Processing by Centrifugation

Processing by Centrifugation

Edited by

Liya L. Regel and William R. Wilcox
Clarkson University
Potsdam, New York

Kluwer Academic / Plenum Publishers
New York, Boston, Dordrecht, London, Moscow

Library of Congress Cataloging-in-Publication Data

Processing by centrifugation/edited by Liya Regal and William R. Wilcox.
p. cm.
Includes bibliographical references.
ISBN 0-306-46654-6
1. Materials—Effect of high gravity on—Congresses. I. Regel', L. L. II. Wilcox. William R. III. International Workshop on Materials Processing in High Gravity (4th: 2000: Potsdam, N.Y.)

TA417.7.H53 C47 2001
620.1'12—dc21

2001038547

Proceedings of the Fourth International Workshop on Materials Processing at High Gravity, held May 29–June 2, 2000, at Clarkson University, Potsdam, New York

ISBN 0-306-46654-6

233 Spring Street, New York, New York 10013

http://www.wkap.nl

10 9 8 7 6 5 4 3 2 1

A C.I.P. record for this book is available from the Library of Congress

Printed in the United States of America

"We judge ourselves by what we feel capable of doing, while others judge us by what we have already done." (Henry Wadsworth Longfellow)

PREFACE

Advances in every field of science and technology are no longer measured in centuries or decades, but in years and even months. They seem to surpass in attainments and scope anything that Man has achieved in the past. In the last decade of the 20th century we witnessed an upsurge in our knowledge of the important role played by gravity in materials processing. This knowledge was gained by experiments in orbiting spacecraft (microgravity) and on centrifuges (macrogravity), as well as by theoretical modeling. Our emphasis here is on research done on centrifuges and other rotating systems.

The interest in centrifugal materials processing was largely stimulated by the mysterious results of the first experiments in 1987 with my colleague and friend Huguette Rodot from CNRS in Meudon. From the beginning, we both realized the attractive opportunity presented by the availability of large centrifuges in Russia and in France for crystal growth and other materials processing. We both found it difficult to deal with the skeptical criticism from some of our colleagues, but that didn't deter us. Others began to find high gravity research interesting. Those of like mind gathered to discuss their results and thoughts in 1991 at the First International Workshop on Materials Processing at High Gravity at Dubna in Russia. Subsequently, three other workshops have been held at Clarkson University.

I am also proud to mention here that the proceedings of the Third International Workshop on Materials Processing at High Gravity, published in book form as "Centrifugal Materials Processing" in 1997 by Plenum, received the Best Book Award from the International Academy of Astronautics in 1998 in Melbourne. This was wonderful recognition for all the researchers who have been contributed to this fascinating field.

The present volume constitutes the proceedings of the exciting Fourth International Workshop on Materials Processing at High Gravity, held at Clarkson University May 29 to June 2, 2000. There were more than 75 attendees from 16 countries. Since the topics extended well beyond materials processing, it was felt appropriate to name this proceedings "Processing by Centrifugation." It includes applications of the traditional bench-scale centrifuges, but also all rotating systems utilizing the centrifugal and Coriolis forces to provide unique performance. Centrifugation on a bench-scale apparatus led to the formation of sticky, porous Teflon membranes, as well as improved polymeric solar cells. Centrifugation on large equipment improved the chemical vapor deposition of diamond films, influenced the growth and dissolution of semiconductor crystals, and elucidated the influence of gravity on coagulation of colloidal Teflon. A million-*g* centrifuge was constructed and used to study sedimentation in solids and to prepare compositionally graded materials and new phases. Rotation of a pipe about its axis allowed the casting of large-diameter metal alloy pipes as well as coating the interior of pipes with a cermet utilizing self-propagating high-temperature synthesis. Such coatings are highly corrosion and erosion resistant. Flow on a rotating disk was shown to be useful for process intensification, e.g. large-scale manufacturing of nano-particles, polymerization reactions, and heat & mass transfer. Several theoretical studies dealt with the influence of rotation on fluid convection on surfaces and in pipes, tubes, and porous media. These have applications to integrated-circuit chip manufacturing, alloy casting, oil production, crystal growth, and the operation of rotating machinery.

While I was researching, writing, planning, conceptualizing, organizing, and talking to all contributors, many friends were at my side. The close professional association with all contributors provided the most rewarding and the richest experience arising from hosting this successful fourth international conference and following up with this fourth proceedings. In editing the papers, we made every effort to retain the authors' syntactic and verbal texture, while seeking consistency and readability. All authors strove to convey the flavor of their excellent presentations. I owe them a special thanks for their skill and insight, and for making the collaboration such a pleasure. This book is theirs. All aspects of the work on this volume were done equally by Bill Wilcox and me; the order of our names as editors is purely alphabetical.

We greatly appreciate the financial support of the U.S. National Science Foundation, the Petroleum Research Fund of the American Chemical Society, Corning Corporation, and the New York State Center for Advanced Materials Processing at Clarkson University. We are grateful to those at Clarkson University who so competently handled meeting facilities, housing, etc. Our graduate students particularly deserve awards for helping attendees with travel, setup of projection equipment, entertainment after the banquet, and innumerable other tasks. Our apologies to those whose contributions we have overlooked. The continual encouragement and support of my family and my friends are deeply and sincerely appreciated. My heartfelt thanks to my mother for everything she did for me. She was more than the usual patient and understanding parent. I'm very sorry she couldn't see this new book.

Liya L. Regel, Chair of Workshops and Co-Editor

International Center for Gravity Materials Science and Applications
Clarkson University
Potsdam NY 13699-5814 USA

CONTENTS

VIBRATION DURING CENTRIFUGATION

Liya L. Regel, William R. Wilcox, Ramnath Derebail,[1]
and Peter V. Skudarnov[2]

International Center for Gravity Materials Science and Applications
Clarkson University, Potsdam, NY 13699-5814, USA

INTRODUCTION

Vibration is of concern in many experiments. For example, low-frequency acceleration influences buoyancy-driven convection. If there are two fluid phases present, vibration can cause oscillation of the interface between them, thereby pumping the adjacent fluids. High frequency vibrations can be transmitted through the container walls to cause streaming of the fluid and, in addition, move second-phase particles relative to the fluid.

There are several sources of vibration in centrifuges. Here, we are concerned with the experiment package on large centrifuges operating in the open atmosphere. We offer as an example our experience on the Clarkson centrifuge HIRB.[1,2] A tube furnace was placed in a box, 30.5 cm × 30.5 cm × 25.4 cm, that was attached by a hinge to the centrifuge arm so as to maintain the net acceleration vector approximately aligned with the axis of the furnace. A triaxial accelerometer was placed within this 20 kg package. The attachment hinge was 1.65 m from the center of rotation. At accelerations from 2 to 5 *g*, vibration was observed mainly at 17 Hz in the direction of rotation, with a maximum amplitude at 5 *g*. This was most likely due to periodic vortex shedding that became locked-in to the mechanical resonance frequency (17 Hz) of the system.

SOURCES OF VIBRATION

Laboratory Vibrations

Vibrations are present in all laboratories and these can be transmitted through the centrifuge to the experiment package. Figure 1 shows accelerometer readings taken while our centrifuge was stationary. On the left are the raw data for the three axes. The corresponding spectral power densities are on the right. Vibrations of almost 0.1 *g* occurred in the y and z directions primarily at ~225 Hz and secondarily at ~420 Hz.

[1] Current address: Agilent Technology, San Jose, CA 95131-1008.
[2] Current address: Florida International University, Miami, FL 33174.

Processing by Centrifugation, Edited by Regel and Wilcox
Kluwer Academic/Plenum Publishers, New York, 2001

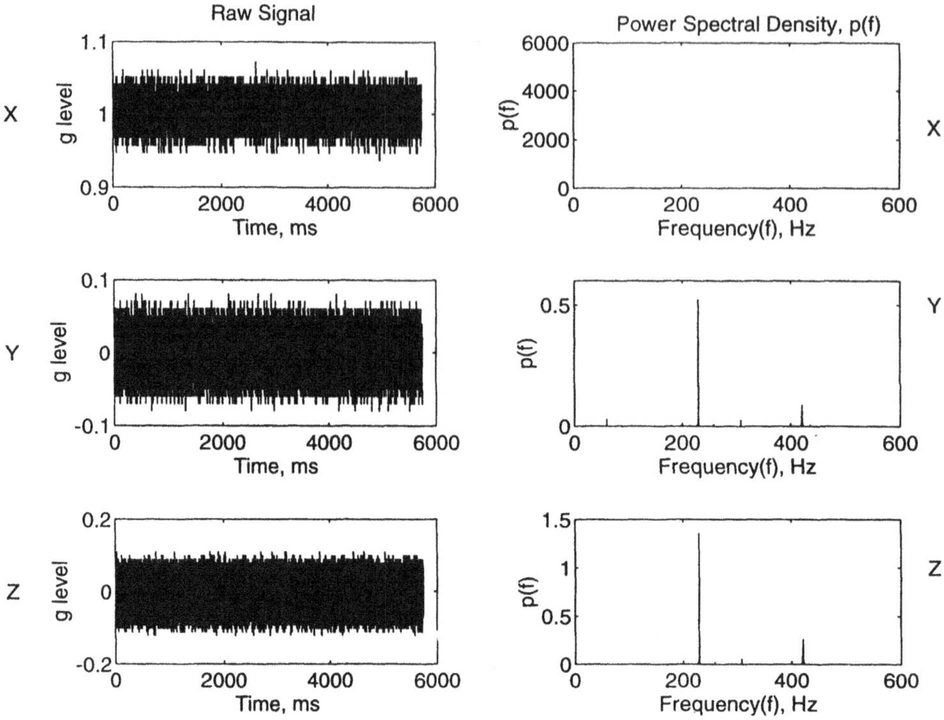

Figure 1. Accelerometer data without centrifuge rotation.[2]
Here, x was along the axis of the furnace (vertical without rotation),
y was circumferential, and z normal to x and y (radial without rotation).

These arose from a co-generation facility in the adjacent room, and could be felt with the hand. They also caused shaking of an interferometer image produced by Dr. Yoshiyuki Inatomi in 1999.

Unbalanced Operation

If a centrifuge is unbalanced, the center of rotation moves in the direction of the center of mass by an amount that depends on the mechanical stiffness of the mounting. While this movement increases slightly the effective radius of the experiment package, and causes oscillation of the mounting at the frequency of rotation, it does not normally lead to vibration of the package itself. HIRB, which was constructed from a large boring machine designed for unbalanced loads, did not noticeably shake its mounting even though it was not well balanced.

Mechanical Noise of the Centrifuge Drive System

As noted by Garnier and Cottineau,[3] mechanical noise in the centrifuge drive system can be transmitted to the experiment package. A common cause is worn bearings. One would expect the frequency of such vibration to be proportional to the centrifuge rotation rate. We did not observe vibration of this type on HIRB.

Aerodynamic Vibrations

When HIRB was rotated, the accelerometer showed that the machine-room vibrations illustrated in Fig. 1 diminished and finally vanished as the resultant acceleration was increased from 2 *g* to 5 *g* in 1 *g* increments. At the same time, a ~17-Hz vibration appeared in the direction of rotation to reach the maximum shown in Fig. 2 at 5 *g*. A lower-amplitude 8.5-Hz vibration also occurred in the direction of rotation. At 3 *g*, the 8.5-Hz vibration reached a maximum and the 17-Hz vibration became a minimum (while remaining stronger than that at 8.5-Hz). These vibrations were driven by the wind flowing over the system when the centrifuge was rotated.

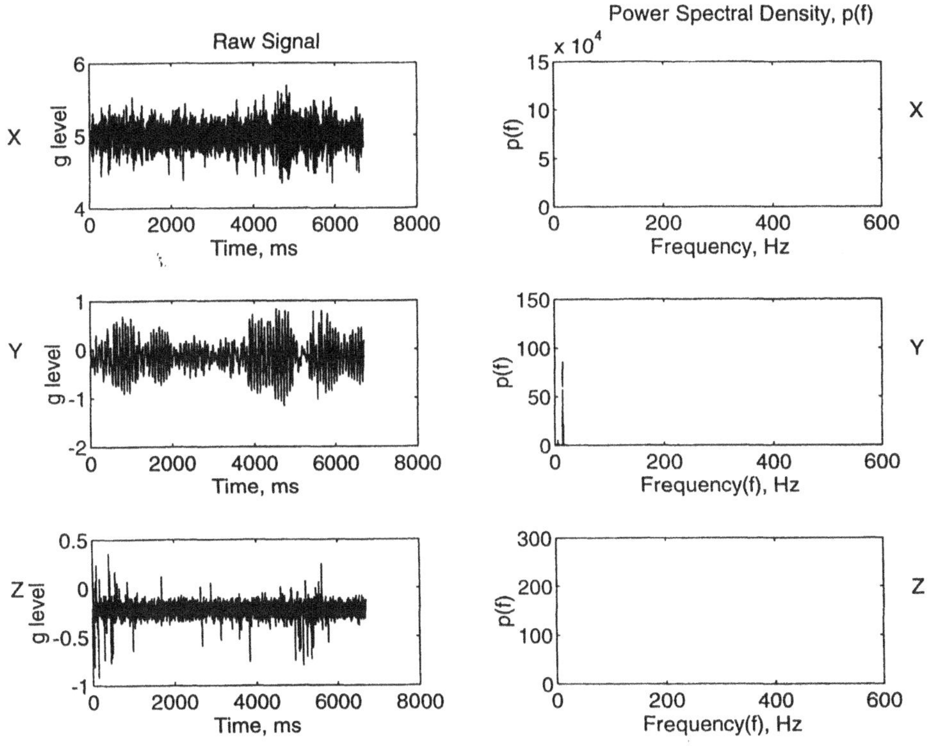

Figure 2. Accelerometer data at 5 *g*. Here, x was along the axis of the furnace (almost radial at 5*g*), y was circumferential, and z normal to x and y.

The natural mechanical vibration frequency of the system was determined to be 17 Hz circumferentially by striking the swing basket with a hammer, as shown in Fig. 3. There is an extensive literature for vibrations caused by flow across long circular cylinders.[4-6] For rigid cylinders, vortices are shed into the wake alternately from one side of the cylinder and then the other for a Reynolds number $Re = DU\rho/\mu$ from 45 to about 10^5. (Here, U is the velocity of the flow, D is the diameter of the cylinder, ρ is the density of the fluid, and μ its viscosity.) The frequency ω_s of vortex shedding is characterized by the Strouhal number $St = \omega_s D/U$, which increases slowly with increasing Re above 300 to reach about 0.28 at $Re = 10^7$. As Re is increased beyond 10^5 the flow becomes increasingly irregular.

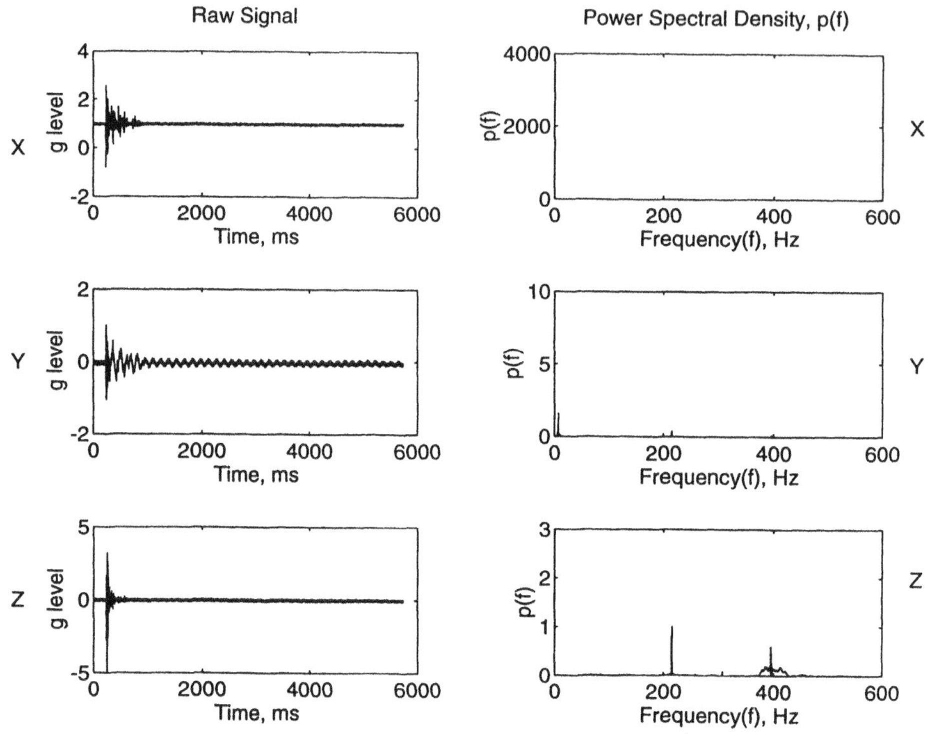

Figure 3. Accelerometer data following impact test.

The alternate vortex shedding from the two sides of a cylinder causes an oscillatory lift force normal to the flow direction. If the cylinder can move in the direction of this force, then it will experience vibration which, in turn, interacts with the flow and the vortex shedding. When ω_s approaches the natural resonant vibration frequency ω_r of the cylinder, "lock-in" occurs with the vortex shedding frequency becoming equal to the mechanical resonance frequency. The vibration amplitude is increased. On the other hand, if the cylinder can move only in the direction of flow then lock-in occurs when $\omega_r \approx 2\omega_s$ and vortices are then shed simultaneously from the two sides of the cylinder. Objects more complex that a cylinder or sphere result in more complex vortex shedding even at low Re. Bluff objects, such as our experimental package, can also interact with wind to give rise to galloping and an amplitude that increases steadily with velocity. Galloping can happen when $\omega_r << \omega_s$ with lightweight, flexible structures.[6]

We now estimate Re and ω_s for the HIRB centrifuge at 5 *g*. The rotation rate N of the centrifuge for 5 *g* was 47.5 rpm, the radius r was about 1.7 m, and D = 30.48 cm.[2] We assume that the wind velocity equaled the velocity of the experiment package, or $U = 2\pi rN = 8.46$ m/s = 19 miles/h. The Handbook of Chemistry and Physics lists the viscosity μ of dry air as 183 micro-poise at 18°C and the density ρ as 1.205 g/liter at 20°C. This gives for the Reynolds number $Re = 1.7 \times 10^5$, which puts the flow in the transition region where it would be complex even for flow over a rigid cylinder. Using 0.28 for the Strouhal number St, we calculate a frequency ω_s = 7.8 Hz, or 2 ω_s = 15.5 Hz.

Thus we are now able to interpret the experimental results as follows. Without centrifuge rotation, we experienced the ~225-Hz vibration generated in the adjacent machine room along with its harmonic of ~420 Hz. The wind caused by centrifuge rotation excited the natural vibration frequency of 17 Hz in the direction of rotation, probably as a twisting motion about the hinge that attaches the swing basket to the centrifuge arm. This indicates locked-in simultaneous vortex shedding from both sides of the box at the same time. It is interesting to note that no vibration was observed normal to the rotation direction, probably because the hinge permits easy movement in this direction with no elastic energy stored.

CONVECTION INDUCED BY VIBRATION

In order to study convection in the gradient-freeze technique, a transparent cell was mounted on HIRB and filled with water containing neutrally-buoyant particles.[7-10] Without a temperature gradient, vibration caused detectable convection only when an air space was present above the water. The surface of the water visibly shook, which vigorously pumped the water beneath. Some shaking of the air-water interface was observed even at 5 rpm, probably due to the machine vibration in the adjacent room. The intensity of the shaking increased as rotation rate increased, with a maximum at about 55 rpm (~6.5 g).

RECOMMENDATIONS

- Construct the experimental package with an aerodynamic shape. This will have the added benefit of reducing the drag, and hence the drive power for the centrifuge and wind buffeting of surrounding structures.
- Completely fill cells to avoid a free liquid surface.
- Locate the centrifuge in a laboratory that is relatively free of external vibration sources.
- Install an accelerometer adjacent to the experimental apparatus inside the swing basket.
- If the vibration level remains a problem, install vibration damping components at appropriate locations, e.g. under the centrifuge and around the experimental apparatus.

Acknowledgement

This research was supported by the National Aeronautics and Space Administration through a CCDS grant to the Consortium for Commercial Crystal Growth at Clarkson University and by the National Science Foundation via grant number DMR-9414304.

REFERENCES

1. R. Derebail, W.A. Arnold, G.J. Rosen, W.R. Wilcox, and L.L. Regel, HIRB – the centrifuge facility at Clarkson, in: *Materials Processing at High Gravity,* L.L. Regel and W.R. Wilcox, eds., Plenum Press, New York (1994), pp 203-211.
2. R. Derebail, *Directional Solidification of Indium Antimonide under High Gravity in Large Centrifuges*, PhD Thesis, Clarkson University, Potsdam, New York (1994).
3. J. Garnier and L.M. Cottineau, Questions raised about materials processing in a centrifuge: lessons derived from the LCPC's experience, in: *Materials Processing in High Gravity,* L.L. Regel, M. Rodot, and W.R. Wilcox, eds., North-Holland, Amsterdam (1992); *J. Crystal Growth* 119:66 (1992).
4. H.J. Lugt, *Vortex Flow in Nature and Technology,* John Wiley & Sons, New York (1983), pp 101-108.
5. E. Simiu and R.H. Scanlan, *Wind Effects on Structures: An Introduction to Wind Engineering,* John Wiley & Sons, New York (1978), chapter 6.
6. R.D. Blevins, *Flow-Induced Vibration,* Van Nostrand Reinhold, New York (1979).
7. P.V. Skudarnov, Flow Visualization and Numerical Modeling for the Gradient Freeze Configuration during Centrifugation, PhD Thesis, Clarkson University (1999).
8. P.V. Skudarnov, L.L. Regel, and W.R. Wilcox, Visualization of Convection in a Simulated Gradient Freeze Cell during Centrifugation, *J. Crystal Growth* 193:219 (1998).
9. P.V. Skudarnov, L.L. Regel, W.R. Wilcox, and G. Ahmadi, Numerical Modeling and Flow Visualization in the Gradient Freeze Configuration during Centrifugation, *present volume.*

APPLICATION OF THE SPINNING-DISC TECHNOLOGY FOR PROCESS INTENSIFICATION IN THE CHEMICAL PROCESS INDUSTRY

Kamelia V.K. Boodhoo and Roshan J.J. Jachuck

Process Intensification & Innovation Centre
University of Newcastle Upon Tyne, NE1 7RU, United Kingdom

INTRODUCTION

The idea of Process Intensification (PI) has received considerable attention lately from research engineers world-wide as an area which could hold the key for the future development and modernization of the chemical process industry (CPI). PI is considered to be a step-change approach to strengthening the industry's competitive edge in today's global marketplace by allowing it to be more flexible and responsive to market needs.

Process Intensification (PI) is a strategy primarily aimed at significantly reducing the size of process equipment without compromising the product throughput. However, as emphasized in recent publication,[1] PI encompasses not only the development of novel, more compact equipment but also the development of intensified methods of processing.

Smaller equipment and their reduced capital costs are the obvious advantages of implementing PI. The intensification of process plants as a whole is particularly favored for huge savings in installation costs. Even greater than the savings in equipment cost, would be savings from reduced piping, civil engineering and support structure, etc.

The other perhaps not so obvious but even more significant potential benefits of PI are improved intrinsic safety, cleaner technology and increased energy efficiency. Adopting the PI approach can substantially improve the intrinsic safety of a process by having a significantly reduced volume of potentially hazardous chemicals at any time in a smaller intensified unit. In addition, one of the objectives of PI is to move away from batch processing to small continuous reactors, the latter giving more efficient overall operation, especially with highly exothermic reactions because the heat can be removed continuously as it is being released. The inherent safety of PI-based technologies and its role in minimizing hazards in the CPI have been intelligently discussed by Hendershot in a recent article.[2]

Environment-friendly technologies are nowadays much sought after. The inherent nature of the PI approach will enable 'clean technology' to be practiced. For instance, high selectivity in intensified reactors could reduce or eliminate altogether the formation of

Processing by Centrifugation, Edited by Regel and Wilcox
Kluwer Academic/Plenum Publishers, New York, 2001

unwanted by-products which, if not removed from the effluent before discharge, can damage the environment. High-purity product could be obtained without incurring enormous downstream purification costs. On the aesthetic side, PI will have tremendously appealing implications whereby a small, compact, highly intensified plant is more likely to be below the tree line, making it far less of an eyesore for the general public than the unsightly and massive steel works characterizing many present chemical plants.

The improved energy efficiency foreseeable in intensified unit operations constitutes yet another highly attractive benefit of PI. Large enhancements in heat and mass transfer, two of the most fundamental and frequently encountered operations in chemical engineering processes, can be achieved in intensified units. Such improvements could lead to dramatic reductions in processing time and the associated energy consumption.

Interesting case studies of intensified processes have been presented for nitration in a compact heat exchanger, in which a gas-liquid reaction takes place using static mixers and hypochlorous production in a rotating packed bed.[3] These case studies serve to clearly illustrate the benefits of PI achieved in real process situations.

Each of the advantages of PI outlined above falls under one of the "process-engineering themes" put forward by Keller and Bryan[4] which, according to the authors, will dictate the future advancement of the CPI. If the changes predicted are to be realized, then there is a need for novel, sustainable technologies such as PI to replace previous chemical engineering practices.

SPINNING DISC TECHNOLOGY

The use of a centrifugal field as an intensification technique is currently being exploited at the Process Intensification and Innovation Centre at the University of Newcastle. This is an internationally recognized centre for the promotion and development of PI. The pioneering development of an intensified spinning disc reactor (SDR) has enabled research into potential areas of application for this novel reactor technology to be undertaken.

The principle of operation of the SDR is based on the creation of a high acceleration field by rotation of a horizontal disc, the surface of which may be smooth or grooved. Under the influence of the centrifugal force, the reacting fluid fed to the centre of the disc by a stationary feed pipe flows on the disc surface in the form of a thin film covered by numerous surface ripples. Experimental evidence for thin film flow on a stationary inclined plane[5] suggests that these ripples or surface waves are probably responsible for vigorous mixing activity within the film that enhances heat and mass transfer in the fluid.

The important benefits offered by spinning-disc technology can be summed up as follows:

1. The extremely high acceleration thus generated is capable of producing very thin films in which heat transfer, mass transfer and mixing rates are likely to be greatly improved.
2. The effect of very short and controllable residence times achieved under the centrifugal action enable heat sensitive materials to be processed without risk of degradation.
3. The rotational speed of the spinning surface provides an additional degree of freedom that can be readily manipulated for optimization of equipment performance.
4. The thin film on the rotating disc is likely to have a plug-flow character, which is desirable to achieve high product throughput by limiting transfer reactions and improving control of product properties, such as the molecular weight distribution (MWD) of polymers.

The potential applications of such a high-acceleration environment in processes involving separation, heat transfer, gas-liquid reactions, crystallization, precipitation of nanoparticles, etc. have been extensively reviewed by Ramshaw.[6] The rotating disc surface would appear to be a particularly attractive choice for systems which, in conventional reactors, are heat or mass-transfer limited. Two main categories of reactions that exhibit such characteristics are inherently fast exothermic reactions and polymerization reactions.

The Novel Spinning Disc Reactor (SDR)

The main design features of a typical SDR are shown in a schematic diagram in Fig. 1. The disc surface, which may be smooth or grooved, is mounted on an internal chamber where the heating or cooling fluid circulates. A channel plate positioned inside the chamber at about 1-3 mm from the underside of the reaction surface allows the heating or cooling medium to be in close contact with the disc, thereby accurately controlling the reaction temperature on the top surface of the disc. The heating/cooling fluid is delivered to the centre of the bottom surface of the disc through an internal pipe and passes out to the edge of the disc in the small channel between the channel plate and the disc surface. After it exits the chamber through the annular space between the drive shaft and the internal pipe, the fluid is recirculated through the system via a temperature controlled oil or water bath. The recirculated fluid thus helps to control the disc surface temperature at a set value. The internal chamber arrangement is supported on the drive shaft, which is driven by a motor. The internal pipe within the drive shaft may or may not rotate depending on the design.

The cover of the reactor is made of stainless steel or UV-transparent glass and has provisions for up to two liquid feed inlet pipes and a gas inlet. The liquid feed pipes are positioned so that they deliver the feeds very close to the centre of the disc, usually in a small recess about 0.5 mm deep. As the disc rotates, this recess acts as a reservoir and helps to distribute the feed uniformly across the surface of the disc under the action of the centrifugal force. When a gas-liquid reaction is to be performed, the turbulence promoter helps to induce turbulence in the gas stream and therefore enhance mass transfer between the gas and the liquid. The base of the reactor has a facility for purging the system with an inert gas such as nitrogen.

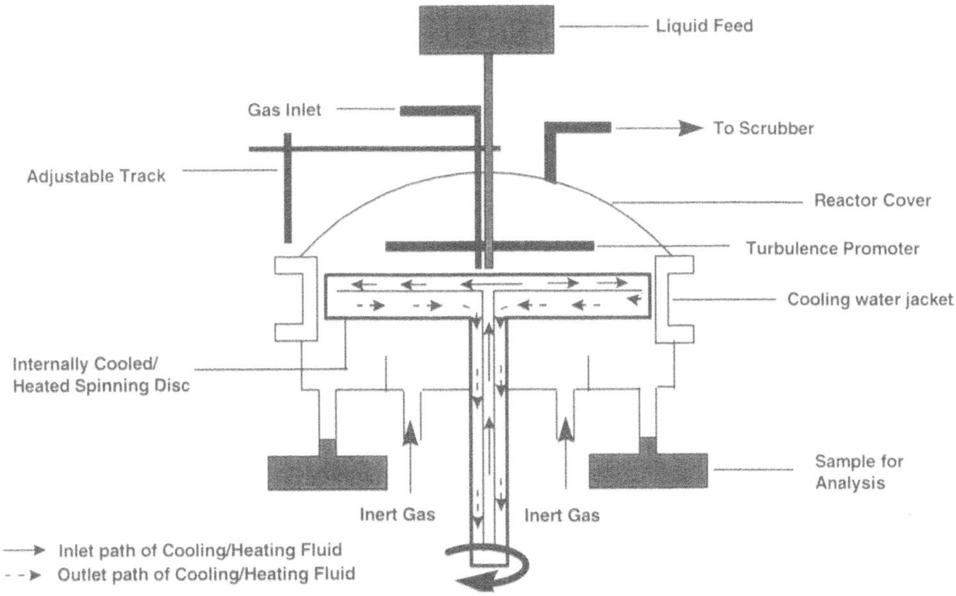

Figure 1. Schematic diagram of the novel spinning disc reactor.

As the fluid is introduced to the centre of the disc at a controlled flow rate, it flows as a thin film on the surface of the rotating disc, assuming that the conditions for film break-up are avoided. After a residence time of the order of a few seconds, the reacting fluid is thrown off the surface and hits the cooled walls of the stationary reactor housing. The product from the disc is thus rapidly cooled to prevent further reaction taking place before it is collected. In some cases, it may also be necessary to add inhibitor to the product as it comes off the disc.

HEAT TRANSFER AND MASS TRANSFER IN SDR

Experimental studies into heat transfer characteristics of thin film flow on a rotating disc system have been performed over many years. These studies have looked into evaporation as well as heat transfer without phase change.[7,8] More recently, the effect of different types of grooves machined in the disc surface on the convective heat transfer characteristics of the disc was investigated using water as a test fluid.[9] Surface instabilities appeared to be significantly increased by grooves and the heat transfer coefficient reached an average value of 18 $kW.m^{-2}.K^{-1}$. This suggests the positive influence of surface waves on heat transfer.

Enhanced mass transfer on a rotating disc have also been obtained. Lim[10] investigated mass transfer rates during the absorption of oxygen in thin liquid films on a rotating disc. He reported mass transfer coefficients much larger than predicted on the basis of molecular diffusion in a smooth laminar liquid film. This result was attributed to vigorous bulk mixing by the numerous ripples formed on the film surface. Moore[11] also compared the effect of rotation on mass transfer using various perforated and smooth plates as the active mass transfer surface. Mass transfer coefficients for the perforated surfaces were much higher than those for a smooth disc due to the many random turbulent waves in the former case.

POLYMERISATION IN SDR

A rather unique characteristic encountered during polymerization reactions is the rapid viscosity build up with increasing monomer conversion. At the resulting high viscosities, complete and efficient agitation throughout the bulk of the monomer-polymer mixture becomes increasingly difficult in stirred-tank reactors and reduces heat transfer. Non-uniformity in temperature broadens the molecular weight distribution and poor product quality. In extreme cases there may be a total loss in temperature control so that a runaway reaction occurs. This has been responsible for a number of accidents in polymer plants.

For equilibrium-controlled reactions or step-growth polymerization, such as polyester production, the polymerization is heavily dependent on water or glycol removal from the reacting mixture. This removal of small molecules requires mass transfer through the liquid polymer and at the liquid-vapor interface. This mass transfer is hindered by a high viscosity so that the rate of polymerization eventually drops significantly. The problem is further aggravated by insufficient surface area available in the usual large polymer reactors. Consequently, drastic operating conditions are applied to achieve the desired degree of polymerization.

The thin-film characteristics of the SDR fulfil the major requirements of an ideal polymer reactor: excellent heat transfer, mass transfer and mixing even in quite viscous reaction media. The benefits that can potentially be achieved by using the SDR for carrying out polymerization processes are:

1. Thin film to allow efficient penetration of UV light for photo-initiated polymerization.
2. Capable of achieving higher conversions due to enhanced heat and mass transfer rates.
3. Significant reduction in reaction time.
4. Improved intrinsic safety of the process.
5. Tighter molecular weight distribution / Improved product quality.
6. Minimal risk of polymer degradation at high temperatures due to short residence times.

Step-Growth Polymerization

The polyesterification reaction between maleic anhydride and ethylene glycol shown below was studied as an example of step-growth polymerization.

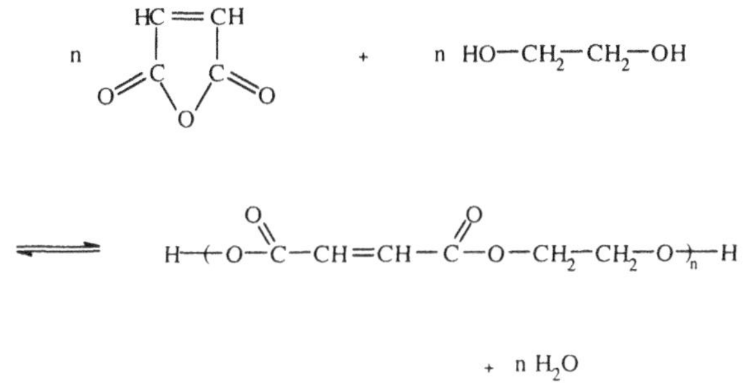

(1)

The reaction was performed on a grooved disc at a temperature of 200°C and rotating at a disc speed of 1000 rpm. Pre-polymer was prepared to known levels of conversion or acid value (AV) in a batch reactor and was fed to the centre of the spinning disc. The runs were carried out in a closed system under a blanket of nitrogen at 1 atmosphere to prevent oxidation and unnecessary side reactions. As soon as the pre-polymer melt came in contact with the disc surface, condensation polymerization started and proceeded rapidly due to the enhanced heat and mass transfer achieved due to thin films and high shear mixing provided by the grooves on the disc. The ease with which water vapor was removed from the reaction system determined the rate of polymerization.

The performance of this SDR was compared to a conventional laboratory batch reactor by measuring the rate of polymerization in both types of reactors. The rate profile for the batch polymerization was obtained from the gradient of the acid value profile for a range of reaction times. In the case of the SDR, the residence time on the rotating disc surface was calculated to be about 2 s, based on an assumed value of 1.5 Pa.s for the average viscosity of the polymer, water-like density and an estimated flow rate of pre-polymer of 50 cc/s on the disc. A very conservative value of 1 min was used to calculate the rate of polymerization in the SDR to take into account the time it takes for the product to flow down the collecting pipe (up to 4 or 5 min), during which some polymerization could have occurred. The experimental rate profiles are compared in Fig. 2. The results indicate that the rate of polymerization in the batch reactor falls off quite rapidly. The rate becomes slower as a result of the increasing viscosity of the melt and the associated drop in the rate removal of water. The SDR rate profile follows a similar trend to that of the batch profile but is shifted up. Furthermore, the drop in rate in the SDR with reaction time is not as drastic as that observed for the batch reactor.

The SDR rate profile approaches the batch-reactor profile in the initial stage of the polymerization, showing that the reaction proceeds equally well in both reactors. In the very viscous melt conditions of the later stage of the reaction, the rate of polymerization in

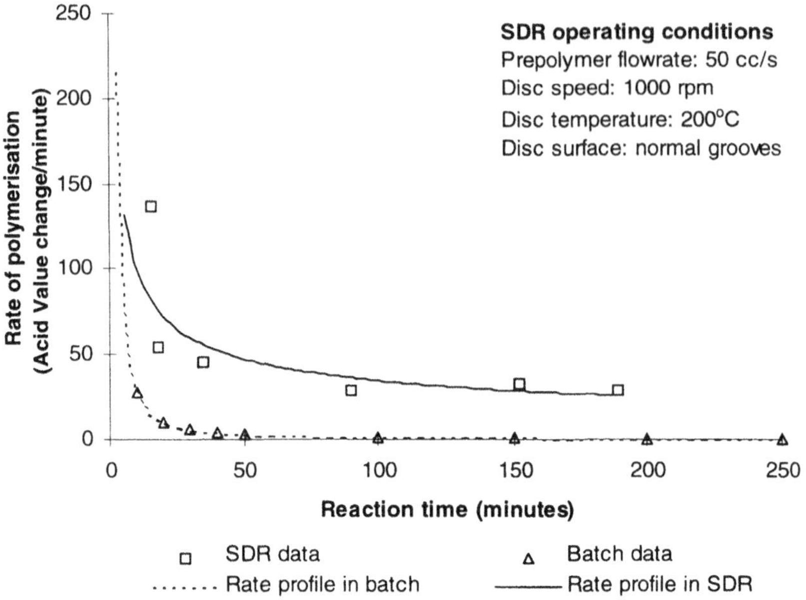

Figure 2. Comparison of experimental rate of change of AV in a batch reactor and in a SDR during step-growth polymerization of maleic anhydride and ethylene glycol (Eq. 1).

the SDR is tremendously larger than that in the batch reactor. This improvement is attributed to the significant increase in mass transfer in the SDR by the intense mixing that occurs both within the highly sheared films and on the film surface. The numerous surface waves also help to constantly renew the exposed surface from where the water evaporates. Performing this reaction in an SDR could reduce processing times by as much as 100 min in the viscous stage of the polymerization. Also, the ability of continuous processing offered by the use of the SDR makes it highly desirable as a replacement unit for batch operations.

Addition Polymerization

In contrast to step-growth polymerization, addition or chain polymerization does not involve removal of any reaction bi-product. Species are added in sequential steps to build up a macromolecule. Free-radical chain polymerization's of styrene and n-butyl acrylate were studied using either thermal initiation or photo-initiation. The chain reaction steps and the associated rate expressions are given elsewhere.[12] Some of the results are presented in the following sections.

Thermal Initiation of Initiator. A study of free-radical polymerization of styrene via thermal initiation of benzoyl peroxide was undertaken to assess the performance of the SDR for a typical free-radical polymerization process. The detailed results are in a thesis.[13] Figure 3 shows that the single-pass conversion in an SDR increased from ~3% to ~8% as the speed of rotation was increased from 300 to 750 rpm, with the other operating conditions held constant.* There was a steep rise in conversion to 17% with a change of

* Each data point is the average of three runs with very little scatter. An optimal value at a given speed was also found for a different set of operating conditions.

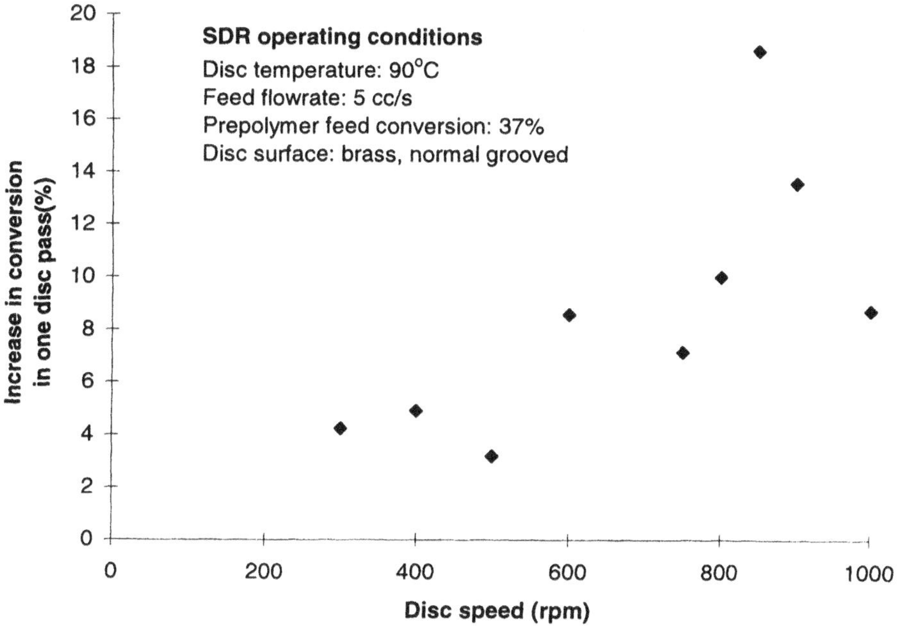

Figure 3. Effect of disc rotational speed on single-pass conversion in SDR.

only 100 rpm in going from 750 to 850 rpm. Beyond 850 rpm, however, the conversion declined rapidly to about 8% at 1000 rpm. An optimal speed of 850 rpm was therefore observed at which the polymerization in the SDR proceeded at its fastest rate. The advantage of SDR can be appreciated by observing that it would take 29 min in a batch reactor to achieve the same conversion realized in only a few seconds on a disk rotating at 850 rpm. The optimal rotation rate depends on the pre-conversion of the feed, and hence on the feed viscosity that affects the film thickness and the residence time on the disc surface.[13] The optimal speed shifted from 850 rpm for a pre-conversion of 37% to 1000 rpm for a pre-conversion of 58%.[13] The increase in film thickness due to higher feed viscosity was counteracted by increasing the speed of rotation and hence the centrifugal acceleration. The same reasoning is applicable to residence time on the spinning disc.

For the range of speeds considered, the product from the disc generally had a lower average molecular weight M_w and lower polydispersity index PDI (i.e., narrower MWD) than those from a batch reactor for the same conversion. However, there did not seem to be any significant trend in the M_w and PDI with disc speed.

The trend of single-pass conversion in an SDR with disc rotational speed may be explained in terms of the film thickness, the residence time and the surface waves covering the fluid film. The film thickness determines the molecular diffusion path length, which is an important consideration when micromixing effects comes into play, especially in viscous reaction media. The mean residence time controls the time of exposure to conditions imposed in the SDR and can thus have a great impact on the productivity and properties of the end-product. The surface waves significantly affect mixing in the fluid on the rotating surface. Expressions for the average film thickness δ and the mean residence time t_{res} of a Newtonian liquid film flowing over a smooth disc have been derived on the basis of a centrifugal model as:[13]

$$\delta = \left(\frac{3}{2\pi}\frac{\nu Q}{\omega^2 r^2}\right)^{1/3} \qquad (2)$$

$$t_{res} = \left(\frac{81\pi^2 \nu}{16\omega^2 Q^2}\right)^{1/3} \left(r_o^{4/3} - r_i^{4/3}\right) \qquad (3)$$

where ν is the kinematic viscosity (= η/ρ), Q is the volumetric flow rate, ω is the angular velocity (= $2\pi N/60$), N is the rotational speed, r is the radial distance, and the subscripts o and i refer to outlet and inlet, respectively. The criterion for which surface waves are clearly distinguished on hydrodynamically smooth surfaces is:

$$Re > 16 \qquad (4)$$

where the Reynolds number Re is defined as $2Q/\pi\nu r$.

Although a grooved-disc configuration was used in the experiments and expressions (2) and (3) apply to a smooth disc, these equations are still useful because similar dependencies on the variables are expected with the inclusion of an appropriate correction factor to account for the surface roughness. From Eqs. (2) and (3), it is seen that increases in the speed of rotation translate into shorter residence times and thinner films. The onset of wave formation seems to be independent of the disc rotational speed (Eq. (4)), but numerous studies have indicated that increasing the rotational speed does influence the type of wave patterns as well as the intensity of the ripples produced.[14-17] Thus, the reduced path length combined with the intense mixing generated by wave motions would encourage faster mass transfer of reacting species in the polymer film, thus enhancing conversion in spite of the increasingly shorter residence time at higher rotational speeds. Beyond the optimal speed, however, the increasing change in conversion cannot be sustained because the time of exposure on the disc becomes the controlling factor. The rise in conversion then falls off from its peak value.

The decrease in molecular weight and tightening of the MWD suggest that polymerization on a rotating surface takes place by mechanisms different from those observed in the gel-effect phenomenon which, in contrast, gives high polymerization rates at the expense of an increase in molecular weight and a severe broadening of its distribution. Again, this effect may generally be attributed to enhanced mixing in the SDR film, especially in highly viscous polymerizing melts.

Photo-initiation. It is well-known that the rate of initiation for photo-initiated polymerization systems is extremely fast due to the low activation energy associated with this process. However, this method is not commercially used for preparing polymers on a continuous basis because of an immense lack of thin film polymer reactor technology. The existing conventional polymer reactors are not suitable since they cannot provide the thin film environment for the monomer and initiator system. The thin film requirement is critically important because the absorption of the UV radiation that directly influences the extent of polymerization or conversion is affected by the depth of the film. The thin-film technology offered by the SDR presents a major advantage for performing photo-induced bulk polymerization's in that it should allow efficient penetration of UV light or any other form of radiation used.

An exploratory study into the polymerization of n-butyl acrylate by UV irradiation in the SDR has recently been undertaken. As seen in Fig. 4, the disc rotational speed is an important parameter controlling the extent of polymerization of n-butyl acrylate in an SDR. Conversions as high as 55% were reached in a single disc pass in the low-speed range, indicating that the disc residence time is the controlling factor. Molecular weights in the range 250,000 to 300,00 have also been achieved at a residence time of less than 3 s on a smooth rotating disc of 200-mm diameter. Increasing the feed flow rate gave a similar trend in conversion, as illustrated by Fig. 5. A rise in the feed rate causes the residence time on the disc surface to be reduced (Eq. (3)) and, therefore, the polymerizing film is exposed to the UV light for less time. It also appears that the corresponding decrease in film thickness (Eq. (2)) and increase in intensity of waves at higher flow rates did not contribute much to counteracting the effects of a shorter residence time.

With regard to the molecular weight distribution (MWD) of the polymer, a low polydispersity index (PDI) in the range between about 1.5 to 2.0 was generally achieved, reflecting the fact that SDR, with its enhanced mixing, has the potential to yield polymer with tightly controlled MWD and hence of better quality.

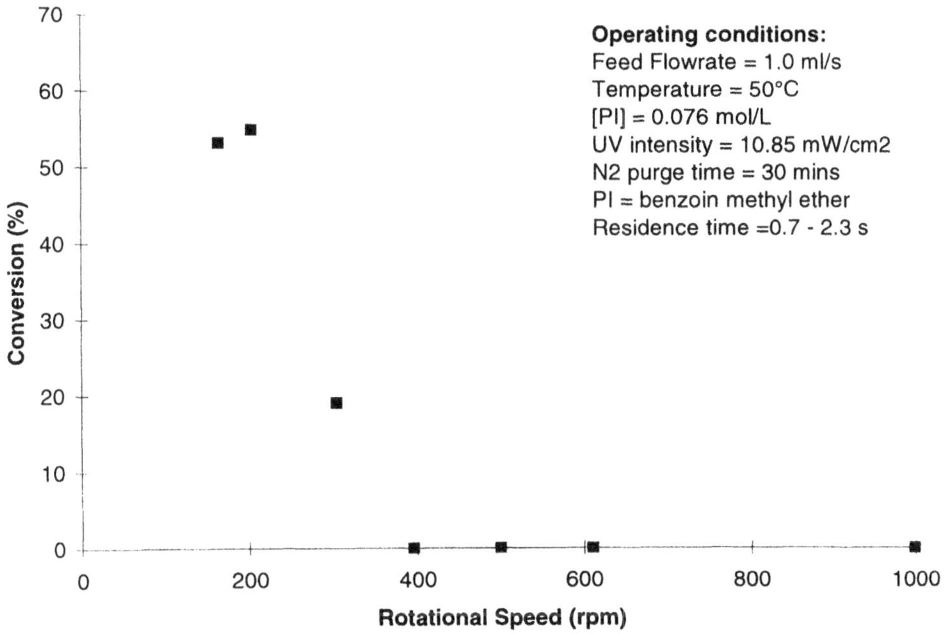

Figure 4. Effect of disc rotational speed on conversion of n-butyl acrylate in SDR in one disc pass.

CRYSTALLISATION IN SDR

Precipitation or crystallization reactions take place in two distinct steps: nucleation followed by crystal growth. Good mixing is very important in the nucleation step. Homogeneous nucleation is required to produce crystals with a very narrow size distribution. An exploratory study evaluating the performance of the SDR for the precipitation of barium sulphate showed that a very uniform and narrow crystal size distribution was produced on a smooth disc rotating at about 900-1000 rpm with an initial supersaturation S_o of 2000. The uniform shape and size (Fig. 6) of the resulting barium sulphate crystals suggest that homogeneous nucleation conditions prevailed.

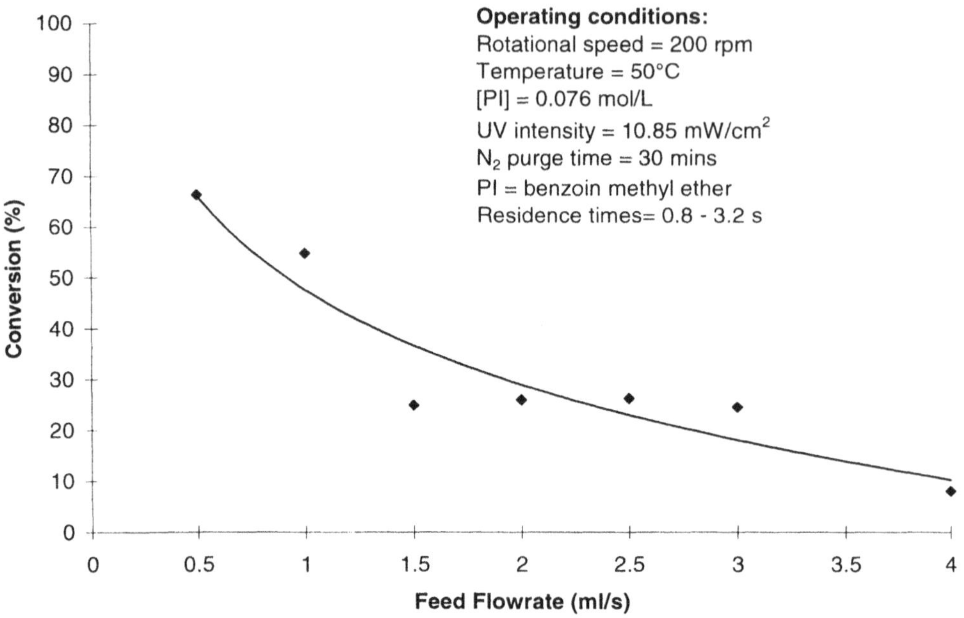

Figure 5. Effect of feed flow rate on n-butyl acrylate conversion in one pass in SDR

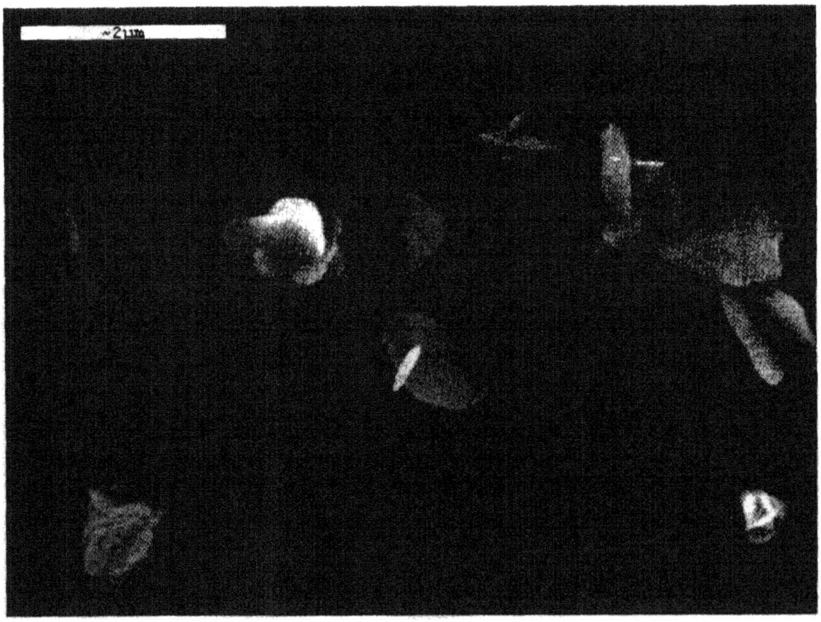

Figure 6. TEM micrograph of barium sulphate crystals produced in SDR.

The crystals from the SDR also compared well with those produced in a tee-mixer,[18] which is known to allow homogenous nucleation to take place. Recent studies at our Centre indicate that nanoparticles with tight crystal size distribution can be produced at a higher supersaturation S_o of 5000.

CONCLUSIONS

Intense mixing and enhanced heat and mass transfer in the SDR make it a very versatile reactor for performing a wide range of reactions, such as polymerization and crystallization. Higher rates of polymerization, reduced processing times and controlled molecular weight distributions (MWD) have been achieved in the SDR for benzoyl-peroxide-initiated styrene polymerization and UV-induced bulk polymerization of n-butyl acrylate. Uniformly shaped crystals of barium sulphate having a tightly controlled crystal size distribution were formed in the spinning disc crystallizer. The SDR has the added flexibility of additional degrees of freedom, such as disc speed and temperature, which can be manipulated by the operator to obtain end-products tailored to meet product requirements.

REFERENCES

1. A.I. Stankiewicz and J.A. Moulijn, Process intensification: transforming chemical engineering, *Chem. Eng. Prog.* 96: 22 (2000).
2. D.C. Hendershot, Process minimisation: making plants safer, *Chem. Eng. Prog.* 96: 35 (2000).
3. A. Green, B. Johnson and A. John, Process intensification magnifies profits, *Chem. Eng.* 106(13): 66 (1999).
4. G.E. Keller and P.F. Bryan, Process engineering: moving in new directions, *Chem. Eng. Prog.* 96(1): 41 (2000).
5. N. Brauner and D.M. Maron, Modelling of wavy flow in inclined thin films, *Chem. Eng. Sci.* 35: 775 (1983).
6. C. Ramshaw, The opportunities for exploiting centrifugal fields, *Heat Recovery Systems & CHP* 13: 493 (1992).
7. C. Bell, *The Hydrodynamics and Heat Transfer Characteristics of Liquid Films on a Rotating Disc*, Ph.D. Thesis, University of Newcastle Upon Tyne, UK (1975).
8. J.R. Khan, *Heat Transfer on a Rotating Disc with and without Phase Change*, Ph.D. Thesis, University of Newcastle Upon Tyne (1986).
9. R.J. Jachuck and C. Ramshaw, Process intensification: heat transfer characteristics of tailored rotating surfaces, *Heat Recovery Systems & CHP* 14: 475 (1994).
10. S.T. Lim, *Hydrodynamics and Mass Transfer Processes associated with the Absorption of Oxygen in Liquid Films Flowing across a Rotating Disc*, Ph.D. Thesis, University of Newcastle Upon Tyne, UK (1980).
11. R.S. Moore, *Mass Transfer to Thin Liquid Films on Rotating Surfaces with and without Chemical Reaction*, Ph.D. Thesis, University of Newcastle Upon Tyne, UK (1986).
12. K.V.K. Boodhoo and R.J. Jachuck, Process intensification: spinning disc reactor for styrene polymerisation, *Appl. Therm. Eng.* 20: 1127 (2000).
13. K.V.K. Boodhoo, *Process Intensification: Spinning Disc Reactor for Styrene Polymerisation*, Ph.D. Thesis, University of Newcastle Upon Tyne, UK (1999).
14. H. Epsig, and R. Hoyle, Waves in a thin liquid layer on a rotating disc, *J. Fluid Mech.* 22: 671 (1965).
15. A.F. Charwat *et al.*, The flow and stability of thin liquid films on a rotating disk, *J. Fluid Mech.* 53: 227 (1972).
16. S. Thomas *et al.*, Experimental analysis and flow visualization of a thin liquid film on a stationary and rotating disc, *ASME J. Fluids Eng.* 113: 73 (1991).
17. W. Woods, *The Hydrodynamics of Thin Liquid Films Flowing over a Rotating Disc*, Ph.D. Thesis, University of Newcastle Upon Tyne, UK (1995).
18. R. Mohanty *et al.*, Characterizing the product crystals from a mixing tee process, *AIChE J.* 34: 2063 (1988).

SYNTHESIS AND APPLICATION OF NANOPARTICLES BY MULTIPHASE REACTIVE PRECIPITATION IN A HIGH-GRAVITY REACTOR: I: EXPERIMENTAL

Jianfeng Chen and Yuhong Wang

Higravitec Center, Department of Chemical Engineering
Beijing University of Chemical Technology, Beijing 100029, China

ABSTRACT

A new technology, "high-gravity multiphase reactive precipitation," was developed for the massive production of nanoparticles. A high-gravity reactor was designed to generate high acceleration. The synthesis of $CaCO_3$ nanoparticles was experimentally investigated. The mean particle size could be controlled in the range of 15~40nm by changing the operating conditions. Nano-sized $CaCO_3$ nanoparticles gave much better performance than micron-sized particles for reinforcement of polymers and as a high-functional rheopexy filler in paint.

INTRODUCTION

Nanoparticles have high potential applications in chemicals, ceramics, micro-electronics, pharmaceuticals, automobiles, and so on. The increasing industrial interest in massive production of nanoparticles is driving research activities toward the development of methods with low cost and high volume. For example, inorganic and organic hybrid nanocomposites are currently emerging in the market. The potential demand of inorganic nanoparticles, such as $CaCO_3$, SiO_2, TiO_2 and Al_2O_3, is on the order of millions of tons per year.

Many methods have been reported for the synthesis of nanoparticles, such as physical vapor deposition, chemical vapor deposition, sol-gel, microemulsion, sonochemical processing, mechano-chemical processing, supercritical processing, etc. However, all of those technologies seem to be limited to small production volumes or bench scale, i.e. a few kg/hr.

Reactive precipitation is a traditional way to produce particles or powders in industry. It features convenient processing, low cost and massive production. However, conventional

Processing by Centrifugation, Edited by Regel and Wilcox
Kluwer Academic/Plenum Publishers, New York, 2001

precipitation is often carried out in a stirred tank or column reactor, and the quality of the product is difficult to control, and the morphology and particle size distribution often change from one batch to another. In addition, reactive precipitation often involves a multiphase reaction, making the processing more complicated and difficult to control.

Here, we describe a new technology called high-gravity multiphase reactive precipitation and its application to the synthesis of $CaCO_3$ nanoparticles involving gas, liquid and solid phases.

ANALYSES OF REACTIVE PRECIPITATION PROCESS

Analysis of Kinetics

Precipitation processes consist of three main steps: chemical reaction, nucleation and crystal growth. If agglomeration and ripening mechanisms are neglected, precipitation mechanisms can be described by:

$$\mathrm{A} + \mathrm{B} \rightarrow \mathrm{P} \quad \text{(Reaction)} \tag{1}$$

$$\mathrm{P} + \mathrm{P} + \cdots + \mathrm{P} \rightarrow \mathrm{P}^* \quad \text{(Nucleation)} \tag{2}$$

$$\left.\begin{array}{l} \mathrm{P}^* + \mathrm{P} \rightarrow \mathrm{P}_1 \\ \mathrm{P}_1 + \mathrm{P} \rightarrow \mathrm{P}_2 \\ \cdots\cdots \\ \mathrm{P}_{n-1} + \mathrm{P} \rightarrow \mathrm{P}_n \end{array}\right\} \quad \text{(Crystal Growth)} \tag{3}$$

where the symbols are defined at the end. These mechanisms are quite similar to that of polymerization. The particle size distribution (PSD) of the product depends on each step. In general, reaction and nucleation are both relatively fast, while crystal growth proceeds slower. The kinetic equations of the three steps are often described as follows:

$$dc_P / dt = k c_A^{\alpha} c_B^{\beta} \tag{4}$$

$$dn / dt = k_N (c_P - c^*)^m \tag{5}$$

$$dl / dt = k_g (c_P - c^*)^{\gamma} \tag{6}$$

The growth order γ is usually in the range of between 1 and 2. In the preparation of ultrafine particles via reactive precipitation, the chemical reaction is sufficiently fast that a high degree of supersaturation of P is achieved in local volumes. This causes the formation of crystals to be dominated by homogenous nucleation,[1] with the nucleation order m in the range of 5 to 18. The values of reaction orders α and β are often <2. Therefore, a high concentration of product c_p favors an increase in the number of nuclei and a decrease in average particle size because nucleation depends more strongly on c_p than do the reaction and crystal growth.

The induction time τ is the time from the first creation of the conditions for homogeneous nucleation to establishment of a steady-state nucleation rate, and can be estimated by the equation given by Dirksen and Ring:[1]

$$\tau = 6d^2n^* /(D\ln S) \quad (7)$$

For homogeneous nucleation, tiny variations of c_p lead to a large change of nucleation rate due to the high value of m (see Eq. 5). As a result, the PSD is broad if the spatial concentration distribution in the reactor is non-uniform. In addition, differences of growth time of the crystals in the reactor broaden the PSD. Therefore, to obtain nanoparticles with a narrow size distribution, one should do as much as possible to meet the requirements of: (a) high degree of supersaturation, (b) spatially uniform concentration, and (c) the same growth time for all crystals.

Analysis of Engineering Factors

Very little attention has been paid in the literature to the effect of engineering factors on PSD in the synthesis of ultrafine particles. But these are very critical when scaling up a process. Experimental results reported by Chen *et al.*[2] and Tosun[3] showed that micromixing and macromixing strongly influence the PSD in reactive precipitation from $BaCl_2$ and Na_2SO_4 solutions. A spatially-uniform concentration on the vessel scale can only be obtained by good macromixing. A spatially-uniform concentration on the molecular scale can only be reached by intense micromixing. Macromixing and micromixing should be going on simultaneously in the vessel. Micromixing is a crucial factor determining the supersaturation of solute and its local spatial distribution. From the viewpoint of chemical reaction engineering, the rate of reaction and subsequent nucleation in precipitation are controlled by the intrinsic kinetics without the influence of micromixing for $t_m < \tau$, and controlled or influenced by micromixing when $t_m > \tau$, where t_m is the characteristic time of micromixing for a species reaching a maximum mixed state at the molecular level. Due to the very strong non-linearity of homogenous nucleation, micromixing should be intense in order to reach $t_m < \tau$ so that the rate of nucleation at different locations in a precipitator is nearly the same, allowing the PSD to be controlled.

The characteristic micromixing time can be expressed by:[4]

$$t_m = k_m (\upsilon / \varepsilon)^{1/2} \quad (8)$$

The constant k_m has been assigned various values in the literature; k_m =16 is used here. The energy dissipation rate ε in a common stirred tank is on the order of 0.1~10 W/kg with a kinematic viscosity υ of 1×10^{-6} m^2/s for aqueous solutions. In this case, the characteristic micromixing time t_m is estimated to be in the order of 5~50 ms. In aqueous solutions, the nucleation induction time is often of the order of 1 ms or less, so that $t_m >> \tau$. This implies that the PSD in a stirred tank cannot be easily controlled, and scale-up is difficult owing to poor micromixing. This corresponds to industrial experience.

Micromixing has little effect on the crystal growth rate because its concentration dependence is small. Thus only the influence of macromixing is important for crystal growth, so that nuclei formed at the same instant should grow to yield a narrow PSD if the macro-environment has a uniform concentration.

High-Gravity Multiphase Reactive Precipitation (HGRP)

High-gravity technology in the form of a rotating packed bed (RPB), also called "Higee," is a novel method to intensify mass transfer between gas-liquid, liquid-liquid, and solid-liquid multiphase systems.[5-8] Recently, our investigation on a Higee reactor revealed that micromixing is also tremendously intensified.[9] Micromixing in a RPB reactor is much faster than in a static mixer, T-type mixer, Y-type mixer, or stirred-tank reactor. The micromixing characteristic time t_m is estimated to be on the order of 10~100μs. This should satisfy the requirement of $t_m < \tau$ to permit good control of PSD. In addition, the higher mass transfer rate in a multiphase system in a RPB reactor would generate a higher supersaturation of the product from gas-liquid and gas-liquid-solid precipitation reactions, thereby yielding smaller particles. As a result, it is predicted that the RPB high-gravity reactor is ideal for synthesis of nanoparticles using multiphase reactive precipitation. Experimental verification of this prediction is discussed in the following.

EXPERIMENTAL

The synthesis of nanoparticles of $CaCO_3$ from the gas-liquid-solid chemical reaction between a lime suspension and CO_2 gas was employed as a demonstration case study. The chemical reaction is:

$$CO_2\ (g) + H_2O\ (l) + Ca(OH)_2\ (s) = CaCO_3\ (s) + 2H_2O\ (l) \tag{9}$$

High-purity industrial-grade quicklime and chemical-grade CO_2 bottled gas were used as raw materials. The quicklime was first hydrated with tap water to yield lime milk.

Figure 1 shows the experimental setup schematically. It consisted of a rotating packed bed and a stirred tank. In the rotating packed bed, a rotator was inside a fixed casing and was rotated at several hundreds to thousands of rpm. The lime slurry was introduced into the eye of the rotator from a liquid inlet pipe and then sprayed by distributor (5) onto the inside edge of the rotator. The liquid entering the bed flowed in the radial direction under centrifugal force, past the packing and outside space between the rotator and shell, and was finally collected and left the equipment through liquid exit (7). The gas was introduced through gas inlet (10) and flowed inward countercurrent to the liquid in the packing of rotator, and finally left through the gas exit under the force of the pressure gradient.

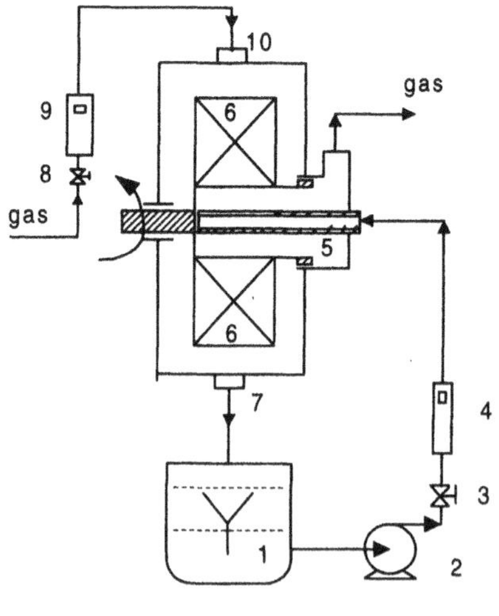

Figure 1. Schematic diagram of experimental setup. 1. Stirred tank; 2. Pump; 3. Valve; 4. Rotor flow meter; 5. Distributor; 6. Rotating packed bed; 7. Outlet; 8. Valve; 9. Rotor flow meter; 10. Inlet.

In operation, 80 liter of lime milk initially in stirred tank (1) was pumped through distributor (5) into rotating packed bed (6), where it reacted with CO_2 to form $CaCO_3$ particles. The mixed slurry of $Ca(OH)_2$ and $CaCO_3$ flowed back into stirred tank 1, where it was re-pumped into the rotating packed bed. This recycle was ended when the pH of the slurry reached around 7. At this time, $Ca(OH)_2$ had been completely converted to $CaCO_3$. The time from the beginning of this process to the end is termed the reaction process time or carbonation time, which is denoted by t_R. Product samples were immediately taken after the process was stopped for transmission electron microscopy (TEM) (mode H-800, Hitachi, Japan).

For experiments on production of polymer composites, the above $CaCO_3$ nanoparticles were surface-treated and then dried. The dried powder was mixed with polymers using the normal blending methods. Standard testing methods were used to determine the physical and mechanical properties of the polymer composites filled with nano-sized $CaCO_3$. Commercial micron-sized $CaCO_3$ particles were also used as a control.

RESULTS AND DISCUSSION

Process Time of Carbonation Reaction

The operating conditions, such as acceleration a, gas flow rate G and liquid flow rate L, influence the batch processing reaction time t_R when the processing volume and solid concentration of lime in the suspension remained fixed. Figure 2 shows the effect of high-gravity level on the carbonation time. The carbonation time decreased sharply with increasing acceleration for $a<500$ m/s^2, but quickly reached a plateau with further increases in a. The batch carbonation time was 12~20min here, while Gu *et al.*[10] reported 70~90 min under similar conditions using a conventional stirred bubbling tank reactor. Comparing these two technologies, it is seen that the high-gravity reactor has a much larger production capacity due to the significant reduction of processing time, which should save in operating costs.

Figure 3 shows the influence of the ratio of liquid-to-gas flow rates, L/G, on the carbonation time t_R when G was held constant. An increase in L/G would be expected to decrease t_R due to an increased mass transfer rate between liquid and solid phases.

Particle Size Distribution

The particle size distribution (PSD) was calculated on the basis of the TEM pictures. Figure 4 shows the effect of acceleration on PSD in terms of the mean particle size d_P and dimensionless variance of the size distribution σ. Both decrease with acceleration.

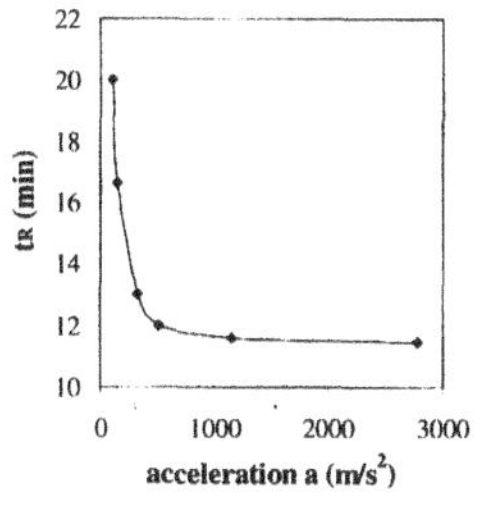

Figure 2. Effect of acceleration on carbonation time t_R.

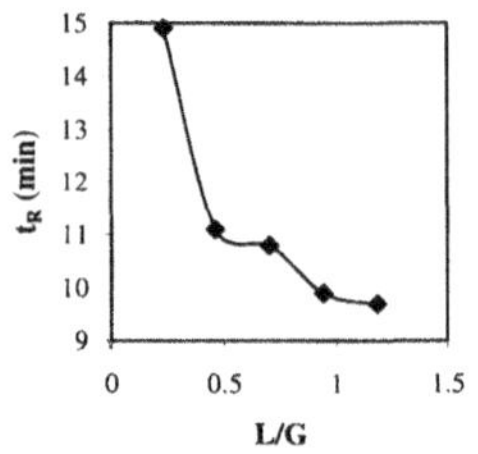

Figure 3. Influence of the ratio of flow rates of liquid to gas (L/G) on carbonation time t_R.

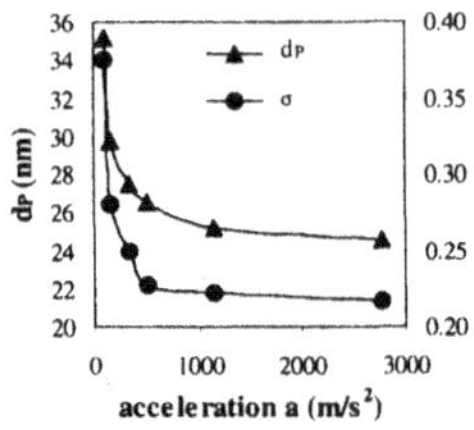

Figure 4. Effect of acceleration a on mean d_P and standard deviation σ of particle size.

Figure 5 shows the influence of L/G with G kept constant. Figure 6 shows a typical TEM picture of $CaCO_3$ nanoparticles synthesized by this HGRP method and a comparison with some commercially-available samples of $CaCO_3$ particles. The nanoparticles of $CaCO_3$ synthesized by HGRP were around 30nm in size, much smaller than the commercial particles which were 70~100nm.

Evaluation of Application Performance of Nanoparticles

Nanoparticles of $CaCO_3$ can be used as reinforcement or rheopexy-functional filler in rubber, adhesives, plastics, paints, coatings, etc. Table 1 gives experimental test results for the reinforcement performance of 30nm $CaCO_3$ particles compared with that for μm sized.

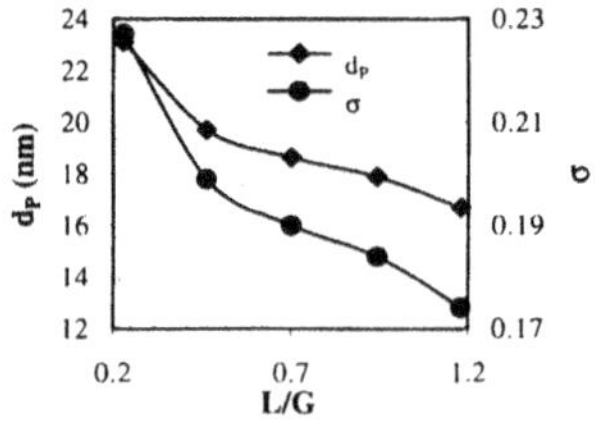

Figure 5. Influence of L/G on mean and standard deviation of particles size.

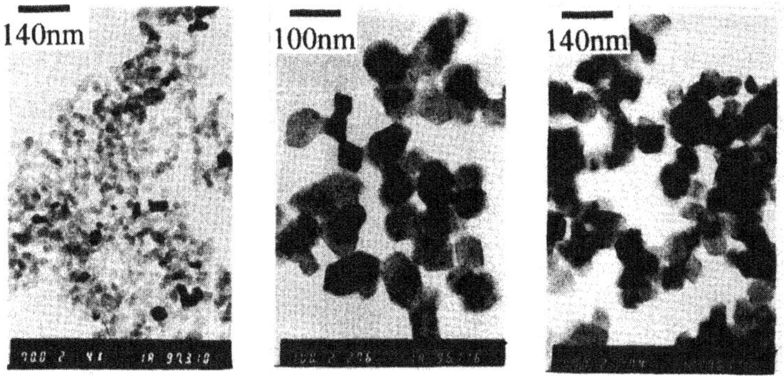

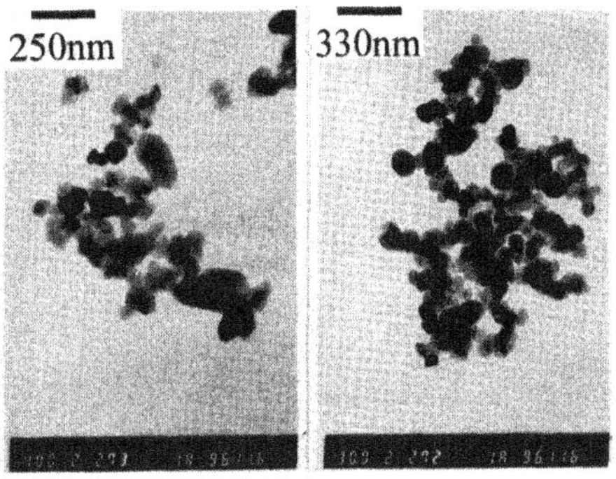

Sources	mean size (nm)
HGRP	30
U.S.A.	75
Japan	80
France	90
England	100

Figure 6. Comparison of $CaCO_3$ nanoparticles synthesized by HGRP with commercial samples from various sources.

Table 1. Comparison of reinforcement of polyvinyl chloride (PVC) with nano-sized $CaCO_3$ particles and μm-sized $CaCO_3$ fillers (polymerization degree=1000).

Material	Tensile stress	Elongation at rupture
Pure PVC	7 MPa	350 %
PVC with μm $CaCO_3$	7 MPa	330 %
PVC with 30nm $CaCO_3$	11 MPa	480 %

The μm-sized particles did not provide any reinforcement function while nano-sized particles exhibited significant reinforcement with greater tensile stress and higher elongation ratio.

Figure 7 shows the effect of particle size on the rheology of PVC filled with $CaCO_3$ particles. The 30nm $CaCO_3$ nanoparticles caused a significant rheopexy character while the μm-sized particles did not. Note that rheopexy plays a key role in painting and coating applications. Figure 8 shows that nanoparticles promoted more rapid thermal curing of polymer composites.

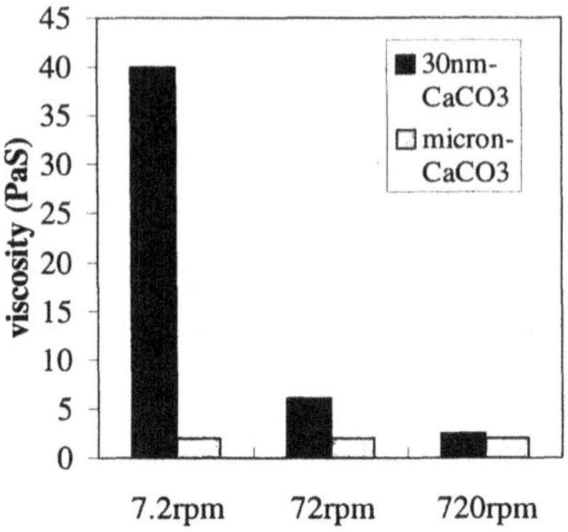

Figure 7. Effect of particle size on the rheology character of PVC.

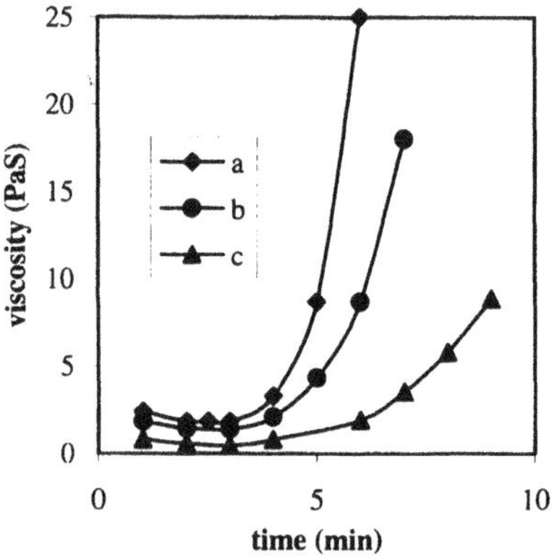

Figure 8. Change of viscosity of PVC plastisol with and without being filled with 30%w $CaCO_3$ during thermal curing at 85 ^0C and 720rpm: (a) 30nm$CaCO_3$; (b) ~1μm$CaCO_3$; (c) no filling.

CONCLUSIONS

A new technology, high-gravity multiphase reactive precipitation, was presented here for synthesis of nanoparticles with mass production and low cost. Nanoparticles of $CaCO_3$ were synthesized from lime and CO_2 to give a controllable mean size ranging from 15nm to 40nm. Such nanoparticles have unique functional applications in the polymer industry compared to μm-sized particles. Therefore, it has a strong potential for commercialization.

Acknowledgements

This work was supported by the National Science Foundation of China via grant 29606001 and by the National High Technology Program "863" via grant 715-009-0080.

NOTATION

A, B = reactants
a = acceleration, m/s^2
c = concentration, in weight % or mol/m^3
c^* = supersaturation concentration, mol/m^3
D = coefficient of diffusion, m^2/s
d = diameter of molecules, or the average particles size, m
G = gas volume flow rate, m^3/s
g = gas phase
k = constant of reaction rate
k_g = constant of crystal growth rate
k_m = constant
k_N = constant of nucleation rate
l = characteristic size of crystal, m; or liquid phase
L = liquid/slurry volume flow rate, m^3/s
m = nucleation rate order
n = number of nuclei per volume, m^{-3}
n^* = number of ions embodied in a crystal embryo
P = product P
P^* = crystal embryo
$P_1 \cdots P_n$ = crystals of different sizes
S = ratio of super-saturation concentration to saturated concentration, c/c^*
s = solid phase
t_R = carbonation time, s
t_m = micromixing characteristic time, s

Greek letters
α, β = orders of reaction
γ = order of nucleation
ε = energy dissipation rate per mass, w/kg
τ = induction time, s
υ = kinematic viscosity, m^2/s

Abbreviations
HGRP = high-gravity multiphase reactive precipitation
PSD = particle size distribution

REFERENCES

1. J.A. Dirksen and T.A. Ring, Fundamentals of crystallization -kinetic effects on particle size distributions and morphology, *Chem. Eng. Sci.* 46: 2389 (1991).
2. J.F. Chen, C. Zheng, and G.T. Chen, Interaction of macro- and micromixing on particle size distribution in reactive precipitation, *Chem. Eng. Sci.* 51:1957 (1996).
3. G. Tosun, An experimental study of the effect of mixing on the particle size distribution in $BaSO_4$ precipitation reaction, in: *Proceedings of 6th European Conference on Mixing,* Cranfield, England: BHRA (1988) p 161.
4. X. Li, J.F. Chen, and G.T. Chen, Morphological configurations of material elements during turbulent mixing - experimental study and modeling, *Acta Mechanic Sinica.* 26: 266 (1994).
5. R. Mallinson and C. Ramshaw, *European Patent* 2568B(1979), *US Patent* 4263255 (1981).
6. C. Ramshaw, Higee distillation - an example of process intensification, *The Chem. Engr.* 389:13 (1983).
7. R. Fowler, Higee- a status report, *The Chem. Engr.* 456: 35 (1989).
8. C. Zheng, X.M. Zhou, D.G. Ai, K. Gao, J.F. Chen, J.S. Zhu, Y.D. Feng and N. Gardner, *Advances in Rotating Packed Bed Technology. 5th World Congress of Chemical Engineering*, San Diego, California (July 14-18, 1996).
9. J. Liu, J.F. Chen, Y.H. Song and C. Zheng, Experimental study on the micromixing in a novel rotating packed bed reactor, *Chemical Reaction Engineering and Technology,* 15(3): 327-332 (1999).
10. Y. F. Gu, S. Wang, L. M. Hu and A. I. Zhang, Synthesis of ultrafine $CaCO_3$ particles by reactive precipitation. *J. East China Inst. Chem. Techn.* 15: 550 (1993).

SYNTHESIS AND APPLICATION OF NANOPARTICLES BY MULTIPHASE REACTIVE PRECIPITATION IN A HIGH-GRAVITY REACTOR: II: MODELING

Yuhong Wang and Jianfeng Chen

Higravitec Center, Department of Chemical Engineering
Beijing University of Chemical Technology, Beijing 100029, China

ABSTRACT

A mathematical model was developed to describe the particle size distribution of nanoparticles (e.g., $CaCO_3$) synthesized by multiphase reactive precipitation in a high-gravity reactor reported in part I. This model utilizes a material balance, population balance, kinetics of crystallization, and mass transfer equations. The agreement between predictions and experimental results is satisfactory.

INTRODUCTION

In part I of this work, we reported promising results for the experimental synthesis of nano-$CaCO_3$ with the novel high-gravity multiphase reactive precipitation. A mathematical model would be helpful for process scale-up, engineering and control, and would also be useful to understand the process mechanism. Here, we present a mathematical model to describe the particle size distribution resulting from the synthesis of nanoparticles with the high-gravity reactive precipitation technology.

The synthesis of nano-$CaCO_3$ particles is used as a demonstration system. It is known that the carbonation reaction between CO_2 gas and a $Ca(OH)_2$ slurry mainly involves gas-liquid mass transfer, chemical reaction and crystallization. Juvekar and Sharma,[1] Capuder and Koloini,[2] and Sada[3, 4] studied the chemical absorption of CO_2 gas into an aqueous $Ca(OH)_2$ solution using the double-film model. The difference between the predictions of the double-film model, penetration model and surface-renewal model was less than 12%.

We have investigated experimentally the chemical absorption of CO_2 gas into a $Ca(OH)_2$ solution in a high-gravity reactor.[5] The liquid-side mass transfer coefficient k_L, effective specific interfacial area a_E, liquid-side volumetric mass transfer coefficient k_La_E, and surface renewal frequency s were obtained. These data are used here.

PROCESS MODELING

Physical Model

The experimental flowsheet and the procedures were reported in part I. Figure 1 is a schematic diagram of the carbonation process with high-gravity reactive precipitation. PFR denotes the high-gravity reactor (rotating-packed bed reactor), CSR the stirred feed and recycle tank. The flow direction of the fluid in the packing of the high-gravity reactor is parallel to the radius direction of the rotator.

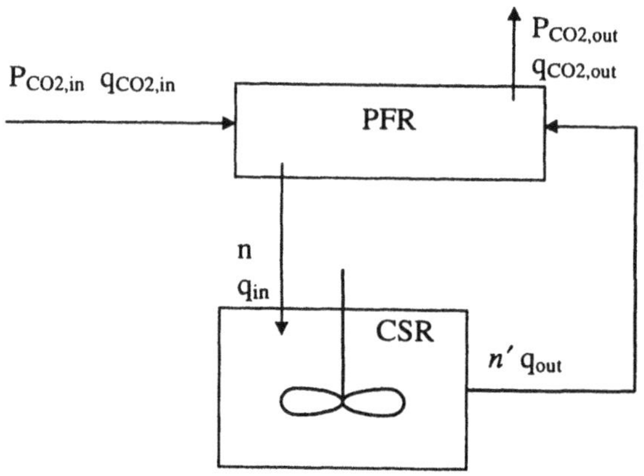

Figure 1. Flow diagram for high-gravity reactive precipitation of $CaCO_3$.

The following assumptions are made:

1. The absorption reaction and crystal nucleation mainly take place in the packed section of the high-gravity reactor.
2. No reaction takes place in the CSR.
3. Plug flow in the high-gravity reactor.
4. Continuous well-mixed flow in the stirred tank.
5. The crystal growth obeys the ΔL law, that is, the growth rate of each particle is independent of its size ($dG/dL=0$).
6. The rates of the nucleation and growth of the particles mainly depend on the supersaturation of solute.
7. Neglect aggregation and break-up of particles.
8. Neglect interaction between reactant particles (here, $Ca(OH)_2$) and product particles ($CaCO_3$).
9. The volume of the system remains constant.
10. Completely mixed at the micro-scale in both PFR and CSR.
11. In the high-gravity reactor, the surface of the liquid is renewed continuously and its volume remains constant ($dV_A/dt=0$).
12. The dissolution of $Ca(OH)_2$ particles is faster than the rate of the CO_2 absorption.

Mathematical Model

The mechanism of the carbonation reaction can be expressed as:

$$CO_{2(G)} + H_2O_{(L)} \Leftrightarrow H_2CO_{3(L)} \Leftrightarrow H^+{}_{(L)} + HCO^-_{3(L)} \Leftrightarrow 2H^+{}_{(L)} + CO_3^{2-}{}_{(L)} \quad (1)$$

$$Ca(OH)_{2(S)} + H_2O_{(L)} \Leftrightarrow Ca(OH)_{2(AQ.)} \Leftrightarrow Ca^{2+}{}_{(L)} + 2OH^-{}_{(L)} \quad (2)$$

$$Ca^{2+}{}_{(L)} + HCO^-_{3(L)} \Leftrightarrow Ca(HCO_3)_{2(L)} \Leftrightarrow CaCO_{3(S)} \quad (3)$$

$$Ca^{2+}{}_{(L)} + CO_3^{2-}{}_{(L)} \Leftrightarrow CaCO_{3(S)} \quad (4)$$

$$H^+{}_{(L)} + OH^-{}_{(L)} \Leftrightarrow H_2O_{(L)} \quad (5)$$

The carbonation reaction and the reaction rate are:

$$A + B \rightarrow P\downarrow \quad (6)$$

$$r_A = -\frac{dc_A}{dt} = k_r c_A c_B = k_I c_A \quad (7)$$

$$k_I = k_r c_B \quad (8)$$

where the symbols are defined at the end. The model is given below for each unit separately.

PRF. The material balance equations are:

$$\frac{dc_A}{dt} = k'_L a_E (HP_A - c_A) - k_r c_A c_B \quad (9)$$

$$\frac{dc_B}{dt} = -k_r c_A c_B \quad (10)$$

$$\frac{dc_P}{dt} = k_r c_A c_B - \frac{\rho_P k_V}{M_P}(3G_q \mu_2 + B_0 L_0^3) \quad (11)$$

The moment equations of the population balance are:

$$\frac{d\mu_0}{dt} = B_0 \quad \text{(0 order moment)} \quad (12)$$

$$\frac{d\mu_i}{dt} = iG_q \mu_{i-1} + B_0 L_0^i \quad (i \geq 1, \text{ integers}) \quad (13)$$

The kinetics equations for crystallization are:

$$B_0 = k_N (c_p - c_{ps})^{m_1} \quad (14)$$

$$G_q = k_q (c_p - c_{ps})^{q_1} \quad (15)$$

The coefficients for gas-liquid mass transfer rate are:[5]

$$k'_L a_E = E k_L a_E \quad (16)$$

$$k_L a_E = 0.69042 L_R^{0.15971} (\omega^2 R)^{0.17096} \quad (17)$$

$$E = \sqrt{1 + 9.285 \times 10^5 D_A k_2 c_{OH^-} L_R^{-0.10702} (\omega^2 R)^{-0.07452}} \quad (18)$$

The initial conditions at t = 0 are:

$$c_j = c_j' \quad \text{(j=A, B, P)} \quad (19)$$

$$\mu_i = \mu_i' \quad \text{(i=0, 1, 2, 3, ...)} \quad (20)$$

CSR. The material balance equations are:

$$\frac{dc_A'}{d\theta} = \frac{q}{V_{CSTR}}(c_A - c_A') \quad (21)$$

$$\frac{dc_B'}{d\theta} = \frac{q}{V_{CSTR}}(c_B - c_B') \quad (22)$$

$$\frac{dc_P'}{d\theta} = \frac{q}{V_{CSTR}}(c_P - c_P') - \frac{\rho_p k_V}{M_P}(3G_q'\mu_2' + B_0' L_{01}^3) \quad (23)$$

The moment equations of the population balance are:

$$\frac{d\mu_0'}{d\theta} = B_0' + \frac{q}{V_{CSTR}}(\mu_0 - \mu_0') \quad \text{(0 order moment)} \quad (24)$$

$$\frac{d\mu_i'}{d\theta} = iG_q'\mu_{i-1}' + B_0' L_{01}^i + \frac{q}{V_{CSTR}}(\mu_i - \mu_i') \quad \text{(i≥1, integers)} \quad (25)$$

The kinetics equations for crystallization are:

$$B_0' = k_N'(c_P' - c_{PS})^{m_2} \quad (26)$$

$$G_q' = k_q'(c_P' - c_{PS})^{q_2} \quad (27)$$

The initial conditions for $\theta = 0$ are:

$$c_j' = 0 \quad \text{(j=A, B, P)} \quad (28)$$

$$\mu_i' = 0 \quad \text{(i=0, 1, 2, 3, ...)} \quad (29)$$

Here, μ_i is the ith order moment of the population density n, and is defined as:

$$\mu_i = \int_0^\infty n L^i dL \quad (30)$$

Several methods can be used to characterize the particle size and particle size distribution for non-spherical particles using the equations below to calculate the mean particle size and the dimensionless variance of the particle size distribution.

- Mean particle size and variance based on the number:

$$d_{10} = \mu_1 / \mu_0 \tag{31}$$

$$\sigma_{10} = \sqrt{(\mu_2 / \mu_0) - (\mu_1 / \mu_0)^2} \tag{32}$$

- Mean particle size and variance based the particle area:

$$d_{32} = \mu_3 / \mu_2 \tag{33}$$

$$\sigma_{32} = \sqrt{(\mu_4 / \mu_2) - (\mu_3 / \mu_2)^2} \tag{34}$$

- Mean particle size and variance based on the volume or mass:

$$d_{43} = \mu_4 / \mu_3 \tag{35}$$

$$\sigma_{43} = \sqrt{(\mu_5 / \mu_3) - (\mu_4 / \mu_3)^2} \tag{36}$$

Thus the mean and variance of the particle size distribution can be calculated if the n moments of the population density are known. In industry, the particle size and particle size distribution for non-spherical particles are often expressed as the volumetric mean particle size and variance, for which we need up to the fifth order moment (Eqs. (35) and (36)).

The experimental measurement of L_0 and L_{01} is very difficult. For $CaCO_3$, the Kelvin equation gives L_0=10 nm for primary homogeneous nucleation.[5] Here, we assume that L_0 in the PFR is 1 nm and L_{01} in the CSR is 10 nm.

Solution of Mathematical Equations

Generally speaking, if we know the kinetics of crystallization in both PFR and CSR as well as the mass transfer coefficients in PFR, the above equations do not have any unknown parameters, and so the particle size distribution of product can be predicted for any operating conditions. If the kinetics of crystallization are not known, there are 8 parameters that must be assumed: k_N, m_1, kq, q_1, k'_N, m_2, k'_q, q_2. Fortunately, the rates of nucleation and growth of $CaCO_3$ during the carbonation reaction in a stirred stank. have been measured. Shun and Alan,[6] for example, obtained the following typical results:

$$\text{Nucleation rate:} \quad J_N = 1.0 \times 10^7 \left(c_p - c_{ps}\right)^{4.2} \tag{37}$$

$$\text{Growth rate:} \quad J_G = 8.06 \times 10^{-9} \left(c_p - c_{ps}\right)^2 \tag{38}$$

We used these expressions here and set:

$$B'_0 = J_N \tag{39}$$

$$G'_q = J_G \tag{40}$$

The above equations were numerically by the Gear method, with auto-adjustment of the step and order during the calculation.[5]

By fitting the experimental data, the crystallization kinetics in PFR were found to be:

$$\text{nucleation rate: } B_0 = 1.08 \times 10^{24} (c_p - c_{ps})^{6.7} \tag{41}$$

$$\text{growth rate: } G_q = 1.16 \times 10^{-7} (c_p - c_{ps})^2 \tag{42}$$

COMPARISON OF MODEL PREDICTIONS WITH EXPERIMENTS

Figure 2 shows typical results comparing experimental results with theoretical predictions. The average difference was 7.9% for mean particle size and 11.4% for the variance of particle size. The agreement between model and experiment is satisfactory.

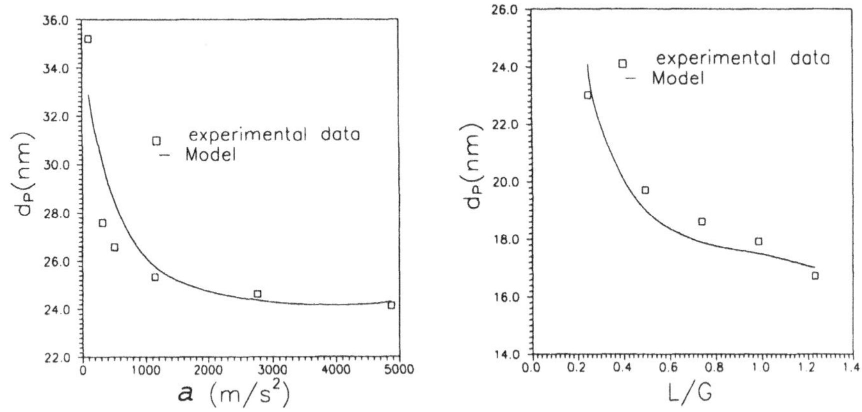

Figure 2. Comparison of model calculations and experiments for average particle size d_p.

Acknowledgements

This work was supported by the National Science Foundation of China under grant 29606001 and the National High Technology Program "863" under grant 715-009-0080.

NOMENCLATURE

A = reactant (here, CO_2)
a = high-gravity acceleration
a_E = effective specific interfacial area • m^2/m^3
B = reactant (here, $Ca(OH)_2$)
B_0 = crystal nucleation rate in PFR, number/m^4/s
B_0' = crystal nucleation rate in CSTR, number/m^4/s
c_A = concentration of A in PFR, mol/m^3
c_A' = concentration of A in CSR, mol/m^3
c_B = concentration of B in PFR, mol/m^3
c_B' = concentration of B in CSR, mol/m^3
c_{OH^-} = concentration of OH^-, mol/m^3
c_P = concentration of P in PFR, mol/m^3
c_P' = concentration of B in CSR, mol/m^3
c_{PS} = saturation concentration of P, mol/m^3
D_A = diffusion coefficient of A in the solution, m^2/s
d_{10} = mean particle size of number basis, m
d_{32} = mean particle size of area basis, m
d_{43} = volumetric mean particle size, m
d_p = mean particle size of product particle, m
E = chemical absorption factor
G = volumetric flow rate of gas, m^3/s
G_q = crystal growth rate in PFR, m/s
G'_q = crystal growth rate in CSR, m/s
H = coefficient of dissolution, mol/m^3/Pa
J_G = growth rate, m/s
J_N = nucleation rate, number/m^4/s
k_2 = chemical reaction rate constant
k_I = chemical reaction rate constant
k_N = crystallization rate constant in PFR
k_q = crystallization rate constant in PFR
k_V = shape factor of volume
k_J = frequency factor
k_L = mass transfer coefficient of physical absorption
k_N' = crystallization rate constant in CSR
k_q' = crystallization rate constant in CSR
k_r = chemical reaction rate constant
k_L' = mass transfer coefficient of chemical absorption
k_0 = coefficient
k_0' = coefficient
L = volumetric flow rate of recycled liquid, m^3/s
L_0 = primary particle size of crystal of $CaCO_3$ in PFR, m
L_{01} = primary particle size of crystal of $CaCO_3$ in CSR, m
L_R = liquid density in packing at radius R, $m^3/m^2/s$
M_p = molecular weight of P
m_1 = unknown parameter
m_2 = unknown parameter
n = number density of product particles ($CaCO_3$) in outlet of PFR, number/m^3/m

n' = number density of product particles ($CaCO_3$) in inlet of PFR, number/m^3/m
P = product (here, $CaCO_3$)
P_A = pressure of A in gas phase, Pa
$P_{CO2.in}$ = pressure of A in inlet of gas, Pa
$P_{CO2.out}$ = pressure of A in outlet of gas, Pa
q = volumetric flow rate, m^3/s
q_1 = unknown parameter
q_2 = unknown parameter
q_{in} = volumetric flow rate in inlet of CSR, m^3/s
q_{out} = volumetric flow rate in outlet of CSR, m^3/s
$q_{CO2.in}$ = volumetric flow rate of CO_2 in inlet of gas, m^3/s
$q_{CO2.out}$ = volumetric flow rate of CO_2 in outlet of gas, m^3/s
R = radius of packing, m
r_A = chemical reaction rate of reactant A, mol/m^3/s
t = time, s
V_{CSR} = volume of liquid in CSR, m^3

μ_i = ith order moment in PFR
μ_i' = ith order moment in CSR
θ = time, s

REFERENCES

1. V. A. Juvekar and M. M. Sharma, Absorption of CO_2 in a suspension of lime, *Chem. Eng. Sci.* 28: 825 (1973).
2. E. Capuder and T. Koloini, *Chem. Eng. Res. Des.* 62: 110 (1984).
3. E. Sada, H. Kumazawa and C. H. Lee, *Chem. Eng. Sci.* 38: 79 (1983).
4. E. Sada, H. Kumazawa, C. H. Lee, and N. Fujiwara, *Ind. Eng. Chem. Proc. Des.* 24: 16 (1983).
5. Y.H. Wang, *Study of Synthesis and Scale-Up of Nanoparticles of Calcium Carbonate by High-Gravity Reactive Precipitation*, Ph.D. dissertation, Beijing University of Chemical Technology (1998).
6. W. Shun and G.J. Alan, Mass transfer with chemical reaction and precipitation, *Chem. Eng. Sci.* 46: 1027 (1991).

SPREADING OF A THIN LIQUID DROP UNDER THE INFLUENCE OF GRAVITY, ROTATION AND NON-UNIFORM SURFACE TENSION

Ebrahim Momoniat

Department of Computational and Applied Mathematics, University of the Witwatersrand, Wits, 2050, Johannesburg, South Africa

INTRODUCTION

Figure 1 shows an axisymmetric thin liquid drop that is spreading under the influence of gravity, rotation and non-uniform surface tension. This process is modeled using the Lie group method. Functional forms of the non-uniform surface tension $\Sigma(x)$ are obtained. A fourth-order nonlinear partial differential equation describing the evolution of the free surface, $u(t,x)$, of the liquid drop is derived by imposing the thin-film approximation on the Navier-Stokes equations and solving the resulting system of equations subject to boundary conditions on the surface of the rotating disk and on the free surface of the liquid drop. Exact and approximate Lie point symmetry generators admitted by the free surface equation are determined. These symmetry generators are imposed on the moving boundary, $x=R(t)$, where $R(t)$ is the radius of the foot of the liquid drop at time t and x is the radial coordinate fixed on the rotating disk. Functional forms of $R(t)$ are determined. The rate of spreading $dR(t)/dt$ can then be easily calculated.

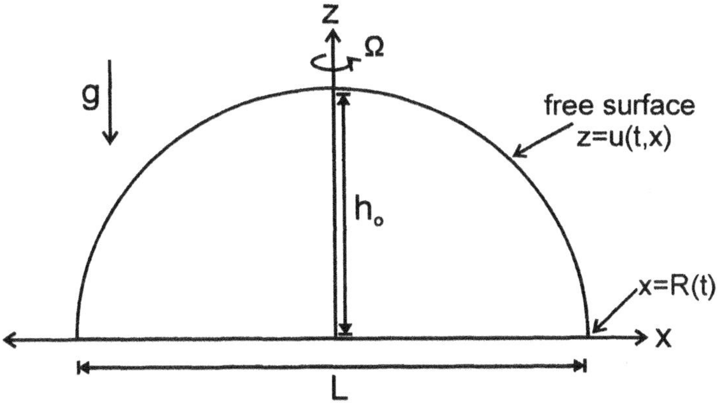

Figure 1. Schematic diagram of the spreading axisymmetric drop.

Processing by Centrifugation, Edited by Regel and Wilcox
Kluwer Academic/Plenum Publishers, New York, 2001

The free surface equation, which is derived in this paper, is important in the modeling of coating flows,[1-3] the manufacture of optical disks, and the spread of molten lava. The functional form of $R(t)$ that is determined can be used to represent the industrial process of spin coating to determine how far or how fast a liquid drop will spread. The function $R(t)$ can also be used as an ansatz in a curve-fitting algorithm when analyzing experimental observation of thin film flow.

Derivation of Equations

We introduce the following characteristic lengths: let L be the characteristic length of the liquid drop, h_o the characteristic height and U the characteristic horizontal velocity. The thin film approximation implies that:[4]

$$\frac{h_o}{L} << 1, \qquad \mathrm{Re}\left(\frac{h_o}{L}\right)^2 << 1 \tag{1}$$

where Re is the Reynolds number defined as UL/ν where ν is the kinematic viscosity. We introduce the following dimensionless variables:

$$\begin{gathered}\overline{x}=\frac{x}{L}, \quad \overline{z}=\frac{z}{L}, \quad \overline{v_x}=\frac{v_x}{U}, \quad \overline{v_\theta}=\frac{v_\theta}{U}, \\ \overline{v_z}=\frac{Lv_z}{h_o U}, \quad \bar{t}=\frac{Ut}{L}, \quad \overline{p}=\frac{l}{\mu U}\left(\frac{h_o}{L}\right)^2 p, \quad \overline{\underline{b}}=\frac{L}{U^2}\underline{b},\end{gathered} \tag{2}$$

where x is the coordinate in the radial direction, z is the coordinate in the vertical direction and θ is the angular coordinate. The velocity vector $\underline{v}$ has components (v_x, v_z, v_θ), which represent the velocity components in the horizontal, vertical and angular directions, respectively. The quantity p is the pressure and $\underline{b}$ is the body force. The only body force present in the problem under consideration is gravity, thus $\underline{b}=(0,0,-g)$ where g is the gravitational constant.

We need to determine an estimate for the characteristic horizontal velocity U in terms of known parameters. We first balance the pressure gradient force per unit volume in the horizontal plane with the viscous force per unit volume to obtain:

$$p=\frac{\mu UL}{h_o^2} \tag{3}$$

where μ is the shear viscosity of the fluid. By balancing the pressure gradient force per unit volume that is normal to the plane with the force of gravity per unit volume we obtain:

$$p=\rho g h_o \tag{4}$$

where ρ is the fluid density. By equating Eqs. (4) and (5) we note that the characteristic horizontal velocity in the horizontal direction can be given as:

$$U=\frac{gh_o^3}{\nu L}, \tag{5}$$

where $\nu=\mu/\rho$. We introduce the thickness, E, of the Ekman boundary layer in a rotating fluid:[4,5]

$$E = \left(\frac{\nu}{\Omega}\right)^{\frac{1}{2}}. \tag{6}$$

The thin film approximation given by Eq. (1) then simplifies to:

$$\frac{h_o}{L} << 1, \quad \left(\frac{h_o}{E}\right)^4 << 1. \tag{7}$$

We substitute the dimensionless variables in Eq. (2) with the characteristic horizontal velocity U given by Eq. (5) into the Navier-Stokes equation for a rotating frame:[5]

$$\frac{D\underline{v}}{Dt} = -\frac{1}{\rho}\text{grad}\left(p - \frac{1}{2}\rho(\underline{\Omega}\times\underline{x})^2\right) - 2\underline{\Omega}\times\underline{v} + \underline{b} + \nu\nabla^2\underline{v} \tag{8}$$

and the continuity equation:

$$\text{div}\,\underline{v} = 0. \tag{9}$$

Ignoring terms with coefficients satisfying Eq. (7), we obtain the system:

$$\frac{\partial p}{\partial x} = \left(\frac{F}{Ro}\right)^2 x + \frac{\partial^2 v_x}{\partial z^2}, \tag{10}$$

$$\frac{1}{x}\frac{\partial p}{\partial \theta} = \frac{\partial^2 v_\theta}{\partial z^2}, \tag{11}$$

$$\frac{\partial p}{\partial z} = -1, \tag{12}$$

$$\frac{1}{x}\frac{\partial}{\partial x}(x v_x) + \frac{1}{x}\frac{\partial v_\theta}{\partial \theta} + \frac{\partial v_z}{\partial z} = 0. \tag{13}$$

Here, F is the dimensionless Froude number defined as the ratio $U/(gh_o)^{1/2}$ and Ro is the Rossby number given by the ratio $U/L\Omega$. The system Eqs. (10)-(13) is solved subject to the following boundary conditions on the surface of the disk ($z=0$):

$$v_x(x,0,t) = 0, \text{ (a)} \quad v_\theta(x,0,t) = 0, \text{ (b)} \quad v_z(x,0,t) = 0. \text{ (c)} \tag{14}$$

The boundary conditions at the free surface of the liquid drop, $z=u(x,t)$, are:

$$p(x,u,t) = p_o - \frac{1}{Bo}\Sigma(x)\left(u_{xx} + \frac{1}{x}u_x\right) \tag{15}$$

$$\frac{\partial v_x}{\partial z}(x,u,t) = \left(\frac{L}{h_o}\right)^2 \frac{1}{Bo}\frac{d\Sigma(x)}{dx} \tag{16}$$

$$\frac{\partial v_\theta}{\partial z}(x,u,t)=0, \tag{17}$$

$$v_z(x,u,t)=u_t+u_x v_x(x,u,t) \tag{18}$$

Here, Bo is a dimensionless Bond number defined as $\rho g L^2/\sigma$, where σ is the characteristic surface tension at the boundary $x=R(t)$ when $t=0$. The boundary conditions Eq. (14a) and Eq. (14b) imply that there is no slip between the drop and the rotating disk. Boundary condition (14c) implies that no cavities form in the drop. Conditions (15), (16) and (17) result from the presence of a non-uniform surface tension, $\Sigma(x)$, at the free surface of the liquid drop. Atmospheric pressure is p_o is. Condition (18) occurs as a result of the fact that any fluid element on the free surface remains on the free surface throughout the motion of the liquid drop.[4] Solving the system of Eqs. (10)-(13) subject to (14)-(18), we obtain the following fourth-order nonlinear evolution-type partial-differential equation that describes the evolution of the free surface of the liquid drop:

$$u_t=-\frac{1}{3x}\frac{\partial}{\partial x}\left(xu^3u_x-\frac{1}{Bo}xu^3\frac{\partial}{\partial x}\left(\Sigma(x)\left(u_{xx}+\frac{1}{x}u_x\right)\right)-\frac{3}{2}\left(\frac{L}{h_o}\right)^2\frac{1}{Bo}xu^2\Sigma_x(x)\right)$$

$$-\left(\frac{F}{Ro}\right)^2\frac{1}{3x}\frac{\partial}{\partial x}\left(x^2u^3\right) \tag{19}$$

In deriving Eq. (19) we assumed that the flow is axisymmetric, i.e. all quantities are independent of θ. Myers[6] reviewed fourth-order nonlinear partial differential equations that model surface-tension phenomena. Huppert[7] obtained solutions for spreading under gravity alone. Middleman[8] showed how a similarity solution for spreading under gravity alone can be determined. Emslie *et al.*[9] considered spreading under rotation only, ignoring the effects of gravity, surface tension and Coriolis force. Momoniat and Mason[10] considered the effects of Coriolis force on spreading with rotation only. The effects of gravity, rotation and non-uniform surface tension are included in Eq. (19), although the effects of Coriolis force are not included.

EXACT LIE GROUP METHOD

Now we use the Lie group method to analyze Eq. (19). The Lie group method is used extensively to determine analytic solutions to differential equations. The interested reader is referred to references[11-17] for details of the Lie group method. Hill[18] and Hill and Hill[19] showed how to determine solutions to evolution-type equations using group methods.

To use the Lie group method, we need to determine the functions ξ^1,ξ^2 and η such that:

$$X=\xi^1(t,x,u)\partial_t+\xi^2(t,x,u)\partial_x+\eta(t,x,u)\partial_u \tag{20}$$

is a Lie point symmetry generator admitted by Eq. (19). Details of the required calculation can be found in the references listed above. Alternatively, the ansatz proposed by Momoniat and Mahomed[20] can be used to determine Lie point symmetry generators admitted by Eq. (19). Algebraic computer packages are available to perform this calculation. Using MATHEMATICA,[21] for example, we find that Eq. (19) admits the Lie point symmetry generator:

$$X = \partial_t \tag{21}$$

Group invariant solutions corresponding to the Lie point symmetry generator Eq. (20) can be determined by solving:

$$X\left(u - \Phi(t,x)\right)\Big|_{u=\Phi(t,x)} = 0. \tag{22}$$

Substituting Eq. (21) into Eq. (22) we obtain the stationary solution:

$$u = F(x), \tag{23}$$

where F is an arbitrary function of x. We also find that for Eq. (19) to admit Eq. (21), we must have that:

$$\Sigma(x) = x^2. \tag{24}$$

A non-uniform surface tension may be caused by the presence of surfactants on the free surface of the liquid of the drop.[22] A surfactant can be thought of as something whose concentration at the surface of the drop exceeds that in the adjacent solution. This is because the surfactant lowers the free energy (surface tension) of the surface. When the liquid drop moves, the surface excess deviates from equilibrium with the adjacent solution. it tends to exceed the equilibrium value at the end of the moving surface, leading to transport of surfactant back into the solution. The reverse is true at the beginning of the moving surface. The net effect is to retard the movement of the surface. Equation (24) implies that the deviation from equilibrium caused by surfactants on the surface of the liquid drop increases as a square of the distance from the center of the liquid drop. This non-uniform surface tension may be strong enough to prevent the liquid drop from moving, causing the drop to behave as if it were rigid.

APPROXIMATE LIE GROUP METHOD

The approximate group method has been introduced to determine group invariant solutions for equations that admit a small parameter.[23-27] Here, we adopt the approach described by Gazizov[28] to determine approximate Lie point symmetry generators admitted by Eq. (19). We consider the case when:

$$\left(\frac{F}{Ro}\right)^2 = \varepsilon << 1. \tag{25}$$

This assumption implies that the disk is rotating slowly. The dimensionless angular velocity is then given by:

$$\Omega = \sqrt{\frac{gh_o\varepsilon}{L^2}} \tag{26}$$

Multiplying Eq. (26) by $\sqrt{h_o / h_o}$ we find that:

$$\Omega = \delta\sqrt{\frac{g}{h_o}} \tag{27}$$

where $\delta = (h_o / L)\sqrt{\varepsilon}$ is very small since Eq. (1) holds and $\varepsilon << 1$. We note that the magnitude of the angular velocity depends on the ratio of g to the characteristic height of the liquid drop. We can assume that as the value of g increases, the characteristic height of the liquid drop, which we can take as the initial height of the drop, decreases. Hence:

$$h_0 = \frac{\kappa}{g}, \tag{28}$$

where κ is a proportionality constant that depends on the physical properties of the liquid drop. We do not investigate the effect of chemical composition on the spreading of the liquid drop. We can, however, conclude that the approximation Eq. (25) and the results we obtain by imposing this approximation are dependent in a direct way on the chemical composition of the liquid drop and the value of g acting on the liquid drop. Substituting Eq. (28) into Eq. (27) we observe that:

$$\Omega = \left(\frac{\delta}{\sqrt{\kappa}}\right)g \tag{29}$$

We now investigate approximate symmetry generators of the form:

$$X \approx X_0 + \varepsilon X_1 \tag{30}$$

where X_0 is the symmetry generator admitted by the unperturbed part of Eq. (19). We find that Eq. (19) admits the approximate Lie point symmetry generator:

$$X \approx (a_2 + \varepsilon(tb_1 + b_2))\partial_t - \varepsilon b_1 x\partial_x - \varepsilon b_1 u\partial_u. \tag{31}$$

where a_2, b_1 and b_2 are arbitrary constants.

'EXACT' RATE OF SPREAD

Instead of using the approximate symmetry generator Eq. (31) to determine group invariant solutions to Eq. (19) that may not be physical, we now show how Eq. (31) can be used to determine the radius of the liquid drop. The behaviour at the interfaces of the liquid drop with the rotating disk and with the air has been of considerable interest to researchers for some time,[29-34] In this section, we show how to overcome the physics of this interface by treating it as a moving boundary and imposing an invariance principle from mathematical physics.[11-13,17]

Since the liquid drop is incompressible and the flow is axisymmetric, the total volume, V, of the liquid drop is conserved and is given by:

$$V = 2\pi\int_0^{R(t)} xu(t,x)dx. \tag{32}$$

Conservation of volume implies that:

$$\frac{dV}{dt} = \int_0^{R(t)} x u_t \, dx + R'(t) x u \Big|_{x=R(t)} = 0. \tag{33}$$

By an invariance principle,[11-13,17] we impose the approximate Lie point symmetry generator Eq. (31) on the moving boundary $x=R(t)$ and on Eq. (33). Imposing Eq. (31) on the boundary $x=R(t)$ as follows:

$$X\left(x - R(t)\right)\Big|_{x=R(t)} = 0 \tag{34}$$

and solving the resulting ordinary differential equation for $R(t)$ we find that:

$$R(t) = \frac{a_2}{a_2 + \varepsilon b_1 t}. \tag{35}$$

In this way, the equation obtained by imposing Eq. (31) on Eq. (33), with u_t replaced by Eq. (19) and $R(t)$ by Eq. (35), is satisfied exactly. Equation (35) gives us an expression for the radius of a liquid drop spreading under the influence of gravity and rotation.

CONCLUDING REMARKS

In this paper we derived a fourth-order nonlinear diffusion equation describing the evolution of the free surface of a thin liquid drop spreading axisymmetrically under the influence of gravity, rotation and non-uniform surface tension. We used the approximate Lie group method to determine the rate at which a liquid drop spreads under the influence of a quadratic non-uniform surface tension. Experiments now need to be conducted to verify that Eq. (35) is a good model for describing the radius of the liquid drop at time t. The parameters a_2 and b_1 can be determined by fitting experimental data to Eq. (35). For example, Tanner[35] has derived:

$$R(t) \approx 0.94\left(\frac{\sigma V^3 t}{\mu}\right)^{0.1} \tag{36}$$

to model the spreading of a silicone oil drop on a horizontal surface under the influence of gravity and constant surface tension. Tanner's model is a good fit to the experimental data given by Chen.[36]

The reader is referred to a recent paper by Wilson *et al.*[37] for a discussion of the effects of weak surface tension on the centrifugally driven spreading of a liquid drop.

REFERENCES

1. D.E. Bornside, C.W. Macosko, and L.E. Scriven, On the modeling of spin coating, *J. Imaging Tech.* 13: 122 (1987).
2. W.W. Flack, D.S. Soong, A.T. Bell and D.W. Hess, A mathematical model for spin coating of polymer resists, *J. Appl. Phys.* 56: 1199 (1984).
3. C.J. Lawrence, The mechanics of spin coating of polymer films, *Phys. Fluids,* 31:2786 (1988).

4. D.J. Acheson, *Elementary Fluid Dynamics*, Clarendon Press, Oxford (1990).
5. G. K. Batchelor, *An Introduction to Fluid Mechanics*, Cambridge University Press, Cambridge (1967).
6. T.G. Myers, Thin films with high surface tension, *SIAM Review*, 40:441 (1998).
7. H.E. Huppert, The propagation of two-dimensional and axisymmetric viscous gravity currents over a rigid horizontal surface, *J. Fluid Mech.* 121:43 (1982).
8. S. Middleman, *Modeling axisymmetric flows: dynamics of films, jets and drops*, Academic Press, San Diego (1995).
9. A.G. Emslie, F.T. Bonner, and L.G. Peck, Flow of a viscous liquid on a rotating disk, *J. Appl. Phys.* 29:858 (1958).
10. E. Momoniat and D.P. Mason, Investigation of the effect of the Coriolis force on a thin fluid film on a rotating disk, *Int. J. Non-Linear Mech.* 33:1069 (1998).
11. L.V. Ovsiannikov, *Group Properties of Differential Equations*, Izdat. Sibirsk. Otdel. ANSSSR, Novosibirsk (1962). English translation by G. Bluman (1967).
12. L.V. Ovsiannikov, *Group Analysis of Differential Equations*, Academic Press, New York (1982).
13. P.J. Olver, *Applications of Lie Groups to Differential Equations*, Springer-Verlag, New York (1986).
14. H. Stephani, *Differential Equations: Their Solution Using Symmetries*, Cambridge University Press, Cambridge (1989).
15. G.W. Bluman, and S. Kumei, *Symmetries and Differential Equations*, Springer-Verlag, Berlin (1989).
16. N.H. Ibragimov, *Lie Group Analysis of Differential Equations, Vol. 1*, CRC Press Inc, Boca Raton, Florida (1994).
17. N.H. Ibragimov, *Lie Group Analysis of Differential Equations, Vol. 2*, CRC Press Inc, Boca Raton, Florida (1995).
18. J.M. Hill, Similarity solutions for nonlinear diffusion-A new integration procedure, *J. Eng. Math.* 23:141 (1989).
19. D.L. Hill and J.M. Hill, Similarity solutions for nonlinear diffusion-Further exact solutions, *J. Eng. Math.* 24:109 (1990).
20. E. Momoniat and F. M. Mahomed, The existence of contact transformations for evolution-type equations, *J. Phys. A: Math. Gen.* 32: 8721 (1999).
21. G. Baumann, Symmetry analysis with Mathematica, in: *Modern Group Analysis VI*, N.H. Ibragimov and F.M. Mahomed, eds., New Age International Publishers, New Delhi (1996).
22. F.S. Sherman, *Viscous Flow*, McGraw-Hill, New-York (1990).
23. V.A. Baikov, R.K. Gazizov, and N.H. Ibragimov, Approximate symmetries of equations with a small parameter, *Mat. Sbornik.* 136: 435 (1988). English trans. in Math. USSR Sbornik, 64: 427.
24. W.I. Fushchich, and W. M. Shtelen, On approximate symmetry and approximate solutions on the non-linear wave equation with a small parameter, *J. Phys. A: Math. Gen.* 22: L887 (1989).
25. M. Euler, N. Euler, and A. Köhler, On the construction of approximate solutions for a multi-dimensional nonlinear heat equation, *J. Phys. A: Math. Gen*, 27:2083 (1994).
26. N. Euler, M.W. Shul'ga, and W.H. Steeb, Approximate symmetries and approximate solutions for a multi-dimensional Landau-Ginzburg equation, *J. Phys. A: Math. Gen.* 25:L1095 (1992).
27. N. Euler, and M. Euler, Symmetry properties of the approximations of multi-dimensional generalized Van der Pol equations, *Nonlinear Mathematical Physics*, 1:41 (1994).
28. R.K. Gazizov, Symmetry of equations with a small parameter, in: *Proceedings on Modern Group Analysis VII*, N.H. Ibragimov, R.K. Naqvi and E. Straume, eds., MARS Publishers, Norway (1999).
29. E.V. Dussan, and S.H. Davis, On the motion of a fluid-fluid interface along a solid surface, *J. Fluid Mech.* 65:71 (1974).
30. E. V. Dussan, On the spreading of liquids on solid surfaces: static and dynamic contact lines, *Ann. Rev. Fluid Mech.* 11:371 (1979).
31. L.M. Hocking, A moving fluid interface on a rough surface, *J. Fluid Mech.* 76: 801 (1976).
32. L.M. Hocking, A moving fluid interface. Part 2. The removal of the force singularity by a slip flow, *J. Fluid Mech.* 79: 209 (1977).
33. L.M. Hocking and A.D. Rivers, The spreading of a drop by capillary action, *J. Fluid Mech.* 121: 425 (1982).
34. L.M. Hocking, The spreading of a thin drop by gravity and capillarity, *Q. J. Mech. Appl. Math.* 36: 55 (1983).
35. L.H. Tanner, The spreading of silicone oil drops on horizontal surfaces, J. *Phys. D Appl. Phys.* 12: 1473 (1979).
36. J.D. Chen, Experiments on a spreading drop and its contact angle on a solid, *J. Coll. Interf. Sci.* 122: 60 (1988).
37. S.K. Wilson, R. Hunt and B.R. Duffy, The rate of spreading in spin coating, *J. Fluid. Mech.* 413: 65 (2000).

RECOVERY OF DILUTED COMPONENTS FROM LARGE AQUEOUS STREAMS BY AN EXTREMELY COMPACT, CENTRIFUGAL COUNTERCURRENT ADSORPTION SYSTEM

Luuk A.M. van der Wielen, Marc A.T. Bisschops and Karel Ch.A.M. Luyben

Kluyver Laboratory for Biotechnology, Delft University of Technology, Julianalaan 67, 2628 BC, Delft, the Netherlands

INTRODUCTION

The development of efficient and generally applicable recovery techniques for dilute, non-volatile species from large industrial streams or waste flows has been a technological challenge for many decades. The generality of this problem can be illustrated by many examples, such as the purification of diluted biopharmaceutical proteins from fermentation broth, removal of dissolved organic contaminants from petrochemical waste streams and the recovery of precious metals such as uranium from leach streams. The use of adsorbent or ion exchange agents is usually preferred over other separation techniques, because of their high selectivities towards the desired product, the possibility of recycling the auxiliary (sorbent) material, the relatively low regeneration cost, and the variety of possible, efficient processes. A typical example of this category of equipment was described by Cloete and Streat in 1962,[1] and was considered breakthrough technology at that time. However, the available equipment is without exception inherently voluminous, and may contain tens to hundreds of cubic meters of adsorbent material for processing flows in the order 10 to 100 m^3 per hour. In this work, we present a novel technique, Centrifugal Adsorption Technology (CAT), to process similar liquid streams with a high recovery efficiency, but in very compact equipment, typically 100 to 1000 times smaller than the previous equipment.

The usual technological problem in countercurrent adsorption is the inherent conflict of two counteracting phenomena, namely that a high processing capacity requires adsorbent particles with very small dimensions to maximise the interfacial area and minimise diffusional distances, whereas hydrodynamics requires large adsorbent particles to minimise pressure drop over the equipment or prevent washing-out of the adsorbent particles. Optimised processes compromise generally at particles diameters in the range of millimeters. In CAT, a substantial reduction in equipment volume is achieved, by using adsorbent particles with diameters typically between 1 and 10 μm (microsorbents). A centrifugal field is used to establish a stable flow of the tiny particles in countercurrent contact with a large liquid flow. This is shown schematically in Fig. 1 for dense particles. Because the construction does not rely on close tolerances, the liquid flow may contain some solids, without hampering continued operation of CAT-systems.

Four liquid and sorbent streams enter and leave the CAT via the central axis, as shown in Fig. 2. The special central axis has a nozzle-discharge for microsorbents into the contacting zone, whereas the liquid stream to be processed is fed to the contacting zone at the rim of the bowl. The liquid flow, from which the dissolved components are removed, leaves the contacting zone near the axis of rotation. The loaded microsorbents leave the contacting zone via a special distributor-collector head that is mounted at the rim of the centrifuge bowl. Careful adjustment of the pressure balance of the various solid and liquid streams enables the transport of the suspension of microsorbents from the rim of the centrifugal chamber toward the axis and subsequent withdrawal from the equipment for further processing. The transport of the microsorbents as well as that of the liquid stream is fully continuous, which can be considered to be an additional advantage when compared to existing semi-continuous techniques such as the multistage fluidized bed (MFB) technology,[1] as well as the simulated moving bed (SMB) technology.[2]

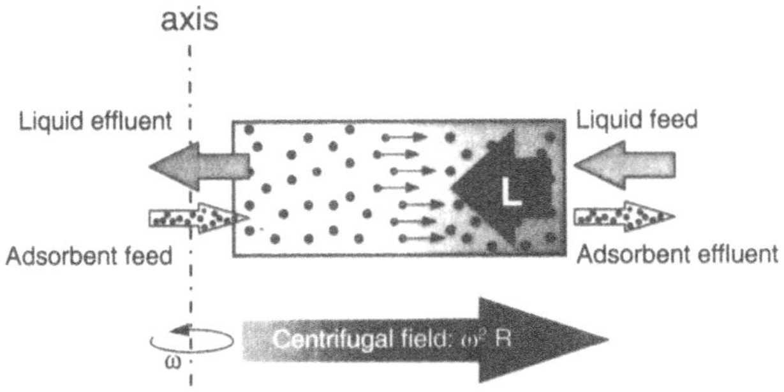

Figure 1. The CAT-concept for dense microsorbent particles.

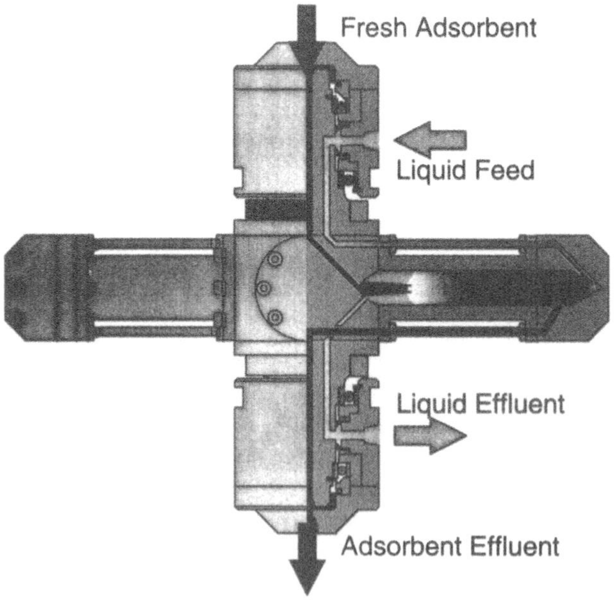

Figure 2. Schematic overview of the construction of the laboratory scale CAT-apparatus.

MODEL SYSTEM

To perform hydrodynamic and adsorption experiments in centrifugal fields (with accelerations ranging from 100g to 1000g), a laboratory scale CAT-apparatus with two straight parallel contacting columns was constructed from stainless steel. Each contained a cylindrical contacting zone 125 mm in length and 25 mm in diameter (contacting volume: 61 mL). In the hydrodynamic experiments, ambient water and commercial cation exchange resin particles Dowex 50WX8 (polystyrene matrix crosslinked with 8% divinylbenzene) in the Na^+ - form were used. The particle size was determined by Image Analysis at 71 ± 14 μm and the density was 1350 kg m^{-3}.

Experiments were performed at 500 and 700 rpm, a liquid flow rate of approximately 20 L/hr and a sorbent flow rate of 0.7-2 L/hr. The residence-time distribution of the liquid phase was determined by pulse-response measurements using NaCl and differential-conductivity measurements, whereas the solids residence time was measured by differential pulse responses of coloured particles, which were monitored with a Sony CCD video camera connected to an Image Analyser.

Under these conditions, the liquid residence time was approximately 10 s, and the sorbent residence time was in the range of 30-75 s. Mixing in the liquid phase corresponded to approximately 5 mixers (Peclet number, Pe, ~10), but mixing in the sorbent phase was relatively severe with Pe ~2. As a simple model system, the exchange of Na^+ for Ca^{2+} from the liquid phase was used, and the liquid effluent was analysed by means of ICP. The efficiency of Ca^{2+} removal was better than 99 % for separation factors in excess of 1.5. To achieve such high removal efficiencies, the Numbers of Transfer Units had to be at least 20, which coincides with predicted values.

OBVIOUS AND NON-OBVIOUS PHENOMENA IN CAT

Compactness

CAT reduces length scales (particle diameters, contacting zones) by several orders of magnitude relative to those in previous adsorption equipment. This miniaturisation of process equipment is expected to lead to several obvious and non-obvious phenomena that are related to the very small dimensions. The residence time of the microsorbent particles is on the order of 1 min, whereas the residence time of the liquid phase is typically in the order of 10 s. In the previous process equipment, liquid residence times were on the order of tens of minutes up to one or several hours, whereas sorbent residence times may be as large as several hours up to days. It should be noted that the residence times of the phases in CAT can be adjusted independently, as long as the flooding conditions are not violated. Therefore, CAT systems are particularly advantageous for relatively unstable products (e.g., some biopharmaceuticals). The compactness of the CAT systems also leads to minimal equipment volumes, which is advantageous when products are hazardous (e.g., radioactive metals, toxic species, and products from some categories of recombinant micro-organisms).

Flooding in a Centrifugal Field

Flooding limits determine the relative flow rates of the liquid and microsorbent phases that a countercurrent CAT system can handle without rejecting one of the phases at their respective entrance. The flooding conditions are clearly determined at the location of the smallest centrifugal force, which is at the particle entrance close to the axis of rotation. However, a classical, one-dimensional analysis of the flooding limits following Wallis's drift flux theory[3] leads to severe underpredictions. This appears to be due to the Coriolis force, possibly leading to a spiralling trajectory of the microsorbent particles in the centrifugal field despite the small dimensions of the microsorbent particles and the high rates of revolution of CAT systems. It is nearly impossible to verify these trajectories in a rotating suspension and to determine the particle velocity's radial component, which determines the flooding condition, as well as the component along the particle trajectory, which sets the processing capacity.

Effect of Dispersion

The high efficiency of mass transfer results in a steep decline in the concentration of the species to be adsorbed over the small distance between the liquid feed to the microsorbent entrance. In CAT, this distance is typically 0.1 m. In conventional equipment, comparable distances are on the order of 1 m (SMB-equipment) or 10 m (MFB-equipment). In large-scale equipment, backmixing is usually not too important. The small distances in CAT imply that each disturbance to plug-flow behaviour of the microsorbent and liquid phase will affect the performance negatively. In our laboratory scale CAT-equipment, the geometry of the internals had not yet been optimized, leading to Pe between 5-20 for the liquid phase and 1-5 for the microsorbent phase. We anticipate that a more optimal design of the internals should also take the spiralling particle trajectories into account.

Radial Variation of Particle Properties and Hold-up

Another non-obvious phenomenon that may seriously affect the separation performance of a CAT-system is the relatively strong variation of the properties of the microsorbent with the radial position. For instance, when heavy metal ions are adsorbed from a liquid stream by a microsorbent ion-exchange resin in the protonated form, the density of the microsorbent particles may increase several hundreds $kg.m^{-3}$ upon complete loading. Similarly, microsorbents composed of hydrophobic or hydrogel networks can swell significantly due to the uptake of sorbates. It was recently shown that the swelling kinetics of gel-type sorbents can be very fast due to rapid expansion of the network.[4-6] The time constants corresponding to these variations in density and diameter of adsorbing particles are the same order of magnitude as the residence time of the microsorbent particles. The effects on the separation performance depend on the actual process. For instance, sorbent particles accelerate in a radial direction due to the increasing strength of the centrifugal field while moving towards the rim of the bowl. Increasing particle densities magnifies this effect even further and the solid's residence time is expected to decrease. This may adversely affect the amount adsorbed and reduce recovery performance. When particles swell due to adsorbate uptake, the particle diameters increase but the average particle density decreases. Whether the particles accelerate or decelerate is determined by the relative variation of the particle diameter and the difference of particle and suspension densities. For most practical cases, it is expected that the particles increasingly accelerate while moving from the central axis to the rim of the bowl.

Adsorption Kinetics and Mass Transfer

The last phenomenon that may set CAT apart from "conventional" processes is the virtual unimportance of mass transfer due to the small dimensions of the microsorbent particles. In most other situations, the actual "reaction" kinetics of the sorption event at the adsorption or ion exchange sites can safely be neglected relative to the (slower) diffusion processes, maybe with the exception of some chelating resins.[7] This is, however, not guaranteed for microsorbents and adsorption kinetics may become limiting.

CONCLUSION

CAT appears to be a promising new technique that combines compactness, high recovery efficiencies, and very broad applicability. The general features can be understood, and the technique seems sufficiently mature for industrial application and to provide serious competition and an attractive alternative to existing techniques.

Acknowledgement

We thank the Dutch Ministry of Economic Affairs for financial support of this work through its programme IOP MT Prevention.

REFERENCES

1. F.L.D. Cloete and M. Streat, A new continuous solid-fluid contacting technique, *Nature* 200: 1199-1200 (1962).
2. D.M. Ruthven and C.B. Ching, Review article number 31: Countercurrent and simulated countercurrent adsorption separation processes, *Chem. Eng. Sci.* 44: 1011-1038 (1989).
3. G.B. Wallis, *One-Dimensional Two-Phase Flow*, McGraw-Hill, New York (1969).
4. M.A.T. Bisschops, L.A.M. van der Wielen and K.Ch.A.M. Luyben, Werktuig voor het uitvoeren van een behandeling in aanwezigheid van een centrifugale kracht en inrichting daarvoor. *Dutch Patent Application* 1002569 (1996).
5. M.A.T. Bisschops, L.A.M. van der Wielen and K.Ch.A.M. Luyben, Method for carrying out a treatment in the presence of a centrifugal field and an apparatus therefor, *PCT Application* WO97/33687 (1997).
6. M.A.T. Bisschops, L.A.M. van der Wielen and K.Ch.A.M. Luyben, Swelling kinetics of gel-type particles using Maxwell Stefan theory, *Ind. Eng. Chem. Res.* 37: 3312-3322 (1998).
7. Turse and Rieman, *J. Phys. Chem.* 65: (1961).

GAS-PHASE MASS TRANSFER IN A CENTRIFUGAL GAS-LIQUID CONTACTOR WITH A STACK OF FLEXIBLE DISKS AS PACKING

Rajib Mukherjee, Deepanjan Deb, Pavitra Sandilya and D. P. Rao

Department of Chemical Engineering
Indian Institute of Technology, Kanpur, 208016, India

ABSTRACT

Rotating packed beds have been proposed for carrying out distillation and absorption as an alternative to conventional columns. It had been claimed that the mass transfer rates are one to two orders of magnitude higher than for conventional columns, permitting a volume reduction of a similar order of magnitude. However, we found that the gas undergoes 'solid-body' like rotation, implying that there should be no enhancement of the gas-side transfer coefficient. To verify this contention, mass transfer studies were carried out on the absorption of SO_2 from air into an NaOH solution using wire-mesh packing. The *volumetric* mass-transfer coefficient was in the range of that for conventional packed beds. The *gas-side* mass-transfer coefficient was far lower.

With a view to promote tangential slip and thereby enhance the transfer rate, we used a stack of flexible disks of cloth with a spacing of 1.5 mm as packing. Studies were carried out on gas-phase pressure drop and mass transfer. From the total pressure drop, the local friction factor in the rotor was deduced. The gas-side transfer coefficient was much lower than for conventional beds, but in close agreement with those reported for flow through stationary annular disks. The experimental values for gas-side mass transfer coefficient were an order-of-magnitude lower than those estimated from theory considering the flow to be laminar. This discrepancy was attributed to severe liquid maldistribution.

INTRODUCTION

In conventional distillation and absorption columns, the liquid flows down under the influence of gravity. The allowable gas and liquid flow rates through the column, the intensity of contact of the phases and the upper limit on the specific area of the packing that could be used, and hence the mass transfer rates, are controlled by gravity. In 1983 Ramshaw and coworkers at Imperial Chemical Industries in the U.K. introduced a rotating packed bed as a means to overcome the limitations imposed by gravity.[1] In the rotating bed, the liquid flows outward due to centrifugal force whereas the gas flows inward under an imposed pressure gradient. A high centrifugal acceleration was used, up to 1000*g*. The liquid flowed over the packing as a thin film. As a consequence, a packing with a surface area as high as 4000 m^2/m^3 could be used, which was about 10 times that of packing in conventional columns. In addition, the liquid-side mass transfer coefficient was about 4 to 6 times that of conventional columns. They claimed that a volume reduction of about one to two orders-of-magnitude can be obtained with a rotating packed bed. However, this concept did not find ready commercial acceptance by the chemical process industry. This was partly due to the general resistance to employ moving parts (in preference to conventional stationary columns), and, more importantly, to a lack of understanding of the hydrodynamics and mass transfer in the contactor.

Several studies on hydrodynamics and liquid-side mass transfer coefficient in rotating contactors have been reported. Kelleher and Fair[2] and Mukherjee[3] reviewed the literature. Recently, attention has been focused on the nature of the flow over the packing and liquid maldistribution in the rotor.[4] Kelleher and Fair[2] reported studies on distillation in a centrifugal contactor. They deduced the gas-side mass-transfer coefficient from the overall volumetric coefficient based on available correlations for the interfacial area and liquid-side transfer coefficients. Recently, Sandilya[5] analyzed the flow of gas in the rotor and showed that the gas rotates with the packing, that is, the tangential slip between the gas and packing is negligible. Therefore, rotation is expected to have no effect on the gas-side mass transfer coefficient and it should be in the same range as that for conventional packed beds. To verify this contention, he obtained experimental data employing the absorption of SO_2 from air into NaOH with wire-mesh packing. Rotation had little effect on the gas-side transfer coefficient.

With a view to promote tangential slip and to enhance the gas-side transfer rate, we carried out studies using a stack of flexible disks as packing. The objectives were to determine the local friction factor in the rotor, the gas-side mass-transfer coefficient, and the potential for enhancement of the transfer coefficient.

EXPERIMENTAL METHODS

Figure 1 shows the centrifugal contactor. It consisted of an annular rotor (1) encased in a cylindrical casing with flat ends (2). The casing was made of flexiglass™ for visual observations. The rotor was made up of two flexiglass disks held in position by spacers and bolts placed 90° apart near the outer radius. The upper one was an annular disk with 6 elliptic holes (20 mm major axis and 10 mm minor axis) closer to the inner radius. A cylindrical housing (3) was provided on the top disc to direct the gas flow out of the

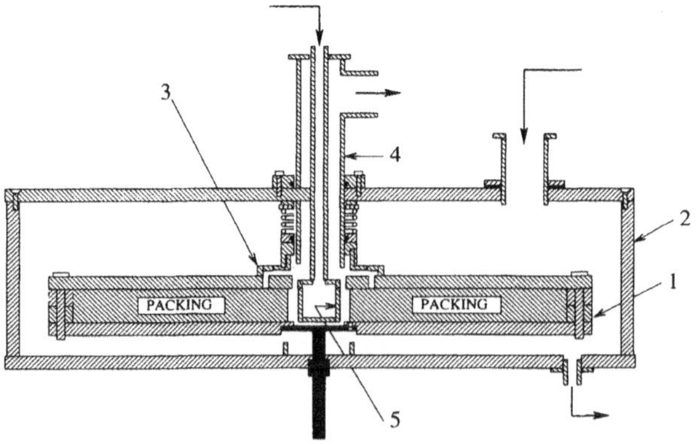

Figure 1. Schematic diagram of the rotating packed bed.
1 – rotor, 2 – cylindrical casing, 3 – cylindrical housing, 4 – gas outlet, 5 – liquid distributor.

packing into the gas outlet pipe (4) to eliminate hindrance to the entry of the liquid into the packing by the outgoing gas. A cylindrical liquid distributor (5) was fixed concentric to the disks. To ensure the liquid entered the rotor as jets, six rows of 0.5-mm holes 60° apart were provided on the distributor wall.

The lower disk of the rotor was fitted to a circular flange, which, in turn, was fixed on a vertical steel shaft as shown in Fig 1. The shaft was connected to an AC induction motor. The speed of this motor was varied using an inverter drive.

Annular disks of the shape shown in Fig. 2 were cut out of thick cotton cloth. Stainless-steel wire-mesh rings were fixed to their inner edges to impart rigidity and to provide spacing when they were stacked one over another. A stack of 12 disks of 18-mm height was stitched together at the wire mesh and fixed inside the rotor. The disks were stacked such that the elliptical holes aligned to form channels for gas flow. When this assembly was rotated at about 1000 rpm, the disks became stiff and flat. Details of this packing are given in Table1.

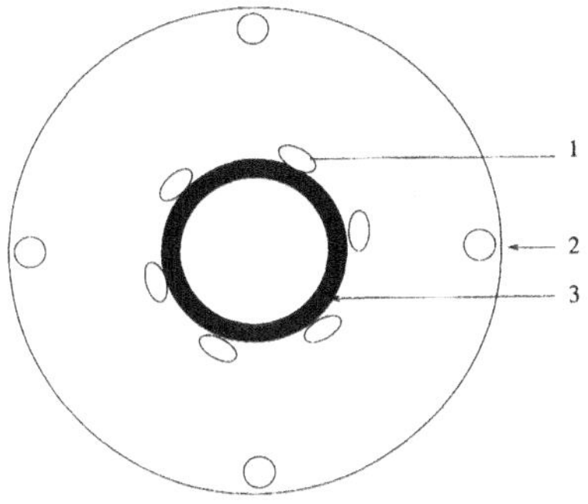

Figure 2. Details of the rotating packed bed.
1 – gas passage, 2 – hole for spacer, 3 – wire-mesh support.

Table 1. Details of the rotor assembly

Parameter	Dimension
Rotor	
Thickness of top and bottom disks	1.8 cm
Inner diameter	6 cm
Outer diameter	31cm
Packing depth	1.8 cm
Cloth packing	
Number of cloth disks	12
Bed porosity	0.70
Specific area	1333 m^2/m^3

Figure 3 shows the complete experimental setup. Using the air-water system, we measured the pressure drop across the casing and the gas outlet for different gas and liquid flow rates. The absorption of SO_2 from air into an aqueous solution of NaOH was employed to study the gas-phase mass transfer. Significant absorption of SO_2 took place in the casing. To reduce this absorption, we covered the bottom of the casing with polyurethane sheet to reduce direct contact of the liquid with the gas. The concentration of SO_2 in the inlet air-SO_2 mixture was kept in the range of 2 to 7%m. After steady state had been attained, samples were drawn at the inlet of the casing, from the casing and at the gas outlet. These samples were analyzed for SO_2 using iodometric titration.[5]

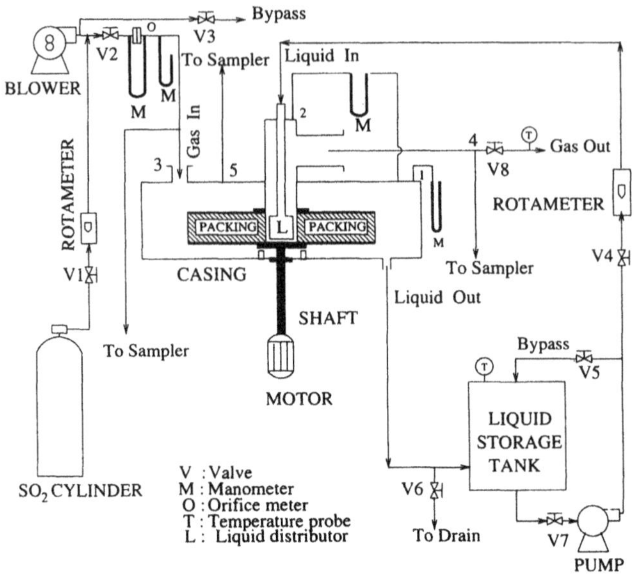

Figure 3. Schematic diagram of experimental set-up.

RESULTS AND DISCUSSION

The total pressure drop, ΔP_t, between the gas inlet and outlet was analyzed to determine the friction factor in the rotor, which was, in turn, related to mass transfer in the rotor. Figure 4 shows that ΔP_t increased with increasing rotor speed.

Let us consider the variation of ΔP_t with rotor speed without gas flow ($Q_g = 0$). At steady state, the gas inside the rotor is expected to rotate with the packing like a solid body. Therefore, ΔP_t under this condition corresponds to the pressure difference induced across the rotor by the centrifugal force field. It can be seen from Fig. 4 that the measured ΔP_t for $Q_g = 0$ is lower than the theoretical value, which was computed assuming the gas rotates with the rotor. On the other hand, analysis of the Keyvani and Gardner[6] and Kelleher and Fair[2] data shows that the experimental values were greater by about 2.5 and 2 times the theoretical ones, respectively, at 900 rpm. If the packing is '*neutral*', that is, if the rotor does not act as a blower or as a compressor, there is no apparent reason for such a significant discrepancy between experimental and theoretical values.

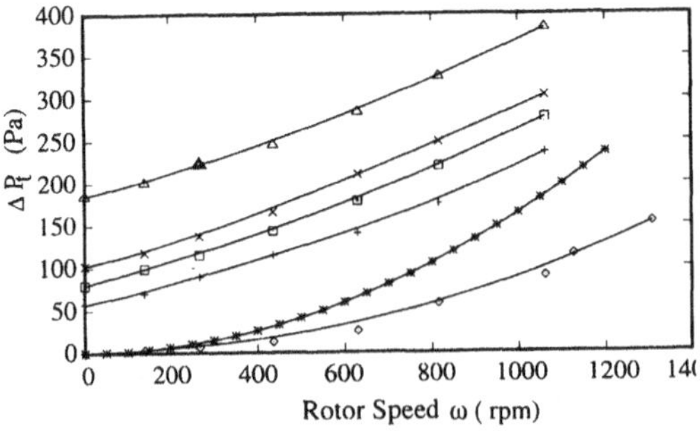

Figure 4. Variation of total pressure drop with rotor speed and gas velocity Q_g in $10^3 m^3/s$.
◇ $Q_g = 0$, + 3.6, □ 4.8, ×5.6, △ 7.1, ✱ theoretical.

This aspect is of significance in understanding the intriguing behavior of pressure drop with liquid flow reported by Keyvani and Gardner.[6] Figure 4 also shows that the liquid flow had no effect on the total pressure drop, which indicates that the thickness of the liquid film in the rotor was negligible compared to the spacing between the disks.

Different Pressure Drops

The total pressure drop, ΔP_t , is considered to consist of components due to the centrifugal force, ΔP_c , the momentum gain because of the change in gas velocity, ΔP_m, and the pressure drops due to bends, ΔP_b , and to friction in the rotor, ΔP_f :

$$\Delta P_t = \Delta P_c + \Delta P_m + \Delta P_b + \Delta P_f \tag{1}$$

The centrifugal pressure difference was evaluated from the tangential velocity component of the gas, V_θ , from:

$$\Delta P_c = \rho_g \int_{r_i}^{r_o} \frac{V_\theta^2}{r} dr \tag{2}$$

where ρ_g is the density of the gas and r is the radius. Sandilya showed that the gas attains the rotor velocity within a few mm after entering the rotor due to drag by the packing.[5] Although the gas attains a higher velocity than the rotor due to the Coriolis effect, this overshoot is small compared to the tangential velocity. Nonetheless, we use V_θ to evaluate ΔP_c. We find the momentum pressure drop from:

$$\Delta P_m = \rho_g \int_{r_i}^{r_o} \frac{V_r^2}{r} dr \tag{3}$$

where V_r is the radial velocity component. The gas undergoes three 90° bends in its path from the rotor to the outlet pipe. The pressure drop was estimated for each of these bends by multiplying the kinetic energy of the gas by a loss coefficient of 0.9.[3] Thus the frictional pressure drop was found from Eq. 1 because the other pressure drops are known.

Table 2 shows a typical set of pressure drops. It can be seen that ΔP_c and ΔP_f accounted for 90% of the total pressure drop. There was a marginal increase in ΔP_c with gas velocity due to the Coriolis effect. With increasing gas velocity, the frictional drop became dominant. It may be noted that the uncertainty in estimation of the pressure drop due to bends would affect the value computed for the frictional drop.

Table 2. Various pressure drops at 830 rpm and no liquid flow.

Gas flow 10^2x m^2/s	**Gas velocity** m/s	**Re**	**Pressure drop** (Pa) ΔP_t	ΔP_c	ΔP_m	ΔP_b	ΔP_f
0.184	0.54	68	123	109	0.4	1.7	12
0.437	1.29	161	210	115	2.0	9.5	83
0.708	2.09	261	326	121	5.2	24.8	175
0.887	2.62	327	448	124	8.2	39.0	227
1.04	3.05	382	531	126	11.1	53.1	341

Friction Factor

The radial velocity of the liquid at the gas-liquid interface is negligible compared to the gas velocity,[5] and hence the friction factor f is related to the frictional drop by:

$$\Delta P_f = \frac{1}{2}\int_{r_i}^{r_o} (4f)\rho_g \frac{V_r^2}{d_h} dr \tag{4}$$

where f *is* the local friction factor and d_h is the hydraulic diameter of the channel. For parallel plates of infinite width, d_h is related to the channel height h by:

$$d_h = 2h \tag{5}$$

The friction factor varies with the radius due to the variation of gas velocity. A functional form based on the Ergun equation is chosen to relate f to the Reynolds number Re:

$$f = \frac{\alpha}{\text{Re}} + \beta \tag{6}$$

where α and β are constants. Replacing V_r in Eq. (4) in terms of the volumetric flow of gas per channel, $Q_{g,ch}$, and integrating, we get:

$$\Delta P_f = \frac{\rho_g}{h}\left(\frac{Q_{g,ch}}{2\pi h}\right)^2 \left[\alpha \frac{\pi \vartheta_g}{Q_{g,ch}} \ln \frac{r_o}{r_i} + \beta\left(\frac{1}{r_i} - \frac{1}{r_o}\right)\right] \tag{7}$$

The constants α and β were evaluated using a regression analysis of the data for frictional pressure drop versus rotational speeds. From theory, the friction factor for gas flow through a channel of infinite width is:[7]

$$f = \frac{24}{\text{Re}} \tag{8}$$

Figure 5 shows the variation of friction factor with Reynolds number. Note that the friction factor for a stationary rotor (0 rpm) cannot be obtained because the packing collapses. Therefore, the frictional pressure drop was deduced for 0 rpm by extrapolation. The theoretical value is considerably lower than the present results. It seen that the friction factor for different velocities is nearly constant for Re greater than 250 and is nearly independent of rotational speed. The latter observation is consistent with our finding that the gas undergoes solid-body like rotation with the rotor.

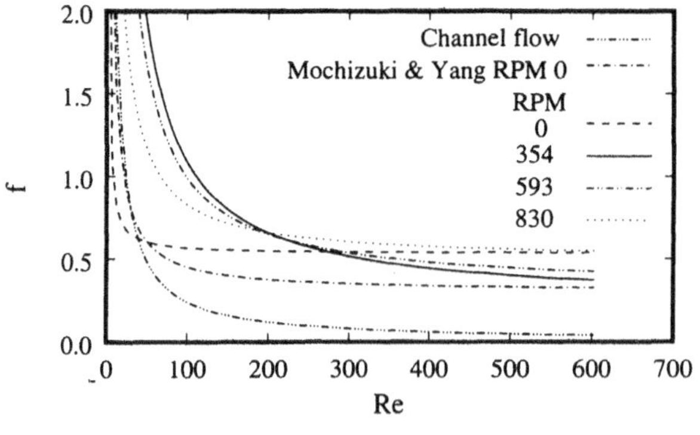

Figure 5. Variation of friction factor f with Reynolds number Re.

Therefore, we fitted a single correlation for all data to yield:

$$f = \frac{45}{\mathrm{Re}} + 0.4 \tag{9}$$

The ΔP_f estimated from Eq. 9 is within ± 12.5% of the experimental values.

Mochizuki and Yang[8] presented the following correlation for average friction factor, f_{av}, for flow through a stationary annular channel:

$$f_{av} = \frac{1.4}{\mathrm{Re}^{0.222}} \tag{10}$$

We evaluated α and β for their experimental conditions to obtain 15.1 and 0.3, respectively.

Mass Transfer. We measured the rate of absorption at different gas velocities keeping the liquid flow rate at 25 ml/s and the rotational speed at 1000 rpm. We assumed that the gas inside the casing was well mixed and the concentration of SO_2 entering the rotor was the same as in the casing. Further, it was considered that the concentration of SO_2 was zero at the gas-liquid interface because the reaction between SO_2 and NaOH is instantaneous and irreversible.[9] Based on these assumptions, the average volumetric mass-transfer coefficient, $k_g a$, was calculated from the following equation:

$$k_g a = \frac{Q_g \ln(C_C / C_o)}{\pi\, h(r_o^2 - r_i^2)} \tag{11}$$

Figure 6 shows the present results and a correlation proposed by Kelleher and Fair[2] for $k_g a$ based on distillation studies in a rotating packed bed employing a packing with specific area of 2500 m^2/m^3. The hash lines show the range covered in their experiments. Though there is no overlap of the data, the overall trend appears to be in agreement. Figure 7 shows the comparison of the coefficient obtained in the present work with that for two types of conventional packing, Berl saddles (1 inch, specific area 250m^2/m^3) and Sulzer BX (specific area 492 m^2/m^3). At higher gas flow rates, the coefficient was higher than for conventional packing. Figure 8 shows the gas-side mass transfer coefficients, obtained assuming that the wetted area was the same as the specific area.

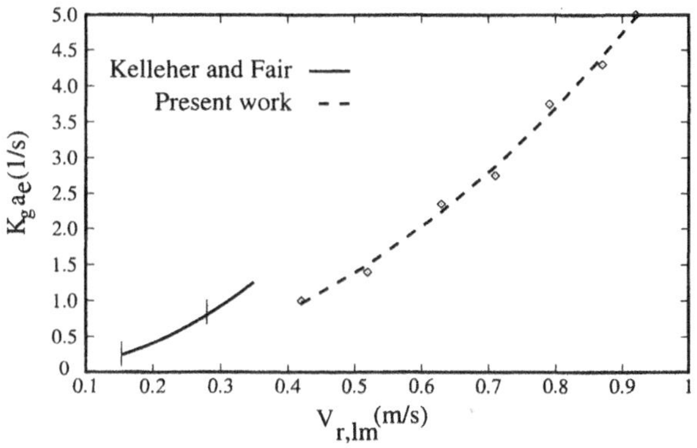

Figure 6. Comparison of volumetric mass transfer coefficients from the present results with those by Kelleher and Fair.[2]

The coefficient obtained in the present work is much less than those for conventional packing. Thus it appears that the wetted area was far lower than the specific area of the packing. Recent studies showed that the liquid flows radially outward and does not spread in a tangential direction.[4] The holes provided for the disengagement of gas could have led to the lower wetted area.

To examine the potential of this contactor, we estimated the average gas-side mass transfer coefficient from the analogy between momentum, heat and mass transfer for laminar flow through conduits.[10] We also estimated the mass transfer coefficient from a j_h-factor correlation for outward gas flow through annular disks obtained by Mochizuki and Yang.[8] These coefficients are compared in Fig. 9. It appears that there is a possibility to obtain higher transfer coefficients by ensuring proper liquid distribution in the rotor.

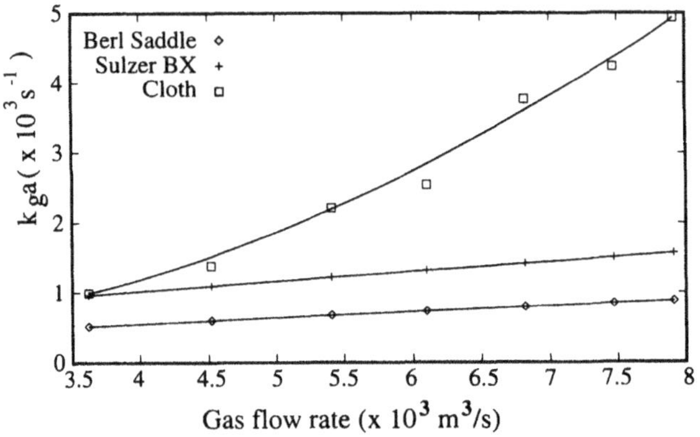

Figure 7. Comparison of volumetric coefficients of rotating and conventional packed beds.

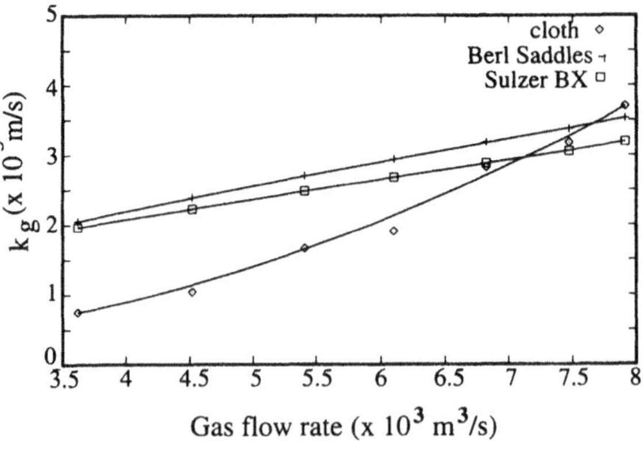

Figure 8. Comparison of mass transfer coefficients of rotating and conventional packed beds.

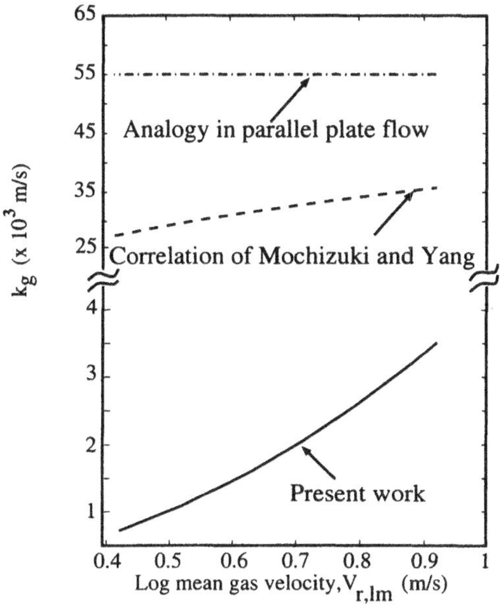

Figure 9. Comparison of experimental mass transfer coefficient with that estimated by the heat and mass transfer analogy.

NOMENCLATURE

a_p	specific area of packing m^2/m^3
C_c	SO_2 concentration in casing, $kmol/m^3$
C_o	SO_2 concentration at gas outlet, $kmol/m^3$
d_h	hydraulic diameter, m
f	local friction factor
h	disc spacing, m
k_g	mass transfer coefficient, m/s
k_ga	volumetric mass transfer coefficient, 1/s
L	liquid mass flow rate, k_g/m^2s
ΔP	pressure drop, Pa
Re	Reynolds number, $(=Vd_h/\upsilon)$
Q_g	volumetric flow rate of gas, m^3/s
r_o	outer radius of rotor, m
r_i	inner radius of rotor, m
V	gas velocity, m/s
υ	kinematic viscosity, m^2/s

Subscripts

b	bend
c	centrifugal
f	frictional
m	momentum
r	radial
t	total
θ	tangential

REFERENCES

1. C. Ramshaw, Higee distillation- An example of process Intensification, *The Chemical Engineer,* February, 13 (1983).
2. T. Kelleher and J.R. Fair, Distillation studies in a high-gravity contactor, *Ind. Eng. Chem. Res.* 35: 4646 (1996).
3. R. Mukherjee, Mass transfer in Higee contactor: Rotor with a stack of flexible disks, M. Tech. thesis, I.I.T. Kanpur, India (1999).
4. J.R. Burns and C. Ramshaw, Process intensification: Visual study of liquid maldistribution in rotating packed beds, *Chem. Eng. Sci.* 51:1347 (1996).
5. P. Sandilya, Studies on gas-phase mass transfer in a centrifugal contactor, Ph. D. thesis, I. I. T. Kanpur, India (1999).
6. M. Keyvani and N. Gardner, Operating characteristics of rotating beds, *Chem. Eng. Prog.* 85: 48 (1989).
7. J.G. Knudsen and D.L. Katz, *Fluid Dynamics and Heat Transfer,* McGraw-Hill Book Company, Inc. New York (1958).
8. S. Mochizuki and W.J. Yang, Heat transfer and friction loss in laminar radial flows through rotating annular disks, *Tran. ASME, J. Heat Transfer* 103: 112 (1981).
9. P.V. Danckwerts, *Gas-Liquid Reactions*, McGraw-Hill Book Company, Inc. New York (1970).
10. W.M. Kays and M.E. Crawford, *Convective Heat and Mass Transfer,* 3rd ed, McGraw-Hill Book Company, Inc., New York (1993).

ANALYTICAL ULTRACENTRIFUGATION: A VALUABLE TOOL TO RECOGNIZE CRYSTALLIZATION CONDITIONS OF PROTEINS

Joachim Behlke and Otto Ristau

Max Delbrück Center for Molecular Medicine, D-13092 Berlin, FRG

SUMMARY

The use of analytical ultracentrifugation to elucidate crystallization conditions for proteins is discussed. The methods are based on careful enhancement of the protein concentration by sedimentation, leading to a partial formation of associates. The results can be monitored either from the broadening of the moving boundary in a Schlieren pattern or from the considerable concentration increase near the cell base. The conditions under which highly charged proteins are able to undergo a self-association process (nucleation) as prerequisite for crystallization can be determined from the second virial coefficient, to which the excluded volume and the net charge mainly contribute. Crystallization experiments are successful when the repulsive forces of charged proteins are screened by addition of neutral salts and the net charge contribution becomes smaller than that of the excluded volume. Protein crystallization can be considered as an open association event with binding constants in the millimolar range. An exact description of single steps by using solutions of the Lamm differential equation is not possible because of the unknown association kinetics and the large number of parameters that must be estimated. However, based on fitting of simulated concentration profiles we can exclude a statistical binding mode for nucleation.

INTRODUCTION

The growth of crystals suitable for X-ray diffraction has always limited protein structure analysis. The main problem is that there are at least a dozen different solution variables that can influence crystallization. Therefore, it is difficult to predict crystallization conditions for proteins. Rather, trial and error methods have been employed to grow protein crystals.[1] To conclude whether or not such experiments have been successful very often requires several weeks or months.

There is broad interest in developing pre-crystallization assays to analyze concentrated

Processing by Centrifugation, Edited by Regel and Wilcox
Kluwer Academic/Plenum Publishers, New York, 2001

protein solutions with respect to forming either crystals or precipitates. Such procedures would be helpful to reduce the number of screening experiments required.

Protein crystallization can be considered as a molecular assembly process that is complex in detail. A high supersaturation is required to induce nucleation. A rather low supersaturation is desirable for crystal growth to avoid formation of crystals of poor quality.[2] Several different methods have been employed to analyze the mechanisms responsible for nucleation and crystal growth of proteins (see Table 1).

Table 1. Methods used to study protein crystal growth and nucleation

Methods	Subject	Authors
electron microscopy	crystal growth	Durbin and Feher[3]
atomic force microscopy	crystal growth	Durbin and Carlson[4]
fluorescence anisotropy	nucleation	Jullien *et al.*[5]
light scattering	nucleation	Pusey[6]
light scattering	nucleation	Skouri *et al.*[7]
light scattering	nucleation	Georgalis *et al.*[8,9]
light scattering	nucleation	Malkin *et al.*[10]
light scattering	nucleation	Malkin and McPherson[11]
light scattering	nucleation	George and Wilson,[12]
light scattering	nucleation	Rosenbaum and Zukowski[13]
neutron scattering	nucleation	Boue *et al.*[14]
neutron scattering	nucleation	Niimura *et al.*[15]
neutron scattering	nucleation	Minezaki *et al.*[16]
osmosis	nucleation	Todd *et al.*[17]
centrifugation	nucleation	Behlke and Knespel[18]
centrifugation	nucleation	Lenhoff *et al.*[19]

These methods are sensitive to associates that are formed during the nucleation process. Most of them only permit measurement of average values of the macromolecular size, and the published data differ in most cases by orders of magnitude. In contrast to the different techniques mentioned in Table 1, analytical ultracentrifugation also allows measurement of the size distribution of different species. The size distribution can be derived either from observations of i) the moving boundary in sedimentation velocity experiments or ii) the concentration distribution near the cell base. Another methodical variant that allows prediction of crystallization conditions will be discussed. It is based on the consideration of the thermodynamic nonideality measured at sedimentation equilibrium.

METHODS

A Schlieren-optical system is suitable to study association events with the analytical ultracentrifuge. It allows us to observe the moving boundary of homogenous substances in a sedimentation velocity experiment as a bell curve, which becomes asymmetric when substances of higher molecular mass substances are present. The size of the dissolved proteins is found by an approximate solution of the Lamm differential equation:

$$\left(\frac{\partial c}{\partial t}\right)_r = \frac{1}{r}\frac{\partial}{\partial r}\left[rD\frac{\partial c}{\partial r} - s\omega^2 r^2 c\right] \tag{1}$$

where D is the diffusion constant, s the sedimentation constant, ω the angular velocity of the rotor, r the distance from the center of revolution, and c the concentration. Two different variants are known; the approximate analytical solutions of Fujita,[20] Holladay,[21] Philo,[22,23] and Behlke and Ristau[24,25]; and the numerical finite element method developed by Claverie.[26] These have been applied by Demeler and Saber[27] and Schuck.[28] A program named CRYSTAL fits the model function of Fujita and McCosham[29] that had been improved by us. Since the experimental curves are concentration gradients, the differentiated form of the model function is used:

$$\frac{dc}{dr}=\frac{c_m e^{-\tau}}{r_m}\left[\frac{e^{-p_1^2}}{\sqrt{\pi\varepsilon(e^{\tau}-1)}}-\frac{be^{\frac{b(w-2)}{\varepsilon}}}{(1-b)\varepsilon}\mathrm{erfc}(p_2)+\frac{(2-b)e^{\frac{w-2+\tau(1-b)}{\varepsilon}}}{(1-b)\varepsilon}\mathrm{erfc}(p_3)\right.$$
$$\left.+\frac{e^{-p_2^2}}{(1-b)\sqrt{\pi\varepsilon\tau}}e^{\frac{b(w-2)}{\varepsilon}}-\frac{(2-b)e^{-p_3^2}}{(1-b)\sqrt{\pi\varepsilon\tau}}e^{\frac{w-2+\tau(1-b)}{\varepsilon}}\right] \tag{2}$$

with

$$\varepsilon=\frac{2D}{s\omega^2 r_m^2},\quad x=\left(\frac{r}{r_m}\right)^2,\quad w=2\frac{r}{r_m},\quad \tau=2s\omega^2 t$$
$$w_0=2[1+0.5(0.5-0.25\varepsilon)\tau],\qquad b=\frac{w_0^2-2\varepsilon}{2w_0} \tag{3}$$
$$p_1=\frac{\sqrt{e^{\tau}}-\sqrt{x}+0.5\varepsilon\left(1-\sqrt{e^{\tau}}\right)}{\sqrt{\varepsilon(e^{\tau}-1)}};\quad p_2=\frac{w-2+b\tau}{2\sqrt{\varepsilon\tau}};\quad p_3=\frac{w-2+(2-b)\tau}{2\sqrt{\varepsilon\tau}}$$

To describe associates that appear during the nucleation process at concentrations used in our experiments one can present bands (fractions) of up to 60 single gradients of equal running time. Assuming spherical associates, the sedimentation coefficients s or diffusion coefficients D of the associates can be calculated by the following relations:

$$s_n=s_1 n^{2/3},\quad D_n=D_1 n^{-1/3},\quad \varepsilon_n=\varepsilon_1/n \tag{4}$$

The program fits the sedimentation constant s_1 and the parameter ε_1 of the monomers. The association number n is a pre-set value. It is possible to fit simultaneously several gradients obtained at different run times (global fitting). Furthermore, a base line and if necessary a rotation parameter can be fitted additionally (global).

In the sedimentation equilibrium technique the following equation, which includes consideration of nonideality, is valid:

$$c_r=c_0\exp\left(\frac{M(1-\bar{v}\rho_0)\omega^2(r^2-r_0^2)}{2RT}-2B_2(c_r-c_0)-\frac{3}{2}B_3(c_r^2-c_0^2)\ldots\right) \tag{5}$$

where c_0 is a reference concentration, M is the molecular mass, $(1-\bar{v}\rho_0)$ the buoyancy term, R the gas constant, T the absolute temperature, and B_2 and B_3 the second and third virial coefficients, respectively. The accuracy of the determined parameters, B_2 and B_3, is considerably increased when M is not estimated but rather introduced as the theoretical value of the monomeric protein. This allows us to estimate B_2 with an error less than 5 %. For experiments with a loading concentration up to 3 mg/ml, B_3 is nearly zero and can be neglected.

To calculate the second virial coefficient and, additionally, the contribution of the net charge a procedure developed by Wills and Winzor[30] was used. Based on the Debye Hückel theory the second virial coefficient, given in the dimension of a weight concentration, can be obtained by:

$$B_2 = \frac{16\pi}{3M} N_A R_M^3 + \frac{Z^2[1+2\kappa R_M]}{4MI(1+\kappa R_M)^2} \tag{6}$$

The first term of this equation represents the excluded volume derived from the hydrodynamic radius R_M of the macromolecule and the second part describes the net charge. Here, I denote the ionic strength and κ the inverse Debye Hückel radius. For a known hydrodynamic radius of the macro ion or the excluded volume, the net number of charges Z can be derived directly from B_2.

RESULTS AND DISCUSSION

Analysis of the Moving Boundary

In order to analyze the appearance of protein associates under crystallization conditions, sedimentation velocity experiments were carried out on lysozyme at 40,000 rpm using an analytical ultracentrifuge Spinco E (Beckman, Palo Alto, CA) with Schlieren optics. The time-dependent Schlieren patterns shown in Fig. 1 were recorded by means of a mounted CCD-camera. The curves represent differentiated radial concentration distributions at different running times. After 30 min the bell curve appears nearly symmetrical, and later becomes slightly asymmetric.

Recently, from the types of curves shown in Fig. 1, we estimated the partial concentrations of monomeric lysozyme and the amount of oligomers, such as dimers, trimers, tetramers, etc.[18] Based on calculations of the association constants and the free energy data or $\Delta(\Delta G)$ values we could assume that the first steps of association are at equilibrium and formed by unfavorable (positive) energy. From a certain size (larger than dodecamers) the uptake of monomers by the existing oligomers is accompanied by a release of energy. Such complexes can be considered as nuclei that grow spontaneously to crystals. When observing this state we interrupted the sedimentation velocity experiments and put the centrifuge cells in a refrigerator. After two or three days we obtained crystals of lysozyme. In other experiments, papain crystals were obtained. The advantage of this procedure seems to be a cautious increase of the protein concentration during the sedimentation run. This slow increase favors oligomerization and nucleation of proteins. Although it is difficult to determine precisely the various species involved, the association steps proceed most likely in the manner shown in Fig. 2.

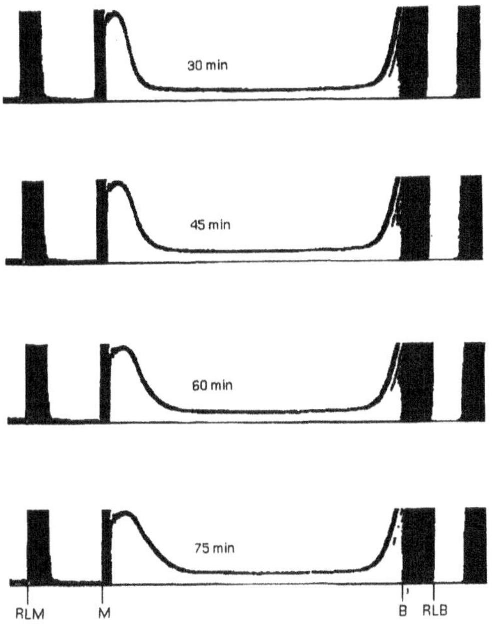

Figure 1. Representative ultracentrifuge patterns from a sedimentation velocity experiment on lysozyme (10 mg/ml) dissolved in 50 mM sodium acetate buffer, pH 4.5, containing 5% NaCl at 40,000 rpm and 20°C. A conventional single sector cell was used. The curves derived by the program TIFVIEW demonstrate the sedimentation from left to right obtained at different times after reaching speed. M and B denote the meniscus and bottom positions of the cell. RLM and RLB are the reference lines at meniscus and bottom. The angle of the Schlieren diaphragm was 70°.

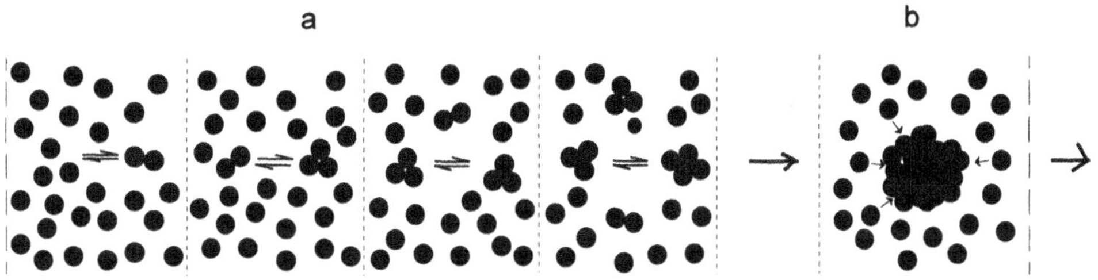

Figure 2. Proceeding from open association (a) to nucleation (b) of proteins under crystallization conditions.

Observations at the Cell Bottom

In earlier sedimentation experiments on catalase using a preparative ultracentrifuge Barynin and Melik-Adamyan[31] obtained crystals suitable for X-ray diffraction and structure analysis. The crystals were found on the bottom of the centrifuge tubes. Recently Lenhoff *et al.*[19] successfully repeated these studies. In order to study the expected association events we have used an analytical ultracentrifuge XL-A which, in contrast to the preparative device, allowed us to observe the radial concentration distribution at any time. Figure 3 demonstrates such records obtained at three different wavelengths after 20 h at 18,000 rpm. At that time most of the protein was found near the cell base. A molecular mass of about 240 kDa was found for the native tetrameric catalase by analyzing those data that could be treated as sedimentation equilibrium, from the slow concentration increase up to a radius of

7.136 cm. Beyond this radius the curves fell off strongly, indicating a considerably higher molecular mass of about 1.5 million or, on average, hexamers of the 240-kDa catalase. This seems to be a transition between both forms and is the first step of nucleation. Unfortunately, the bottom of the analytical cells was planar and the column height of analytical cells was incomparably shorter than those from preparative tubes. Therefore, it was only possible to recognize a dark brown oil-like drop at the cell base; possibly as a precursor of crystals or nuclei still soluble in buffer.

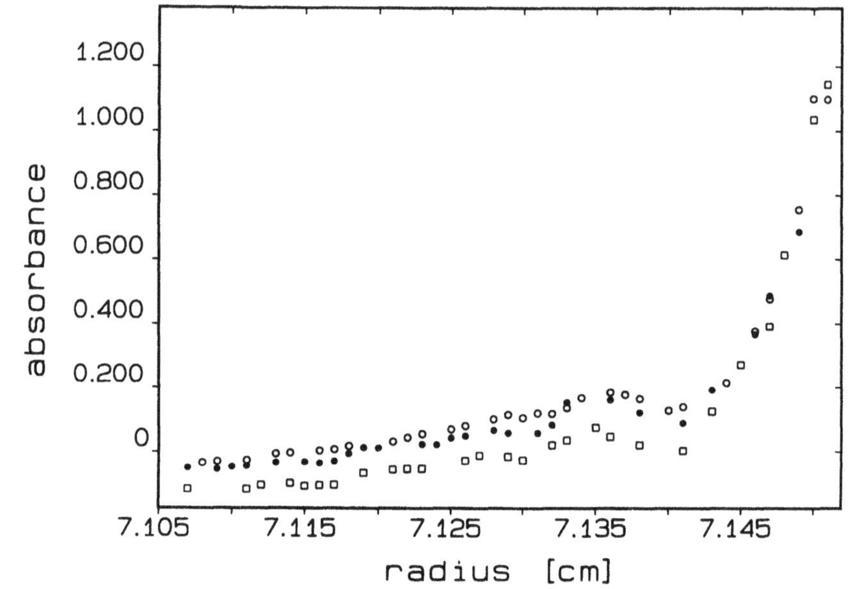

Figure 3. Sedimentation run on bovine liver catalase at 18,000 rpm and 10°C. Loading concentration 2 mg/ml in K-phosphate buffer, pH 6.2.
The radial absorbance was recorded at 490 nm (○), 500 nm (●) and 510 nm ( ).

Crystallization Conditions Derived from the Thermodynamic Nonideality

All of the experiments above were performed without considering the thermodynamic nonideality expressed by the virial coefficients. The second virial coefficient B_2 arises from at least two different properties of macromolecules, charge and the excluded volume. High net charge is observed far from the isoelectric point of a protein. It implies a very high solubility. Net charge and excluded volume provoke a positive value of B_2. This behavior is typical for proteins in the monomeric state because of the large repulsion forces.[32] It is more difficult to interpret B_2 for monomer oligomer equilibria because of the unknown association model and size of excluded volume.

Recently, George and Wilson[12] and Rosenbaum and Zukowski[13] have discussed the advantage of the second virial coefficient to recognize crystallization conditions. Negative values for B_2 were found to be indicators for suitable crystallization conditions. However, we believe such data are too complex to be considered as an implicit predictor because of the contributors (charge and excluded volume) are of different sign and size. Therefore, we have developed a method for the exact determination of the virial coefficients based on sedimentation equilibrium technique using an analytical ultracentrifuge (Behlke and

Ristau[32]). For this procedure a protein solution e. g. the basic lysozyme dissolved in 100 mM Na-acetate, pH 4.5, was centrifuged simultaneously at seven different loading concentrations among 1.1 and 2.7 g/L and using an eight hole rotor until reaching sedimentation equilibrium. The radial concentration distributions c_r representing e-functions of the seven curves were fitted simultaneously according to equation (5). The accuracy of the determined parameters B_2 and B_3 is considerably increased when M is not estimated but introduced as the theoretical value of the monomeric protein. This allows us to estimate B_2 with an error of smaller than 5 %. In these experiments B_3 is nearly zero and can be neglected. The second virial coefficient $B_2 = 24 \pm 1$ ml/g (see Fig. 4) is typical for a high net charge of the protein that prevents an association reaction due to repulsive forces between the macromolecules when assuming the charges to be spread on the surface of proteins. As can be seen from Fig. 4 the second virial coefficient depends strongly on the ionic strength. By adding NaCl or other neutral salts the B_2 values decrease nearly to the amount of the excluded volume that is calculated by the first part of Eq. 6 to be 3.57 ml/g when considering R_M = 1.72 nm (Mikol *et al.*[33]). At conditions where the charge-dependent part of B_2 becomes smaller than the excluded volume, presumably the attractive forces overcome the repulsive ones and association as well as nucleation is possible as can be seen from the appearance of crystals. Only a reduction of B_2 by increasing the pH is not advantageous for crystal formation.

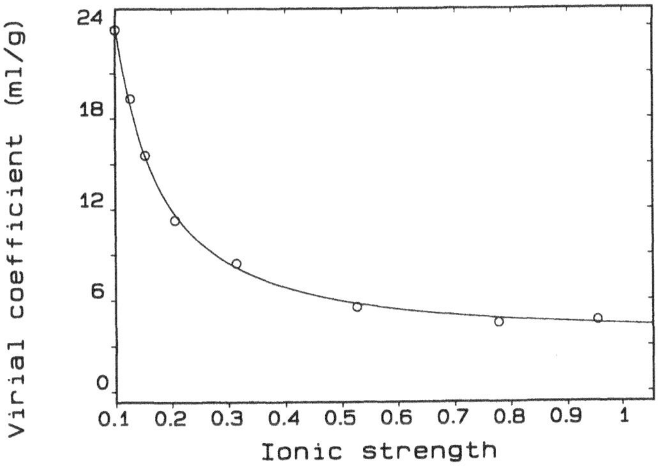

Figure 4. Influence of ionic strength on the second virial coefficient B_2 of lysozyme dissolved in 100 mM Na-acetate at pH 4.5.

Approach to Describe the Nucleation Process using Analytical Ultracentrifugation

All of the above calculations on sedimentation velocity experiments were done assuming i) that the association equilibria are established slowly compared to the moving boundary and ii) thermodynamic nonideality is not be considered. With respect to the association kinetics it is difficult to give a correct answer. Surely, direct formation of trimers, tetramers or higher oligomers from monomers requires more time than the uptake of monomers by existing oligomers. This is also due to the low binding constant in the millimolar range.

The associates formed during nucleation can be considered either as stable complexes or as complexes which are in instant equilibrium with monomers. The approximate analytical solutions are suitable to study integral concentration distribution curves of stable complexes but not of association equilibria. Simple equilibria (monomer-dimer) also can be analyzed by numerical methods. However, at present it seems to be impossible to fit concentration gradient profiles with numerous equilibria by using this method. This is due to the large number of parameters and by the less structured concentration gradients. Furthermore, there is no knowledge about the kinetics of single association steps. This may depend on the size of associates. It can be supposed that the association kinetics become slower with increasing association number. In order to obtain first rough information about the association reactions and to save computing time we used approximate solutions of the normal Lamm equation, making visible hints of associate formation became visible (Fig. 5).

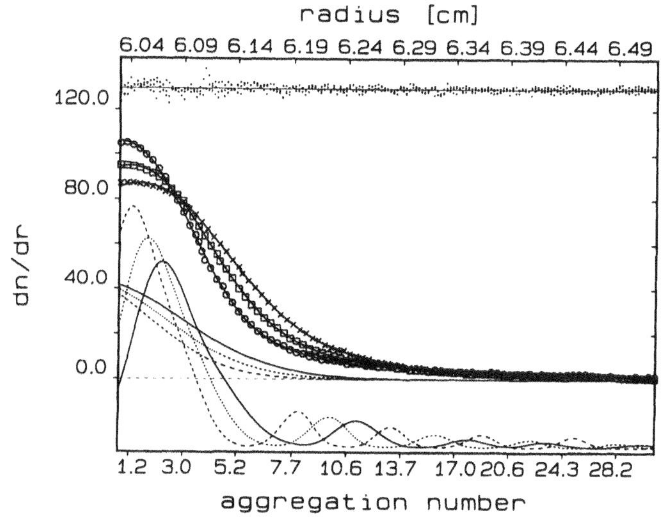

Figure 5. Sedimentation velocity run on lysozyme (10 mg/ml) under the conditions described in Fig. 1. The moving boundary, given as dn/dr values versus the radial position (upper abscissa), was recorded at 40 min (O), 50 min ( ) and 60 min (x) after reaching full rotation speed. These data were fitted globally using Eq. 2 to yield the partial concentrations of the different species. The monomer fraction of lysozyme is given by the dashed, dotted or solid lines derived from the three experimental curves in the upper part. A deconvolution of the remaining concentration area between each experimental curve and those of the monomers yielded the oligomer fractions, which are plotted in arbitrary units versus the aggregation number. The corresponding scale (lower abscissa) is for the last experimental curve (here the solid line). Residuals are amplified fourfold.

To get an overview about their appearance, Claverie simulations of concentration gradients with explicit consideration of fast chemical equilibria were done. A program based on the finite element method was written to simulate sedimentation velocity events with rapid self-association equilibria according to Schuck.[18] To get gradient curves the concentration profiles were differentiated numerically. Curves similar to the experimentally ones were obtained for very low equilibrium constants. When analyzing these synthetic curves with our program CRYSTAL we obtain many aggregates, but the aggregation numbers differ somewhat from the expected ones. Therefore, we can assume that the analysis of the experimental curves by our program CRYSTAL allows recognition of aggregates, although their sizes cannot be estimated with certainty. In addition, radial concentration gradients were generated for an open association reaction according to Schuck[18] with consideration of the statistical binding model of Wyman and Gill.[34] In this way complexes (aggregates) up to an arbitrary size (in our case 24) were simulated (see Fig. 6). However, these gradients show a sharpening at the right wing of the bell curve rather than a broadening as observed in the experimental curves. Therefore, the binding forces according to one site for increasing aggregates seem to be much stronger than expected for the statistical binding model. This is reasonable and explains the formation of stable complexes or nuclei and finally crystallization.

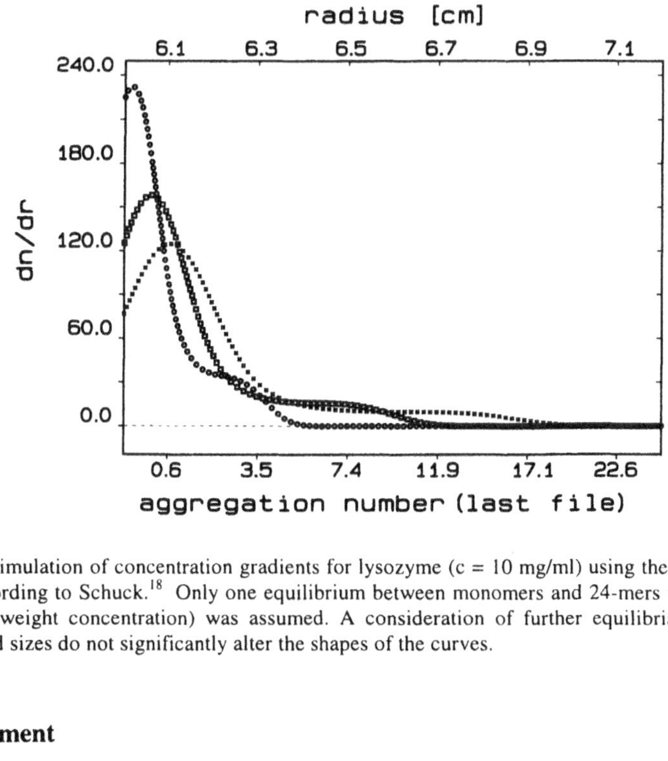

Figure 6. Simulation of concentration gradients for lysozyme (c = 10 mg/ml) using the finite element method according to Schuck.[18] Only one equilibrium between monomers and 24-mers with $K = 10^{-23}$ (defined in weight concentration) was assumed. A consideration of further equilibria with similar affinities and sizes do not significantly alter the shapes of the curves.

Acknowledgment

The financial support of the Deutsche Forschungsgemeinschaft (HE 1318/15 -1/2) is gratefully acknowledged.

REFERENCES

1. J. Jankaric and S.-H. Kim, Sparse matrix sampling: a screening method for crystallization of proteins, *J. Appl. Crystal.* 24:409 (1991).
2. E.A. Stura and I.A. Wilson, Applications of the streak seeding technique in protein crystallization, *J. Crystal Growth* 110:270 (1991).
3. S.D. Durbin and G. Feher, Studies of crystal-growth mechanisms of proteins by electron microscopy, *J. Mol. Biol.* 212:763 (1990).
4. S.D. Durbin and W.E. Carlson, Lysozyme crystal-growth studied by atomic force microscopy, *J. Crystal Growth* 122:71 (1992).
5. M. Jullien, M.P. Crosio, S. Baudet-Nessler, F. Merola, and J.C. Brochon, Evidence for a dimeric

intermediate on the crystallization pathway of ribonuclease-A, *Acta Crystallographica* D 50:398 (1994).
6. M.L. Pusey, Estimation of the initial equilibrium-constants in the formation of tetragonal lysozyme nuclei, *J. Crystal Growth* 110:60 (1991).
7. M. Skouri, M. Delsanti, J.P. Munch, B. Lorber, and R. Giegé, Dynamic light scattering-studies of the aggregation of lysozyme under crystallization conditions, *Febs- Letters* 295:84 (1991).
8. Y. Georgalis, A. Zouni, and W. Saenger, Dynamics of protein precrystallization cluster formation, *J. Crystal Growth* 118:360(1992).
9. Y. Georgalis, A. Zouni, W. Eberstein, and W. Saenger, Formation dynamics of protein precrystallization fractal clusters, *J. Crystal Growth* 126:245 (1993).
10. A.J. Malkin, J. Cheung, and A. McPherson, Crystallization of satellite tobacco mosaic-virus. 1. Nucleation phenomena, *J. Crystal Growth* 126:544 (1993).
11. A.J. Malkin and A. McPherson, Crystallization of satellite tobacco mosaic virus. 2. Postnucleation events, *J. Crystal Growth* 126:555 (1993).
12. A. George and W.W. Wilson, Predicting protein crystallization from a dilute solution property, *Acta Cryst.* D 50:361 (1994).
13. D.F. Rosenbaum and C.F. Zukowski, Protein interactions and crystallization, *J. Cryst. Growth* 169:752 (1996).
14. F. Boue, F. Lefaucheux, M.C. Robert, and I. Rosenman, Small-angle neutron-scattering study of lysozyme solutions, *J. Crystal Growth* 133:246 (1993).
15. N. Niimura, Y. Minezaki, M. Ataka, and T. Katsura, Aggregation in supersaturated lysozyme solution studied by time-resolved small angle neutron scattering, *J. Crystal Growth* 154:136 (1995).
16. Y. Minezaki, N. Niimura, M. Ataka, and T. Katsura, Small-angle neutron-scattering from lysozyme solutions in unsaturated and supersaturated states, *Biophys. Chem.* 58:355 (1996).
17. P. Todd, S.K. Sikdar, C. Walker, and Z.R. Korszun, Application of dewatering to the controlled crystallization of biological macromolecules and organic-compounds, *J. Crystal Growth* 110:283 (1991).
18. J. Behlke and A. Knespel, Observation of precrystallization aggregation in protein solutions during centrifugation, *J. Crystal Growth* 158:388 (1996).
19. A.M. Lenhoff, P.E. Pjura, J.G. Dilmore, and T.S. Godlewski, Ultracentrifugal crystallization of proteins - transport-kinetic modeling, and experimental behavior of catalase, *J. Crystal Growth* 180:113 (1997).
20. H. Fujita, *Mathematical Theory of Sedimentation Analysis*. Academic Press, New York (1962).
21. L.A. Holladay, An approximate solution of the Lamm equation, *Biophys. Chem.* 10:187 (1979).
22. J.S. Philo, Measuring sedimentation, diffusion, and molecular weights of small molecules by direct fitting of sedimentation velocity concentration profiles, in: *Modern Analytical Ultracentrifugation*, T. M. Schuster and T. M. Laue, eds., Birkhäuser, Boston (1994).
23. J.S. Philo, An improved function for fitting sedimentation velocity data for low-molecular weight solutes, *Biophys. J.* 72:435 (1997).
24. J. Behlke and O. Ristau, Molecular mass determination by sedimentation velocity experiments and direct fitting of the concentration profiles, *Biophys. J.* 72:428 (1997).
25. J. Behlke and O. Ristau, An improved approximate solution of the Lamm equation for the simultaneous estimation of sedimentation and diffusion coefficients from sedimentation velocity experiments, *Biophys. Chem.* 70:133 (1998).
26. J.M. Claverie, H. Dreux, and R. Cohen, Sedimentation of generalized systems of interacting particles. I. Solution of systems of complete Lamm equations, *Biopolymers* 14:1685 (1975).
27. B. Demeler and H. Saber, Determination of molecular-parameters by fitting sedimentation data to finite element solutions of the Lamm equation, *Biophys. J.* 74:444 (1998).
28. P. Schuck, Sedimentation analysis of noninteracting and self-associating solutes using numerical-solutions of the Lamm equation, *Biophys. J.* 75:1503 (1998).
29. H. Fujita and V.J. MacCosham, Extension of sedimentation to molecules of intermediates sizes, *J. Chem. Phys.* 30:291 (1959).
30. P.R. Wills and D.J. Winzor, in: *Analytical Ultracentrifugation in Biochemistry and Polymer Science*, S. E. Harding, A. J. Rowe, J. C. Horton, eds., Royal Society, Cambridge, U.K. (1992).
31. V.V. Barynin and W.R. Melik-Adamyan, The ultracentrifugation protein crystallization mechanism, *Kristallografiya* 27:981 (1982).
32. J. Behlke and O. Ristau, Analysis of the thermodynamic non-ideality by sedimentation equilibrium experiments, *Biophys. Chem.* 76:13 (1999).
33. V. Mikol, E. Hirsch, and R. Giegé, Diagnostic of precipitant for biomacromolecular crystallization by quasi-elastic light-scattering, *J. Mol. Biol.* 213:187 (1990).
34. J. Wyman and S.J. Gill, *Binding and Linkage*, University Science Books, Mill Valley, CA (1990).

THE EFFECT OF CENTRIFUGATION ON THE COMPOSITION, STRUCTURE AND PROPERTIES OF IONIC COMPOUND CRYSTALS

Stanislav P. Nikanorov,[1] Vladimir N. Gurin,[1] Lev I. Derkachenko,[1] Alexei P. Nechitailov,[2] Igor N. Zimkin,[1] Zoya I. Uspenskaya,[3] Vladimir N. Osipov,[1] Maria M. Korsukova,[1] and Liya L. Regel[4]

[1] Ioffe Physical-Technical Institute, St. Petersburg 194021, Russia
[2] All-Russian Aluminum-Magnesium Institute, St. Petersburg 199106, Russia
[3] Institute of Cytology, St. Petersburg 194021, Russia
[4] Clarkson University, Potsdam, NY 13699-5814, USA

INTRODUCTION

Most investigations on centrifugal materials processing dealt either with the processes taking place in various media or with the properties of the resulting materials.[1] Practically no attempts have been made to study in detail the influence of centrifugation on the properties in relation to changes in chemical composition and crystal structure. We demonstrated[2,3] such changes for the first time for centrifugation of supersaturated aqueous solutions of the alkali halide compounds KCl and KBr at accelerations a of 1.3, 6.2 and 11.8×10^3 g. The microhardness was increased by 1.5 to 2.4 times. Further studies of these and some other compounds were carried out in the present work, resulting in new data that confirm and extend the previous results.

EXPERIMENTAL

Crystallization of KCl, KBr, $Ba(NO_3)_2$, $PbCl_2$ and KI from highly supersaturated aqueous solutions was carried out on the centrifuge K24D. All experimental conditions were the same as described elsewhere.[4] In most cases the crystalline sediments strongly stuck to the walls and bottom of the tubes and had to be removed with a scalpel. Crystallization products were in the form of dense 1 - 2 cm crystalline aggregates, which consisted of intergrown single crystals up to 2 mm in size. These crystals were regular cubes in the case of alkali halides and dendrites in the case of $PbCl_2$. For measuring the refractive index (N) of the crystals obtained, the standard set of immersion liquids IZH-1 and their mixtures were used.[5] Expected values of N were estimated by analogy with the dependence of N on lattice parameter in the system KCl-NaCl.[6] The lattice parameter and microhardness (precision ≈ 7%) of some KCl and KBr crystals were re-measured 4 years after they had first been obtained in order to determine the time relaxation, if any.

RESULTS AND DISCUSSION

Figure 1 and Table 1 show the influence of centrifugation on compositions and properties. We concluded in our previous work that the difference in the atomic masses of the ions is responsible for centrifugation during crystallization altering the chemical composition and lattice parameter of the resulting crystals.[2,3] The larger this difference, the more substantial is the alteration, although a proportional dependence is not expected since every kind of ion has its own peculiarities of hydration, which must influence sedimentation and, consequently, centrifugation results.

The atomic mass ratio *i/i* of ions can be employed both to choose systems for investigation and for qualitative interpretation of centrifugation results.[4] It can be seen from Table 1 and Figure 1 that for *i/i* ≈ 1.1 (KCl, $Ba(NO_3)_2$) the alterations of the lattice parameter, microhardness and refractive index were not so pronounced as for *i/i* ≈ 2 and larger (KBr and KI). More or less uniform alterations of the chemical composition, lattice parameter and microhardness with increasing acceleration were observed only for $Ba(NO_3)_2$. For KCl and KBr, the expected increase of the content of heavier ion was observed only at $6.2 \times 10^3 g$ (see Figure 1). The increase of *i/i* to 3 and higher resulted in even more pronounced alteration of the chemical composition and crystal properties (see Table 1).

The chemical analyses show that all crystals displayed some deviation from their stoichiometric compositions. This should have resulted in the change of their color (appearance of so-called F-centers of coloration). However, this was not the case, and all crystals were practically colorless. There are several possible explanations for this colorless state, one being the aggregation of such centers, i.e. vacancies clustering to become voids. Perhaps a small amount of water was trapped in the crystals that grew with centrifugation[2] and prevented the color change.

Figure 1 shows that with increasing acceleration a the alteration of the lattice parameter of KCl and KBr was opposite to the alteration of Cl and Br content in the crystals, while the refractive index varied in accordance with the lattice parameter. The microhardness of both compounds increased with increasing a.

Another interesting and important result of the present work was the detection of 10-70μm micro-inclusions in the crystals. These inclusions were of various shapes: rounded, oval, pole-like and prismatic, as for $Ba(NO_3)_2$. As a rule, they were along the crystal edges. By qualitative estimation, their amount reached 5-10%. These inclusions were revealed only in transmitted light. Such inclusions produced a peculiar composite material with increased hardness, e.g. for KBr the value of H_K increased by 2.4 times (see Table 1). But the microhardness may have increased also due to an increase of the dislocation density with increasing acceleration, due either to plastic deformation or to mismatch as the crystal closed around each inclusion. However, in the present work the dislocation density was not measured. Thus, the possibility to control the size, shape and amount of such inclusions in the crystals of various compounds could be an interesting continuation of these studies. It is necessary to note that gas bubbles, particles and solution inclusions were observed inside RDX single crystals grown with centrifugation[7].

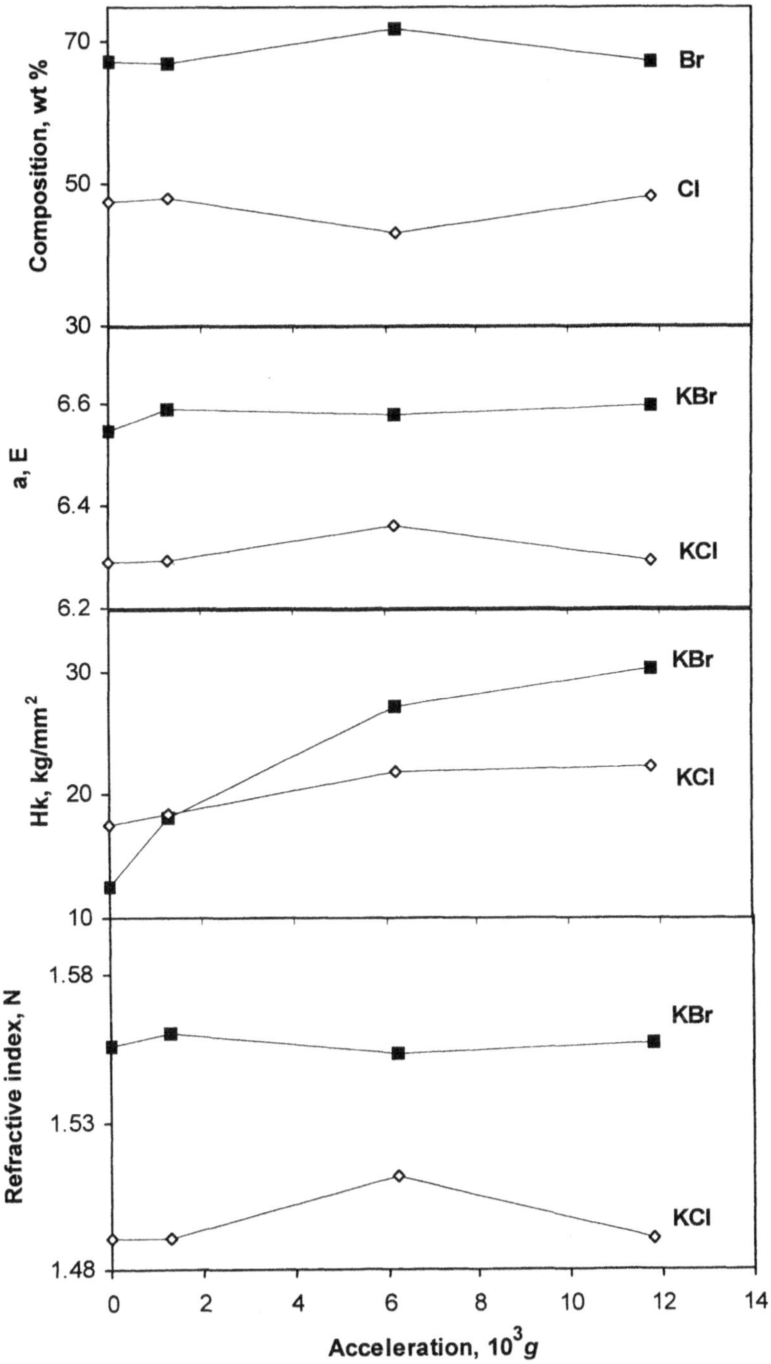

Figure 1. The influence of the centrifugal acceleration on the composition, structure and properties of KCl and KBr crystals.

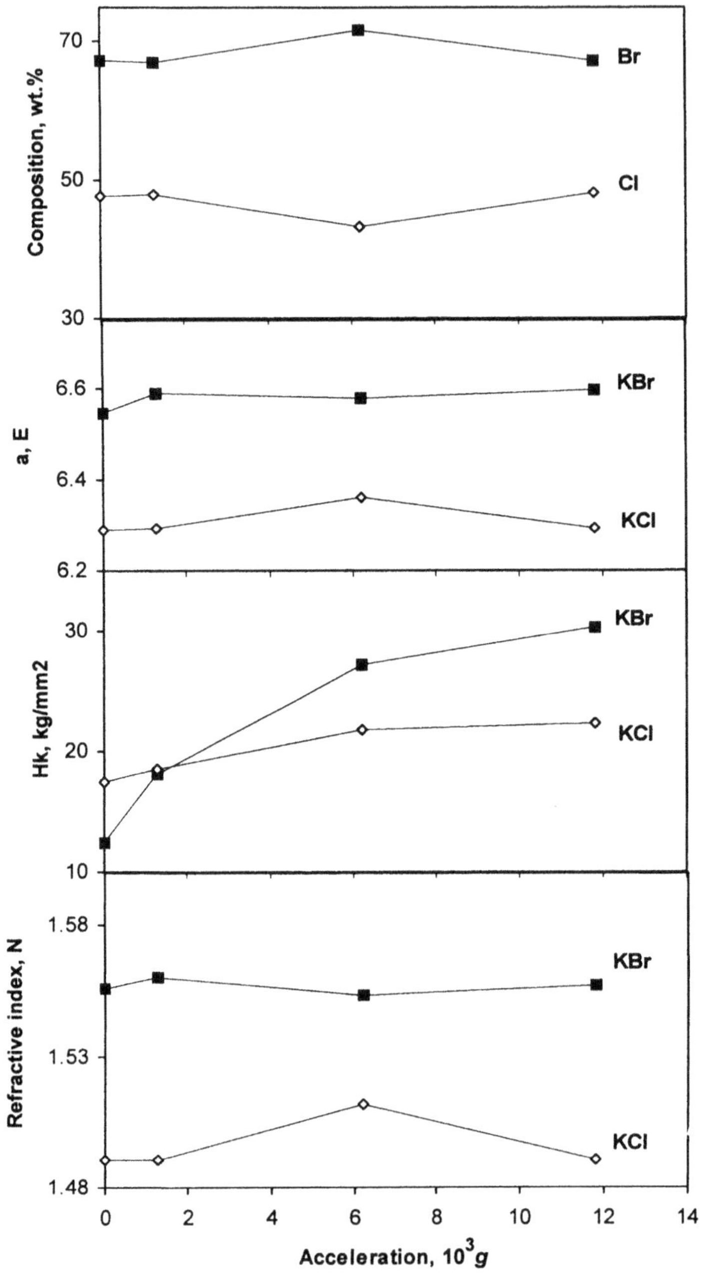

Figure 1. Continued.

Table 1. Alteration of the chemical composition, lattice parameter and properties of ionic compounds by centrifugation during crystallization.

Properties	Acceleration a			
	1 g	1.3 x $10^3 g$	6.2 x $10^3 g$	11.8 x $10^3 g$
		$Ba(NO_3)_2$ (Ba/NO_3 = 1.11)		
Ba, mass.%	51.5	52.1	52.4	52.5
a, Å	8.1166(4)	8.1169(4)	8.1218(4)	8.1224(4)
Refractive index, N	1.5710(3)	1.5712(3)	1.5738(3)	1.5700(3)
Estimated N	1.5711	1.57107	1.57055	1.57049
H_K, kg/mm^2 (P = 2g)	71.7	88.4	103.6	128.0
		$PbCl_2$ (Pb/Cl = 2.92)		
Cl, g/l	6.7	6.3	5.4	6.8
H_K, kg/mm^2 (P = 2g)	-	233 (ǁ {001})	268 (ǁ {001}) 184 (ǁ {100})	229.5 (ǁ {001}) or (ǁ {100}
		KI (I/K = 3.25)		
K, mass%	25.1	25.0	24.6	25.1
a, Å	7.0663(5)	7.0641(5)	7.0666(5)	7.0666(5)
Refractive index, N	1.6656(3)	1.6660(3)	1.6656(3)	1.6658(3)
Estimated N	1.6656	1.6657	1.6656	1.6656
H_K, kg/mm^2 (P = 10g)	-	10.2	11.1	13.6

Table 2. Alteration of microhardness (P = 5g) and lattice parameter of KCl and KBr crystals after 4 years of aging.

	Acceleration, a			
	1 g	1.3 x 10^3 g	6.2 x 10^3 g	11.8 x 10^3 g
		KCl		
H_K, kg/mm^2	17.5	18.5	21.8	22.3
H_K (new), kg/mm^2	-	17.3	18.5	17.4
		KBr		
H_K, kg/mm^2	12.5	18.1	27.1	30.3
H_K (new), kg/mm^2	-	14.7	19.5	16.1
a, Å	6.5478(9)	6.5908(8)	6.5775(9)	6.5974(2)
a, Å (new)	6.5992(4)	6.5998(4)	6.5490(4)	6.5989(4)

For the first time, the microhardness H_K of KCl and KBr crystals was re-measured 4 years after they had been prepared with centrifugation. Aging and relaxation were manifested as a decrease in microhardness (see Table 2), though complete restoration to the value for crystals grown without centrifugation did not occur. The re-measured microhardness values displayed maxima for growth at 6.2 x $10^3 g$. For KCl the new H_K values for crystals obtained at 6.2 x $10^3 g$ and 11.8 x $10^3 g$ practically coincided with data for stoichiometric crystals.[2] For KBr crystals the lattice parameter was re-measured, too. One can see from Table 2 that only the value for 6.2 x $10^3 g$ was restored to the original one.[2]

CONCLUSIONS

Centrifugation at 1.3 to 11.8x$10^3 g$ during crystallization of highly supersaturated aqueous solutions of ionic compounds brought about the alteration of chemical composition, lattice parameter, refractive index and microhardness of the resulting crystals. With the atomic mass ratio of cations and anions in these compounds exceeding one, the changes increased with increasing acceleration. Growth of crystals by rapid cooling of highly supersaturated aqueous solutions both under ordinary conditions and with centrifugation resulted in the formation of microscopic inclusions.

Aging of KCl and KBr crystals for 4 years resulted in considerable alteration of the lattice parameter and microhardness.

REFERENCES

1. L.L. Regel and W.R. Wilcox, Centrifugal materials processing, in: *Centrifugal Materials Processing*, L.L. Regel and W.R. Wilcox, eds., Plenum Press, New York and London (1997) pp. 1-16.
2. V.N. Gurin, S.P. Nikanorov, M.M. Korsukova, L.I. Derkachenko, W. Jung, and L.L. Regel, Alteration of the lattice constant and the composition of KCl and KBr crystals under macrogravity, *Phys. Solid State* 39:1599 (1997).
3. V.N. Gurin, S.P. Nikanorov, L.L. Regel, and L.I. Derkachenko, Influence of gravity on the habit of single crystals of compounds grown from solutions, in: *Centrifugal Materials Processing*, L.L. Regel and W.R. Wilcox, eds., Plenum Press, New York and London (1997) pp. 197-202.
4. V.N. Gurin, S.P. Nikanorov, A.P. Nechitailov, V.N. Osipov, Z.I. Uspenskaya, M.M. Korsukova, and L.L. Regel, Alteration of the chemical composition of solid solutions and eutectics at centrifugation, in: *Centrifugal Materials Processing IV, 4th International Workshop on Materials Processing at High Gravity,* May 29 - June 2, 2000, Clarkson University, Potsdam, NY, USA.
5. V.B. Tatarsky, *Crystal Optics and Immersion Method,* Nedra, Moscow (1965), *in Russian.*
6. A.N. Vinchell, *Optical Properties of Synthetic Minerals,* Mir, Moscow (1967), *in Russian.*
7. M.Y.D. Lanzerotti, J. Autera, L. Borne, and J. Sharma, Crystal Growth of Energetic Materials during high Acceleration, in: *Centrifugal Materials Processing*, L.L. Regel and W.R. Wilcox, eds., Plenum Press, New York and London (1997) pp. 213-219.

ALTERATION OF THE CHEMICAL COMPOSITION OF SOLID SOLUTIONS AND EUTECTICS BY CENTRIFUGATION

Vladimir N. Gurin,[1] Stanislav P. Nikanorov,[1] Alexei P. Nechitailov,[2] Vladimir N. Osipov,[1] Zoya I. Uspenskaya,[3] Maria M. Korsukova,[1] and Liya L. Regel [4]

[1] Ioffe Physical-Technical Institute, St. Petersburg 194021, Russia
[2] All-Russian Aluminum-Magnesium Institute, St. Petersburg 199106
[3] Institute of Cytology, St. Petersburg 194021, Russia
[4]Clarkson University, Potsdam, NY 13699-5814, USA

INTRODUCTION

We have established for the first time that centrifugation alters the chemical composition of binary alkali halide crystals growing from supersaturated aqueous solutions.[1,2] To see if this phenomenon occurs in more complex systems, we investigated solid solution and eutectic systems constituted of two binary alkali halide compounds.

EXPERIMENTAL

Crystallization products were studied that had been obtained from supersaturated aqueous solutions of the systems LiCl-KCl, KCl-KI and LiCl-NaCl during centrifugation. These systems were chosen so as to contain different cations with the same anion and *vice versa*. Experiments were carried out in the centrifuge K24D (arm length 70 mm) at a=1.3g, 6.2g and 11.8x10^3g for 30 min. Initial compositions contained alkali halide mass ratios of 90:10 and *vice versa*. Supersaturation was achieved by cooling down to -10°C solutions that were saturated at 40-50°C and had been heated to 70-80°C. The volume of the centrifuged solutions was 40-60ml. Superheated solutions of two compositions were balanced in two special polyethylene test tubes for centrifugation. The test tubes were closed, placed into the centrifuge and subjected to different accelerations while cooling.

After an experimental run, the undersaturated solutions above the crystalline sediments on the bottom of the centrifuge tubes were poured off into ordinary test tubes. In most cases the sediments strongly stuck to the walls and bottom of the tubes and had to be removed with a scalpel. Crystallization products were in the form of rather big pieces (up to 1-2 cm) which consisted of smaller (up to 1 mm) intergrown crystals. These products were then dried at about 100°C in a drying box and chemically analyzed by standard

Processing by Centrifugation, Edited by Regel and Wilcox
Kluwer Academic/Plenum Publishers, New York, 2001

techniques: gravimetric, titrimetric and spectro- and flame-photometric. Crystalline sediments obtained by centrifuging the systems containing Li, i.e. LiCl-KCl and LiCl-NaCl, were very hygroscopic, which made it practically impossible to study the properties of the crystals.

RESULTS AND DISCUSSION

The results for solid solutions are presented in Tables1-3 and Fig. 1.

Table 1. Composition and lattice parameter of KCl-KI solid solution crystals obtained by crystallization with centrifugation (in mol %; data normalized to 100% are in parentheses).

Acceleration, *a*	KCl – KI (*a/a* = 3.58)					
	Excess KCl			Excess KI		
	KCl content	KI content	Lattice parameter, **a**, Å	KI content	KCl content	Lattice parameter, **a**, Å
1 *g*	(95.07)	(4.93)		(80.15)	(19.85)	
	98.50	1.50		89.06	10.94	
$1.3 \times 10^3 g$	95.24	4.76	6.2925(3)	88.33	16.27	7.0629(3)
$6.2 \times 10^3 g$	96.12	3.88	6.2920(3)	77.94	22.06	7.0636(3)
$11.8 \times 10^3 g$	96.12	3.88	6.2934(3)	98.36	1.64	7.0636(3)

Table 2. Composition and lattice parameter of LiCl-NaCl solid solution crystals obtained by crystallization with centrifugation (in mol %; data normalized to 100% are in parentheses).

Acceleration, *a*	LiCl - NaCl (*c/c* = 3.31)			
	Excess LiCl		Excess NaCl	
	LiCl	NaCl	NaCl	LiCl
1*g*	(92.50)	(7.50)	(86.89)	(13.31)
$1.3 \times 10^3 g$	29.03	70.97	98.30	1.70
$6.2 \times 10^3 g$	14.60	85.40	96.50	3.50
$11.8 \times 10^3 g$	-	-	-	-

It can be seen from Table 1 that crystals formed during centrifugation were enriched in the heavier ions compared to the initial compositions. The samples obtained with an initial KCl excess displayed an increase of the lattice parameter with increasing *a* compared to the initial KCl value of **a** = 6.2830(3) Å.

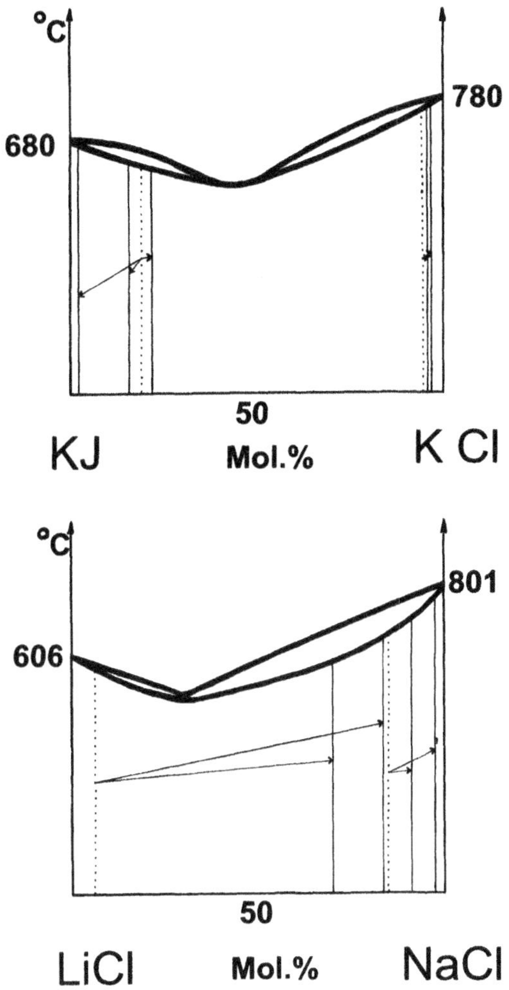

Figure 1. Alteration of the chemical composition of solid solution crystals in the systems LiCl-NaCl (different cations) and KCl- KI (different anions) by crystallization with centrifugation. The equilibrium diagrams[3,4] are denoted by solid lines. The initial compositions and those after crystallization are denoted by dotted lines. The arrows indicate the shifts of composition.

This trend correlates with the alteration of the chemical composition. It should be noted that all samples were constituted of a single-phase, indicating the formation of solid solutions of KCl-KI. The samples obtained with an initial KI excess displayed a small increase of the lattice parameter with increasing *a* compared to the initial KI value of **a**=7.066 Å. X-ray analysis showed that all of these samples contained some KCl. Thus, the initial excess of the heavier component brought about the separation of the solid solution into separate phases.

All of these results may be attributed to the considerable difference of sedimentation rates for ions with different masses. To characterize and predict this effect we propose using the atomic mass ratio of heavier ions to lighter ions, denoted as *c/c* for cations in two compounds, *a/a* for anions, and *a/c* (*c/a*) for anions and cations in the same compound. When the value of such ratio exceeded 2 (e.g., $a/c = 2.04$ for KBr) centrifugation altered both the composition and the properties of the resulting crystals.[1] Therefore, this ratio can be used as an approximate criterion for choosing systems for investigation, providing that other properties are not deleterious, e.g., the hygroscopic nature of Li salts in air and the small solubility of some Ba and Pb salts. Here, in addition, systems were chosen that contain the same cations and different anions and *vice versa* in order to compare the effect of centrifugation with ions of different nature.

As shown in Tables1 and 2, the atomic mass ratios of cations and anions in the systems KCl-KI and LiCl-NaCl are nearly the same. However, the effect of centrifugation was different. In the system KCl-KI (anions with different masses) the composition shift towards the component with the heavier anion, i.e. KI, was manifested with an initial KCl excess for all accelerations only in comparison with the experimental data for $1g$. With an initial KI excess a composition shift can be seen by comparing the data obtained at $1.3g$ and $11.8x10^3g$ with that at $1g$, and by comparing the data obtained at $11.8x10^3g$ with that at $1g$. No systematic variation in the Cl/I ratio with acceleration can be detected. In the system LiCl-NaCl (cations with different masses, see Table 2) the composition shift is confirmed to the full extent, both for initial excesses of lighter LiCl and heavier NaCl. For an initial excess of LiCl the composition shift became more pronounced with increasing acceleration, while for excess NaCl this was not observed.

Figure 1 also shows that the chosen mixtures with different cations and anions displayed different variations with centrifugation. In the system with different anions the tendency for composition shift was rather weak and was manifested mainly as a surplus of the heavier anion. This tendency was much more pronounced in the system with different cations, where the composition shifted by more than 70% towards the heavier cation with an initial excess of the lighter cation.

Table 3 shows that the most pronounced composition shifts were observed in the eutectic system LiCl-KCl, which has a big difference in the cations' masses, c/c=5.63. Two compositions were investigated: the eutectic composition and a hypoeutectic composition with a large excess of lighter component. Both compositions displayed big shifts towards a higher content of the heavier cation with centrifugation. When the acceleration a was increased from $1.3x10^3g$ to $11.8x10^3g$ an unexpected tendency was observed for the eutectic: the Li content increased. For the hypoeutectic composition the alteration of Li content with increasing acceleration showed a minimum at $6.2x10^3g$.

Table 3. Alteration of the composition of the eutectic system LiCl-KCl (c/c = 5.63) by crystallization during centrifugation (in mol %).

Acceleration, a	Eutectic composition (58.25% LiCl and 41.75% KCl)		Hypoeutectic composition (94.03% LiCl and 5.97% KCl)	
	LiCl	KCl	LiCl	KCl
$1\ g$	27.93	72.07	(94.03)	(5.97)
$1.3 \times 10^3\ g$	16.16	83.84	36.48	63.52
$6.2 \times 10^3\ g$	18.71	81.29	28.87	70.13
$11.8 \times 10^3\ g$	20.41	79.59	37.50	62.50

In many respects the crystallization of ionic compounds with centrifugation is far from understood. However, we have established that centrifugation drastically changes the composition of the crystals of solid solution and eutectic systems. The observations can be presented on binary phase diagrams as shifts of the composition points or the so-called figurative composition points[5] towards the compound with a heavier mass (more precisely, with a heavier ion). We note that the alteration of the compositions in these systems was studied by crystallization from aqueous solutions at low temperature and not by using binary melts at high temperature as shown on binary phase diagrams. However, the shift of figurative composition points on a binary phase diagram may be an acceptable representation for crystals formed from solutions, since the chemical composition of crystals obtained from the same initial composition with centrifugation is compared with the chemical composition of crystals obtained without centrifugation.

CONCLUSIONS

It was shown for the first time that centrifugation of supersaturated aqueous solutions of binary alkali halide systems brings about considerable alteration of the chemical composition of the crystals, both for mixtures forming solid solutions and those forming a eutectic. It is proposed to use the ratio of atomic masses of the differing ions to evaluate the influence of centrifugation on the chemical composition of crystals obtained. Binary systems with different anions were less subject to alteration of chemical composition than those with different cations.

REFERENCES

1. V.N. Gurin, S.P. Nikanorov, M.M. Korsukova, L.I. Derkachenko, W. Jung, and L.L. Regel, Alteration of the lattice constant and the composition of KCl and KBr crystals under macrogravity, *Phys. Solid State* 39: 1599 (1997).
2. V.N. Gurin, S.P. Nikanorov, L.L. Regel and L.I. Derkachenko, Influence of gravity on the habit of single crystals of compounds grown from solutions, in: *Centrifuging Materials Processing*, L.L. Regel and W.R. Wilcox, eds., Plenum Press, New York and London (1997) pp 197-202.
3. *Handbook on the Melting of Systems of Waterless Inorganic Salts,* Academy of Sciences of USSR, Moscow-Leningrad (1961) (in Russian).
4. V.I. Posipaiko, E.A. Alekseeva, N.V. Vasina, E.S. Gryslova, V.N. Afonova, N.N. Petrova, I.G. Popova, Z.L. Guseva and V.T. Shemyatenkova, *Melting Diagrams of Salt Systems,* Metallurgiya, Moscow (1977, 1979) (in Russian).
5. V.Ya. Anosov, M.I. Ozerova, and Yu.Ya. Fialkov, *Introduction to Physical-Chemical Analysis*, Nauka, Moscow (1976) (in Russian).

DISSOLUTION RATE OF GaP CRYSTAL ON A CENTRIFUGE

Yuko Inatomi,[1] Kengo Horiuchi,[2] and Kazuhiko Kuribayashi[1]

[1]The Institute of Space and Astronautical Science, 3-1-1 Yoshinodai, Sagamihara, Kanagawa 229-8510, Japan
[2]Faculty of Engineering, The University of Tokyo, Tokyo 113-8654, Japan

INTRODUCTION

The acceleration of gravity is often an important factor influencing a processing procedure, in addition to thermodynamic parameters such as temperature and pressure, especially when a fluid phase is involved. Accordingly, the control of acceleration is important to obtain high quality materials, and to understand the heat and mass transport phenomena during the process. In recent years, materials processing experiments have been performed not only in microgravity but also in hypergravity environments. Utilization of a centrifuge is used to generate hypergravity acceleration on earth. Most centrifugal crystal growth experiments were employed to study the relationship between acceleration and 1) composition and defect distributions in the resulting crystal,[e.g.,1,2] 2) sedimentation during growth,[3] 3) morphological instability of the solid/liquid (S/L) interface,[4,5] 4) growth striations,[6,7] 5) temperature oscillations in the liquid phase,[8,9] and 6) heat transmission.[10]

It has been shown by numerical simulations and by analysis of crystals that convection might be suppressed at a particular rotation rate on a centrifuge. However, a clear understanding of the reason for such behavior has not yet been achieved. Wilcox *et al.*[11] classified the following three simple models to explain the phenomena:

1. Flow transition model. A minimum in convection might occur in the neighborhood of the transition from a low-acceleration flow mode to a high-acceleration flow mode.[12]
2. Thermal stability model. This model stemmed from the observation that there would be no buoyancy-driven convection in a fluid if the net acceleration vector were everywhere exactly parallel to the density gradient, with density decreasing with height.[13, 14]
3. Buoyancy-Coriolis balance model. Scaling analysis suggested that the buoyancy force in the melt is balanced by the Coriolis force and that a minimum in convection might occur at a particular rotation rate.[13, 15]

The Buoyancy-Coriolis balance model may be suitable when the contribution of the Coriolis force to the convection is not negligible due to the small radius of a centrifuge, although a numerical simulation predicted that the Coriolis force may alter the flow modes only slightly in experiments on an 18 m centrifuge.[12] The concentration profile of Ga in a Ge crystal grown by a vertical Bridgman method agreed with results from a computational simulation.[15] The validity of this model should be investigated from the viewpoint of heat

Processing by Centrifugation, Edited by Regel and Wilcox
Kluwer Academic/Plenum Publishers, New York, 2001

and mass transport phenomena, because the radius of the centrifuge in materials processing experiments has been on the order of meters.

In high-gravity materials experiments, the materials are analyzed after the completion of the experiments. *In-situ* simultaneous observation of the growing interface and the environment would directly reveal the influence of supersaturation, temperature and concentration gradients, convection, and surface kinetics on the growth behavior. Some visualization experiments of crystal growth process and fluid behavior with high Prandtl number (*Pr*) liquids have been performed on centrifuges.[*5,16] However, there has been no report of visualization experiments using a low *Pr* fluid, such as a melt or solution of a semiconductor. Accordingly, we concluded that fruitful information on the relationship between the rotation rate of the centrifuge and the fluid behavior could be obtained if the morphological change of a solid-liquid interface during solution growth of a semiconductor were measured in real-time. Thus, we performed *in-situ* observations of the solid-liquid interface during semiconductor solution growth using near-infrared (NIR) microscopic interferometry.[17,18] The dissolution rate of GaP(111)B into a gallium solution was measured on a centrifuge in real-time in the hope of observing a condition where buoyancy-driven convection would be strongly damped.

EXPERIMENTAL PROCEDURE

A centrifuge system with a radius of 1.2 m, TACT,[5] was used for the experiments. The angular velocity of the table was controlled with an inverter from 0 to 16 rad/s. This experimental setup consisted of four components: a goniometer with an NIR microscopic interferometer and an image furnace, a data acquisition system, a power supply for the furnace, and counterweights. Eighteen slip rings were used for primary power supplies of the components and telemetering with computers. The axis of the specimen ampoule was made parallel to the resultant acceleration vector, ***b***, by adjusting the angle of the goniometer. The center of inertia for the ampoule was located at R_0 = 0.85 m, where a maximum acceleration $Ng = |\boldsymbol{b}|$ of 22 *g* could be generated.*

The (111)B faces of S-doped liquid-encapsulated Czochralski GaP single crystals, 7.8mm in diameter and 300μm in thickness, were dissolved into gallium during heating from 973 K to 993 K at 1 K/min. Both sides of each substrate were mechanically and chemically polished to mirror planes. A quartz ampoule, 10.0mm O.D. and 8.0mm I.D., contained the substrate, 0.8 g of gallium solvent, and a graphite crucible whose O.D., I.D., and length were 7.8 mm, 3.6 mm and 26 mm, respectively. The ampoule was evacuated to $\sim 10^{-6}$ mbar and sealed with a mechanical valve. To prevent Marangoni convection, a BN cylinder of 5.9mm diameter and 13mm length was placed on the solution as a lid and was pressed down slightly by quartz-glass wool that filled the remainder of the ampoule. The image furnace with a ring-shaped halogen lamp of maximum power of 1 kW was used to heat the specimen. The lamp power was controlled by a PID system to ± 0.1K using a ceramic-coated chromel-alumel thermocouple whose tip was situated at the bottom of the crucible and attached to the substrate. Figure 1 shows the experimental configuration on the centrifuge.

During dissolution, the substrate developed a depression with a faceted region at the center at a depth of about 60μm in every experiment. The dissolution rate on the faceted region was measured in real-time using NIR microscopic interferometry.[17] Figure 2 shows a typical image of an interference fringe pattern. Measurement of the dissolution rate was more suitable than the growth rate to detect variations in mass transport in the solution, because surface kinetics is much more rapid for dissolution than for growth.[19,20]

---

* See the table of nomenclature at the end.

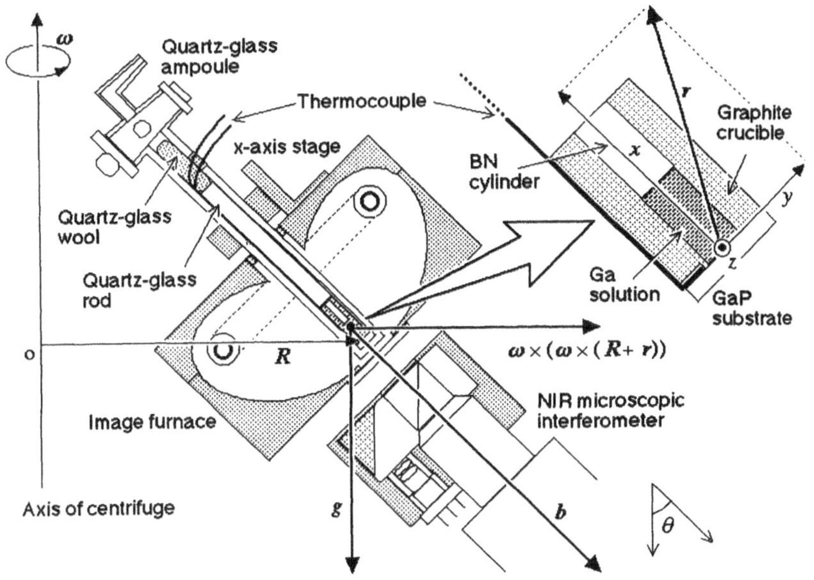

Figure 1. Schematic diagram of the experimental configuration.

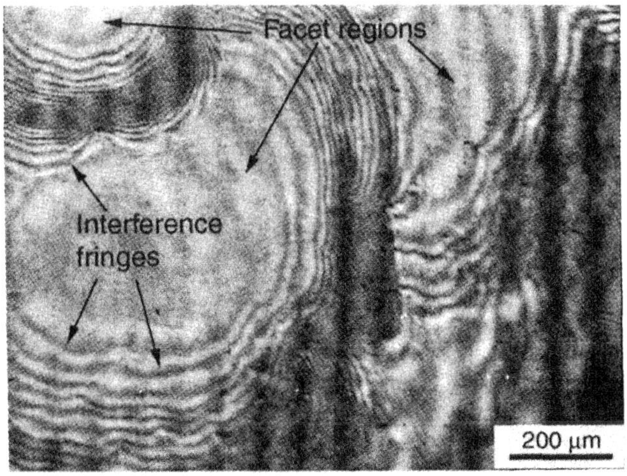

Figure 2. Typical interference fringe pattern on a growing facet on the (111) B face of GaP. The contour lines correspond to 0.120 μm in height difference.

In order to ensure that the solutal driving force for convection was negligible compared to the thermal one, the onset temperature of dissolution was set at 973 K, where the equilibrium concentration of phosphorus in gallium is 0.04 at% (= 2.2×10^{25} m^{-3}).[21] It was also necessary to consider that the refractive index of the GaP depends on temperature. Thus, we measured the refractive index and the thermal expansion of GaP crystals from 300 K to 1200 K.[22] In interpreting the interferograms we assumed that the refractive index of GaP from 973 K to 993 K is 3.283 and neglected the thermal expansion of the substrate. Table 1 shows the physical properties of the Ga solution and Table 2 the experimental conditions.

Table 1. Physical properties of gallium solution.

Physical properties of Ga solution at 973 K	Symbol	Value
Thermal expansion coefficient	β [K^{-1}]	1.0×10^{-4}
Kinematic viscosity	ν [m^2/s]	7.0×10^{-8}
Density	ρ [kg/m^3]	5.6×10^{3}
Diffusion coefficient of phosphorus	D [m^2/s]	7.6×10^{-10}
Thermal diffusivity	κ [m^2/s]	2.8×10^{-5}
Specific heat	C_p [J/kg.K]	3.8×10^{2}

Table 2. Experimental conditions.

Experimental condition	Symbol	Value
Initial phosphorus concentration in liquid	C_o [mol/m^3]	2.2×10^{25}
dCe/dt	m [m^{-3}/s]	6.8×10^{21}
Axial temperature gradient	G_x [K/m]	1.0×10^{3}
Radial temperature gradient	G_{yz} [K/m]	6.9×10^{2}
Melt radius	r_0 [m]	2.0×10^{-3}
Liquid height	h [m]	1.3×10^{-3}
Distance between rotation axis and origin of coordinate system in specimen	R_0 [m]	0.85
Resultant acceleration	Ng [m/s^2]	1.0 – 7.0

THEORETICAL MODEL

Scaling Analysis

The influence of buoyancy force on heat and mass transport was studied using a scaling analysis.[23] Friedrich *et al.*[15] modified this analysis to estimate fluid behavior accounting for both centrifugal force and Coriolis force. They assumed that the temperature field and the velocity field for low *Pr* and Reynolds number *Re* can be treated separately because the heat transport is mainly conductive. At steady state they obtained:

$$Re = -\left(1 + Ta^{0.5}\right)/2 + \sqrt{\left(1 + Ta^{0.5}\right)^2/4 + Gr\left|\boldsymbol{b} \times \nabla T\right| / \{(Ng)(\Delta T/h)\}} \tag{1}$$

Here, we derive the dependence of *Re* on ω_0 from Eq. (1) to find the conditions for minimum convection. We assume that: 1) an axisymmetric ampoule for the specimen is held parallel to the resultant acceleration $\boldsymbol{b}$; 2) heat transport in the solution is conductive; 3) the temperature gradient $\boldsymbol{G}$ in the solution is a constant vector composed of:

$$G_x \approx \Delta T / h = 1.0 \times 10^3 \text{K/m} \ , \ G_{yz} = \sqrt{G_y^2 + G_z^2} \approx 6.9 \times 10^2 \text{K/m} \tag{2}$$

We note that normally $G_y/G_x << r_0/R_0 << 1$ and define a non-dimensional centrifugal force as $\varphi = \omega_0 R^2/g$ so that the components of the acceleration $\boldsymbol{b}$ can be approximated by:

$$b_x = \omega_0^2 (x \sin\theta - y\cos\theta - R_0 \sin\theta) - g\cos\theta \approx -g\sqrt{1+\varphi^2} \tag{3}$$

$$b_y = \omega_0^2 \cos^2\theta(-x\tan\theta + y) \approx -g\,\{\varphi/(1+\varphi^2)\}(y - x\varphi)/R_0 \approx 0 \tag{4}$$

$$b_z = \omega_0^2 z = g\varphi(z/R_0) \approx 0 \tag{5}$$

Then the non-dimensional variables in the square root in Eq. (1) can be written:

$$Ta = \varphi/q^2 \ , \ Gr|\boldsymbol{b} \times \nabla T| / \{(Ng)(\Delta T/h)\} \approx (p/q^2)\sqrt{1+\varphi^2} \tag{6}$$

with

$$p = R_0 G_{yz}\beta \ , \ q = (\nu / r_0^2)\sqrt{R_0/g} \tag{7}$$

By substituting Eq. (6) into Eq. (1) and setting $(dRe/d\varphi)_{\varphi=\varphi m} = 0$ we find that the condition for minimum convection, φ_m, is a solution of:

$$p\varphi_m^3 - (1 - q\varphi_m^{1.5} - \varphi_m^2)\sqrt{1+\varphi_m^2} = 0 \tag{8}$$

Diffusion-Controlled Model

If we assume that the solubility is very small, then we can neglect the motion of the crystal surface during dissolution and the one-dimensional equation for diffusion is:[24]

$$\partial C/\partial t = D\partial^2 C/\partial x^2 \tag{9}$$

We assume that the mass flux of the solute atoms to the BN cylinder and the crucible is negligible, and that the solute concentration at the crystal surface increases with the linear-heating method as follows:

$$\partial C/\partial x|_{x=h} = 0 \ , \ \partial C/\partial x|_{x=0} \approx C_S R(t)/D \tag{10}$$

$$C(0,t) = C_0 + mt \ , \ C(x,0) = C_0 \tag{11}$$

The dissolution rate, $R_{\text{diff}}(t)$, can be expressed as a Fourier series in time by solving Eq. (9) with boundary conditions Eqs. (10) and (11), similar to Tiller:[25]

$$R_{\text{diff}}(t) = (hm/C_S)\left[1 - 2\sum_{n=1}^{\infty}(1/a_n^2)\exp(-a_n^2 Dt/h^2)\right] \tag{12}$$

where h is the length of the solution, $C_S = 2.5 \times 10^{28}$ m^{-3} is the concentration of solute in the solid, and $a_n = (2n-1)\pi/2$. If we substitute $1/(1+a_n^2 Dt/h^2)$ for $\exp(-a_n^2 Dt/h^2)$ in Eq. (12), a simple approximation can be derived as:

$$R_{\text{diff}}(t) \approx (m/C_S)\sqrt{Dt}\tanh\left(h/\sqrt{Dt}\right) \tag{13}$$

RESULTS AND DISCUSSION

By substituting the constants in Tables 1 and 2 into Eq. (1), we obtain Re as a function of φ, as shown in Fig. 3, with Re exhibiting a minimum at $\varphi = \varphi_m \sim 1.0$. Figure 4 shows the measured dissolution rate with linear-heating. The dissolution rate was lower for accelerations from 1.1 g_a to 7.0 g_a than without centrifugation. Moreover, the curve at 1.4 g_a was relatively close to the calculated values of $R_{\text{diff}}(t)$ based on Eq. (13) with the parameters shown in Tables 1 and 2. Thus the experimental data appear to support the Buoyancy-Coriolis balance model.

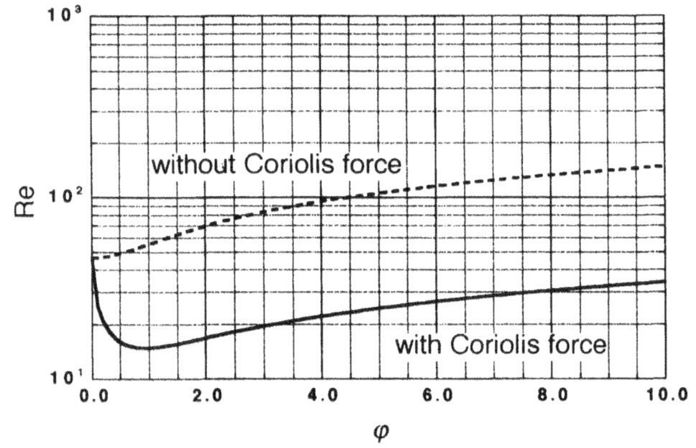

Figure 3. Values of Re versus φ calculated from Eq. (1) with and without Coriolis force.

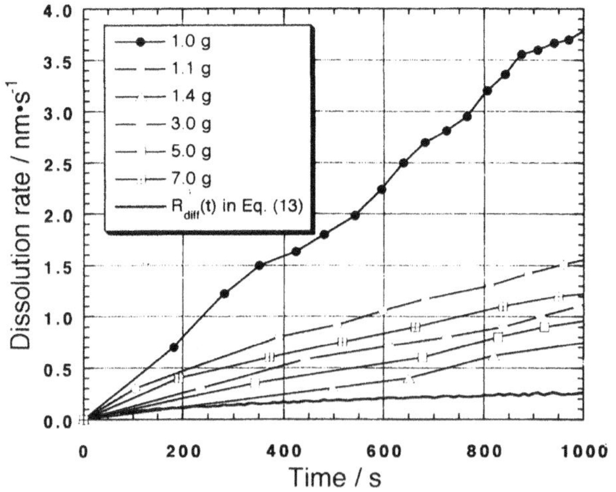

Figure 4. Measured dissolution rate as a function of time.

Figure 5 shows φ_m as a function of log p and log q from Eq. (8). Note that φ_m depends appreciably on p only in the range of q = 10^{-5} - 5.0×10^{-1}. For semiconductor and metal liquids, $p=10^{-2}$ to10^{-1} and q = 10^{-4} to 10^{-1} since $\nu=10^{-7}$ to 10^{-6} m^2/s, $\beta=10^{-4}$ to 10^{-3} K^{-1}, and $G=10^2$ to 10^4 K/m, while $r_0=10^{-3}$ to 10^{-1} m and R_0=1 to 10 m for the usual large centrifuge. This result shows that the minimum of the characteristic flow velocity can be defined by the product of R_0 and $G_{yz}\beta$, where the former and the latter are, respectively, caused by the Coriolis force and the buoyancy force induced by the temperature gradient. Thus the non-dimensional parameters, p and q, characterize the behavior of thermal convection on a centrifuge.

If $Re(\varphi_m)$ is assumed to result from a decrease of acceleration from g to Mg with M<1, the parameter M can be derived as:

$$M \approx \left(R_0 G_{yz} \beta\right)\varphi_m^3 / \left(1+\varphi_m^2\right) \leq R_0 G_{yz} \beta / 2 = p/2 \tag{14}$$

By substituting physical properties into Eq. (14) we conclude that the fluid behavior of a low gravity environment, e.g. 0.01 g, could be obtained on a centrifuge with a short radius.

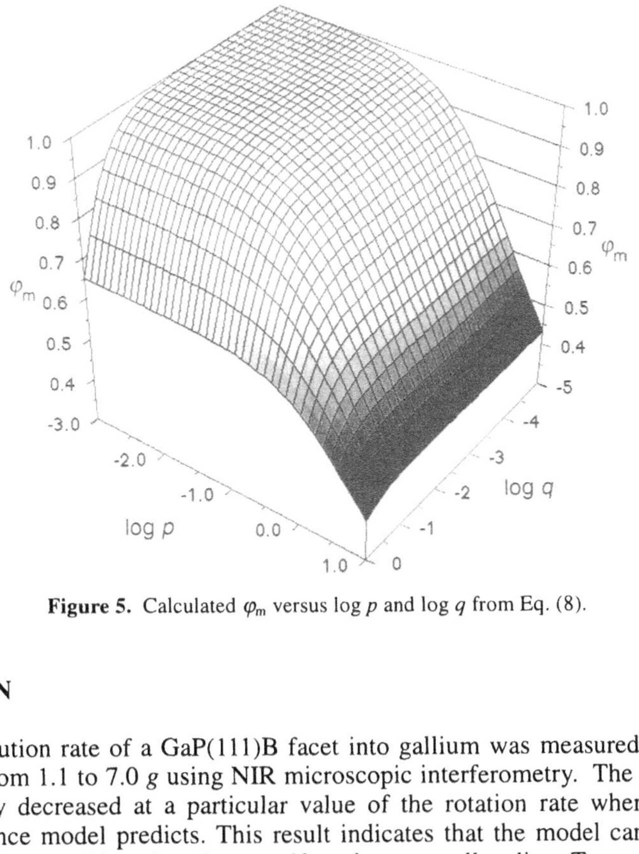

Figure 5. Calculated φ_m versus log p and log q from Eq. (8).

CONCLUSION

The dissolution rate of a GaP(111)B facet into gallium was measured in real time at accelerations from 1.1 to 7.0 g using NIR microscopic interferometry. The dissolution rate was remarkably decreased at a particular value of the rotation rate where the Coriolis-Buoyancy balance model predicts. This result indicates that the model can describe fluid behavior with the condition that the centrifuge has a small radius. Two non-dimensional parameters were derived from the model that characterize the behavior of low-Pr fluids such as liquid metals and semiconductors.

NOMENCLATURE

$\boldsymbol{b}$ - vector of resultant acceleration, = (b_x, b_y, b_z) $\boldsymbol{g}$ - $(\boldsymbol{\omega} \times (\boldsymbol{\omega} \times (\boldsymbol{r} + \boldsymbol{R}))$ (m/s^2)
C_0 - initial concentration in liquid (atom/m^3)
C_e - equilibrium concentration in solution at the solid-liquid interface (atom/m^3)
C_S - solute concentration in solid (atom/m^3)
D - diffusion coefficient of solute atoms in the solution (m^2/s)
$\boldsymbol{G}$ - characteristic temperature gradient in the solution, = (G_x, G_y, G_z) (K/m)
Gr - Grashof number, $\beta (\Delta T/h)(Ng)r^4/\nu^2$
$\boldsymbol{g}$ - vector of acceleration of earth's gravity, $g = |\boldsymbol{g}| = 9.81$ m/s^2
m - dC_e/dt; defined to be constant (atom/m^3.s)
Ng - absolute value of resultant acceleration ($|\boldsymbol{b}|$) (m/s^2)
P - pressure (Pa)
$\boldsymbol{R}$ - vector from the rotation axis to the origin of the local coordinate system in the specimen, $R_0 = |\boldsymbol{R}|$ (m)
Re - Reynolds number, $V_{ref}\, r/\nu$
$\boldsymbol{r}$ - local vector in the specimen, = (x, y, z) (m)
r_0 - radius of the specimen (m)
T - temperature (K)
Ta - Taylor number, $4R_0^4\omega_0/\nu^2$
t - time (s)
V_{ref} - characteristic velocity in liquid (m/s)

β - thermal expansion coefficient (K^{-1})
θ - angle of specimen to gravity vector
ΔT - temperature difference between top and bottom in the solution (K)
ν - kinematic viscosity (m^2/s)
$\boldsymbol{\omega}$ - vector of angular velocity of the centrifuge, $\omega_0 = |\boldsymbol{\omega}|$ (s^{-1})

Acknowledgements

The authors thank T. Maki, N. Yamada and A. Yamamoto of Olympus Optical Co., Ltd. for manufacturing and adjusting the NIR microscopic interferometer. The authors also thank L.L. Regel and W.R. Wilcox for helpful discussions and advice. This work was supported by grant No. 9JSF-283 from the Japan Space Forum.

REFERENCES

1. P.J. Shlichta and R.E. Knox, Growth of crystals by centrifugation, *J. Crystal Growth* 3/4: 808 (1968).
2. H. Rodot, L.L. Regel, and A.M. Turtchaninov, Crystal growth of IV-VI semiconductors in a centrifuge, *J. Crystal Growth* 104: 280 (1990).
3. R.S. Sokolowsky and M.E. Glicksman, Gravitational influence on eutectic solidification, *J. Crystal Growth* 119: 126 (1992).
4. G. Müller and G. Neumann, Tenfold growth rates in the travelling heater method of GaSb crystals by forced convection on a centrifuge, *J. Crystal Growth* 63:58 (1983).
5. Y. Inatomi, O. Kitajima, W. Huang, and K. Kuribayashi, Directional solidification of faceting material in a centrifuge, *Cryst. Res. Technol.* 32: 947 (1997).
6. W. Weber, G. Neumann and G. Müller, Stabilizing influence of the Coriolis force during melt growth on a centrifuge, *J. Crystal Growth* 100: 145 (1990).
7. G. Müller, G. Neumann, and W. Weber, The growth of homogeneous semiconductor crystals in a centrifuge by the stabilizing influence of the Coriolis force, *J. Crystal Growth* 129: 8 (1992).

8. F. Tao, Y. Zheng, W.J. Ma and M.L. Xue, Unsteady thermal convection of melts in a 2-D horizontal boat in a centrifugal field with consideration of the Coriolis effect, in: *Materials Processing in High Gravity*, L.L. Regel and W.R. Wilcox, eds., Plenum, New York (1994) p 67.
9. L. Bergelin and A. Chevy, Analysis of thermal convection in molten tin under centrifugal conditions, in: *Centrifugal Materials Processing*, L.L. Regel and W.R. Wilcox, eds., Plenum Press, New York (1997) p 45.
10. T. Hibiya, S. Nakamura, K.W. Yi and K. Kakimoto, Coriolis effect on heat transfer experiment using hot-wire technique on centrifuge, in: *Materials Processing under High-Gravity*, L.L. Regel and W.R. Wilcox, eds., Plenum Press, New York (1994) p 171.
11. W.R. Wilcox, L.L. Regel and W.A. Arnold, Convection and segregation during vertical Bridgman growth with centrifugation, *J. Crystal Growth* 187: 543 (1998).
12. W.A. Arnold, W.R. Wilcox, F. Carlson, A. Chait, and L.L. Regel, Transport mode during crystal growth in a centrifuge, *J. Crystal Growth* 129: 24 (1992).
13. V.A. Urpin, Convective flows during crystal growth in a centrifuge, in: *Materials Processing in High Gravity*, L.L. Regel and W.R. Wilcox, eds., Plenum Press, New York (1994) p 35.
14. W.A. Arnold and L.L. Regel, Thermal stability and the suppression of convection in a rotating fluid on earth, in: *Materials Processing in High Gravity*, L.L. Regel and W.R. Wilcox, eds., Plenum, New York (1994) p 17.
15. J. Friedrich, J. Baumgartl, H.-J. Leister, and G. Müller, Experimental and theoretical analysis of convection and segregation in vertical Bridgman under high gravity on a centrifuge, *J. Crystal Growth* 167: 45 (1996).
16. P.V. Skudarnov, L.L. Regel, and W.R. Wilcox, Flow visualization study of convection on a centrifuge, in: *Centrifugal Materials Processing*, L.L. Regel and W.R. Wilcox, eds., Plenum, New York, (1997)
17. Y. Inatomi and K. Kuribayashi, Real-time measurement of LPE growth rate in GaP, *J. Crystal Growth* 114: 380 (1991).
18. Y. Inatomi, P. Dold, A.N. Danilewsky, and K.W. Benz, In-situ observation setup for semiconductor growth interface from solution in a magnetic Field, *Cryst. Res. Technol.* 32: 759 (1997).
19. K. Onuma, K. Tsukamoto and I. Sunagawa, Dissolution kinetics of K-alum crystals as judged from the measurements of surface undersaturations, *J. Crystal Growth* 110: 724 (1991).
20. Y. Inatomi, S. Aoki, K. Kuribayashi, and Y. Sawada, Growth and dissolution rates on GaP(111)B facet surface during solution growth under a transverse static magnetic field, *Cryst. Res. Technol.* 33: 857 (1998).
21. M.B. Panish, A thermodynamic evaluation of the simple solution treatment of the Ga-P, In-P and Ga-As systems, *J. Electrochem. Soc.* 121: 556 (1974).
22. D. C. Yin and Y. Inatomi, Measurement of refractive index of GaP crystal over a large temperature range using interferometry, *Cryst. Res. Technol.* 35: 221 (2000).
23. D. Camel and J.J. Favier, Scaling analysis of convective solute transport and segregation in Bridgman crystal growth form the doped melt, *J. Physique* 47: 1001 (1986).
24. W.R. Wilcox, A Generalized Treatment of Mass Transfer in Crystal Growth, *Preparation and Properties of Solid State Materials: Vol. 2*, W.R. Wilcox, ed., Marcel Dekker, New York (1977).
25. W.A. Tiller and C. Kang, On the growth rate of crystals from solution, *J. Crystal Growth* 2: 345 (1968).

INFLUENCE OF CENTRIFUGATION ON THE CONDUCTIVITY OF POLY-N-EPOXYPROPYLCARBAZOLE - COPPER PHTHALOCYANINE FILMS DEPOSITED FROM SOLUTION

Khasan Karimov,[1] Khakim Akhmedov,[1] Mohammad Mahroof-Tahir,[2] Rizwan M.Gul,[3] and Abduakhad Ashurov

[1]Physical Technical Institute, Dushanbe 734063 ,Tajikistan
[2]St. Cloud State University, St. Cloud, MN 56301, USA
[3]GIK Institute of Engineering Sciences and Technology, Topi, Pakistan
[4]Firm DORO, Aini 70-35, Dushanbe 734001, Tajikistan

ABSTRACT

We investigated the electrical conductivity of films composed of two photosensitive organic semiconductors, poly-N-epoxipropylcarbazole (PEPC) and copper phthalocyanine (CuPc, $C_{32}H_{16}N_8Cu$), which had been deposited from solution at different gravity conditions from 1*g* to 1107*g*. Visually, the films grown with centrifugation were more homogeneous. The X-ray analysis of the deposited films indicated a small shift of the characteristic peak for copper phthalocyanine (at $2\theta = 7.10^o$) as acceleration was increased. The electrical conductivity at 1000 Hz was measured in the range from 300 K to 410 K using a sandwich-type cell. There was a 1.5-fold increase in the conductivity and 2-fold decrease of activation energy with increased acceleration. Scanning electron microscopy (SEM) showed more uniform films at higher accelerations and with larger size of agglomerates. Analysis of the conductivity measurements indicated the presence of a compensative effect that manifested itself as a linear increase of activation energy with the logarithm of the pre-exponential factor.

INTRODUCTION

The conductivity of organic semiconductors has both fundamental and applied importance.[1,2] For example, the efficiency of organic solar cells is limited mainly by the low conductivity of photosensitive materials.[3,4] We investigated films made of two photosensitive organic semiconductors, PEPC and CuPc, which have potential application as organic photoelectric converters or solar cells.[4-6]

Organic materials have a large variety of structures and properties that depend on processing conditions. Centrifugal materials processing had given many interesting results with different materials.[e.g.,7] Centrifugal processing of organic semiconductors was considered because of their large molecular weights and weak van der Waal's intermolecular bonding. For example, copper phthalocyanine thin films deposited by physical vapor transport at microgravity were more homogeneous, smoother and denser than those made at unit-gravity conditions.[8-10] At high-gravity conditions we expect the centrifugal force to influence the deposition process and the structure and properties of the resulting films. We present here experimental results on investigation of the conductivity of PEPC-CuPc films deposited from solution at different gravity conditions.

EXPERIMENTAL

We chose PEPC because of its strong adhesion to substrates and its high photosensitivity. The electrical conductivity of PEPC should be increased by doping with CuPc. We believe the PEPC and CuPc form a charge-transfer complex, because PEPC is an electron donor in many organic low molecular weight materials.[11] So this deposited matter should not be a simple mixture.

For these experiments, we used PEPC synthesized by us[11] and CuPc produced in industry. Figure 1 shows the molecular structures of PEPC (here, "n" was about 4-6) and CuPc. The molecular weight of the PEPC was about 1000 and CuPc 576 amu. The films were deposited from 1%w solutions of PEPC-CuPc (90:10 by weight) in benzol during centrifugation at $1g$ to $1107g$. A table-top centrifuge (Fig. 2) was used for deposition on 15-mm diameter thin (0.1-0.2 mm) aluminum substrates at standard room temperature. The substrate was put inside an aluminum vessel. The centrifuge had an 11-cm arm and could provide an acceleration up to $1107g$ at a maximum rotation speed of 3000 rpm.. For every experiment, two symmetrically installed aluminum vessels were used that had been filled with an equal volume solution to within 0.5 ml.

During an experiment the solution was evaporated at room temperature (300 K) without heating. At accelerations of $1g$, $277g$ and $1107g$ processing times of 30, 20 and 10 min, respectively, were sufficient to evaporate the solvent completely and to deposit the films. The film thickness was determined by weighing the deposited matter.[12] AC electrical conductivity at 1000 Hz was determined by measuring the resistance from 300K to 410K using a conventional measuring bridge and a sandwich-type cell. The cell's structure was the following: aluminum (as first electrode) - organic film - soot (as second electrode).

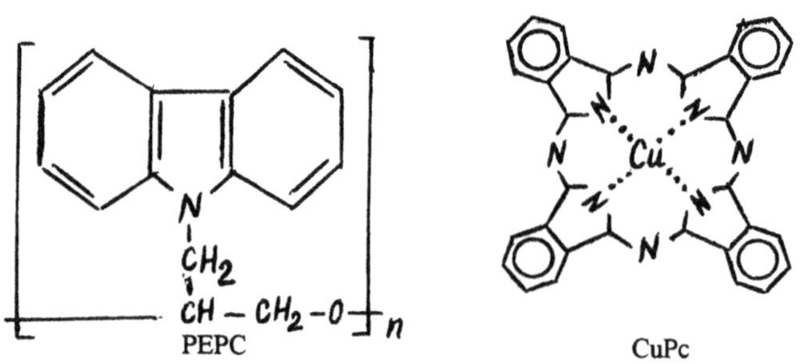

PEPC CuPc

Figure 1. Molecular structures of PEPC and CuPc .

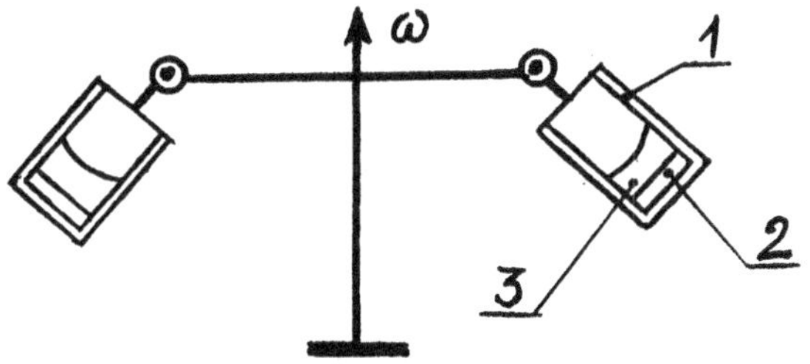

Figure 2. Schematic diagram of table-top centrifuge
1- aluminum cylinder for vessel, 2- aluminum substrate, 3- solution .

RESULTS AND DISCUSSION

Visually, the blue CuPc color seemed more uniform in films grown at higher gravity. The film thickness determined by weighing decreased with increasing acceleration: 11.1μm for 1*g*, 3.5 μm for 277*g*, and 2.6 μm for 1107*g*. In calculating thickness we assumed that the density of PEPC is 1.3 g/cm^3 and CuPc 0.8 g/cm^3, to give an average density of 1.25 g/cm^3 for the films. It is possible that the apparent decrease in thickness was due to an increase in density with increasing acceleration. Another possibility is that acceleration increased the area inside the vessel that was covered by the film.

Figure 3 shows x-ray 2θ scattering curves measured from the deposited films, giving $2\theta = 6.86^o$ for the α-form of phthalocyanine.[13] From the Al_2O_3 peaks of $2\theta = 38.5^o$ and 67.6^o,[13] it was possible to identify a shift in the CuPc peak as the acceleration was increased. The samples prepared at 1*g*, 277*g* and 1107*g* showed the CuPc peak at 7.10°, 7.16° and 7.20°, respectively. This means that this material may have became denser via decreasing distance for the interstacking of CuPc molecules with increasing acceleration.[9] At present, at least seven distinguishable crystalline polymorphs are known.[9] At the same time, the β-CuPc form is probably more stable at normal conditions. For example, α-CuPc is metastable and can be converted to the β-CuPc form by heating or by solvent evaporation.[9]

Figure 4 shows the dependence of electrical conductivity on temperature. Usually such behavior can be described by:[1,2]

$$\sigma=\sigma_o \exp(-E/kT) \qquad (1)$$

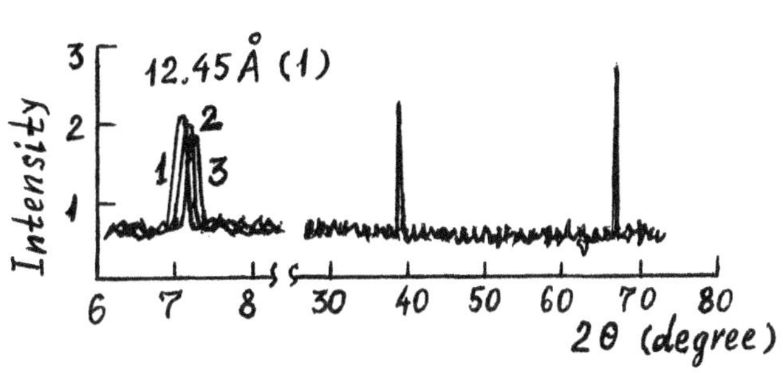

Figure 3. Typical 2θ x-ray diffraction scan of films deposited at:
1 – 1*g*, 2 – 277*g* and 3 – 1107*g*.

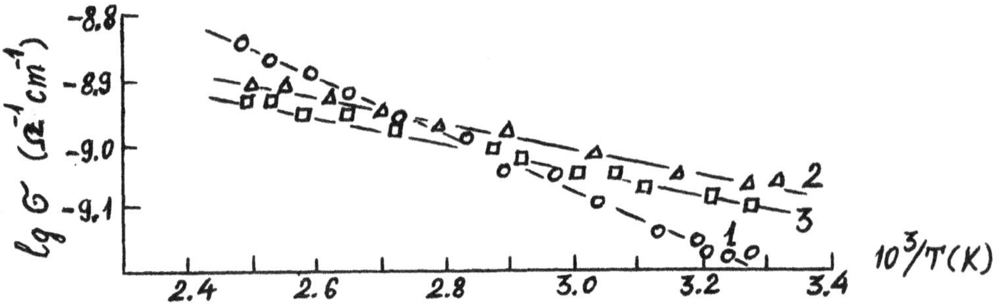

Figure 4. Dependence of the PEPC-CuPc films' conductivity on temperature. 1, 2 and 3 represent films deposited at 1*g*, 277*g* and 1107*g*, respectively.

where σ is conductivity, E is activation energy, k is Boltzmann constant, T is absolute temperature, and σ_o is the pre-exponential factor. The room temperature conductivity of the films deposited at 1*g*, 277*g* and 1107*g* was 0.36×10^{-9}, 0.54×10^{-9} and 0.50×10^{-9}/Ohm.cm, whereas the activation energy E in Eq. (1) was 0.10, 0.04 and 0.04 eV, respectively. This indicates a 1.5-fold increase in conductivity and a 2-fold decrease in activation energy with increased acceleration.

We inspected the films by SEM in order to help interpret the conductivity results. Figure 5 shows that the films consisted of different sized agglomerates. The film deposited at 277g seemed to be more uniform with the smallest agglomerates. Larger agglomerates can be seen in films deposited at 1107g. Higher electrical conductivity would be expected for films with greater uniformity and larger agglomerates.

Figure 6 shows the relation between the activation energy E and the pre-exponential factor σ_0. There appears to have been a compensative effect, indicating that the generation and transfer of charge carriers is similar to that found in several groups of organic semiconductors with different compositions.[14] Here, this effect was found on organic semiconductors of similar composition but deposited at different accelerations.

Acknowledgments

We highly appreciate the efforts of L.L. Regel to initiate and facilitate this research. We express our sincere thanks as well to Y. Abe for helping this work to materialize.

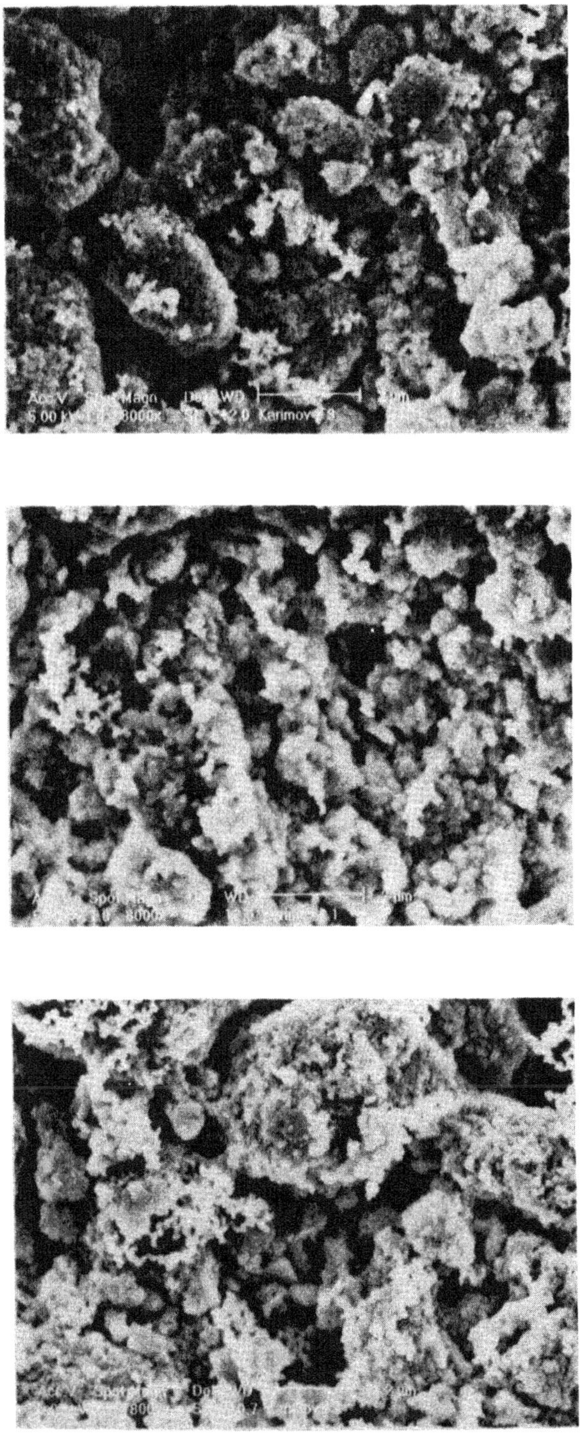

Figure 5. Scanning electron micrographs of PEPC-CuPc films deposited at: top: 1*g*, middle: 277*g*, bottom: 1107*g*.

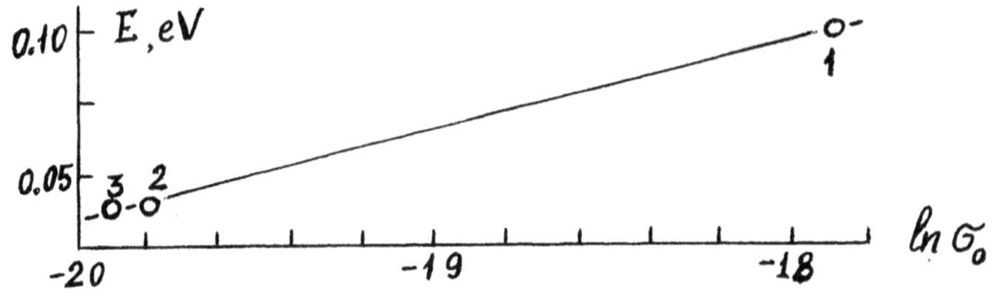

Figure 6. Dependence of activation energy E of conductivity on pre-exponential factor σ_o: 1 - 1g, 2 - 277g and 3 - 1107g.

REFERENCES

1. F. Gutman and L.E. Lyons, *Organic Semiconductors*, Part A, Robert Krieger Publishing Company, Malabar, Florida (1981).
2. F. Gutman, H. Keyzer, L.E. Lyons, and R.B. Somoano, *Organic Semiconductors*, Part B, Robert E. Krieger Publishing Company, Malabar, Florida (1983).
3. G.A. Chamberlain, Organic solar cells: a review, *Solar Cells* 8: 47 (1983).
4. Kh.M. Akhmedov, Kh.S. Karimov, M.I. Fiodorov, Organic solar cells, *Geliotekhnika* 1-3: 178 (1995).
5. A. Al-Mohamad, M. Soukieh, Solar cells and high efficiency photo-diodes having metal-organic thin film-semiconductor structures, *Thin Solid Films* 27: 132 (1995).
6. H. Yonehara and C. Pac, Photoelectrical properties of double-layer organic solar cells using C_{60} and phthalocyanines, *Thin Solid Films* 278:108 (1996).
7. L.L. Regel and W.R. Wilcox, eds., Centrifugal Materials Processing, Plenum, New York (1997).
8. M.K. Debe, R.J. Poirier, D.D. Erickson, T.N. Tommet, D.R. Field and K.M. White, Effect of gravity on copper phthalocyanine thin films. I: Enhanced smoothness and local homogeneity of copper phthalocyanine thin films grown in a microgravity environment, *Thin Solid Films* 186: 257 (1990).
9. M.K. Debe and K.K. Kam, Effect of gravity on copper phthalocyanine thin films. II: Spectroscopic evidence for a new oriented thin film polymorph of copper phthalocyanine grown in a microgravity environment, *Thin Solid Films* 186: 289 (1990).
10. M.K. Debe and R.J. Poirier, Effect of gravity on copper phthalocyanine thin films. III: microstructure comparisons of copper phthalocyanine thin films grown in microgravity and unit gravity, *Thin Solid Films* 186: 327 (1990).
11. Kh.M. Akhmedov, Kh.S. Karimov, I.M. Sherbakova, and M.I. Cherkashin, Elektroprovodnost novikh karbazolsoderzhashikh organicheskikh poluprovodnikov, *Doklady RAN* 354: 625 (1997).
12. L.I. Maissel and R. Glang, eds., *Handbook of Thin Film Technology*, McGraw Hill Hook Company, New York (1970) v 1.
13. S. Rudiono, M. Okazaki, and M. Takeuchi, Effects of NO_2 on photovoltaic performance of phthalocyanine thin film solar cells, *Thin Solid Films* 334: 187 (1998).
14. L.I. Boguslavski and A.V. Vannikov, *Organic Semiconductors and Biopolymers*, Moscow, Nauka (1968) (in Russian).

PHYSICAL VAPOR TRANSPORT OF CuPc ORGANIC THIN FILMS IN HIGH GRAVITY

Khasan Karimov,[1] Stefano Bellingeri,[2] and Yoshiyuki Abe[2]

[1]Physical Technical Institute, Dushanbe, 73406, Tajikistan
[2]Electrotechnical Laboratory, Tsukuba, Japan

INTRODUCTION

Changes in processing conditions can yield different molecular structures and properties of organic materials.[1,2] Photosensitivity is a property that characterizes several organic semiconductors, but their application as photoelectric converters or solar cells is limited because of the low efficiency of most of them.[3,4] In fact their solar-cell efficiency is strongly influenced by their conductivity, particularly by the low mobility of the charge carriers (electrons and holes). In order to increase the mobility, denser and more effective electron intermolecular transfer structures should be developed. Centrifugal processing[5] of these materials should be very promising because of the weak molecular bonding that characterizes organic materials, and because of the high molecular weight of organic semiconductors. Because of the low sublimation temperatures of many organic semiconductors,[1,2] physical vapor transport (PVT) is a reasonable process to investigate.

For our experiments we used the photosensitive organic semiconductor copper phthalocyanine (CuPc). This organic compound is considered as one of the most promising organic semiconductors for solar cells because of the modest efficiency (2%) already achieved[2] for CuPc-based solar cells. Investigations on CuPc thin films deposited by PVT in microgravity and normal gravity conditions had showed that in former case the films were more homogeneous, smoother and denser.[6,7,8]

Here, we present experimental results on PVT of CuPc thin films deposited at different gravity conditions, the structure of the film as revealed by XRD and SEM, and the absorbance characteristics of the films. We also present the results of our numerical Direct Simulation Monte Carlo (DSMC) of the gas phase and gas-substrate interface.

EXPERIMENTAL

In our experiments, industrially produced CuPc ($C_{32}H_{16}N_8Cu$) was used. Figure 1 shows the molecular structure of CuPc.[7] Its molecular weight is 576 atomic mass units. There are at least seven crystalline polymorphic states of CuPc; α, β, γ, $\delta_{1,}$ ε. The α-CuPc form is metastable and at 165°C can be thermally or solvent converted to the β-form.[1] The α- and β-forms are those most frequently encountered.[7] Because of thermal conversion we can say that our CuPc was of the β-form. The crystal structure of the β-form is monoclinic $P2_1/a$ with a=120.93Å, b=4.79 Å, c= 14.628 Å and β=120.93°.[7] It has a gap in its electronic spectrum of about 1.6 eV and a conductivity at 300 K of 5.10/ohm.cm.[1,2] Its sublimation temperature varies from 400°C[9] at 10^{-4} Pa pressure to 580°C[10] at 10^{-4} Pa.

Figure 2 shows the PVT chamber, which was installed inside the rotating drum that constitutes the high gravity experimental facility described elsewhere.[11] In the PVT chamber there was 40mm between the heater and the substrate . The substrate was a BK-7 glass plate 40mm in diameter and 2mm thick. A circular CuPc tablet 30 mm in diameter was used as a sublimation source. During deposition, rotary and turbo pumps were used to evacuate the chamber to 0.3-0.4 Pa. The temperatures of the heater and the substrates were 570°C and 50-70°C, respectively. The depositions had a duration of 2-2.5 hours at accelerations of -50 and 50 g.

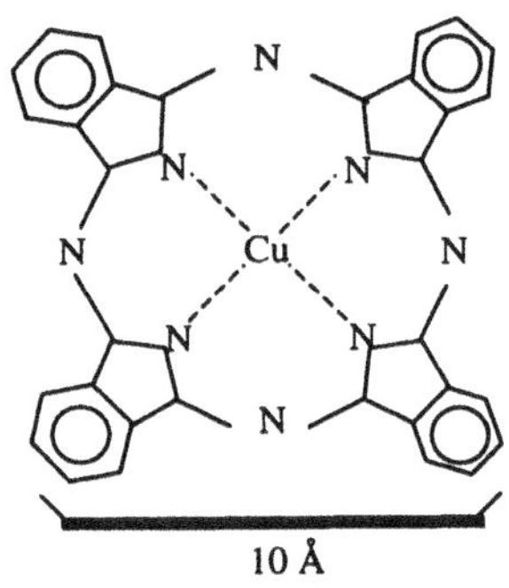

Figure 1. Molecular structure of copper phthalocyanine (CuPc).

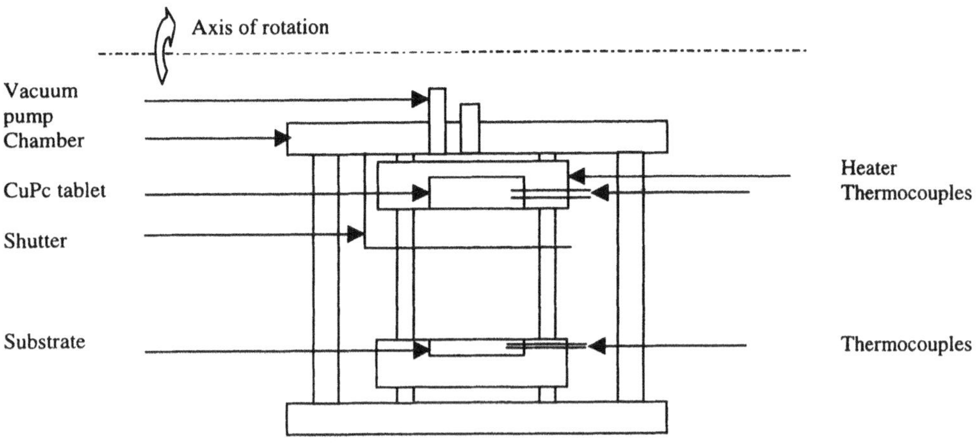

Figure 2. Schematic representation of the physical vapor transport chamber. For negative acceleration *a*, the substrate was located on the top of the chamber and the heater on the bottom (farther from the axis of rotation of the drum). For positive *a*, the heater was on the top and the substrate on the bottom.

EXPERIMENTAL RESULTS AND DISCUSSION

The resulting CuPc films were visually homogeneous, thin, and blue-colored. The film thickness was estimated using two independent measurements. As first, we measured optical absorbance and used the Lambert-Buger formula,[12] $I=I_0exp(-bx)$, where I_0 and I are beam inlet and outlet light intensities, respectively, b is the absorption coefficient and x is the thickness of the film. During the calculations, it was taken into account that approximately 10% of the incoming light was reflected from the glass surface so that the light's intensity at the inlet (I_0) was reduced by 10%. Film thickness was also determined by measuring the weight[12] and using $x=m/\rho S$, where m is the mass of the film, ρ is the density of CuPc (0.8g/cm^3) and S is the surface of the substrate covered by the film. By measuring the optical absorbance and the weight, it was possible to estimate the film thickness in the interval of 15-100 nm.

Figure 3 shows the deposition rate for negative accelerations. Figure 4 shows the acceleration-dependence of the wavelengths of the maxima and minima in the absorption spectra between -50 g and + 50 g. Note that these spectra are characterized by a peak around 600nm, which corresponds to the orange region of the visible spectrum.[9,10]

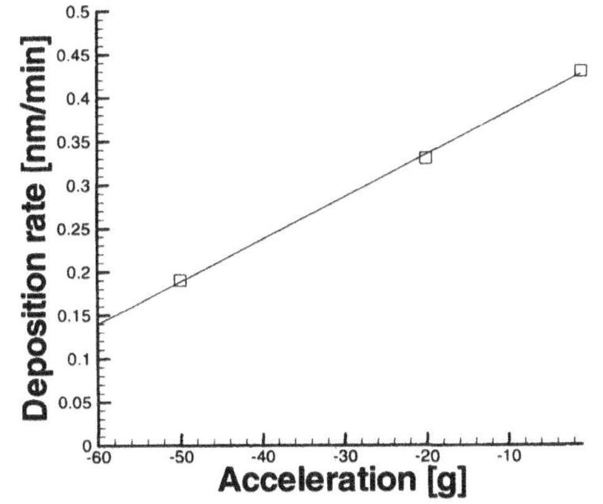

Figure 3. Influence of negative acceleration on the deposition rate.

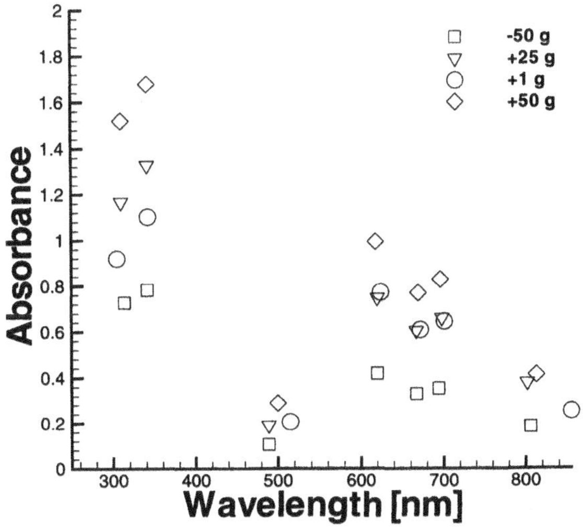

Figure 4. Maxima and minima in absorption spectra for films deposited at four accelerations.

These conditions allowed us to see how acceleration worked, either pulling (negative *a*) or pushing the gas from (to) the substrate. Figure 5 shows the dependence of the absorbance of the CuPc films on gravity. The maximum and the minimum points of the spectrum are plotted. Note that the absorbance increased with increasing acceleration, with the dependence being a function of wavelength.

Table 1 gives a detailed analysis of the behavior of films deposited at +1 *g* and +50 *g*. There was little shift in some of the peaks and minima.

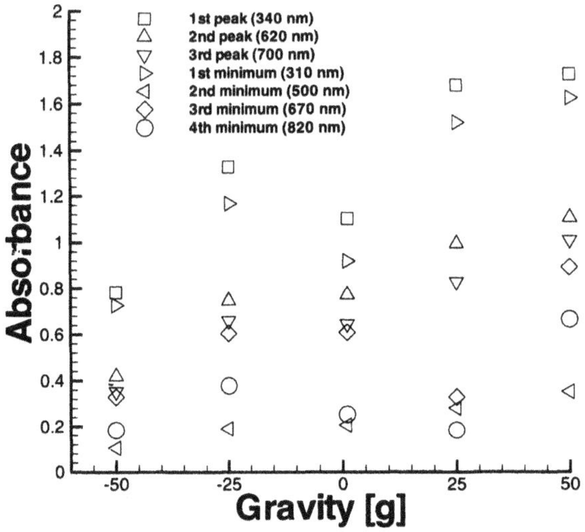

Figure 5. Dependence of absorbance on acceleration.

Table 1. Positions of absorbance peaks (peak mode wavelength, in nm) of CuPc films deposited at +1 *g* and +50 *g*.

+1 *g*		+50 *g*	
Maxima	Minima	Maxima	Minima
701.0	855.0	694.0	836.0
624.0	672.0	623.0	670.0
342.0	515.0	341.0	507.0
	305.0		301.0

Figure 6 shows the dependence of wavelength ($\Delta\lambda$) on acceleration for some spectral peaks. As can be seen from Fig. 6, several spectral peaks of the CuPc films were shifted to a lower wavelength by increasing the acceleration, with the shift difference for each peak.

These results indicate that acceleration influenced the molecular packing in the film. More precisely, it appears that acceleration caused the density to increase. This behavior is in agreement with that observed for the spectrum of organic semiconductors subjected to hydrostatic pressure.[1,13,14]

Scanning electron microscopy was used to compare the structure and growth morphology of films deposited in different gravity conditions. Figure 7 shows micrographs of CuPc films deposited on glass substrates at three difference accelerations. It appears that the crystal grain size increased with increasing acceleration. This could be attributed to the density increasing with *g*. Moreover, the morphology of the CuPc crystals seems to have changed with increasing *g*. This may have been caused by a change in crystal structure. Unfortunately, at the moment, the reproducibility of these results seem to be difficult. The same experiments were repeated using an aluminum-coated glass substrate and gave somewhat different results.

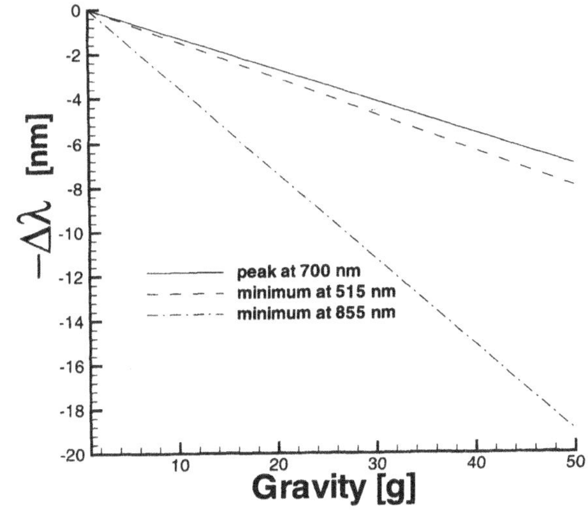

Figure 6. Dependence of the wavelength shift on the acceleration used in preparing CuPc films.

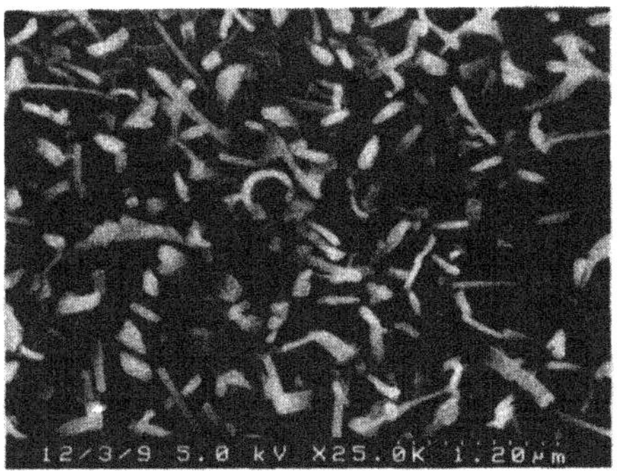

CuPc deposited at –50 g

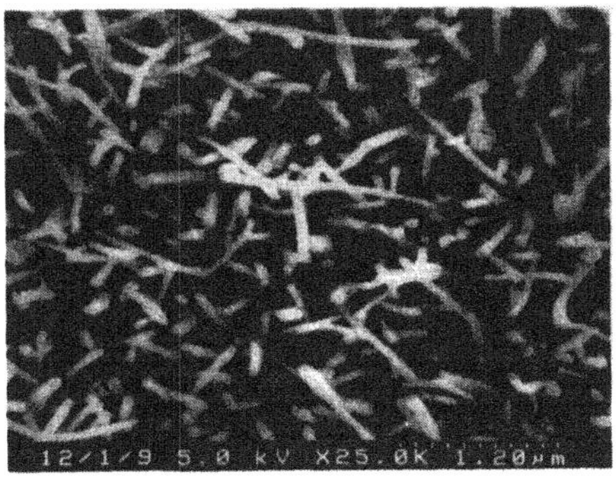

CuPc deposited at 1 g

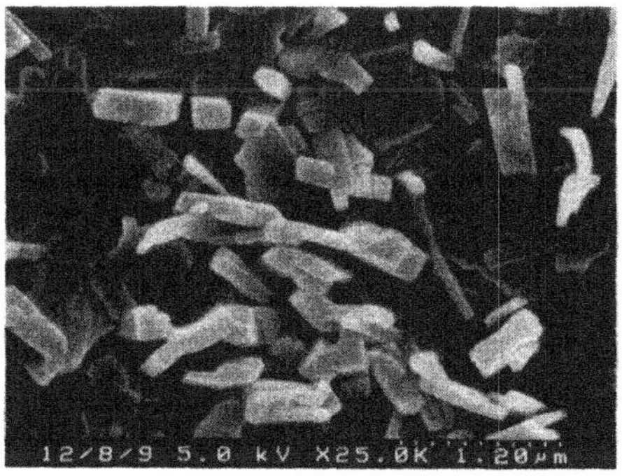

CuPc deposited at 50 g

Figure 7. Scanning electron micrographs of the CuPc films deposited at different accelerations.

Table 2 shows x-ray diffraction data for samples prepared under different accelerations. The distance d between molecular planes was calculated using Bragg's law, $2d \sin \vartheta = n\lambda$, where n is the order, λ the wavelength and θ the glancing angle of the X-ray beam. The glancing angles of the two peaks with wavelengths of 600 and 700 nm are indicated by θ_1, θ_2, with intensities I_1 and I_2. The presence of these two peaks suggests that the α form of CuPc was obtained. The estimated values, from the literature,[6] for the distance between two crystal planes for this form of CuPc is d_1= 12.96 Å and d_2 = 11.96 Å, while the β form has d_1= 9.703 Å and d_2= 7.314 Å. From Table 2 we also note that I_2/ I_2, increased with acceleration. This may be due to the "turning" of the molecule with acceleration. Also increasing with acceleration were d_1 and d_2, suggesting that the molecular spacing increased in the plane perpendicular to the axis of the CuPc stacks. In other words, the CuPc structure that had formed with centrifugal force was strained .

Table 2. X-ray diffraction results for CuPc films prepared with different accelerations.

a	**-50 g**	**-25 g**	**1 g**	**25 g**	**50 g**
d_1 [Å]	12.8747	12.9882	12.9694	12.9314	13.0142
d_2 [Å]	11.7150	11.6841	11.7992	11.7000	11.9050
d_1/ d_2	1.0990	1.1116	1.0993	1.10525	1.0932
$2\theta_1$	6.8600	6.8000	6.8100	6.8300	6.7876
$2\theta_2$	7.5400	7.5600	7.4875	7.5500	7.4200
I_1	82.000	400.00	236.00	10727	773
I_2	91.000	527.00	230.5	5217.5	1793
I_1/I_2	0.90100	0.75901	2.3564	1.9885	3.59

THEORETICAL MODELING

The Direct Simulation Monte Carlo method (DSMC) is a computational particle based technique for gas dynamics modeling. It allows simulation of a wide range of flows from rarefied to the near-continuum region, by tracing sample particles and statistically deducing the flow properties. It has been used to simulate gas flow at high altitude, gas-solid multiphase flows, flow-field and surface interactions for aerospace applications, and nano-scale flow behavior near a solid surface.[15] Since a large number of simulated particles have to be treated, large computer CPU resources and memories are required.

Application of DSMC to an ideal ultracentrifuge provides insight into the behavior of a gas in a chamber for different accelerations. Figure 8 shows the results of such simulations 1s after beginning the chamber rotation. Here, 8a shows the number density of gas particles and 8b the temperature versus radial position for different accelerations. It is particularly important to note from 8c that acceleration increases the number of collisions of the particles with the substrate surface.

CONCLUSIONS

A new experimental PVT chamber was constructed and used to deposit CuPc thin films with different accelerations. The optical absorption increased with increasing acceleration during film growth. Moreover, the peaks shifted to lower wavelengths values as the acceleration was increased. Different morphologies were also produced at the different conditions. X-ray diffraction indicated that the centrifugal force caused the crystals constituting the films to be strained.

From the DSMC calculations for an ideal centrifuge, it was possible to appreciate that even at those low pressures the acceleration influences the gas flux behavior and the gas surface interaction.

Acknowledgements

The authors would like to express sincere thanks to JISTEC for financing the research work with S.B. and K.K.

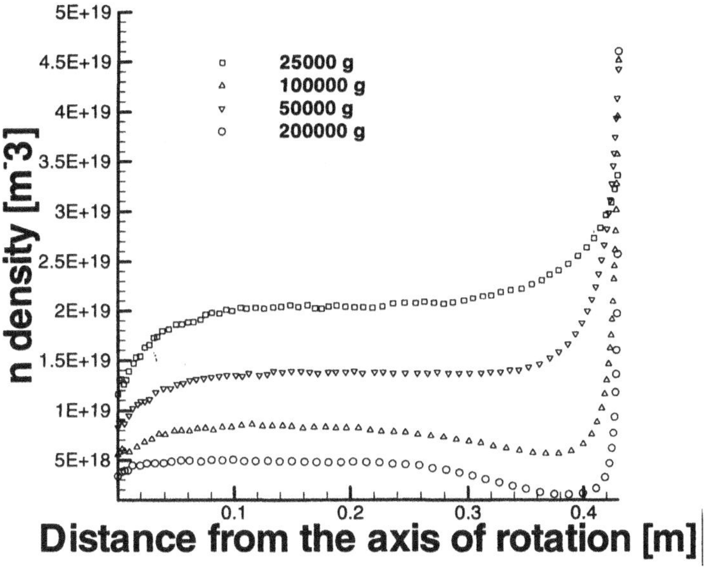

Figure 8a. Simulation results for gas density profile in the PVT chamber at different accelerations.

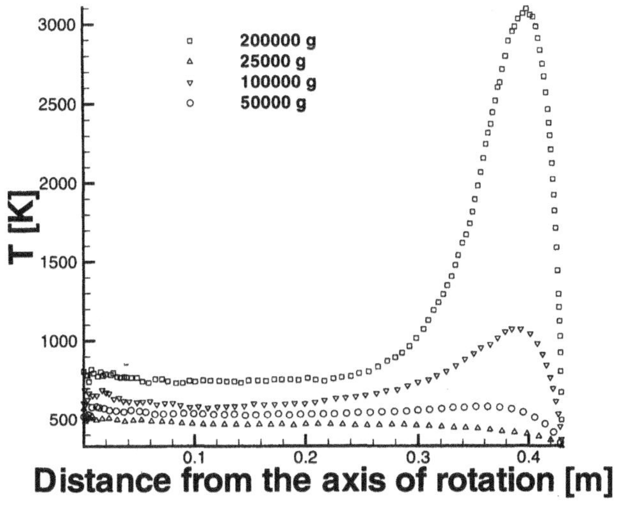

Figure 8b. Simulation results for gas temperature profile in the PVT chamber at different accelerations.

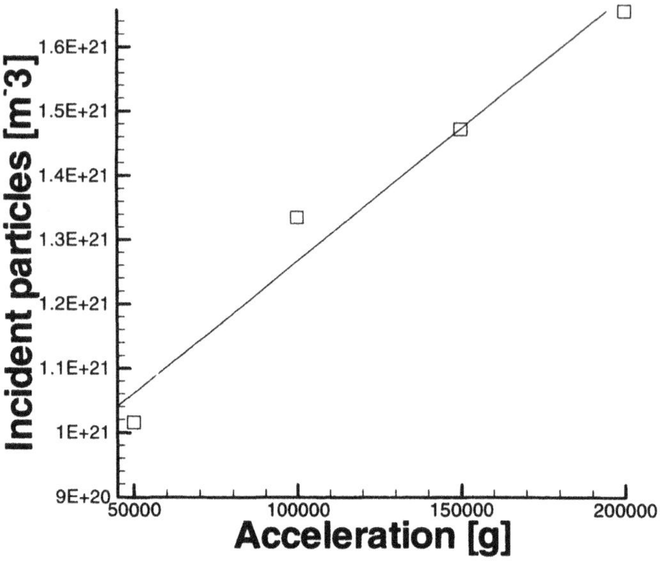

Figure 8c. Simulation results for the number of particle colliding with the substrate at different accelerations.

REFERENCES

1. F. Gutman and L.E. Lyons. *Organic Semiconductors, Part A*, Robert E. Krieger Publishing Company, Malabar, Florida (1981).
2. F. Gutman, H. Keyzer, L. E. Lyons, and R. B. Somoano. *Organic Semiconductors. Part B*, Robert E. Krieger Publishing Company, Malabar, Florida (1983).
3. G.A. Chamberlain, *Solar Cell* 8: 47-83 (1983).
4. Kh.M. Akhmedov, Kh.S. Karimov, M.I. Fiodorov, *Gelioteknika* 1-3: 178-183 (1995).
5. L.L. Regel and W.R. Wilcox, eds., *Centrifugal Material Processing*, Plenum Press, New York (1997).
6. M.K. Debe, R.I. Poirier, D.D. Erickson, T.N. Tommet, D.R. Field, and K.M. White, *Thin Solid Films* 186: 257-288 (1990).
7. M.K. Debe and K.K. Kam, *Thin Solid Films* 186: 289-325 (1990)
8. M. K. Debe and R. I. Poirier, *Thin Solid Films* 186: 327-347 (1990).
9. M.I. Fiodorov, *Influence of Doping on Conductivity and Photoconductivity of Phthalocyanines*, Ph.D. Thesis, Chernogolovka, Russia (1973).
10. A. Al-Mohamad and M. Soukieh, Solar cells and high efficiency photo-diodes having metal-organic thin film-semiconductor structures, *Thin Solid Films* 271: 132-137 (1995).
11. Y. Abe, G. Maizza, N. Sone, Y. Nagasaka and T. Suzuki, A high gravity chemical vapour deposition apparatus, *Rev. Sci. Instr.* 68: 4225-4231 (1997).
12. L.I. Maissel and R. Glang, eds., *Handbook of Thin Film Technology*, McGraw Hill Book Company, New York (1970).
13. J. Shirotani, N. Sakai, H. Inokuchi, and S. Minomura, The spectral blue shift of Li(TCNQ) at high pressures, *Bull. Chem. Soc. Japan* 42: 2087 (1969).
14. N. Sakai, J. Shirotani, and S. Minomura, The effect of pressure on the electronic absorption spectra of TCNQ ion-radical salts, *Bull. Chem. Soc. Japan* 43: 57 (1970).
15. G.A Bird, *Molecular Gas Dynamics and the Direct Simulation of Gas Flow*, Oxford Science Publication (1994).

DIAMOND SYNTHESIZED BY DC-PLASMA CVD AT HIGH GRAVITY

Motoi Nakazato,[1] Tetsuya Suzuki,[1] Yuji Nagasaka,[1] Yoshiyuki Abe,[2]
Stefano Bellingeri,[2] and Giovanni Maizza[3]

[1]Keio University, Yokohama 223-8522, Japan
[2]Electrotechonical Laboratory, Tsukuba 305-8568, Japan
[3]Politecnico di Torino, 10129 Torino, Italy

ABSTRACT

We grew diamond films on molybdenum and silicon substrates at ~900°C between 1 and 100 *g* using a Direct Current (DC) plasma Chemical Vapor Deposition (CVD) apparatus placed in a centrifuge. We characterized these films by scanning electron microscopy, X-ray diffraction method and Raman spectroscopy. We determined the influence of acceleration and gas composition on the nucleation density, morphology, and microstructure of the diamond.

INTRODUCTION

Gravity is one of the most fundamental and important parameters governing physical phenomena, but it had not been regarded as an experimental parameter until the present space era. A number of space and terrestrial experiments have been carried out in the last two decades to synthesize novel materials.[1] Because of technical restrictions and difficulties, especially in manned spacecraft, the number of microgravity experiments on gas-phase materials processing has been limited compared to those on liquid phase.[2] Recently, the number of gas-phase microgravity experiments has increased. For instance, in the Japanese unmanned space-experiment module, the so-called SFU (Space Flyer Unit), four experiments out of nine space materials processing experiments were conducted from the gas phase.[3]

In contrast to microgravity, high gravity conditions are easily obtained in a centrifuge and have been widely used for separation and casting processes. Research on modern materials processing at high gravity was stimulated by the gradient-freeze experiments reported by Regel and Rodot.[4] Subsequent detailed theoretical and experimental examinations revealed that the Coriolis force can stabilize oscillatory natural convection in the melt.[5-7] The flow velocity was sometimes depressed by centrifugation.[8] On the other hand, the role of gravity on the growth mechanism in gas-phase materials processing has been totally veiled. With this in mind, we designed and built a centrifuge facility especially for gas-phase materials processing several years ago.

Several CVD techniques have been invented for preparation of diamond thin films from the gas phase at low pressure, including hot-filament, microwave-plasma, RF-plasma, combustion flame, and DC-plasma CVD.[9] Each method has advantages and disadvantages, but all use atomic hydrogen with an energy to decompose the gases. In the DC-plasma method used in the SFU project, the size of the substrate was limited, so the area on which diamond deposited was consequently small. Although a major disadvantage of DC-plasma CVD had been a limited deposition area, a large deposition area up to 100-mm diameter by a multi-cathode method was recently reported.[10] Generally, the growth rate with the DC-plasma method is faster than with other methods. In this project, we selected the DC-plasma method for diamond synthesis and investigation of gravitational effects. Here, we report the details of the facility and the effects of high gravity on the nucleation density and microstructure of diamond films.

EXPERIMENTAL

Setting up the high gravitational facility

This facility was installed in a pit under the laboratory floor, 1.6 m deep. The total length was 4.22 m with the main part being a rotating drum made of 304 stainless steel, 1.06 m in diameter and 0.97 m long. The maximum rotation speed of this facility was ~600 rpm, with 100 *g* corresponding to 495 rpm. The rotational balance was adjusted by counterweights so that the drum could be operated in any orientation. A detailed description of this facility appears elsewhere.[11]

CVD chamber

Figure 1 shows the CVD chamber, for which the inner diameter and height were 200 mm and 250 mm, respectively. The chamber had two opposing windows 140 mm in diameter. The temperatures at the anode and the substrate were measured with a pyrometer through a window. The cylindrical anode was 18 mm in diameter and attached to the chamber wall, which was electrically grounded. The cathode was hung from the center of the chamber flange using four bolts 50 mm long. The distance between the cathode and the anode was adjusted by the vertical position of the cathode. A molybdenum disc 24 mm in diameter and 4 mm thick was screwed into the end of the cathode. The molybdenum substrate used in the present study was a protuberance 6 mm in diameter and 4 mm high, with a 1.8 mm diameter hole for a thermocouple.

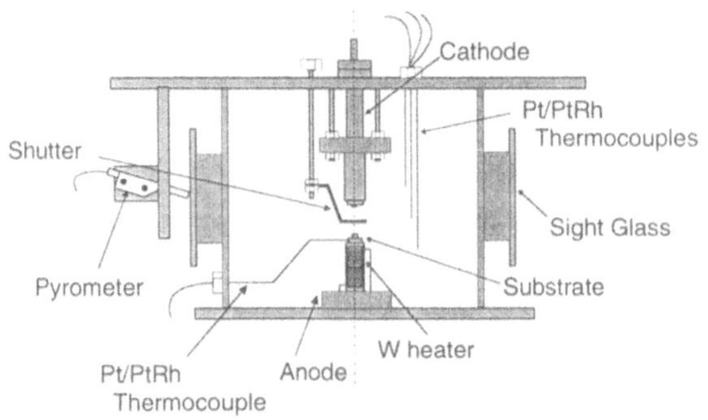

Figure 1. CVD chamber

The heater system (Fig. 2) consisted of a tungsten coil heater surrounding the anode, which was connected to an external independent temperature controller. This anode could be heated by the tungsten coil heater.

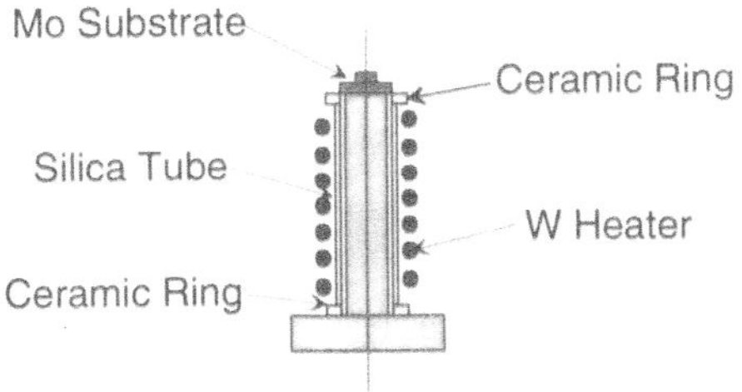

Figure 2. The structure of the anode

Diamond growth

Diamond thin films were deposited at eight different gravitational conditions: normal gravity with the substrate facing up (1*g*), upside down with the substrate facing down (-1*g*), 5*g*,10*g*, 25*g*, 50*g*, 75*g* and 100*g*. The growth conditions were as follows.

- Distance between cathode and substrate: 10 mm - 15mm
- Voltage and electric current: 600-650 V with 0.3-0.4 A for Si and 1 A for Mo
- Gas pressure: 27 kPa
- Gas flow rate: 50 sccm
- Growth time: 2 hr
- Substrate temperature : 900°C
- Methane concentration (in hydrogen): 0.5%, 1%, 2%
- Substrates: molybdenum (polycrystalline), p-type silicon (100)

A silicon substrate, 3 mm square in size, was dipped in 50% hydrofluoric acid for 15 min to removed silicon dioxide. It was fixed on the anode with a carbon paste and heated in a vacuum for drying before deposition. No surface treatment was made on either molybdenum and silicon substrates to enhance nucleation density.

RESULTS AND DISCUSSION

Morphology of CVD diamond

Figure 3 shows scanning electron micrographs and Fig. 4 shows the Raman spectra of diamond deposited on molybdenum and silicon. The nucleation density on molybdenum was larger than that on silicon. Next, we used smaller silicon substrates to obtain a higher plasma density (from 15 mV/mm^2 to 38 mV/mm^2). As expected, the nucleation density became larger; compare Fig. 5 (b) to Fig. 5 (a).

Effect of centrifugation on nucleation and morphology

Figure 6 shows scanning electron micrographs of diamond particles on molybdenum and silicon substrates. In both cases, the nucleation density increased with increasing acceleration. On molybdenum substrates, (100) facets dominated when the deposition was carried out at 50*g* while (111) facets dominated at 1*g*. The average grain size was nearly three times larger at 50*g*. Similarly, the nucleation density was enhanced at 75*g* even when silicon substrates were used.

In the preceding experiments, the methane concentration was 1%. Similar results were obtained with 0.5% methane, as shown in Fig. 7. With 2% methane, the resulting films were rather amorphous, as shown in Fig. 8.

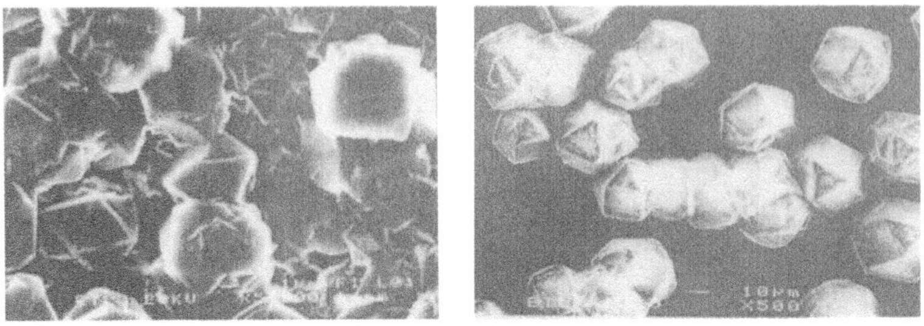

(a) on molybdenum (×5000) (b) on silicon (×500)

Figure 3. Diamond on molybdenum and silicon substrates

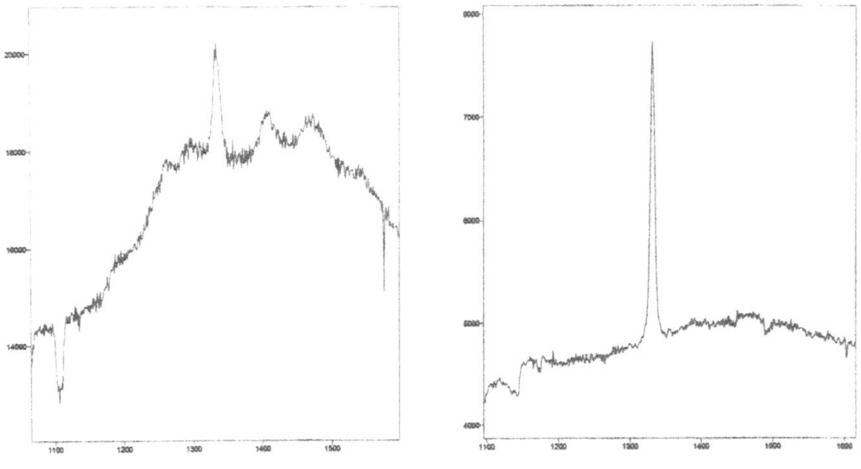

(a) on molybdenum (b) on silicon

Figure 4. Raman spectra of diamond on molybdenum and silicon substrates

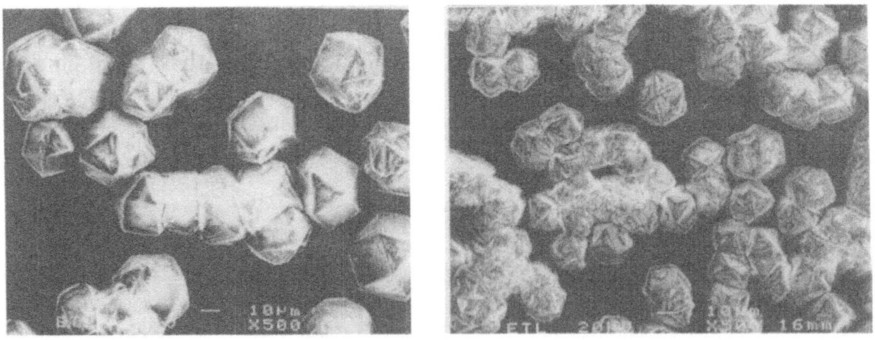

(a) 5mm square (×500) (b) 3mm square (×500)

Figure 5. Effect of silicon substrate size on nucleation density.

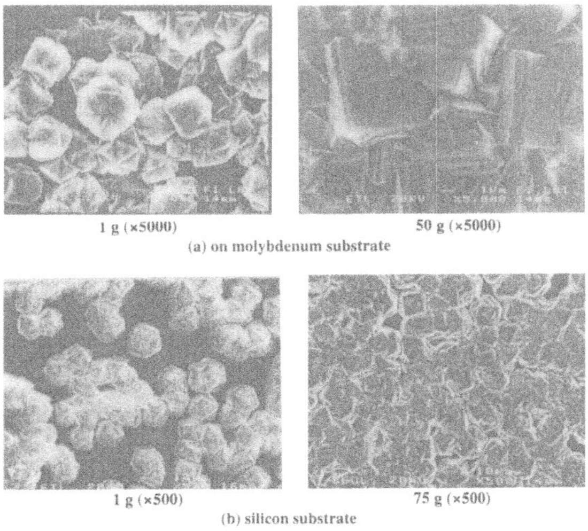

Figure 6. Effect of acceleration on nucleation and morphology with a methane concentration of 1%.

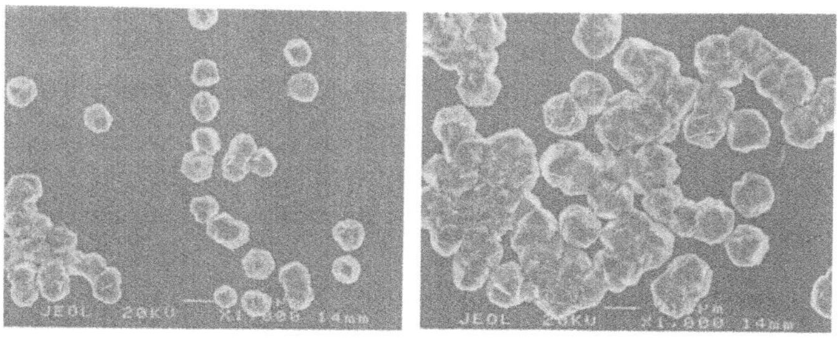

1 g (×1000) 50 g (×1000)

Figure 7. Effect of acceleration on nucleation and morphology with a methane concentration of 0.5%.

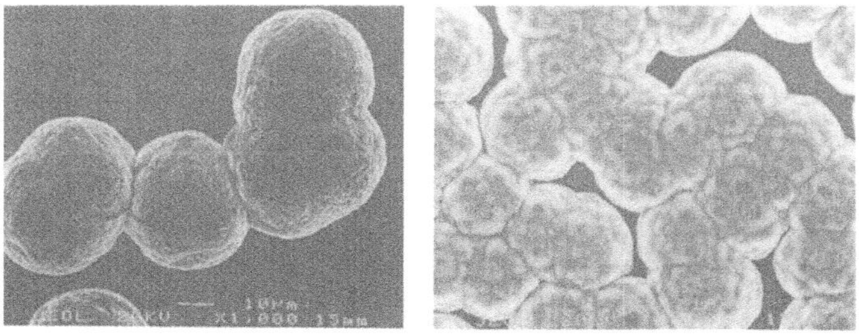

1 g (×1000) 50 g (×1000)

Figure 8. Effect of acceleration on nucleation and morphology with a methane concentration of 2%.

The enhancement of nucleation density had been attributed to a lower substrate temperature at high gravity due to enhanced convective cooling.[12] In the present experiments, however, the substrate temperature was controlled at 900°C with the tungsten heater at all conditions. Thus, the increase in nucleation density and grain size must have originated from acceleration effects. Heat and mass transfer has been simulated for different deposition conditions using this CVD chamber.[13]

CONCLUSIONS

A DC-plasma CVD apparatus was attached to a centrifuge to investigate the effects of centrifugation on the growth behavior of diamond thin films. The centrifuge reached up to 100*g*, where diamond films were successfully synthesized. The results for molybdenum and silicon substrates were:

a) Nucleation density was enhanced with increasing gravity.
b) Grain size became larger with increasing gravity.
c) The (100) facets were more often seen at high gravity.

REFERENCES

1. G. Müller and G. Neumann, *J. Cryst. Growth* 59: 548 (1982).
2. H. Rodot, L.L. Regel, G. Sarafanov, M. Hamidi, I.V. Videvski, and A.M. Turtchaninov, *J. Cryst. Growth* 79: 77 (1986).
3. W.A. Arnold, W.R. Wilcox, F. Carlson, A. Chait, and L.L. Regel, *J. Cryst. Growth* 119: 24 (1992).
4. G. Müller, G. Neumann, and W. Weber, *J. Cryst. Growth* 119: 8 (1992).
5. F. Tao, Y. Zheng, W. J. Ma, and M. L. Xue, in: *Materials Processing in High Gravity*, L.L. Regel and W.R. Wilcox, eds., Plenum, New York (1994) pp 67-79.
6. W.A. Arnold and L.L. Regel, in: *Materials Processing in High Gravity*, L.L. Regel and W.R. Wilcox, eds., Plenum, New York (1994) pp 17-34.
7. J. Friedrich, J. Baumgartl, H. J. Leister, and G. Müller, *J. Cryst. Growth* 167:45 (1996).
8. M. Ishikawa, S. Kamei, Y. Sato, T. Ando, N. Fujimori, N. Koshikawa, K. Murakami and T. Suzuki, *IAA.-97-IAA*.12.2.09, International Astronautical Congress (1997).
9. K. Suzuki, A. Sawabe, H. Yasuda and T. Inuzuka, *Appl. Phys. Lettr.* 50: 728 (1987).
10. Y.J. Baik, J.K. Lee, W.S. Lee, and K.Y. Eun, *Thin Solid Films* 341: 202-206 (1999).
11. Y. Abe, G. Maizza, N. Sone, Y. Nagasaka and T. Suzuki, *Rev. Sci. Instr.* 68: 4225-4230 (1997).
12. N.V. Samokhvalov, V.E. Strel'nitskij, V.A. Belous, V.P. Zubar, A.G. Timchuk, E.D. Obraztsova and V.G. Ral'chenko, *Diamond Rel. Mater.* 4: 964 (1995).
13. G. Maizza, G. Saracco and Y. Abe, *IAF-97-IAA*..12.1.03, International Astronautical Congress (1997).

INFLUENCE OF CENTRIFUGATION ON DIAMOND FILM DEPOSITION BY CHEMICAL VAPOR TRANSPORT

Liya L. Regel and William R. Wilcox

International Center for Gravity Materials Science and Applications
Clarkson University, Potsdam, NY 13699-5814, USA

INTRODUCTION

Chemical vapor deposition (CVD) of diamond films at high gravity was first suggested by Regel in 1990. However, the prior methods for diamond CVD were unsuitable for mounting on a large centrifuge. All required a flowing gas system for hydrogen-hydrocarbon mixtures at low pressure, with the attendant gas cylinders, flow controllers, vacuum pump, and waste gas disposal system. Many required a plasma generator. In 1996, we described a new closed Chemical Vapor Transport (CVT) reactor that could deposit diamond films without continual addition of gas.[1] Hot graphite is used as a carbon source in the presence of hydrogen at low pressure, without a metal filament or plasma. Under proper conditions, this technique has provided diamond growth rates an order of magnitude larger than the ~1 μm/h typical for diamond CVD processes. High-quality polycrystalline diamond was deposited on a wide variety of substrates, including silicon, glass, copper, molybdenum, gold, graphite and carbon felt.[2] The deposits were not removed by the "scotch-tape test." Patterned selective deposition was achieved on a copper pattern on oxidized silicon.[3]

EXPERIMENTAL

The first apparatus utilizing this new technique has been described elsewhere.[2] Since then, two new and larger deposition chambers have been constructed with improved cooling and diagnostics. Basically, a graphite rod is held a few mm away from a substrate in a water-cooled stainless-steel chamber containing about 0.1 atm H_2. The graphite rod is heated to by passing electric current through it. The substrate is heated by the graphite rod and its temperature is measured by a bare K-type thermocouple in contact with its back side.

For centrifuge experiments, the apparatus was mounted on Clarkson's materials processing centrifuge HIRB.[4] In every case investigated, centrifugation resulted in an increase in deposition area, nucleation density, coverage, and crystal size.

Processing by Centrifugation, Edited by Regel and Wilcox
Kluwer Academic/Plenum Publishers, New York, 2001

The first experiments were performed on 2-mm thick single crystal silicon that was 2-cm long and either 1.5 or 2-cm wide. Following standard practice, each substrate was roughened by a diamond slurry in an ultrasonic bath. Later, we found that this roughening is not necessary, although a slightly longer deposition time is then required. Each substrate was rinsed in an ultrasonic bath alternating between acetone and ethanol, and then air-dried for several hours. We obtained a wide variety of microstructures, ranging from isolated crystals to a continuous polycrystalline coating complete coverage over 2 cm^2 area and larger. The best deposits on roughened substrates were obtained with a starting gas pressure in the chamber of 50 torr hydrogen, a reaction time of at least 30 min, a substrate temperature of 738°C, and the substrate located 2.5 mm below the graphite. Figure 1a shows a typical deposit obtained without centrifugation. With centrifugation to a net acceleration of 2.5 times earth's gravity (2.5 *g*), nucleation, growth kinetics, and the area covered were all strongly increased (see, for example, Figure 1b).Without centrifugation, the diamond crystals were on about 50% of the 3 cm^2 surface of the graphite substrate, while with centrifugation the crystals were on about 90%. The size of the crystals was about 55% larger when deposition was carried out at 2 *g* under the same conditions as without centrifugation.

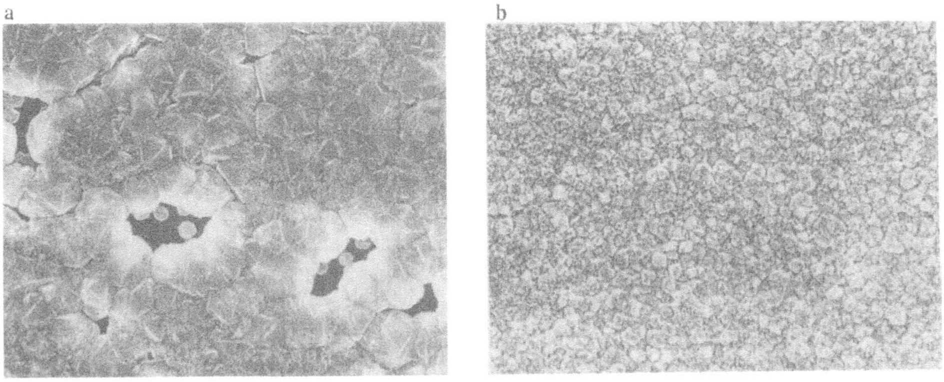

Figure 1. Scanning electron micrographs of diamond deposited on roughened (111) silicon substrates for 40 minutes. Left: without centrifugation. Right: with centrifugation at 2.5 *g* , under otherwise identical conditions. (Magnification same.)

The influence of centrifugation was investigated for deposition of diamond on carbon felt.[3] For each carbon felt experiment, a slab of felt ~1 x 10 x 15 mm was cut and gently rinsed in ethanol and acetone. It was placed on a support made of a 0.5 mm thick (100) silicon wafer that had been cut to 10 x 15 mm and ultrasonically cleaned in ethanol and acetone. Neither the silicon nor the carbon felt were contacted with diamond powder at any time. Six one-hour deposition experiments were performed with the graphite heater rod placed 2.5 mm above the felt, an initial hydrogen pressure of 30 to 60 torr, and a silicon support temperature of ~750°C. An initial hydrogen pressure of 50 torr provided the most dense coating. Without centrifugation, the deposit consisted of isolated crystals (Fig. 2, left). With centrifugation at 2 *g*, complete coverage was obtained for the fibers near the top of the felt (Fig. 2, right). Remarkably, diamond deposition was throughout the ~1-mm thickness of the felt and on the silicon support.

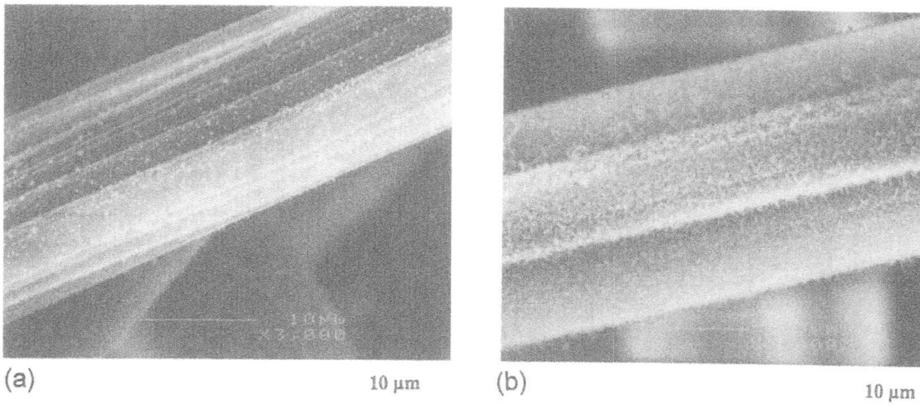

Figure 2. Diamond on fibers from carbon felt after deposition run of 1 h. Left: without centrifugation. Note isolated crystals. Right: with centrifugation at 2 *g*. Note complete coverage.

The influence of centrifugation was also investigated for selective patterned deposition.[2] A silicon wafer was used onto which a thin layer of silicon oxide had been grown thermally. On the silicon oxide was a test pattern of ~1-μm thick copper, under which there was a ~0.05-μm layer of titanium for improved adhesion. This substrate was cleaned in an ultrasonic bath, alternating between ethanol and acetone. The surface was not contacted with diamond powder. Ten deposition experiments were performed with the graphite heater rod placed 2.5 mm above a 2 cm x 2 cm patterned substrate, an initial hydrogen pressure of 30 to 60 torr, and a substrate temperature of 730 to 750°C. Scanning electron microscopy showed the usual faceted diamond crystals, while energy dispersive x-ray spectrometry showed the presence of Cu and Ti only under the diamond. Tape adhesion tests removed none of the deposits. Although selective deposition on copper was obtained in all ten experiments, the best results were obtained at an initial hydrogen gas pressure of 50 torr and a substrate temperature of about 740°C. A deposition time of 60 min was sufficient to grow a continuous film on the copper, on ~70% of the 4-cm^2 area of the substrate. The resolution of the deposited pattern was on the order of 1 μm, being limited by the size of the diamond crystals. Five deposition experiments were also carried out on the HIRB centrifuge. Unlike deposition without centrifugation, during deposition on the centrifuge the pressure and substrate temperature increased dramatically. Again, diamond was deposited selectively on the copper pattern. At an acceleration of 2 *g* under the same conditions used without centrifugation, the portion of the copper pattern on which diamond deposited increased to about 90%.

THEORY

Under the hydrogen pressure and graphite temperature used in our experiments, C_2H_2 and CH_4 are predicted to be the predominant hydrocarbon species at equilibrium,[5-9] with some H_2 dissociated into atomic H.[10] FLUENT[22] was used to calculate the species present at equilibrium in the presence of hot graphite and hydrogen, as shown in Fig. 3. Although many measurements have been made of the reaction products between hydrogen and graphite,[11-21] the definitive study was performed by Balooch and Olander.[11] They used a modulated molecular beam with heated pyrolytic graphite. Up to 2200 K, H_2 molecules neither dissociated on the graphite surface nor reacted with it.

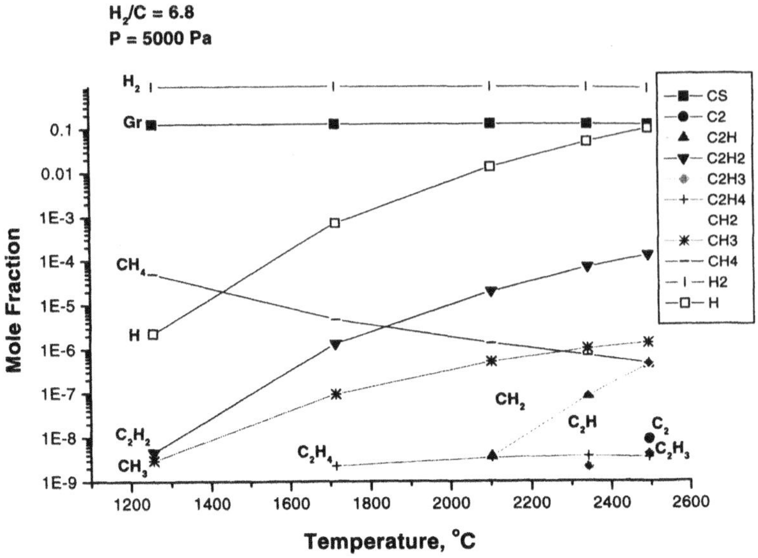

Figure 3. Equilibrium gas composition versus temperature. Here, Gr is graphite.

(This result was confirmed recently.[13]) In subsequent experiments, a beam containing atomic hydrogen was used. For surface temperatures up to 800 K, only CH_4 was produced. Between 800 and 1000 K, H recombined to H_2 but no hydrocarbons formed. From 1000 and 2200 K, C_2H_2 was the sole product.

FLUENT[22] was also used to model numerically the heat transfer and buoyancy-driven convection in our deposition chamber, neglecting chemical reactions and assuming only H_2 is present in the gas. Typical results are shown in Fig. 4. As would be expected at these temperatures and pressures, heat transfer was predominantly by radiation so that buoyancy-driven convection had little effect on the temperature field. The gas flowed upward around the heated rod and substrate, and downward near the cooled wall. The Grashof number (Gr) in the chamber was on the order of 0.005, the Reynolds number 0.3 and the Peclet number for mass transfer 0.2. The flow velocities between the graphite rod and the substrate were about three orders of magnitude smaller. Simulation of the process with centrifugation yielded the same convection pattern, but with increased velocities that are proportional to the total acceleration (as expected for small Gr). See Fig. 5. Similarly, the velocities were proportional to pressure, which is also expected for small Gr. There was no evidence of circulation due to the Coriolis effect as observed in directional solidification.[23]

CONCLUSIONS

From our experimental observations, literature review, and FLUENT modeling we have formed the following preliminary picture of the deposition process. Hydrogen atoms form in the hot gas surrounding the graphite rod. This atomic hydrogen reacts with the graphite to produce acetylene, which reacts further with the surrounding hot gas to produce other hydrocarbons, especially methane. Convection removes the hydrocarbons from the neighborhood of the graphite rod and gently mixes the gas in the chamber. The hydrocarbons diffuse into the gap between the rod and the substrate, where the atomic hydrogen concentration may be near equilibrium. The hydrocarbon concentration in the

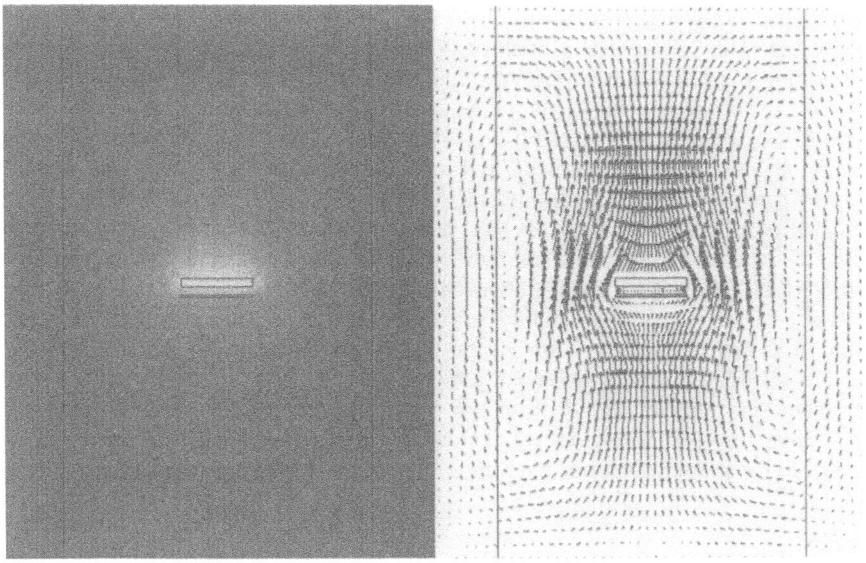

Figure 4. Modeling results for deposition chamber.
Left: Temperature field. The lighter the shade the higher is the temperature.
Graphite T = 1730 K, wall T = 293 K.
Right: Velocity field. Maximum velocity is 0.0609 m/s above and to the sides of the graphite.

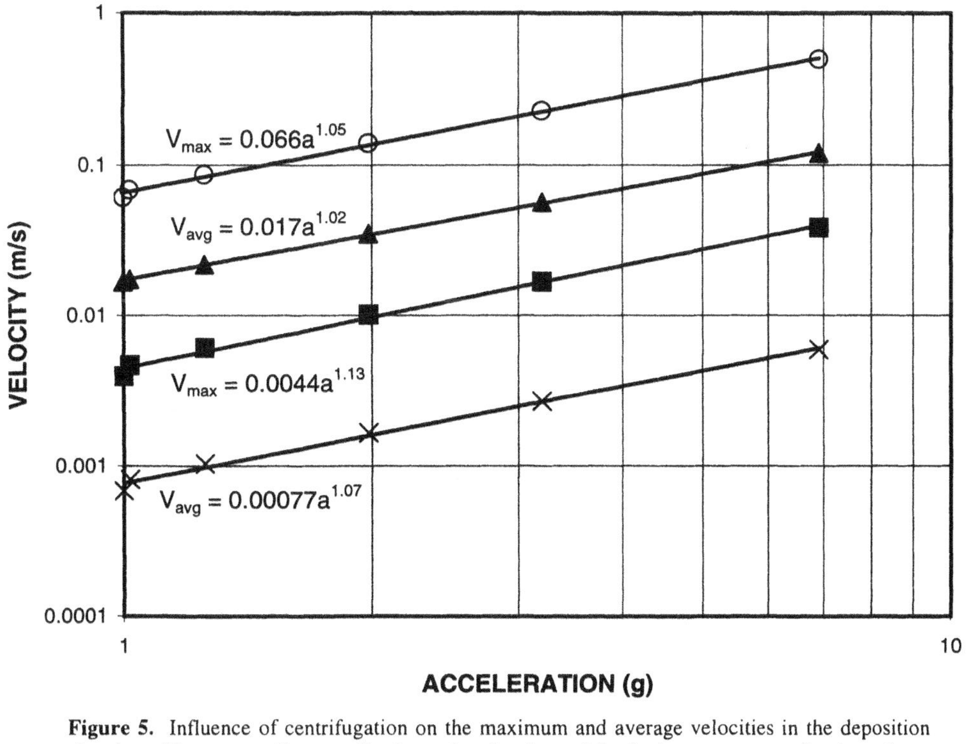

Figure 5. Influence of centrifugation on the maximum and average velocities in the deposition chamber. The top two lines are for the entire chamber, while the bottom two are for the volume between the graphite rod and the substrate.

cell and at the substrate surface gradually increases until it is sufficient to nucleate diamond on the substrate. The subsequent growth of diamond consumes hydrocarbons, decreasing the hydrocarbon concentration at the substrate surface so that nucleation decreases and finally stops. Eventually, a steady state is reached between production of acetylene at the surface of the hot graphite and deposition of diamond. Thus, one advantage of this new technique is that the concentrations of reactants at the substrate surface tend to be self regulating --- they automatically adjust to the levels required for nucleation and then growth. The traditional CVD processes have a flowing gas of fixed composition with rapid generation of growth species that are fixed in concentration during the nucleation and growth processes. Thus, if the reactant concentration is high enough to cause nucleation, it will be too high to yield high quality diamond growth. Consequently, the methane concentration is held deliberately low in the traditional processes in order to yield good diamonds, resulting in low nucleation and growth rates.

The increased nucleation and deposition caused by centrifugation is attributed to an increase in mass transfer rates due to increased convection, enhancing the removal of acetylene from the graphite rod and its eventual transport to the substrate surface.

Acknowledgments

This research was supported by grant DMR-9414304 from the U.S. National Science Foundation and by the New York State Center for Advanced Materials Processing at Clarkson University. We are grateful to Peter Skudarnov and Goodarz Ahmadi for assisting with the FLUENT simulations, and to Craig Burkhard for assisting with the experiments.

REFERENCES

1. L.L. Regel, Y. Takagi and W.R. Wilcox, Centrifugal Diamond Film Processing, in: *Centrifugal Materials Processing*, L.L. Regel and W.R. Wilcox, eds., Plenum, NY (1997) pp 221-228.
2. L.L. Regel and W.R. Wilcox, Deposition of diamond on graphite and carbon felt from graphite heated in hydrogen at low pressure, *J. Mat. Res. Lettr.* 19: 455-457 (2000) plus cover photo.
3. L.L. Regel and W.R. Wilcox, Selective patterned deposition of diamond using a new technique, *J. Mat. Res. Lettr.* 18: 427-430 (1999) plus cover photo.
4. R. Derebail, W.A. Arnold, G.J. Rosen, W.R. Wilcox, and L.L. Regel, HIRB - The centrifuge facility at Clarkson, in: *Materials Processing at High Gravity,* L.L. Regel and W.R. Wilcox, eds., Plenum Press, New York (1994) pp 203-212.
5. B. Lersmacher, H. Lydtin, W.F. Knippenberg, and A.W. Moore, Thermodynamische Betrachtungen zur Kohlensoffabscheidung bei der Pyrolyse Gasförmiger Kohlenstoffverbindungen, *Carbon* 5: 205-217 (1967).
6. R.E. Duff and S.H. Bauer, Equilibrium composition of the C/H system at elevated temperatures, *J. Chem. Phys.* 36: 1754-1767 (1962).
7. R.F. Baddour and J.L. Blanchet, Reactions of carbon vapor with hydrogen and with methane in a high intensity arc, *I&EC Proc. Des. Dev.* 3: 258-265 (1964).
8. P. John, I.C. Drummond, D.K. Milne, M.G. Jubber, and J.I.B. Wilcox, Fundamental limits to growth rates in a methane-hydrogen microwave plasma, *Diamond and Related Materials* 3: 56-60 (1993).
9. M.W. Chase, Jr., *et al.*, editors, *JANAF Thermochemical Tables, J. Phys. Chem. Ref. Data* vol. 14, Supplement 1, Part 2 (1985) p. 1211.
10. R.A. Krakowski and D.R. Olander, Survey of the literature on the carbon-hydrogen system, Lawrence Radiation Laboratory, UCRL-19149 (1970).
11. B. Balooch and D.R. Olander, Reactions of modulated molecular beams with pyrolytic graphite. III. Hydrogen, *J. Chem. Phys.* 63: 4722-4786 (1975).
12. H.F. Winters, J. Seki, R.R. Rye, and M.E. Coltrin, Interaction of hydrogen, methane, ethylene, and cyclopentane with hot tungsten: Implications for the growth of diamond films, *J. Appl. Phys.* 76: 1228-1243 (1994).

13. A. de Koranyi, N.D. Parkyns, and S.J. Peacock, The reactivity of graphite towards hydrogen, in: *Proceedings of the Fifth London International Carbon and Graphite Conference*, Society of Chemical Industry, London (1978) pp 139-149.
14. Z. Szabó, The examination of a system carbon and hydrogen in the temperature range 1100-2600°C, *J. Am. Chem. Soc.* 72: 3497-3502 (1950).
15. S.J. Steck, G. A. Pressley, Jr., S.S. Lin, and F.E. Stafford, Mass-spectrometric investigation of the reaction of hydrogen with graphite at 1900°-2400°K, *J. Chem. Phys.* 50: 3196-3207 (1969).
16. E.A. Gulbransen, K.F. Andrew, and F.A. Brassart, The reaction of hydrogen with graphite at 1200° to 1650°C, *J. Electrochem. Soc.* 112: 49-52 (1965).
17. J.T. Clarke and B.R. Fox, Reaction of graphite filaments with hydrogen above 2000°K, *J. Chem. Phys.* 46: 827-836 (1967).
18. W.L. Hsu, Chemical erosion of graphite by hydrogen impact: A summary of the database relevant to diamond film growth, *J. Vac. Sci. Technol.* A6: 1803-1811 (1988).
19. Z. Pan, Mechanistic studies of uncatalyzed and catalyzed carbon gasification reactions, PhD Dissertation, State University of New York at Buffalo (1996).
20. R.K. Gould, Kinetics of CH_4 formation from the reaction of H with graphite, *J. Chem. Phys.* 63: 1825-1836 (1975).
21. R.R. Rye, Reaction of thermal atomic hydrogen with carbon, *Surf. Sci.* 69: 653-667 (1977).
22. FLUENT (v4.51), *Fluid Flow Modeling*, Fluent Inc., Centerra Resource Park, 10 Cavendish Court, Lebanon, NH 03766, USA.
23. P.V. Skudarnov, L.L. Regel, W.R. Wilcox, and G. Ahmadi, Numerical modeling and flow visualization in the gradient freeze configuration during centrifugation, *present volume*.

POLYMER SYNTHESIS IN A CENTRIFUGAL FIELD

Konstantin G. Kostarev and Andrey V. Shmyrov

Institute of Continuous Media Mechanics, Perm, 614013, Russia

INTRODUCTION

A modern trend in polymeric materials is toward a prescribed variation in properties through the material. With currently available methods, the synthesis of such polymers is a complicated and tedious process, which, in addition, has the deficiency of generating the property distribution in a stepwise manner. We introduce here the perspectives and physical grounds for a new method of obtaining polymeric materials with controlled variations in properties by using a centrifugal field. The key to this method is the formation of the polymer in an inhomogeneous gravitational field, which leads to a change of its properties along the force gradient. Both characteristics of a centrifugal field are useful, namely, the magnitude of the force and the force gradient.

The high sensitivity of polymerization processes to gravity has been revealed in terrestrial and space experiments.[1,2] Experiments on the Mir orbital station showed that eliminating the influence of gravity can yield highly homogeneous samples. Tests on a centrifuge allowed us to detect and understand the mechanisms for the action of gravity on the processes of synthesis and formation of a polymer structure.[3,4] These mechanisms are convection of the liquid monomer, sedimentation of the initial gel microparticles (globules) in the liquid phase of the reaction, and deformation of the polymeric water-permeable sponge generated by combining globules that is forming the solid phase. To gain a better understanding of these mechanisms, we performed a new series of experiments, in which a specially designed setup was used to visualize the flows and conversion fields in a polymerizing monomer directly in the rotating reactor.

EXPERIMENTAL METHODS

A schematic diagram of the experimental setup is given in Fig. 1. An experimental cuvette (3) was mounted on a horizontal supporting plate (2) with a vertical rotating axis and was set in rotation by an electric motor (1). Depending on the experiment's objective, we used either a dc motor with a rotational velocity ranging from 0 to 25 rev/s or an ac motor with a constant velocity of 50 rev/s. The angular error of the horizontal plate was not more than 2 angular minutes. A semi-transparent mirror (4) was located over the

Processing by Centrifugation, Edited by Regel and Wilcox
Kluwer Academic/Plenum Publishers, New York, 2001

cuvette at an angle of 45° to its surface. Coupled with a helium-neon laser (5) and collimators (7 & 8), the cuvette (3) and the mirror (4) formed a Fizeau interferometer providing a beam of coherent light 70 mm in diameter. In order to obtain stationary interferograms of the rotating cuvette, the optical system of the interferometer was fitted with a system of mirrors and diaphragms (6), which allowed synchronization of the exposure and rotation frequencies. The interferogram was recorded by a video camera in real time or by a still photo camera (9).

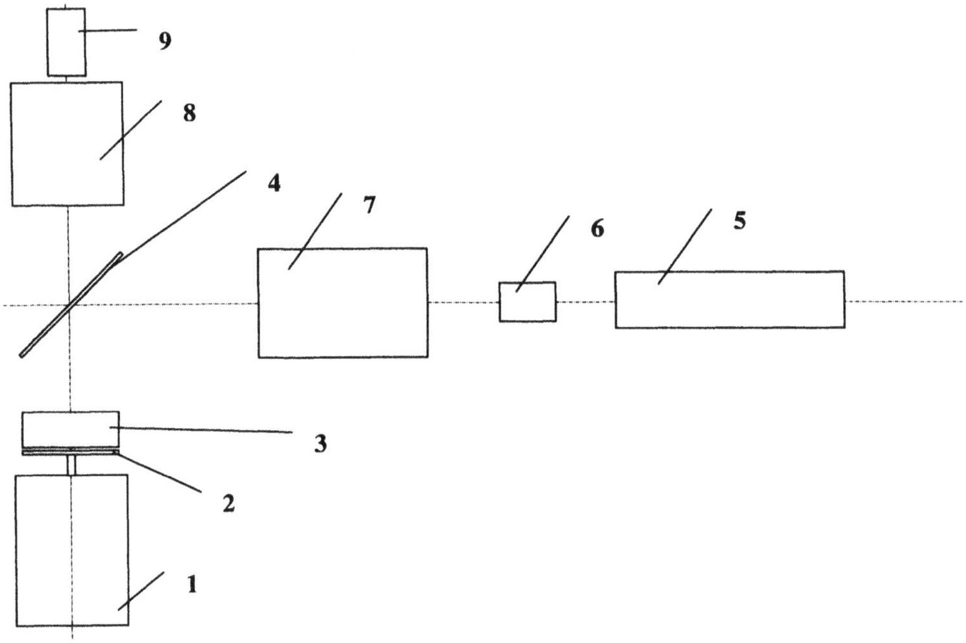

Figure 1. Schematic diagram of setup for polymerization experiments.

The experimental cuvette (Fig. 2) consisted of a demountable interferometric cell and protective case. The cell was formed by two circular optical glasses (1) with semi-transparent mirror coatings and a ring-shaped gasket (2) made of organic glass. The cell cavity was 4 mm in height and from 25 to 30 mm in radius (depending on the inner diameter of the gasket). To obtain interferograms, the glasses were positioned strictly parallel to each other. The cuvette was filled with the reaction mixture through a filler hole at the center of the upper glass. Before an experiment, the cell was assembled and adjusted with the help of a special frame. Then the glasses were cemented together, sealed and housed in a case, which served to align and protect the cell from accidental damage as it was mounted on the supporting plate (9). The cell case consisted of units 5, 6 and 7, which were also bonded together with sealing compound (8). After installation, the position of the cuvette mirrors with respect to the rotation axis (and thus to the direction of illumination) was adjusted with the help of screws (10).

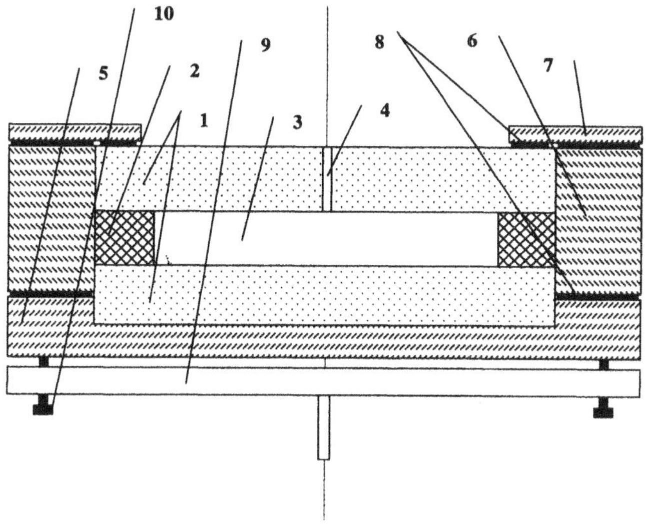

Figure 2. Diagram of cuvette for polymerization experiments.

The above setup was fitted with a video camera having separable lenses that allowed us to differentiate 5 interference bands per mm of the object, provided that the whole radius of the working cell was in the view of the camera. This number could be increased by decreasing the observation area. Use of a photo camera increased the resolution up to 20 bands per mm.

At a rotation frequency of 50 rev/s the maximum acceleration in the cuvette at 25 mm radius was 250 *g*. In the cuvette at 30 mm radius it was 300 *g*.

EXPERIMENTAL RESULTS

Model experiments on diffusion and convection in binary liquids demonstrated that the above apparatus had good performance capabilities. One of our first observations was frontal gel formation in reactive systems with a homogeneous distribution of thermal initiator. This result supported the validity of the hypothesis that the gel structure forms on the basis of initially generated polymer globules.

The first experiments on thermally initiated polymerization of PAG in a centrifugal field demonstrated that rotation for even a short time at the beginning of the reaction gives rise to a gradient in Young's modulus.[4] This phenomenon may be a consequence of sedimentation of primary gel microparticles (globules). The role played by globule sedimentation in gel formation has long been widely debated, but still remains an open question.

To fully understand the influence of globule sedimentation, one needs to exclude the possibility of at least two gravitationally sensitive polymerization mechanisms - convection in the liquid reaction mixture and deformation of the weakly cross-linked polymeric sponge under the action of the centrifugal forces. Buoyancy-driven convection has been thought to be unlikely because the reaction in the monomer proceeds everywhere with the same rate when the initiator is uniformly distributed. Flow visualization using light-scattering particles showed that the spin-up flows in centrifuge experiments vanished during the induction stage of the reaction and therefore were not involved in the formation of gel structure. The possibility of the long-term action of deformation was also thought to be negligible because rotation of the cuvette had been stopped when condensate appeared on the glass surface in the region of an air bubble in the center of the cuvette. (The appearance of condensate corresponds to maximum heat generated in forming the unique polymeric sponge.)

Real-time visualization of the polymerization process was developed in order to better understand the physical mechanisms for gel formation. This technique was based on the hypothesis that if individual gel particles are generated at the beginning of the reaction, sedimentation strengthened by a centrifugal force should increase their concentration at the cuvette boundary. The resulting inhomogeneous distribution of conversion should lead to acceleration of gel formation at the boundary, causing the reaction in the bulk of the monomer to follow a frontal pattern. If so, the process of polymerization should exhibit some other characteristic features - the loss of distribution stability and initiation of convection.

Figure 3 shows a series of interferograms of gel polymerization in a 25 mm radius cuvette rotating at 25 rev/s (63 *g*) with thermal initiation. Figure 3a shows an interferogram made shortly after the beginning of cuvette rotation. It is seen that sedimentation had generated a conversion distribution at the cuvette boundary. (No sedimentation was observed when the reaction was carried out without initiator.) The growth of this conversion boundary initiated gel formation in the bulk of the monomer accompanied by generation of heat. This, in turn, caused a loss of distribution stability (Fig.3b), favoring the development of convection (Fig. 3c). Due to thermal initiation, the difference in the rates of reaction in the bulk of the monomer and at the front was small, so that the mixture rapidly lost its fluidity, except for the region very close to the center where the microparticle concentration was reduced by sedimentation and the reaction mixture remained in a liquid state for a long time (Fig. 3d).

A short duration and low intensity of convection indicate the possibility of the evolution of the process without convection. While short-term convection (when it does arise) is a well-defined phenomenon usually observed in a synthesis reaction with thermal initiation at normal gravity, the occurrence of low-intensity convection was unexpected. The convection observed at 1*g* is normally characterized by high intensity, suggesting that it would increase with increasing acceleration. Possibly the flux of particles due to sedimentation interfered with the convection. The sedimentation may also have created a stratification of density that inhibited buoyancy-driven convection.

The influence of centrifugation on the formation of the three-dimensional water permeable sponge was also unexpectedly large. After generation of a uniform polymeric sponge (Fig. 3e) corresponding to 30-40% conversion, the formation of gel continued. Figures 3f and 3g show the expansion of the gradient gel macrostructure. This expansion was probably related to sponge deformation as well as to diffusion of the low-molecular components of the reaction mixture.

Figures 3h and 3i give interferograms of the gel samples at the end of reaction (before rotation was stopped) and two days later. These show a region with a low gradient of optical properties connected with the plateau on a plot of elastic modulus as a function of cuvette radius.[4] Thus interferometric observation of polymerization allows one to follow the evolution of the distribution of properties.

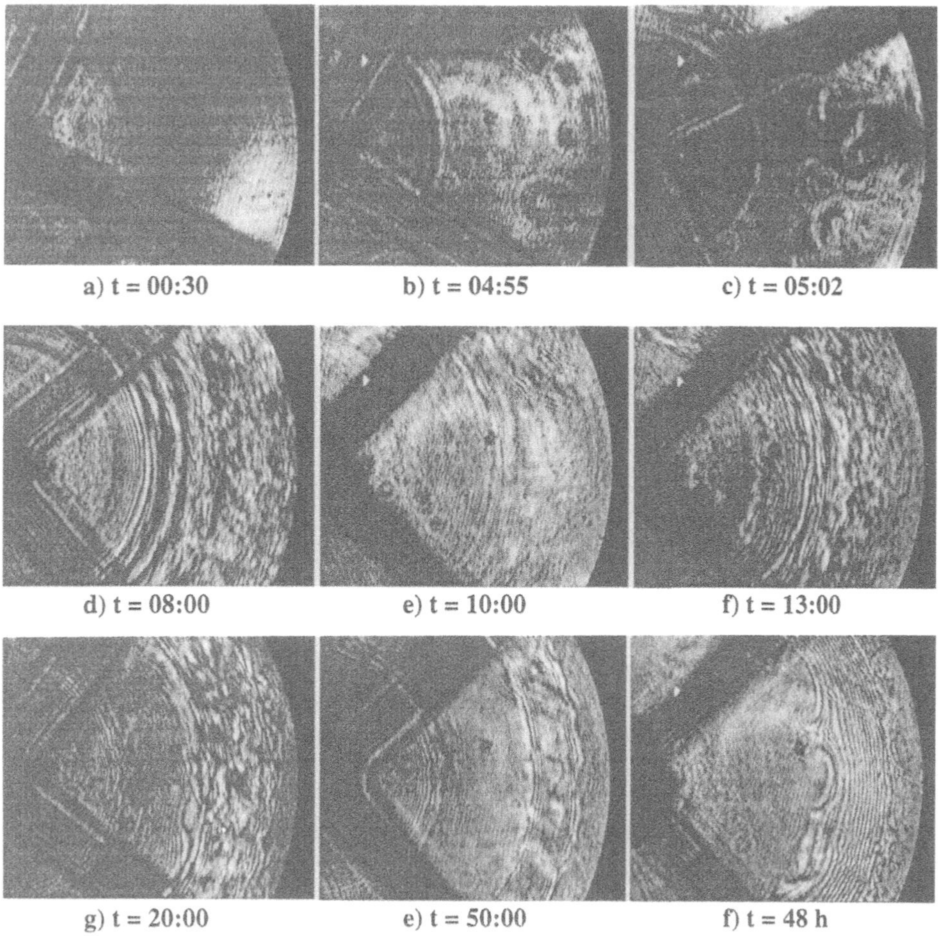

Figure 3. Polyacrylamide gel (PAG) formation in a rotating cuvette with thermal initiation of the polymerization reaction. Rotation rate 25 rev/s. Maximum acceleration of 63 *g* at 25 mm radius. Times are in min:sec.

REFERENCES

1. V. Briskman, K. Kostarev, V. Levtov, V. Romanov, and T. Yudina, Comparative experimental research of polymerization on the "MIR" orbital station and on the Earth, AIAA 95-0263, 33rd Aerospace Sciences Meeting and Exhibit, Reno (1995).
2. V. Briskman, K. Kostarev, V. Levtov, T. Lyubimova, A. Mashinsky, G. Nechitailo, and V. Romanov, Polymerization under different gravity conditions, *Acta Astronautica* 39: 395-402 (1996).
3. V. Briskman, K. Kostarev, and T. Lyubimova, Gel polymerization under high gravity conditions, in: *Materials Processing in High Gravity*, L.L. Regel and W.R. Wilcox, eds., Plenum Press, NY (1994) pp 185-192.
4. V. Briskman, K. Kostarev, and T. Yudina, Comparative investigations of thermal and photopolymerization under action of centrifugal forces. Basic mechanisms of heat/mass transfer, in: *Centrifugal Materials Processing*, L.L. Regel and W.R. Wilcox, eds., Plenum Press, NY (1997) pp 247-255.

UTILIZATION OF MICROGRAVITY AND HIGH GRAVITY TO PREPARE MATERIALS WITH CONTROLLED PROPERTIES

Vladimir A. Briskman, Konstantin G. Kostarev, and Andrey V. Shmyrov

Institute of Continuous Media Mechanics, RAS, Perm, 614013, Russia

INTRODUCTION

The most promising way to extend the scope of properties of known materials is to increase the number of controlled external factors involved in the manufacturing process. These factors include gravitational and centrifugal forces. In particular, in polymer processing many advantages can be gained from changing the level of the mass forces. This conclusion relies on the fact that the density of the reaction mixture during polymerization varies due to heat liberation and to conversion of a monomer to a polymer. The more predictable gravitational effect is convection in a liquid monomer in the early stages of reaction. Convection stirs the reaction mixture, leading to a more uniform temperature and a more uniform degree of conversion. While participating in convective motion, the mixture continues polymerizing and may lose fluidity before homogenization is complete. The gradients of temperature and degree of conversion leads to a variation in the reaction rate and, as a result, to an inhomogeneous polymeric microstructure and a nonuniform distribution of properties in finished products.

INHOMOGENEOUS POLYMERS DUE TO CONVECTION

The first evidence for a possible convection contribution to the formation of inhomogeneous polymers was obtained by a team of researchers from the Institute of Bioorganic Chemistry in Uzbekistan when they compared the results of protein electrophoresis on the basis of polyacrylamide gel matrices prepared under different gravitational conditions - on earth and in space.[1] The results of our recent experiments have supported their important findings. By using optical methods of investigation we showed that the onset of convection entails the formation of inhomogeneities in a polymer synthesized in terrestrial conditions.[2-4] The processes of frontal polymerization exhibits the highest gravitational sensitivity. Here, the reaction proceeds in the form of a relatively local zone in the bulk of the monomer, facilitating the development of an non-uniform density. This type of reaction occurring at normal gravity was found to involve one more mechanism of a gravitational nature – polymerization front instability, which may provoke convective motion or lead to polymer inhomogeneities even in the absence of convection.[5-8]

Figure1 shows a series of frontal polymerization interferograms for a 15% water solution of acrylamide. The reaction was photoinitiated from below in laboratory conditions ($a = 1g$). A gravitationally stable distribution of conversion was first created in the vertical cuvette, which was made of parallel glass plates, 10 mm thick, with inside dimensions 70×30×10mm. Then followed gel formation on the lower illuminated wall of the cuvette. As the thickness of the layer increased, the overall heat release was enhanced, and the temperature front penetrated into the monomer. This caused the generation of unstable density distribution in liquid reaction mixture. After that the mechanical equlibrium was disturbed (Fig.1a) and convection arose in the form of two symmetric counter-rotating rolls (Fig. 1d), with the rotational axes coinciding with the direction of observation. The maximum flow rate was 1-2 mm/s. Convection stirred the liquid, averaging the conversion degree of the monomer (Fig.1c). So the polymer-liquid front remained practically motionless until conversion in the whole volume of the liquid mixture reached the amount necessary for gel formation. As can be seen from Fig.1c the resulting gel structure was inhomogeneous.

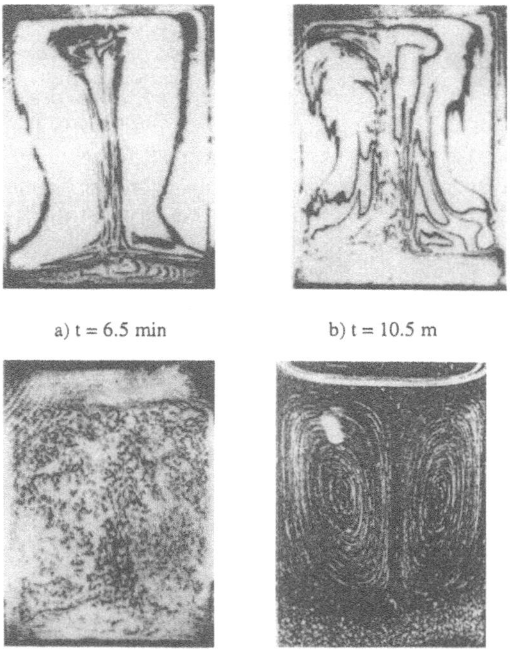

a) t = 6.5 min   b) t = 10.5 m

c) t = 50.0 m   d) flow pattern

Figure 1. Polymerization in a vertical layer illuminated from below; a, b and c are interferograms, and d is the flow pattern

MICROGRAVITY EXPERIMENTS

One would expect a homogeneous polymeric structure when the polymerization is performed with convection suppressed, for example by reducing the acceleration a to $10^{-5}g$. Thus the polymerization experiment "Gel-1" was carried out on the Mir space station in 1992 to investigate the influence of the absence of convection on the homogeneity of the resulting polymer.[8,9]

The working liquid used in our experiment was an aqueous solution of acrylamide doped with methylene-bis-acrylamide as a cross-linking agent (to obtain a three-dimensional cross-linked polyacrylamide gel, PAG), and riboflavine as a photoinitiating agent. The reaction was run in a specially designed cuvette with inner dimensions of 70x30x10mm. The cuvette had a number of viewing windows, which allowed visualization of the reaction in a monomer with the help of the Pion-M shadow device. The structure of the conversion field developed during polymerization was examined more thoroughly on earth using a Fizeau interferometer. Gas bubbles introduced into the reaction mixture were supposed to act as detectors for any convective motion. Video recording of the polymerization process demonstrated that the position of the bubbles remained invariable (Fig. 2). The propagation of the polymerization front was traced with the help of the

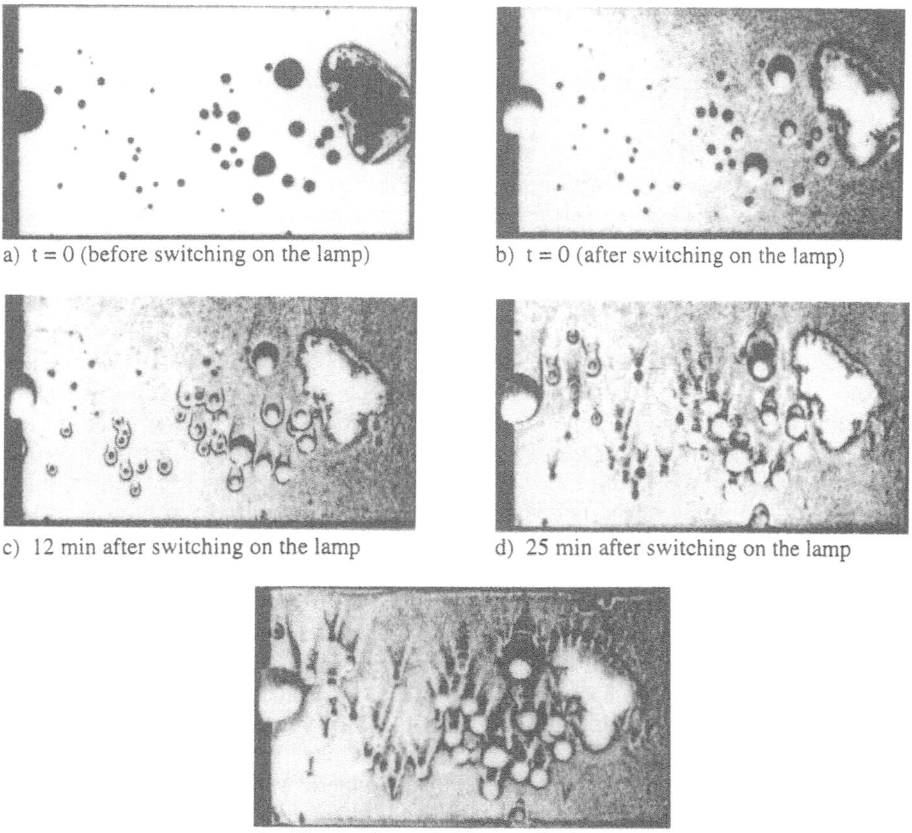

a) t = 0 (before switching on the lamp)

b) t = 0 (after switching on the lamp)

c) 12 min after switching on the lamp

d) 25 min after switching on the lamp

e) 47 min after switching on the lamp

Figure 2. Photographs taken during gelation in microgravity.

bubbles' shadows, which appeared at a certain degree of polymerization.

A combination of theory[10] and characterization of the gel samples on earth showed that the samples had highly homogeneous optical and mechanical properties, except in the vicinity of the bubbles.

CENTRIFUGE EXPERIMENTS

Convection and front instability are not the only mechanisms for a mass force effect on polymer structure and properties. Our investigations at 300*g* of gel formation (PAG) revealed other effects that are attributed to at least two independent mechanisms - sedimentation of primary polymeric microparticles (globules), and deformation of the forming polymeric network under the action of the mass forces.[11,12] The resulting gel samples demonstrated a radial dependence of the mechanical properties (Fig. 3).

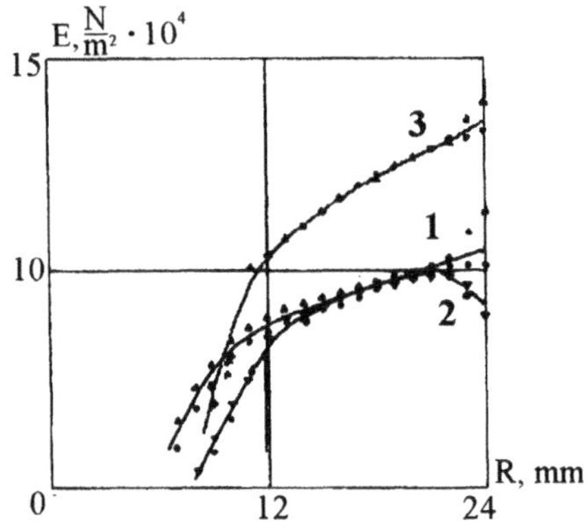

Figure 3. Elastic modulus distribution after polymerization of polyacrylamide gel during centrifugation at 300*g*. Separate and subsequent action of sedimentation (1) and deformation (2).

Several gravity-sensitive mechanisms can operate simultaneously at different stages of polymerization. To gain a deeper understanding of their interaction we performed visualization experiments on the velocity and monomer conversion fields using a specially designed setup. Figure 4 is a series of photographs showing the development of the convective motion during photopolymerization of PAG at 75*g*. The illumination was from above and the cuvette radius was 30 mm. Convective motion was first initiated at the cuvette periphery in the form of several jets and then moved toward the center, stretching over the whole volume of the reaction mixture and disturbing the variation in conversion caused by sedimentation. In contrast to the expected evolution of the process, the excitation of convection occurred immediately before the formation of the gel network.

At the same rotation frequency but with a smaller cuvette radius of 25mm to give 63*g*, the convective motion had time to spread only through only about 1/5 of the cuvette radius by the time the reaction mixture lost its fluidity. Figure 5 shows an interferogram of the resulting polymerized PAG specimen. Here, we observe a well-retained conversion field that gives a radial distribution of the gel properties, including the optical density.[13] Note that the conversion field altered by convection was later subjected to the action of the deformation mechanism during formation of the polymeric network.

The above examples show that our technique of process visualization allowed us to trace the sequence of stages leading to the formation of graded properties in polymers, and to define the role of each gravitationally sensitive mechanism.

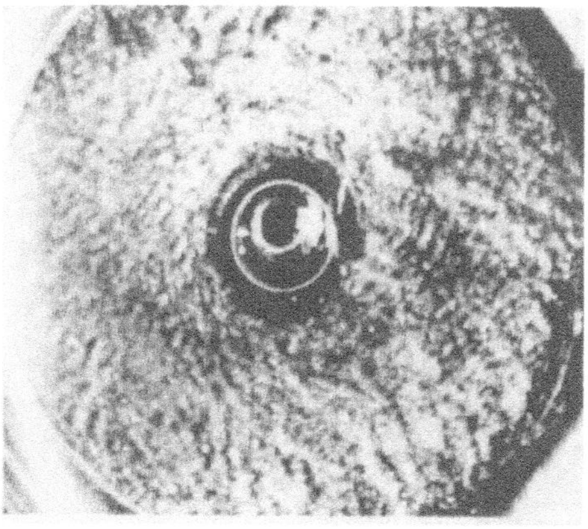

a) t =1 min 25 sec

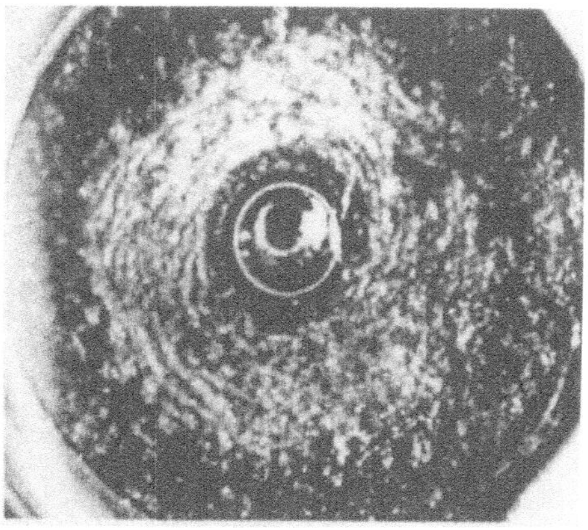

d) t = 1 min 40 sec

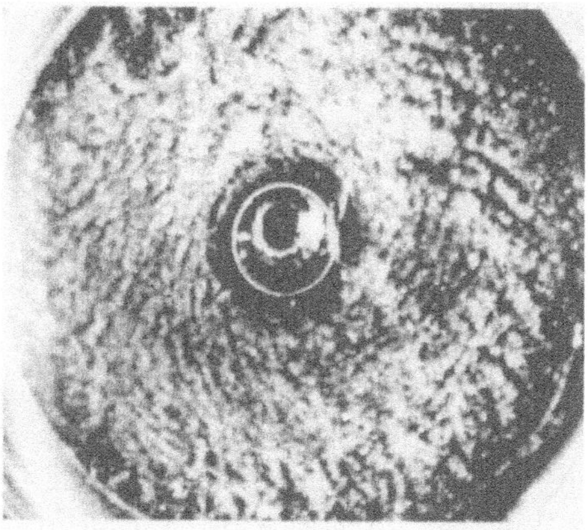

b) t =1 min 30 sec

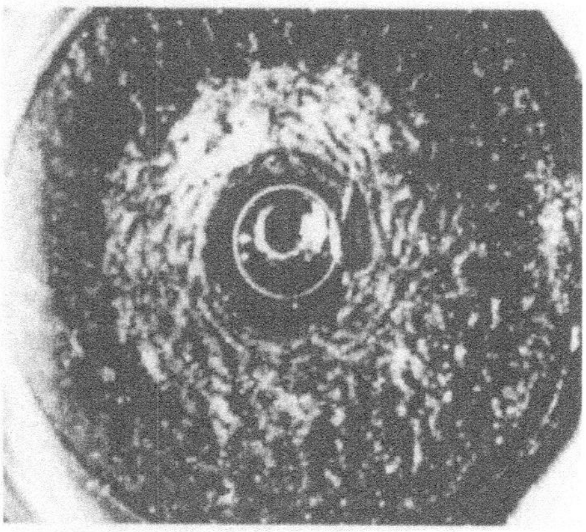

e) t =1 min 45 sec

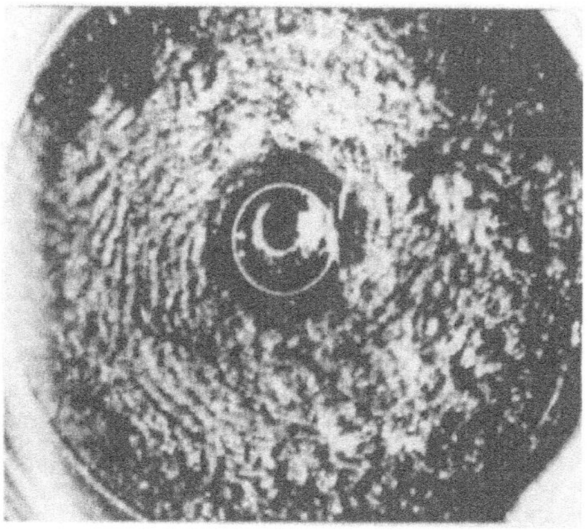

c) t =1 min 35 sec

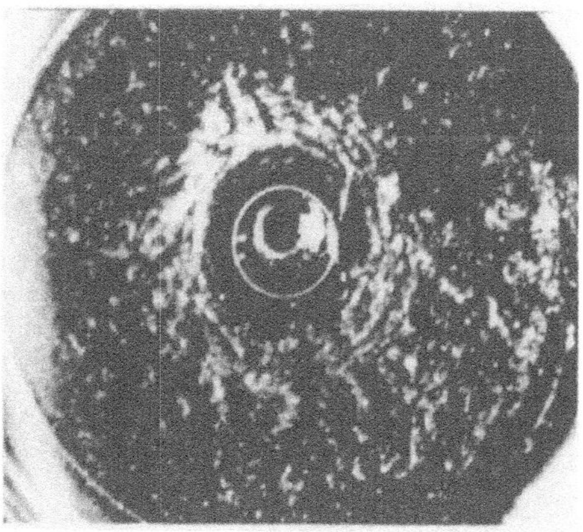

f) t =1 min 50 sec

Figure 4. Convection during polymerization on the centrifuge.

Figure 5. Interferogram of PAG polymerized at 63*g*.

REFERENCES

1. A. Sadykov, V. Leontyev, Y. Mangutova, G. Grechko, G. Nechitailo, and A. Mashinsky, Systems of chemical reactions depending on gravity, *Akademiia Nauk SSSR Doklady* 303: 1004-1007 (1988).
2. Sh. Abdurakhmanov, L. Bogatyreva, V. Briskman, M. Levkovich *et al.*, On polyacrylamide gel formation by photoinitiation under terrestrial and orbital conditions, in: *Numerical and Experimental Modeling of Hydrodynamic Phenomena under Weightlessness*, Sverdlovsk (1988) pp.120-128.
3. L. Bogatyreva, V. Briskman, K. Kostarev, V. Leontyev, *et al.*, Heat/mass transfer mechanisms of the polymerization under terrestrial and microgravity conditions, in: *Proceedings of the VIII European Symposium on Material and Fluid Sciences in Microgravity*, European Space Agency, Brussels, ESA SP-333, 1:173-178 (1992).
4. V. Golubev, D. Gromov, L. Guseva, B. Korolev, K. Kostarev, and T. Lyubimova, Free convective heat and mass transfer in processing polymer materials, *Heat Transfer Research* 25: 888-893 (1993).
5. A. Malkin, V. Begishev, L. Guseva, and K. Kostarev, Cross-linking inhomogeneity of oligomers caused by convection, *J. Polymer Science* A36: 625-630 (1994).
6. J. Pojman, Traveling front of methacrylic acid polymerization, *J. Am. Chem. Soc.* 113: 6284-6286 (1991).
7. V. Briskman, K. Kostarev, V. Moshev, V. Begishev, and G. Nechitailo, Production of polymer materials with prescribed properties in microgravity, STAG Research Number BT-10, NASA-RSA Conference Abstracts, Moscow, 136 (1996).
8. S. Solovyov, V. Ilyashenko, and J. Pojman, Numerical modeling of self-propagating polymerization fronts: The role of kinetics on front stability, *CHAOS* 7: 331-340 (1997).
9. V. Briskman, K. Kostarev, V. Levtov, V. Romanov, and T. Yudina, Comparative experimental research of polymerization on the "MIR" orbital station and on the Earth, AIAA 95-0263, American Institute of Aeronautics and Astronautics, 33rd Aerospace Sciences Meeting and Exhibit, Reno (1995).
10. V. Briskman, K. Kostarev, V. Levtov, T. Lyubimova, A. Mashinsky, G. Nechitailo, and V. Romanov, Polymerization under different gravity conditions, *Acta Astronautica* 39: 395-402 (1996).
11. V. Ivanov, V. Begishev, L. Guseva, and K. Kostarev, Front photopolymerization at constant extinction and limited mass transfer, *J. Polymer Science* B37: 293-296 (1995).
12. V. Briskman, K. Kostarev, and T. Lyubimova. Gel polymerization under high gravity conditions. in: *Materials Processing in High Gravity*, L.L. Regel and W.R. Wilcox, eds., Plenum Press, NY (1994) pp 185-192.
13. V. Briskman, K. Kostarev, and T. Yudina, Comparative investigations of thermal and photopolymerization under action of centrifugal forces. Basic mechanisms of heat/mass transfer, in: *Centrifugal Materials Processing*, L.L. Regel and W.R. Wilcox, eds., Plenum Press. NY (1997) pp 247-255.
14. K. Kostarev and A. Shmyrov, Investigations of flow structures and conversion distributions during photopolymerization in centrifugal fields, *Winter School of Continuous Media Mechanics*, Perm, (1999) 197.

INFLUENCE OF CENTRIFUGATION ON COAGULATION OF COLLOIDAL DISPERSIONS OF TEFLON™

J. Simmons,[1*] L.L. Regel,[1] W.R. Wilcox,[1] and R. Partch[2]

[1]International Center for Gravity Materials Science and Applications
[2]Chemistry Department, Clarkson University, Potsdam, NY 13699-5814, USA

INTRODUCTION

The influence of gravity on coagulation of colloidal dispersions has scarcely been investigated. The objective of the present project was to determine how centrifugation influences coagulation of polytetrafluoroethylene (PTFE) colloidal dispersions and any differences in the coagulated product. PTFE was chosen because it is a unique polymer used in many consumer and industrial applications. Despite being hydrophobic, PTFE remains stable in aqueous dispersions probably due to charge on the particles.[1] PTFE dispersions are usually coagulated by some form of mechanical agitation. This phenomenon has been attributed to the PTFE particles becoming irreversibly surrounded by air, because this material is poorly wet by the liquid.

EXPERIMENTAL METHODS

The experimental methods are described in detail elsewhere.[2] Two Teflon PTFE dispersions from E.I. du Pont de Nemours and Company were used: one with 6%w Triton X-100 surfactant and one without added surfactant. The dispersion with added surfactant was very stable and remained dispersed even with strong agitation, while the other dispersion could easily be coagulated by shaking. The Teflon particles in both dispersions were about 0.2 μm in diameter.

One set of experiments was performed using a Sorval Superspeed Centrifuge at 3000 rpm (705 *g*) for 15 and 60 minutes, 5000 rpm (1958 *g*) for 15 minutes, and 10000 rpm (7833 *g*) for 15 minutes. In several experiments, 20 ml of the dispersion with added surfactant was centrifuged to form a solid sediment on the bottom of the centrifuge tube. The supernatant solution was gently poured off. The sediment was redispersed by stirring

* Present address: Boehringer Ingelheim Pharmaceuticals, Inc., Ridgefield, CT 06877, USA

with 20 ml of deionized water. Any of the solid that couldn't be redispersed was collected, dried and weighed. The new dispersion was centrifuged again and the process repeated until no more Teflon could be redispersed.

Centrifugation of 10 ml of dispersion without added surfactant also resulted in a sedimented solid at the bottom of the centrifuge tube. When an attempt was made to redisperse this sample in 10 ml of deionized water, most of the solid continued to adhere to the centrifuge tube. This solid was collected and the remaining dispersion centrifuged again. A large part of the resulting sedimented solid could not be redispersed and was collected again. After a third centrifugation, all of the originally dispersed Teflon had been collected as non-dispersible solid.

Two studies were performed in the swing bucket of Clarkson materials processing centrifuge, HIRB.[3] Dispersion with no surfactant was used in one set of experiments. A centrifuge tube was secured inside the swing bucket and covered so that the wind created by the centrifuge would not affect the experiment. We tested 10 ml dispersions with 100%, 50% and 25% of the Teflon dispersion plus deionized water. Experiments were performed at 2 *g* and 4 *g* for 2, 5, 7.5 and 10 hours.

For the other study on HIRB, a laboratory mixer with a rotation rate of 350 rpm was mounted to the top inside of the swing bucket. A glass mixing cell with a diameter of 5.5 cm and a height of 11.5 cm was attached to the bottom plate of the swing bucket. The impeller was a 45° pitched-blade turbine with a diameter of 3.7 cm. The four turbine blades were 1.2 cm in length. Figure 1 shows a schematic drawing of this mixing apparatus. Experiments were performed using 80 ml of mixtures of the non-surfactant dispersion with deionized water, and contained either 25% or 50% of the dispersion. Coagulated material was collected after stirring the dispersion for two or three hours at 1, 2, 4 and 6 *g*.

In all experiments, the mass of coagulated material was determined by weighing after drying at 140°C for 8-12 hours. In order to obtain the percent coagulated, this mass was divided by the available amount of Teflon, which was estimated by multiplying the given concentration by the original volume of the dispersion.

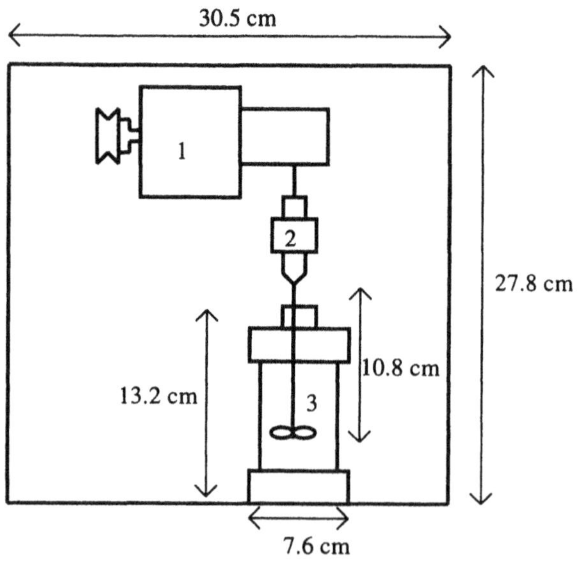

Figure 1. Experimental setup for stirring experiments on HIRB centrifuge. (1) Stirring motor and gear box; (2) Chuck to secure impeller shaft; (3) Mixing cell.

RESULTS

Experiments using Sorval Centrifuge

The non-dispersible solids created by centrifugation of the two Teflon dispersions in the Sorval centrifuge were quite different. The sediments from the dispersion with surfactant were dense and so strong that separation into pieces required them to be torn apart with a spatula. They sank in water. The sediments from the non-surfactant dispersion were less dense, with water and air captured within the structure. Compression of the solid caused water to flow out of it. These solids floated on water and were not wet by it. They adhered to surfaces they contacted, such as the spatula used to collect them and the sides of the vials used to store them. X-ray diffraction showed that all of these solids were Teflon. Scanning electron microscopy of samples from these two experiments revealed that the Teflon particles were fused together differently than a Teflon solid created by coagulation using agitation, where individual particles could clearly be seen. Figure 2 shows non-dispersible solid obtained after four centrifugations of the Teflon dispersion with surfactant. Figure 3 shows solid created by one centrifugation of the non-surfactant Teflon dispersion. Figure 4 shows the sticky membrane resulting from five centrifugations of the surfactant dispersion at 7833 *g*. Figure 5 shows that the amount of non-dispersible solid from the dispersion with added surfactant increased with increasing acceleration, increasing time, and increasing number of centrifugations. Figure 6 shows that the addition of surfactant greatly decreased the amount of sediment that could not be redispersed.

Centrifugation on HIRB without stirring

Centrifugation at 4 *g* of dispersion without surfactant resulted in coagulation only at the air/dispersion interface. Figure 7 shows that dilution with deionized water decreased the rate of coagulation and the amount coagulated. For undiluted dispersion, the coagulated product covered almost the entire area of the air/dispersion interface. At 2 *g*, no coagulation was observed for any mixture, even after 10 hours of centrifugation.

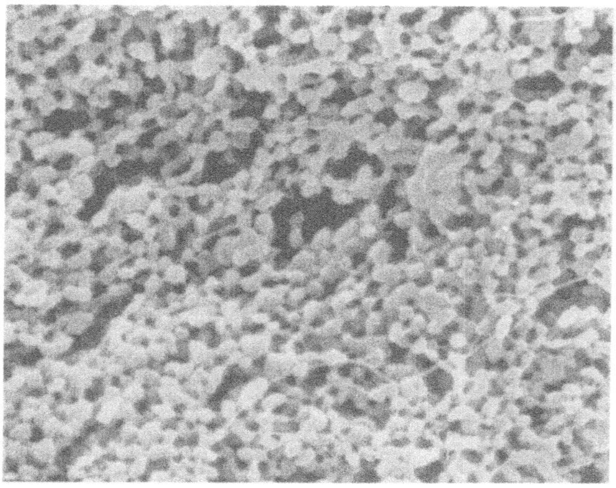

Figure 2. Scanning electron micrograph of solid created by centrifugation of Teflon dispersion with surfactant at 705g. Note the 1 μm white line near the bottom.

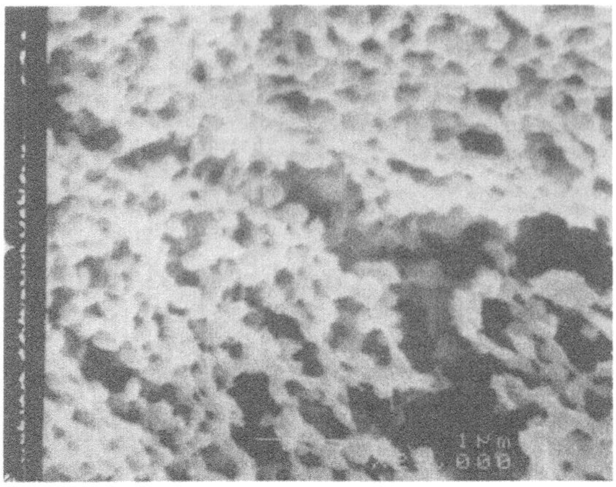

Figure 3. Scanning electron micrograph of solid created by centrifugation of Teflon dispersion without surfactant at 705 *g*.

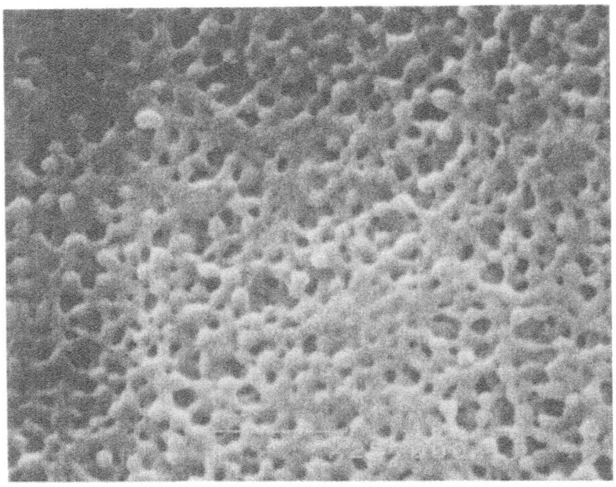

Figure 4. Scanning electron micrograph of the sticky membrane resulting from five centrifugations of the surfactant dispersion at 7833 *g*.

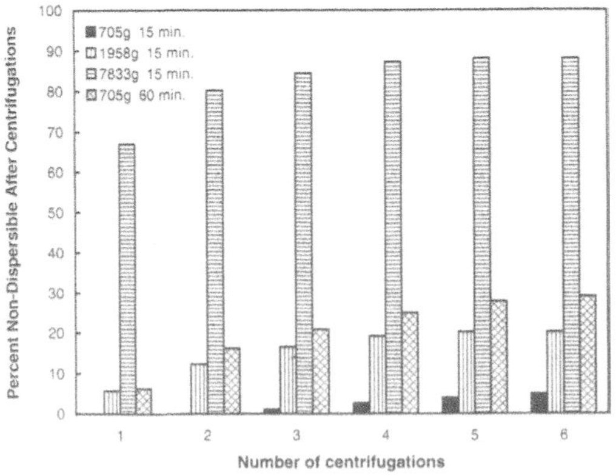

Figure 5. Cumulative non-dispersible solid created by centrifugation of the Teflon dispersion containing added surfactant.

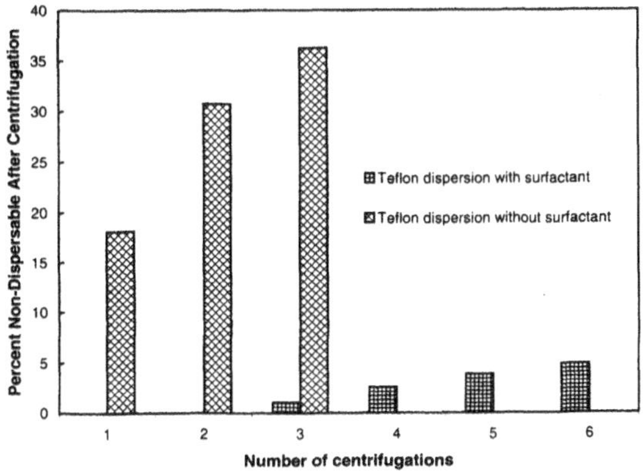

Figure 6. Cumulative non-dispersible solid created by centrifugation at 705g for 15 min of 20 ml of Teflon dispersion containing surfactant and 10 ml of Teflon dispersion without added surfactant.

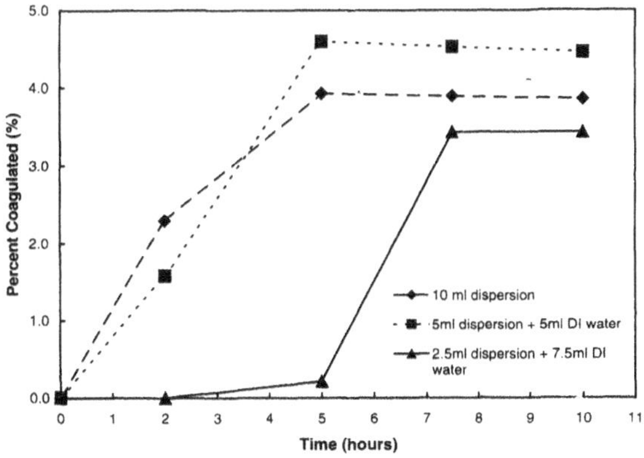

Figure 7. Amount of coagulated solid created after centrifugation of different concentrations of the Teflon dispersion without surfactant at 4g.

Stirring without Centrifugation

Figure 8 summarizes the result of stirring a 25% dispersion/water mixture without centrifugation for 2 hours. This resulted in about 75% of the air/liquid surface being covered by flat, oval-shaped agglomerates with diameters of about 2-5 mm. These non-wettable agglomerates had a dense surrounding layer supporting a less dense, very thin middle section. Attached to the impeller were many porous spherical agglomerates with diameters of 1-2 mm, and some agglomerates similar to those found on the air/water surface. A ring of coagulated Teflon surrounded the impeller shaft where the air/water interface had been located during mixing. Stirring for 3 hours without centrifugation (1g) yielded similar results, except that the agglomerates on the surface of the liquid had diameters of 4-7 mm and the amount of coagulation on the surface and impeller had increased.

Figure 9 shows that stirring the 50% dispersion/water mixture at 1*g* for 2 hours resulted in much more coagulation than the dilute mixture. Agglomerates covered almost the whole air/liquid surface. The agglomerates on the surface were similar to those created with the more dilute dispersion, except that they were spherical in shape. The non-wettable agglomerates had a dense layer surrounding the less dense, thin middle section. Attached to the impeller were many porous spherical agglomerates like those created with the less dilute dispersion, and some agglomerates similar to those found on the air/liquid surface. Unlike the experiments on the more dilute dispersion, there was coagulated solid on the bottom of the mixing cell. Similar results were obtained after stirring for 3 hours.

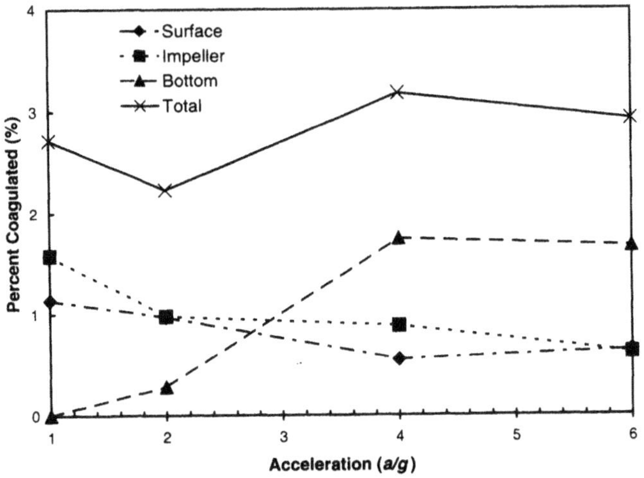

Figure 8. Average amount of coagulated solid created by mixing 20 ml of Teflon dispersion without surfactant plus 60 ml deionized water for 2 hours.

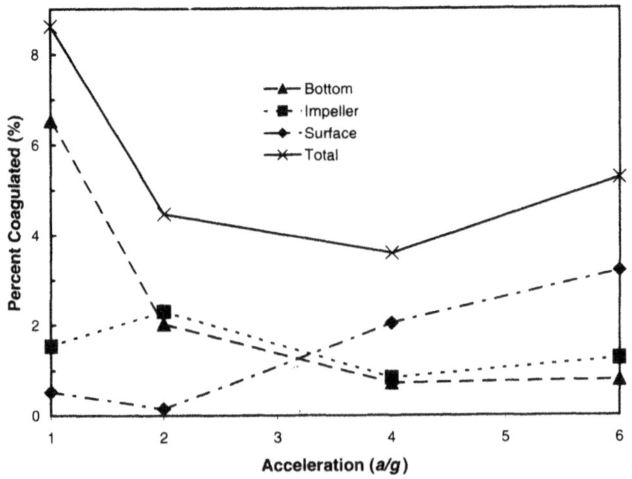

Figure 9. Average amount of coagulated solid created by mixing 40 ml of Teflon dispersion without surfactant plus 40 ml deionized water for 2 hours

Stirring during Centrifugation

Centrifugation significantly affected the coagulation caused by stirring the 25% Teflon dispersion without surfactant. After centrifugation at 2 *g* with stirring for two and three hours, only about 25% of the air/liquid surface was covered by agglomerates. The agglomerates resembled those created without centrifugation except that they were larger (diameters of 5-7 mm), and the edges were not as dense. There were some spherical agglomerates attached to the impeller, but the rest of the coagulation on the impeller was a soft, amorphous solid. A small amount of coagulated Teflon was found on the bottom of the mixing cell.

An acceleration of 4 *g* significantly affected the coagulation caused by stirring. After 2 hours there was much less coagulate on the surface of the liquid, with less defined and weaker agglomerates. The coagulate on the impeller was very soft and not shaped like the well-defined spheres obtained at 1 *g*. Unlike the experiments without centrifugation, significant amounts of coagulated solid were found on the bottom of the mixing cell, including one piece over 1 cm in length. Results were similar for runs at 3 hours, except that more coagulation had occurred than after 2 hours.

An acceleration of 6 *g* while stirring also created less defined, weaker agglomerates. The solid attached to the impeller was similar to that created at 4 *g*, but the amount there was less. Again, there was a significant amount of coagulated solid found at the bottom of the mixing cell, especially after three hours of mixing.

Increasing the acceleration also had a significant effect on the coagulation caused by stirring the more concentrated 50% dispersion. After stirring at 2 *g* for 2 and 3 hours, about 25% of the air/liquid interface was covered by agglomerates. The agglomerates were smaller than those created in the more dilute dispersion at 2 *g*. There was much more coagulated solid attached to the impeller than at 1 *g*, but less on the bottom. The solid on the impeller was either small, spherical coagulates or a mushy, shapeless solid as in the experiments with the less dilute dispersion at 2 *g*.

Stirring at 4 *g* for two and three hours created less coagulate on the surface than at 1g. It resembled the solid that was created at 4 *g* with the more dilute dispersion. There was less coagulate stuck to the impeller than at 2 *g*, but significantly more on the bottom. As with the more dilute dispersion, the solid on the bottom was mostly in one large piece.

Stirring at 6 *g* for two and three hours created less coagulation on the surface than at the other acceleration levels. The coagulate on the surface and attached to the impeller was soft and shapeless like the experiments at 6 *g* with the more dilute dispersion. Again, the amount of solid on the bottom increased with increased acceleration.

CONCLUSIONS AND DISCUSSION

Centrifugation of Teflon dispersions at high acceleration resulted in unique solids with different characteristics than the solid created by coagulation due to shaking or mixing of the dispersion. With repeated centrifugation and resuspension in deionized water, the particles became increasingly fused together, until a porous and sticky membrane was produced at the bottom of the centrifuge tube.

Centrifugation of the Teflon dispersion without surfactant at 4 *g* resulted in coagulation at the air/dispersion interface that reached a constant level at longer run times. The amount of coagulation increased with increasing concentration of the dispersion. At 2*g* no coagulation occurred with run times up to 10 hours.

Centrifugation increased the amount of coagulation caused by agitating the Teflon dispersion without surfactant. With increasing acceleration, the higher-concentration dispersion showed significant decreases of surface coagulation and an increase of coagulated Teflon at the bottom of the mixing cell. With increasing acceleration, the lower-concentration dispersion had a significant increase of coagulated Teflon at the bottom of the mixing cell. The agglomerates on the surface and attached to the impeller changed from distinct shapes with non-wettable properties at 1 *g* to shapeless and soft material at 4 *g* and 6 *g*.

We now speculate on the causes for the behavior observed. Teflon is produced by emulsion polymerization in the presence of a surfactant. We hypothesize that the resulting colloidal particles have a monolayer of surfactant on their surfaces. Adsorbed ions and, possibly, an oil may also be present and play a role in the dispersibility of Teflon. In the usual process of converting this dispersion to a solid product by coagulation, heating and pressing, some adsorbates remain and determine much of the resulting surface properties associated with solid Teflon. Additional surfactant is added to stabilize the Teflon dispersion when it is to be shipped. When a Teflon dispersion is centrifuged at moderately high accelerations, most of the adsorbates are left in the supernatant liquid. By pouring off the supernatant and resuspending in deionized water we reduce the amount of adsorbates on the surface of the Teflon particles. Eventually little adsorbate remains, and the particles are able to merge and flow under the centrifugal force until a smooth membrane is formed that readily adheres to other solids. Although optical microscopy shows a smooth sheet, scanning electron microscopy reveals that this membrane has pores of about the size of the initial Teflon particles.

Centrifugation while stirring shows that gravity plays a role in coagulation by agitation of a dispersion. Coagulant forms not only on the surface but also on the impeller and at the bottom of the vessel. The traditional explanation for coagulation by agitation is that any non-wetting particle that reaches the surface permanently contacts the air. However, our experiments show that agitation alone is sufficient to cause coagulation without the particles reaching the surface. Perhaps the shear field around the impeller causes particles to contact one another and weld together.

Acknowledgements

This research was partially supported by the National Science Foundation under grant DMR-9414304. We are grateful to du Pont for donating the Teflon dispersions and to Egon Matijevic for critiquing the manuscript.

REFERENCES

1. S. Kratohvil and E. Matijevic, Stability of colloidal Teflon dispersions in the presence of surfactants, electrolytes, and macromolecules, *J. Colloid Interface Sci.* 57:104 (1976).
2. J.F. Simmons, *Influence of Centrifugation on the Coagulation of Polytetrafluoroethylene Colloidal Dispersions,* M.S. Thesis, Clarkson University (1996).
3. R. Derebail, W.A. Arnold, G.J. Rosen, W.R. Wilcox, and L.L. Regel, HIRB - The centrifuge facility at Clarkson, in: *Materials Processing at High Gravity,* L.L. Regel and W.R. Wilcox, eds., Plenum Press, New York (1994) pp 203-212.

BOUNDARY CONDITIONS IN DIFFUSIONAL GROWTH AND SEDIMENTATION

Vladimir Privman and Jongsoon Park

Department of Physics and Center for Advanced
Materials Processing, Clarkson University
Potsdam, New York 13699–5820, USA

INTRODUCTION

In many applications, transport of particles can be described by the diffusion equation, or its convective-diffusion generalizations, in part of three-dimensional space. In particular, in surface deposition or in growth of aggregates or sediments, particles in the suspension outside the surface or aggregate boundary can be considered as diffusing under the influence of applied forces, such as the double-layer interactions or gravity. However, the diffusion equation cannot be used *at* the substrate or *inside* the aggregate. One convenient and widely used approach has been to impose boundary conditions at the surface, to supplement and completely define the diffusion problem outside it.[1–11] This short survey describes some theoretical results on the use of such boundary conditions and their physical interpretation.

In order to define the geometry of the problem, let us consider Figure 1. We will assume that the surface is planar, at $x = 0$, and for simplicity use the one-dimensional version of the diffusion equation whenever this causes no confusion. Thus, for $x > 0$, the coordinate x measures the distance away from the substrate or growing aggregate. Some important problems lack planar symmetry. For instance, in the growth of colloid particles from small sub-units, the aggregates are typically spherical, and the radial diffusion equation must be used.

Here, we assume that suspended particles at $x > 0$ move according to the diffusion equation:

$$\frac{\partial c}{\partial t} = D_0 \frac{\partial}{\partial x}\left(\frac{\partial c}{\partial x} + c\,\frac{dV}{dx}\right) \tag{1}$$

where D_0 denotes the particle diffusion constant and the subscript 0 refers to the dilute-suspension limit. The suspended particle number concentration is c, so that for spherical particles, for instance, of radius r, the volume filling fraction ϕ is

$$\phi = 4\pi r^3 c/3 \tag{2}$$

Processing by Centrifugation, Edited by Regel and Wilcox
Kluwer Academic/Plenum Publishers, New York, 2001

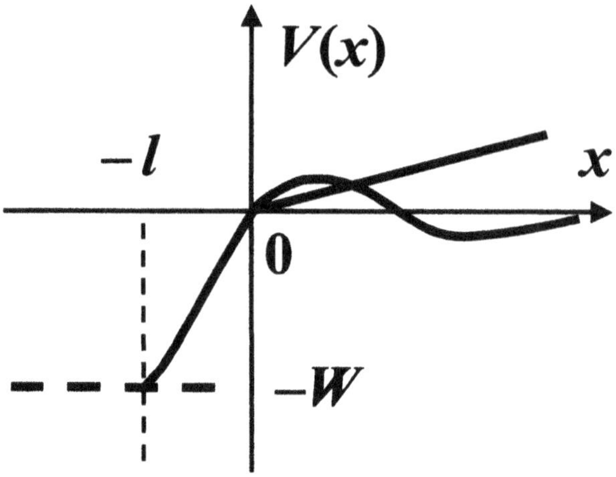

Figure 1. Schematic representation of the dimensionless potential energy $V(x)$ with the typical double-layer and linear forms shown.

It is assumed that for $x > 0$, particles move in the potential energy $U(x)$, here only a function of x to ensure planar symmetry. We define the dimensionless potential energy as $V \equiv U/kT$. The potential energy can be double-layer, for colloid growth, or linear, as in sedimentation; see Figure 1. The dilute-suspension diffusion equation can be used[12] approximately up to $\phi \simeq 0.1$, for particles with relatively short-range interactions as compared to their size, so that they can be regarded similar to hard spheres. Note that some of the longer-range interactions of a given particle with other particles, especially with those in the aggregate, will be included phenomenologically as a contribution to the effective potential energy $U(x)$. The van der Waals dispersion-force component of the double-layer potential energy is one such contribution.

A diffusional description applies for many problems. Specifically, in monolayer surface deposition (or detachment) of particles from solution,[4,7,8] the substrate is fixed at $x = 0$. In other cases, such as growth of submicron colloid particles from nano-size sub-units,[1,9,10] the non-planar, typically spherical, surface actually moves in time. In many instances this motion of the growing surface can be treated as a slow process as compared to particle diffusion. In such cases an approximation can be used whereby the diffusion equation is solved with the fixed boundary at $x = 0$. The solution is then used to calculate the flux of matter to the surface, or away from it in cases of detachment/dissolution.[1,4,7,9,10] The slow motion of the surface itself is then treated separately. Furthermore, particle motion in a dilute solution is much better accessible to experimental probes and understood theoretically than the dynamics of a concentrated system. Therefore it is frequently convenient to lump the effects of the surface and growing, restructuring aggregate in a few-parameter phenomenological boundary condition at $x = 0$.

A typical approach to such problems has been to use boundary conditions that impose a linear relation between the concentration c and its derivatives,

$$c = \alpha \frac{\partial c}{\partial x} + \beta + \gamma \frac{\partial c}{\partial t} \qquad (\text{at } x = 0) \tag{3}$$

The term on the left-hand side is always there. The three terms on the right-hand side vary in their importance depending on the problem at hand. In the next section, we will survey the various forms typically assumed and describe in detail the choice appropriate for growth and sedimentation. In the last section, we consider aspects of diffusion in concentrated suspensions. Finally, when there is no planar symmetry, the coordinate-derivative term in (3) becomes $\alpha \hat{n} \cdot \vec{\nabla} c$, where $\hat{n}$ is the unit vector orthogonal to the surface, pointing into the dispersion.

BOUNDARY CONDITIONS FOR DIFFUSION

The relative importance of the terms on the right-hand side of Eq. (3) depends on the problem at hand. Let us consider the well known example of monolayer surface deposition and detachment.[8] In this case, the boundary condition is usually written as:

$$\frac{\partial c}{\partial t} = -\kappa c + k \qquad (\text{at } x = 0) \tag{4}$$

The first term describes the depletion of the diffusing-particle number at $x = 0$ owing to their adhesion to the surface. The second term corresponds to particle detachment. Within a Langmuirian approach,[8] we take:

$$\kappa = \lambda \left(\theta_{\text{max}} - \theta\right) \qquad \text{and} \qquad k = \omega \theta \tag{5}$$

where θ is the fractional surface coverage and θ_{max} is the maximum coverage, while λ and ω are constants. At equilibrium, when the time-derivative vanishes, the above equations yield the famous Langmuir-isotherm relation between the surface coverage and particle concentration near the surface, and, in the absence of potentials, in the bulk of the solution: $c \propto \theta / \left(\theta_{\text{max}} - \theta\right)$.

The above example illustrates that the coefficients α, β, γ in Eq. (3) can be time-dependent coupled dynamical variables. In another case, that of chemical reactions,[2,3,6] the appropriate relation describes partial reaction probability. It has been established[2,3,6] that the "radiation boundary condition" is applicable: c is proportional to the normal derivative at the reaction radius. In our notation this would correspond to only the coefficient α surviving on the right-hand side of Eq. (3), while $\beta, \gamma = 0$.

In recent work on diffusional growth of spherical colloid aggregates, we explored the form of the boundary condition for the case when the surface is not rigid but rather is itself dynamically evolving.[10] This study was extended to droplet-shaped objects growing on two-dimensional substrates.[11] Here, we survey this approach and offer a different derivation, which is less complicated (at the expense of being somewhat less definitive) but is also more generally applicable. We argue that similar considerations can be applied to sedimentation of submicron or larger suspended particles, provided the growth of the sedimented "cake" layer can be regarded as slow, i.e., the gravitational (centrifugal) acceleration is not too large.

Consider an aggregate or cake-layer slowly growing or dissolving by capture or release of singlet units to/from solution. These "building block" singlets can be solute species or larger suspended particles, depending on the system. Typically in such growth there is no well-defined sharp surface. Instead, there is a region for $x < 0$ in which the particle motion changes from simple dilute-suspension diffusion to whatever restructuring and relaxation dynamical processes are going on deep inside the growing structure. In many experimental situations these restructuring processes are quite fast, leading to dense rather than fractal structures.[9] However, these processes are complicated and cannot be modeled by extensions of the diffusion equation. Still, we assume that for some depth $-\ell < x < 0$, the dilute-limit diffusion Eq. (1) does apply. Such an approximate description deteriorates fast past $x = -\ell$; see Figure 1. The effective potential energy seen by the particles in this region reflects the onset of them being bound within the growing structure and the corresponding increase in the particle concentration. Thus, as shown in Figure 1, we assume that $V(x)$ drops to a large negative value $-W$ at $x = -\ell$, while the particle concentration reaches a large value (as compared to the concentration in the dilute suspension outside, at $x > 0$), denoted c_W.

There is evidence in the literature[12] that the dilute diffusion description applies up to volume filling fractions $\phi_W \simeq 0.1$, for colloid (submicron size) and larger particles which can be approximately treated as hard spheres. We assume that c_W and the potential energy value $U(-\ell) = -kTW$ are properties of the structure of the outer layer of the growing cake and are not dependent on the initial concentration or on time, the latter after a short transient time interval. Furthermore, we assume that the concentration within $-\ell \leq x \leq 0$ is in thermodynamic equilibrium. These assumptions suggest that the values of the concentration at $x = 0$ and $x = -\ell$ are related as follows:

$$c(0) = c_W e^{-W} \tag{6}$$

The above arguments suggest that the boundary condition in this case corresponds to $\alpha, \gamma = 0$ in Eq. (3), and to fixed $\beta = c_W e^{-W}$, i.e., to:

$$c = \beta \left(= c_W e^{-W}\right) \qquad (\text{at } x = 0) \tag{7}$$

A somewhat more detailed study,[10] not presented here, shows that while β is a product of one large parameter, c_W, and one small parameter, e^{-W}, the coefficient α is a product of two small parameters, ℓ and e^{-W}, and that the omission of the time-derivative term $\propto \gamma$ changes only the initial transient behavior of the concentration near $x = 0$. The resulting boundary condition, $c(0) = \beta$, is thus surprisingly simple and involves only a single phenomenological parameter, the value of the concentration at the growing or dissolving aggregate or cake surface. For a completely irreversible capture of particles by the growing aggregate, the effective energy barrier W is very large and $\beta \to 0$, recovering the celebrated boundary condition of Smoluchowski.[1]

CONCENTRATED SUSPENSIONS

One might ask how far can the present approach be taken into the region of the concentrated particle system, $x < 0$, and whether corrections are needed due to nonzero concentration for $-\ell < x < 0$. In order to address this matter, let us survey certain aspects of diffusion in concentrated dispersions.[12–14] Some relations of the ordinary diffusion theory remain unchanged when the particle system is no longer dilute. For instance, if $\vec{v}$ represents the average particle velocity, then the particle number flux $\vec{J}$ is:

$$\vec{J} = c\vec{v} \tag{8}$$

using three-dimensional vector notation. However, the contribution $\vec{J}_F$ to flux, caused by a force field $\vec{F}$, can only be approximately assumed proportional to that force:

$$\vec{J}_F = -c\vec{F}/f(c) \tag{9}$$

Strictly speaking, this relation is only exact in the dilute limit, $c \to 0$, when the friction constant has the value f_0. For denser systems, the relation need no longer be local, and the friction constant $f(c)$, which is the inverse of the mobility, is defined only approximately.

For denser systems, some relations become generally nonlocal and include memory, in that they involve concentration and possibly other properties at nearby points and at earlier times. Perhaps the most familiar such relation has been between the concentration gradient of the suspended particles and their number flux.[13] The dilute-limit-like relation

$$\vec{J} = -D(c)\vec{\nabla}c \qquad (\text{for } V(\vec{R}) = 0) \tag{10}$$

only becomes exact as $c \to 0$ and $D(c) \to D_0$. A more general nonlocal relation with memory is:

$$\vec{J}(\vec{R},t) = -\int d^3\vec{\rho} \int_0^\infty d\tau\, \mathcal{D}(\vec{\rho},\tau)\, \vec{\nabla}c\left(\vec{R}+\vec{\rho},\, t-\tau\right) \tag{11}$$

where the kernel $\mathcal{D}(\vec{\rho},\tau)$ could be a tensor in a more general case. Since relation (11) can be "deconvoluted" to become a product in the Fourier-transformed representation, it can be directly probed by light scattering measurements in some systems.[13,14]

The following two relations remain valid for any concentration:

$$\frac{\partial c}{\partial t} = -\vec{\nabla} \cdot \vec{J} \tag{12}$$

$$\vec{F}_{\text{appl}} = -kT\,\vec{\nabla}V(\vec{R}) \tag{13}$$

Here, $\vec{F}_{\text{appl}}$ is the applied force per particle. However, the expression for the effective thermodynamic driving force per particle, $\vec{F}_{\text{grad}}$, resulting from the particle concentration gradient, again can be only written as an approximate generalization of the dilute-limit expression. Assuming that we can define the osmotic pressure, $\Pi(c)$, for the particles in the suspension, and that local equilibrium applies so that $\Pi(c)$ is a thermodynamic function of c (and of the temperature T, which is assumed constant throughout this work), we have the relation:

$$\vec{F}_{\text{grad}} = -\frac{1}{c}\left(\frac{d\Pi}{dc}\right)\vec{\nabla}c \tag{14}$$

More general formal relations can be written, with this force proportional to the gradient of the chemical potential, accounting for both direct and hydrodynamic particle-particle interactions, as well as applied forces such as gravity.

How far into the region $x < 0$ (see Figure 1) can the approximate Eqs. (9), (10), (14) be used? For hard-sphere-like particle suspensions, the literature[12] seems to suggest that these relations will break down for volume fractions ϕ, see Eq. (2), somewhere between 0.2 and 0.4. There are two common applications of such relations. Firstly, the sedimentation-velocity ratio $K(c)$ can be defined:

$$K(c) \equiv f_0/f(c) \tag{15}$$

and the drag-force relation $\vec{F} = -f_0\vec{v}/K(c)$ probed directly to compare $K(c)$ values to theoretical small-c expansion approximations.[12]

Secondly, all the above relations can be combined to yield a variant of the diffusion equation, with the expression for the flux obtained by adding the external-force and gradient fluxes:

$$\frac{\partial c}{\partial t} = \vec{\nabla} \cdot \left[D(c)\vec{\nabla}c + \frac{kT}{f(c)} c\vec{\nabla}V \right] \tag{16}$$

and the generalized Einstein relation for the gradient-diffusion coefficient:

$$D(c) = \frac{1}{f(c)} \left(\frac{d\Pi}{dc} \right) \tag{17}$$

The latter relation reduces to the familiar $D_0 = kT/f_0$ in the limit $c \to 0$, where we used the ideal equation of state $\Pi = kTc$. The above relations can be used as good approximations for $c > 0$ but with volume fractions below the values at which solid-formation or glassy dynamical effects set in. The diffusion Eq. (16) requires a virial expansion for the osmotic pressure $\Pi(c)$, to derive approximations[12] valid for small positive c.

We assumed that the dynamics of the aggregating system is slow, so that the surface region of the aggregate is in equilibrium. In order to derive the boundary condition at the surface, we used the equilibrium expression $c \propto e^{-V(x)}$ for small negative x. We further assumed that both the effective potential $U(x) = kTV(x)$ felt by a particle and the concentration near $x = -\ell$ do not depend on the concentration profile outside, at $x > 0$, but rather are determined by the structure of the surface of the aggregate. Let us point our that particle-particle (singlet-singlet) interaction effects are accounted for in two different ways in phenomenological relations like Eq. (16). Firstly, short-range interactions are included in thermodynamic properties, such as $\Pi(c)$, or dynamical properties for slow, nearly-steady-state dynamics, such as $K(c)$. Secondly, longer-range interactions contribute to the effective potential $U(\vec{R})$, which also includes the external forces. Avoiding double-counting is therefore a challenge in setting up approximations of this sort.[12–14]

We note that for the modified diffusion-like description given by the equations summarized earlier in this section, the equilibrium assumption of slow dynamics for $x < 0$ can be expressed by the approximate flux balance, equivalent to the force balance $\vec{F}_{\text{grad}} = \vec{F}_{\text{appl}}$, which yields the approximate equation for the (time-independent) concentration, assuming the planar symmetry:

$$\frac{1}{c} \left(\frac{d\Pi}{dc} \right) \frac{dc}{dx} = kT \frac{dV}{dx} \tag{18}$$

In the dilute limit, with $\Pi = kTc$, this relation obviously integrates to $c \propto e^{-V(x)}$. Otherwise, numerical integration will be needed, with a virial expansion for $\Pi(c)$ used for obtaining approximations for small but nonzero c. The latter depend strongly on the details of the particle-particle interactions.

However, no matter the specifics, the modification represented by Eq. (18) does not actually change our basic conclusion. Indeed, it is obvious from its form that Eq. (18) can be integrated from some reference negative x value, $-\ell$, to $x = 0$, and while the expression for β in Eqs. (6)-(7) may change, the form of the boundary condition will not: it will still be without the derivative terms entering. Thus, the phenomenological boundary condition at the aggregate surface remains $c(0) = \beta$.

The authors gratefully acknowledge helpful discussions with Drs. H. L. Frisch, D. Mozyrsky and Y. Shnidman.

REFERENCES

1. M. v. Smoluchowski, Versuch einer mathematischen Theorie der Koagulationskinetik kolloider Lösungen, *Z. Phys. Chem.* 29:129 (1917).
2. F.C. Collins and G.E. Kimball, Diffusion-controlled reaction rates, *J. Coll. Sci.* 4:425 (1949).
3. R.M. Noyes, Effects of diffusion rates on chemical kinetics, *Prog. React. Kin.* 1:129 (1961).
4. S.L. Zimmer and B.E. Dahneke, Resuspension of particles: The range of validity of the quasi-stationary theories, *J. Coll. Int. Sci.* 54:329 (1976).
5. N. Agmon, Diffusion with back reaction, *J. Chem. Phys.* 81:2811 (1984).
6. G.H. Weiss, Overview of theoretical models of reaction rates, *J. Stat. Phys.* 42:3 (1986), and references therein.
7. N. Kallay, E. Barouch and E. Matijević, Diffusional detachment of colloidal particles from solid/solution interfaces, *Adv. Coll. Int. Sci.* 27:1 (1987).
8. V. Privman, H.L. Frisch, N. Ryde and E. Matijević, Particle adhesion in model systems. Part 13. – Theory of multilayer deposition, *J. Chem. Soc. Faraday Trans.* 87:1371 (1991), and references therein.
9. V. Privman, D.V. Goia, J. Park and E. Matijević, Mechanism of formation of monodispersed colloids by aggregation of nanosize precursors, *J. Coll. Int. Sci.* 213:36 (1999), and references therein.
10. D. Mozyrsky and V. Privman, Diffusional growth of colloids, *J. Chem. Phys.* 110:9254 (1999).
11. Z. Tavassoli and G.J. Rodgers, Diffusive growth of a single droplet with three different boundary conditions, *Eur. Phys. J.* B14:139 (2000).
12. S. Bhattacharjee, A.S. Kim and M. Elimelech, Concentration polarization of interacting solute particles in cross-flow membrane filtration, *J. Coll. Int. Sci.* 212:81 (1999), and references therein.
13. J.K.G. Dhont, *An Introduction to Dynamics of Colloids*, Elsevier, Amsterdam (1996).
14. T.G.M. van de Ven, *Colloidal Hydrodynamics*, Academic, London (1999).

SIMULATION OF SEDIMENTATION OF ATOMS IN A BINARY ALLOY IN STRONG GRAVITY

Masao Ono and Tsutomu Mashimo

Shock Wave and Condensed Matter Research Center, Kumamoto University, 2-39-1 Kurokami, Kumamoto 860-8555, Japan

SUMMARY

We obtained atomic-scale graded structures in a solid Bi-Sb (70:30 atom %) alloy by sedimentation of atoms in a mega-gravity field. Simulation of sedimentation in this system was performed using a self-consistent theory. Comparison of experimental results with this simulation suggests that the coefficient for sedimentation is remarkably larger than that for conventional diffusion.

INTRODUCTION

Microgravity has been much used in materials processing research. By comparison, little attention has been given to material science at high gravity, particularly with a very high acceleration. For example, until our work there had been reported only three experiments on sedimentation in solids. Small concentration changes were observed for Au as an interstitial impurity in K, In and Pb.[1,2,3]

To study the sedimentation of atoms or crystal-chemical instability in solids under a strong acceleration field, we developed an ultra-centrifuge apparatus that can generate an acceleration of over one-million *g* for a long time at elevated temperature.[4] Using this apparatus, we obtained atomic-scale graded structures in a Bi-Sb alloy by sedimentation of the constituent atoms.[5] Note that Bi and Sb are completely miscible in the solid phase, with all atoms in substitutional positions.

Here, we simulated sedimentation in the Bi-Sb (70:30 at.%) alloy based on a self-consistent theory[6,7] in both the infinite and finite-energy ranges.

SIMULATION PROCEDURE

We used a self-consistent diffusion equation[6,7] to simulate sedimentation of atoms. The basic self-consistent diffusion equation in a multi-component system,[7] which

Processing by Centrifugation, Edited by Regel and Wilcox
Kluwer Academic/Plenum Publishers, New York, 2001

corresponds to Fick's first law, is:

$$J_i' = -\sum_l^s \left(D_{il1}' \nabla C_l\right) + \sum_j^s \frac{D_{il2}'}{RT} g\left(M_j - M_j^*\right) C_i \tag{1}$$

where J' is the flux relative to the lattice and D_{il1}' and D_{il2}' are the diffusion coefficients for concentration gradients and for sedimentation, respectively. These can be expressed as:

$$D_{il1}' = \sum_l^s \left(L_{ij}' \frac{\partial \mu}{\partial C_l} \right), \quad D_{il2}' = L_{ij}' \frac{kT}{C_i} \tag{2}$$

where C is the concentration of atoms, M_j is the atomic weight of solute j, M^*_j is the effective mass of the surrounding solvent mixture per molar volume of j, R is the gas constant, T is the absolute temperature, L_{ij}' are the phenomenological coefficients, and μ is the chemical potential of j. The self-consistent diffusion equation in a multi-component system,[7] which corresponds to Fick's second law, is:

$$\frac{\partial C_i}{\partial t} = \sum_l^s \left(D_{il1}' \frac{\partial^2 C_l}{\partial r^2} \right) + \sum_l^s \left(D_{il1}' \frac{1}{r} \frac{\partial C_l}{\partial r} \right) + \sum_l^s \sum_n^s \left(\frac{\partial D_{il1}'}{\partial C_n} \frac{\partial C_n}{\partial r} \frac{\partial C_l}{\partial r} \right) - \frac{r\omega^2}{RT} \left[\sum_j^s D_{ij2}' \left(M_j - M_j^*\right) \right] \frac{\partial C_i}{\partial r}$$

$$- \frac{r\omega^2}{RT} \left\{ \sum_j^s \left[\sum_n^s \left(\frac{\partial D_{ij2}'}{\partial C_n} \frac{\partial C_n}{\partial r} \left(M_j - M_j^*\right) \right) - D_{ij2}' \frac{\partial M_j^*}{\partial r} + \frac{2}{r} D_{ij2}' \left(M_j - M_j^*\right) \right] \right\} C_i \tag{3}$$

where r is the radius and ω is the angular speed. By neglecting the non-diagonal elements in the diffusion coefficient matrix of Eq. (3) and putting $Q = D_1 / D_2 = 1 + \partial \ln \gamma / \partial \ln C$ and $D = D_1 = QD_2$, the diffusion equation for a two-component system can be written as:

$$\frac{1}{D} \frac{\partial C}{\partial t} = \frac{\partial^2 C}{\partial r^2} + \left(\frac{1}{r} - \frac{r\omega^2}{QRT} \left[\left(M - M^*\right) - C \cdot \frac{\partial M^*}{\partial C} \right] \right) \frac{\partial C}{\partial r} - \frac{2\omega^2}{QRT} \left(M - M^*\right) C \tag{4}$$

where γ is the activity coefficient. We created a FORTRAN computer program using the one-dimensional finite difference equation derived from Eq. (4) to simulate the sedimentation process in a binary system. For simulation with a finite radius, we used the no-flux boundary condition for both solute and solvent atoms at the two ends of the sample.

SIMULATION RESULTS FOR Bi-Sb (70:30 at%)

Infinite Radius

Figure 1 shows the simulation results for sedimentation of atoms in Bi-Sb with an infinite energy range and an ideal system with $\gamma = 1$, i.e., $Q = 1$. Here, C/C_0 for the vertical axis is the reduced concentration of atoms, and $E = (M_{Bi}-M_{Sb})r^2\omega^2/2RT$ for the horizontal axis is the reduced difference in potential energy between Bi atoms and Sb atoms, where M_{Bi} and M_{Sb} are the atomic weights of Bi and Sb. Basically, this figure shows the evolution of the concentration profiles of the two components, from the uniform starting composition of 70% Bi and 30% Sb to the equilibrium distribution. The ratio of the atomic volumes ($\alpha = V_{Bi}/V_{Sb}$) is 1.44 in this simulation. The calculated concentration profiles change gradually with time. Heavy Bi atoms move in the direction of gravity, and light Sb atoms move in the opposite direction. The Bi concentration first begins to change appreciably at around $E = 1$, where the energy per mol of the external field counterbalances the thermal energy for one-degree of freedom (RT for condensed matter). Therefore, the maximum of the concentration gradient of Sb may be caused by the stagnation of Sb atoms due to the slow diffusion velocity in the small gravity region. Finally, the concentration profiles converge to the respective equilibrium theoretical solutions. This convergence shows that the simulation was successful.

Finite Radius

Figure 2 shows simulation results for sedimentation of the same system in the finite energy range with $r = 1.52\text{-}1.86$ cm, $Q = 1$ (top) and 1/20 (bottom), and $\alpha = 1.44$.

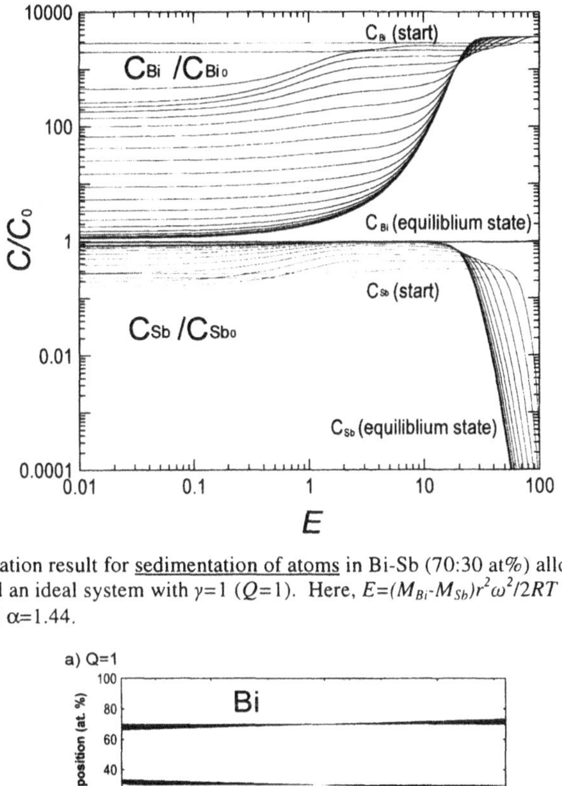

Figure 1. Simulation result for sedimentation of atoms in Bi-Sb (70:30 at%) alloy for an infinite energy range and an ideal system with $\gamma=1$ ($Q=1$). Here, $E=(M_{Bi}\text{-}M_{Sb})r^2\omega^2/2RT$ where $M_{Bi}=208.98$, $M_{Sb}=121.75$, and $\alpha=1.44$.

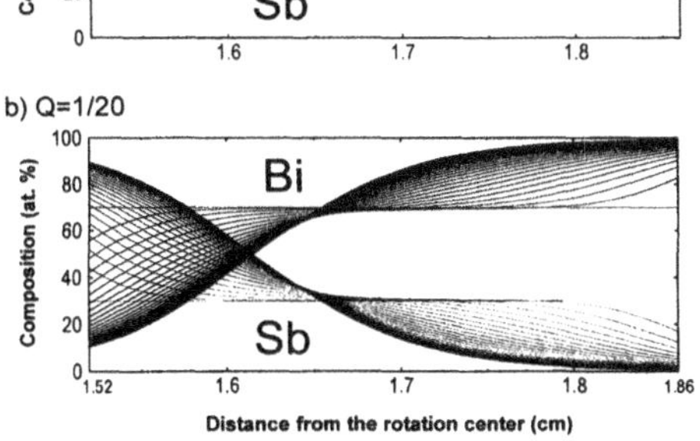

Figure 2. Simulation results for sedimentation of atoms in the Bi-Sb (70:30 at%) system in the finite energy range at $Q=1$ (a) and 1/20 (b). (Simulation conditions: rotation rate = 205,000 rpm; temperature = 503K, and $\alpha = 1.44$)

In Fig. 2 the vertical axis is the composition in atom% and the horizontal axis is distance from the center of rotation. The calculated composition profiles change gradually with time, with the atom % of Bi increasing gradually in the direction of the centrifugal force, and that of Sb increasing in the opposite direction. The composition changes of both elements occur most rapidly near the ends of the sample. Note that at the ends the flux due to the concentration gradient is exactly counterbalanced by that due to sedimentation in order to satisfy the no-flux boundary condition.

Comparison with Experimental Result

Figure 3 shows the experimental composition profiles for this alloy measured by EPMA and X-ray diffraction. The maximum acceleration had been (0.79-0.96)x10^6g. Note that near the outer edge the composition of Bi approached 100% while the composition of Sb approached 0. The simulation in Fig. 2 for Q = 1/20 is similar to this experimental result.

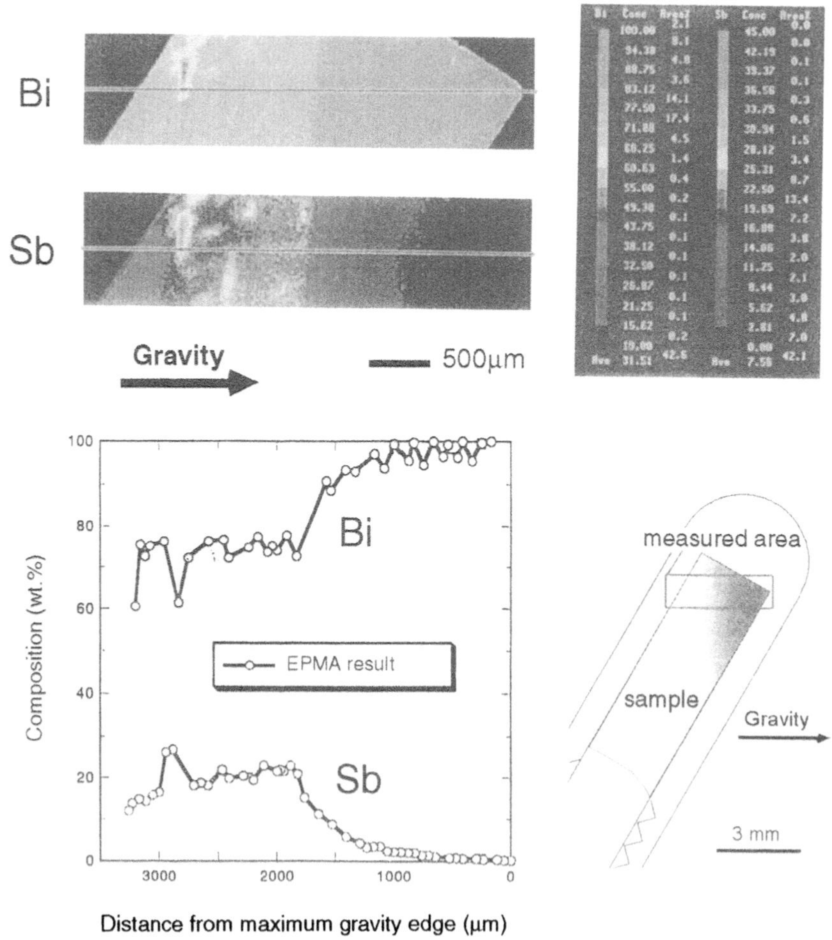

Figure 3. The experimental composition profiles measured by EPMA. (Experimental conditions: rotation rate = 195,000-215,000 rpm, temperature = 493-513 K, duration = 85 h)

Comparison of experimental results with the theoretical simulation suggests that the atomic mobility for sedimentation is remarkably larger than that for conventional diffusion due to a concentration gradient. Since the solute atoms are substitutional, this indicates that the mechanism for sedimentation in this system may differ from the usual vacancy mechanism for diffusion. Note also that the measured concentration change in the lower gravity region was small, although the calculated concentration change was large in this region. This difference may have been caused by the assumption in the simulation that the diffusion coefficients are always constant, whereas the diffusion coefficients may vary with composition or gravity amplitude in this system.

Acknowledgements

The authors would like to thank Prof. K. Nagayama of Kyushu University for useful discussions.

REFERENCES

1. L.W. Barr and F.A. Smith, *Phil. Mag.* 20: 1293-1294 (1969).
2. T.R. Anthony, *Act. Met.* 18: 877-880 (1970).
3. S.J.C. Rushbrook and L.W. Barr: J. Nucl. Materials 68/78: 556-558 (1978).
4. T. Mashimo, S. Okazaki, and S. Shibazaki, Rev. Sci. Instr. 67: 3170-3174 (1996).
5. T. Mashimo, S. Okazaki, and S. Tashiro, Jpn. J. Appl. Phys. 36: 498-500 (1997).
6. T. Mashimo, Phys. Rev. A38: 4149-4154 (1988).
7. T. Mashimo, Phil. Mag. A70: 739-760 (1994).

MEGA-GRAVITY EXPERIMENT ON In-Pb ALLOY

Tsutomu Mashimo, Masao Ono, Takahiro Kinoshita and Hidekazu Kanegae

Shock Wave and Condensed Matter Research Center, Kumamoto University, 2-39-1 Kurokami, Kumamoto 860-8555, Japan.

SUMMARY

A very strong gravity field can cause the sedimentation of atoms even in solids, and may be expected to create a nonequilibrium crystal-chemical state. A mega-gravity field experiment was performed on an In-Pb (80:20%mol) alloy, which forms a three-phase system, to study the sedimentation of atoms and the interaction with phase equilibria. This ultracentrifuge experiment was carried out using a turbine-motor type machine at a rotation rate of 205,000-221,000 rpm and a temperature of 150-160°C for 100 hours. The maximum acceleration was $(0.87\text{-}1.02)\times10^6 g$ at 18.6mm radius. A graded structure was observed on a cut surface of the ultracentrifuged sample by an electron-probe micro-analyzer. It was confirmed by X-ray diffraction that phase transformations occurred under the strong gravity field.

INTRODUCTION

The sedimentation of atoms in condensed matter can be used to control composition based on differences in atomic weight and volume. The sedimentation of component atoms may also form graded structures on the atomic scale in solids. However, sedimentation of atoms should be exceptional in solids because the atomic mobilities are very small. Thus, no sedimentation of component atoms in solid solutions or compounds has been reported, while sedimentation of macro-particles in liquids is commonly used in biochemistry.[e.g.,1,2] There have been only a few experiments[3-5] investigating the sedimentation of Au isotopes in elemental metals with low melting temperatures (K, In, Pb) at a maximum acceleration of $(1\text{-}2)\times10^5 g$. In these studies, the concentrations of Au isotope were, however, at an impurity level, and the resulting concentration changes were small even though these are known as "ultra-fast-diffusion systems" with Au being an interstitial solute. The main reason for the difficulty of such experiments on solids has been that there has been no ultracentrifuge machine that simultaneously generates on the order of 1 million *g* and operates at high temperature. We have developed an ultracentrifuge apparatus that can generate a strong maximum gravity of even 1 million *g* at temperatures of several 100°C

Processing by Centrifugation, Edited by Regel and Wilcox
Kluwer Academic/Plenum Publishers, New York, 2001

for several days.[6] We succeeded in the formation of an atomic-scale graded structure by the sedimentation of substitutional atoms in a Bi-Sb alloy (70:30mol%), which is completely miscible in all proportions.[7]

In the present study, we examined a miscible three-phase system that includes an intermediate phase -- In-Pb alloy (80:20mol%). We performed a mega-gravity experiment on this alloy. The graded structure and crystalline state were investigated by electron-probe micro-analysis (EPMA) and X-ray diffraction (XRD) in order to determine the sedimentation of atoms and the interaction with phase equilibria.

MEGA-GRAVITY EXPERIMENT

For this experiment, we used a turbine motor-type ultracentrifuge apparatus.[6] Figure 1 shows a schematic layout of this apparatus, which used a turbine motor with an oil-floating bearing and a specimen rotor. The turbine motor was driven by hot compressed air supplied by a screw compressor and a combustion system. The specimen capsules were contained in a rotor with an outer diameter of 46 mm made of a titanium alloy. The maximum distance from the rotor axis to the specimen was about 18.6 mm. The specimen could be heated higher than 300°C with hot air.

The In-Pb (80:20atom%) sample was prepared by melting a mixture of In and Pb powder in a capsule at about 223°C for over 20 min under an Ar atmosphere. The powder mixture had been milled before melting to make a more uniform alloy. Both the In and the Pb were 99.99% pure. A titanium alloy capsule with an inner diameter of 3 mm and a depth of about 10 mm was used. It was confirmed that the starting sample was single phase of face-centered tetragonal structure (α-phase).

The ultracentrifuge experiment was performed at a rotation rate of 205,000-221,000 rpm and a temperature of 150-160°C for 100 h. The maximum acceleration field at a 18.6 mm radius was $(0.87\text{-}1.02)\times10^6$ *g*. The composition analysis was carried out using an EPMA apparatus, the JXA-8900R (JEOL Ltd., Tokyo, Japan). Micro-area XRD analysis was carried out using the RINT2500VHF (Rigaku Ltd., Tokyo, Japan).

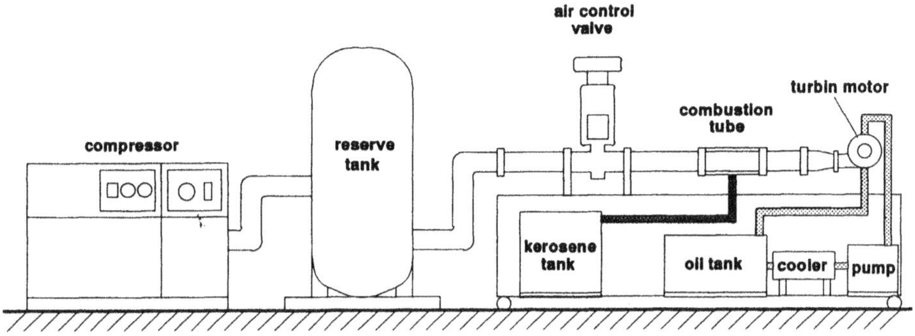

Figure 1. Illustration of the ultracentrifuge apparatus developed at Kumamoto University.

RESULTS

Figure 2 shows an EPMA mapping photograph of In and Pb at a polished surface cut at a plane containing the rotation axis of the ultracentrifuged specimen. The linear composition profiles were obtained along a straight line in the acceleration direction. The Pb composition rate continuously increased in the acceleration direction (arrow direction in the figure) from about 12% to 36%atom, while the In content decreased from about 88% to less than 66%. The composition was homogeneous on a sub-micrometer scale. The slight jaggedness of the composition profiles may have been caused by roughness of the specimen surface.

From the scanning electron microscope image shown in Fig. 2, a clear boundary can be observed at about 0.6 mm from the left edge (weak gravity area) of the capsule. In addition, a very diffused boundary was visible to the naked eye around the center of the sample. The specimen surface can be divided into three regions (areas A, B and C). Figure 3 shows the micro-area XRD patterns (100 μm diameters) at two representative points in the areas A and C (Fig. 2), together with that of the starting sample. The crystal structures in areas A, B and C were tetragonal, face-centered tetragonal (FCT) and face-centered cubic (FCC), respectively. This suggests that the tetragonal phase and the FCC phase appeared at the low- and high-gravity regions, respectively, from the starting FCT phase. In addition, it was found that the lattice parameters changed in each phase. As a result, we concluded that the structure was graded at the atomic scale, and was formed by sedimentation of substitutional atoms.

DISCUSSION AND CONCLUSIONS

In this study, a mega-gravity experiment was performed on an In-Pb alloy. A graded structure was produced with phase transformations. These results are attributed to the sedimentation of atoms.

In previous sedimentation studies on interstitial Au,[3-5] the concentrations of Au were very low so that the concentration changes were small. With sufficiently high acceleration, substitutional solutes also should undergo sedimentation, as in the Bi-Sb system[7] and also the present system. Recently, we performed a simulation of the sedimentation process in these systems by using a self-consistent theory[8,9] for the mechanism. The result indicated that the mobilities for sedimentation were much larger than for conventional molecular diffusion, even for the substitutional solutes in these alloys.

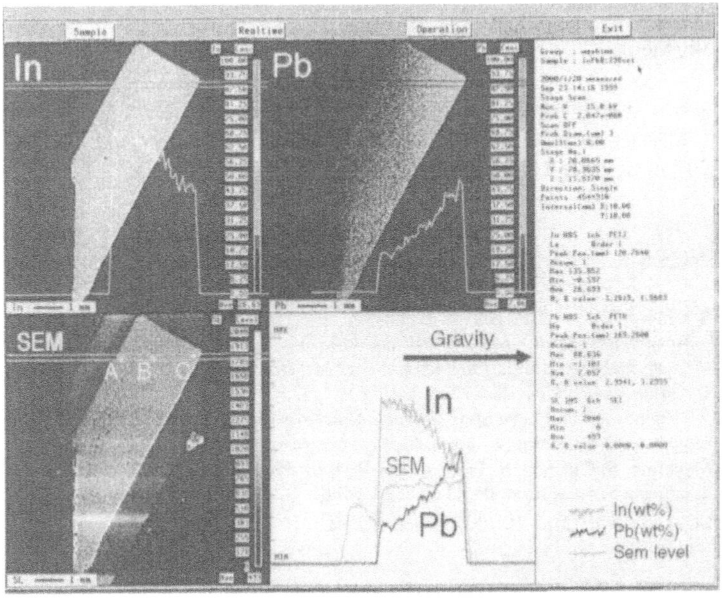

Figure 2. EPMA mapping photograph showing the composition profiles of In and Pb at a polished surface cut from an ultracentrifuged In-Pb alloy sample.

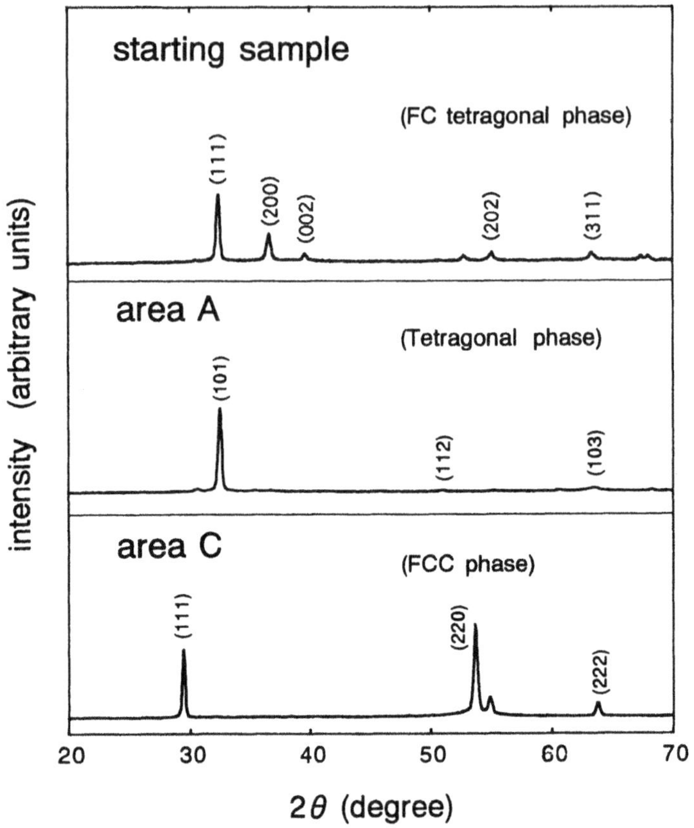

Figure 3. Micro-area XRD patterns of the starting sample and areas A and C (Fig. 2) in the ultracentrifuged sample.

Acknowledgments

This work was partially supported by a Grant-in-Aid from the Japanese Ministry of Education, Science and Culture. This work was partially supported also by a Grant-in-Aid from the Nissan Science Foundation.

REFERENCES

1. T. Svedberg, *Nature* 139: 1051-1062 (1937).
2. J.W. Beams, *Rev. Mod. Phys.* 10: 245-263 (1938).
3. L.W. Barr, and F.A. Smith, *Phil. Mag.* 20: 1293-1294 (1969).
4. T.R. Anthony, *Act. Met.* 18: 877-880 (1970).
5. S.J.C. Rushbrook and L.W. Barr, *J. Nucl. Materials* 68/78: 556-558 (1978).
6. T. Mashimo, S. Okazaki, S. Shibazaki, *Rev. Sci. Instr.* 67: 3170-3174 (1996).
7. T. Mashimo, S. Okazaki, S. Tashiro, *Jpn. J. Appl. Phys.* 36: L498-500 (1997).
8. T. Mashimo, *Phys. Rev.* A 38: 4149-4154 (1988).
9. T. Mashimo, *Phil. Mag.* A 70: 739-960 (1994).

MOLECULAR GRADATION IN GLASS BY USING HIGH GRAVITY: A NOVEL METHOD FOR Si-Ti GRADED-GLASS PREPARATION

Masamichi Nishihara,[1] Hirotaka Ihara,[1] Takashi Sagawa,[2]
Rika Nishimura,[1] Susumu Matsuzaki[3] and Tsutomu Mashimo[4]

[1]Graduate School of Science and Technology
[2]College of Engineering
[3]College of Science
[4]Shock Wave and Condensed Matter Research Center
Kumamoto University, 2-39-1 Kurokami, Kumamoto 860-8555, Japan

SUMMARY

We investigated the preparation of functional graded glass by using the sedimentation of colloid particles under a high gravity field in combination with a sol-gel method. The precursor gel of the graded glass was prepared by centrifuging Si and Ti sols during the conversion from sol to gel. The $SiO_2 - TiO_2$ graded glass was obtained by heating the gel. The optimal conditions for the preparation of the graded glass were determined by changing the times of three steps: homopolymerization, mixing and copolymerization. By lengthening the homopolymerization time and centrifuging quickly after mixing Si and Ti sols, $SiO_2 - TiO_2$ system graded glass could be obtained. The resulting glass had a graded structure at the molecular scale. Efforts are necessary to prevent cracks. We will prepare graded glass containing transition elements and rare earth elements. These are expected to show useful optical properties.

INTRODUCTION

Atomic-scale graded structures have been formed in metal alloys by the sedimentation of atoms under a very high gravity field.[1] It is expected that sedimentation of colloid particles can easily occur in a sol even with a comparatively weak gravity field due to the large molecular weight. By using the sol-gel method, we can prepare glass at a comparatively low temperature.[2] Functional graded glass is expected to show useful optical properties.

The objective of this project was to prepare $SiO_2 - TiO_2$ graded glass by using sedimentation of colloid particles at high gravity in combination with a sol-gel method. The sol mixture consisted of Si and Ti alkoxides and was centrifuged during the conversion from sol to gel. This process formed a graded structure on the molecular scale due to the difference in the rate of sedimentation. The $SiO_2 - TiO_2$ graded glass was prepared by heating the precursor gel.

EXPERIMENTAL

Figure 1 is a schematic diagram of the method used to prepare functional graded glass. To start with, 3.88g (0.0186 mol) of tetraethoxysilane and 6.3 ml of ethanol were mixed and stirred in a vial for an hour. An acid catalyst solution (0.71 ml of water, 6.3 ml of ethanol, 4.5 ml of 99.7% acetic acid and 1.36 g of DMF) were added with stirring to the Si alkoxide for hydrolysis and polymerization. Acetic acid was used in the catalyst to delay the hydrolysis while DMF was added to inhibit crack formation. In another vial, 1.04 g of tetraisopropoxytitan and 1.7 ml of ethanol were mixed and stirred for an hour. An acid catalyst solution (0.19 ml water, 1.7 ml ethanol, 1.2 ml acetic acid, 0.26 g DMF) was added with stirring to the Ti alkoxide. Each sol was kept at 35°C for homopolymerization for time t_A. Then the Ti solution was added to the Si solution and stirred for time t_B. The molar ratios of Si and Ti were 85:15 mol% in this study. The sol mixture was kept at 35°C for copolymerization for time t_C.[3]

The above sol mixture was centrifuged at an acceleration of $(1\text{-}2)\times10^4$ g for 50 hours using an ultracentrifuge, model H-200R from Kokusan Ltd. Because of the ultracentrifuge construction, the apparatus had to be stopped for 30 minutes after every 30 minutes of operation. The temperature of the sample was kept at 40-50°C during centrifugation. At this temperature, Si and Ti alkoxides can be mildly polycondensed to prevent the formation of crystals of SiO_2 and TiO_2. A white wet gel was obtained at the bottom of the centrifuge tube. Water and ethanol were removed from this by evaporation to obtain a transparent dry gel.

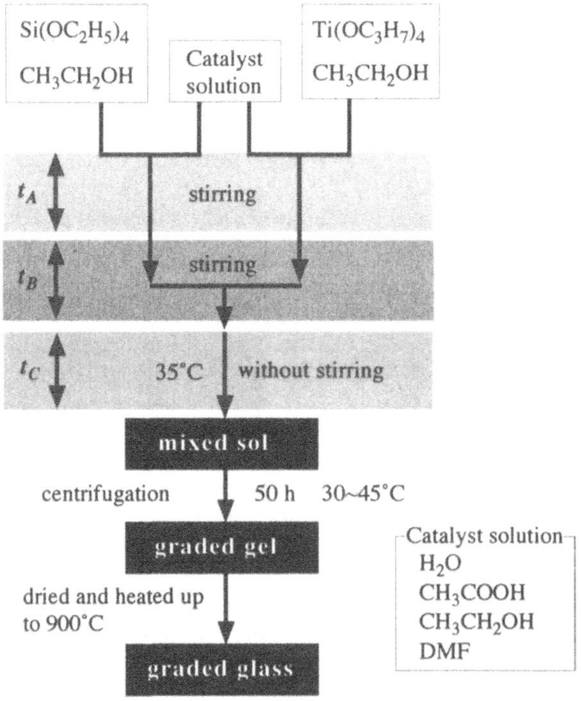

Figure 1. Schematic diagram of sol-gel process used to make graded glass.

Glasses of $SiO_2 - TiO_2$ were prepared by heating the above gel to about 900°C. These glasses were fixed by a polyester resin parallel to the direction of acceleration. These were cut in half and polished with sandpaper and diamond paste for characterization by electron probe micro analyzer (EPMA) and X-ray diffraction (XRD) methods.

RESULTS AND DISCUSSION

We expected that the formation of a graded structure could be controlled by changing the degree of copolymerization and centrifugation. Initially, the Si and Ti concentrations in the glass prepared with various copolymerization times were uniform throughout. Consequently, the homopolymerization time t_A was lengthened with the expectation that the colloidal particles of Si and Ti would grow and sediment at an acceleration of $1 x 10^4 g$. However, the composition remained uniform. Probably the colloids could not appreciably sediment because crosslinked structures formed with the long copolymerization time (sum of mixing time and copolymerization time). Thus, we shortened the mixing time to 0.25 h and eliminated the copolymerization step. We needed at least 15 min to mix the Si and Ti alkoxides solutions, although we desired to shorten the mixing time as much as possible. Then experiments were performed with different homopolymerization times. The EPMA result at t_A = 13 h is shown in Fig. 2. The concentration of Ti continuously increased in the direction of the acceleration, and that of Si component increased in the opposite direction. Some cracks were observed in the glass.

We confirmed by XRD that samples obtained as above were amorphous and did not contain SiO_2 or TiO_2 crystals. We concluded that the structure of this glass was graded at the molecular scale.

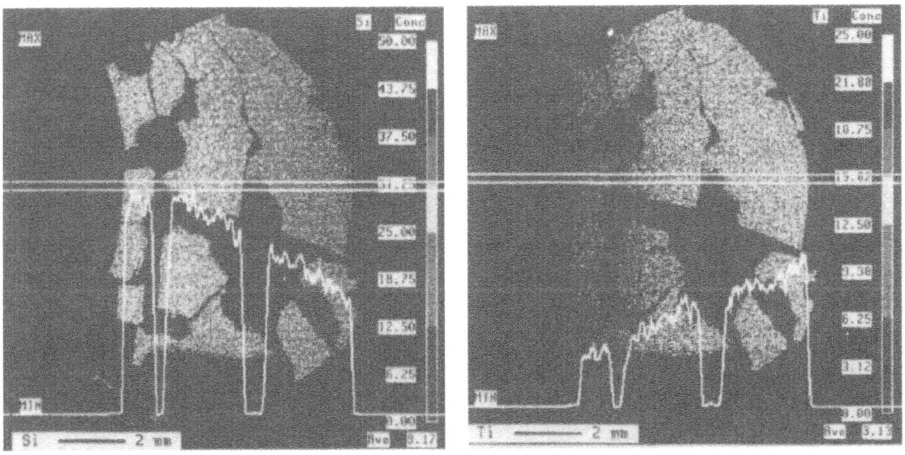

Figure 2. Electron probe micro analysis (EPMA).

REFERENCES

1. T. Mashimo, S. Okazaki, and S. Tashiro, *Jpn. J. App. Phys.* 36: L498-L500 (1997).
2. K. Kamiya and S. Sakka, *J. Mater. Sci.* 15: 2937-2939 (1980).
3. S. Sakka, *Sol-gel Hou no Kagaku* (in Japanese), Agune syohusya (1988).

HIGH - VELOCITY SEDIMENTATION TO DETERMINE CONFORMATIONAL AND HYDRODYNAMIC PARAMETERS OF NITROCELLULOSE MACROMOLECULES

V. A. Malchevsky

Russian Chemical Technological University, Moscow, Russia

INTRODUCTION AND METHODS

The hydrodynamic parameters of linear-polymer macromolecules in solutions, namely the constants for sedimentation S_0, diffusion D_0 and intrinsic viscosity $[\eta]$, are related to their molecular mass M over a wide range of values by:[1-3]

$$S_0 = K_S M^{1-\beta} \tag{1}$$

$$D_0 = K_D M^{1-\beta} \tag{2}$$

$$[\eta] = K_\eta M^{\alpha} \tag{3}$$

where K_S, K_D, K_η, α and β are constants. The correct use of Eqs. (1–3) to determine the molecular masses of cellulose nitro ethers (NC) requires the specifics of the equilibrium rigidity changes of their macromolecules in solutions. The equilibrium rigidity of the NC chain is highly dependent on not only M but also on the substitution degree γ of hydroxyl groups,[1] i.e. the nitrogen content N in the macromolecule: $\gamma = 3.6\ N/(31.1 - N)$, where γ=0–3 and N = 0–14.14%. Especially notable is the impact of γ on $[\eta]$ and the rotational friction coefficient as they are more sensitive to the molecular chain conformation than S_0 and the translational friction coefficient.[2] To exclude those effects while determining M, it is necessary to have specimens with a narrow molecular weight distribution and the same degree of substitution. It is quite challenging to implement this condition. The M for NC with a range of degrees of substitution is normally estimated using empirical equations to reduce $[\eta]$ to that for a standard N value of 13.6%,[3] 13.8%[4] or 14.4%[5] as well as for the constants K_η and α determined for standard N.

The research results obtained for the effect of the NC substitution degree on S_0 are inconsistent. There are data indicating that S_0 does not depend on N.[6] Yet it was shown that increasing N from 10.7% to 13.4% increased the molecular-chain equilibrium rigidity by 2.5 times.[1]

Based on the literature and on new experiments, we attempt here to relate the

Processing by Centrifugation, Edited by Regel and Wilcox
Kluwer Academic/Plenum Publishers, New York, 2001

sedimentation and diffusion constants to the NC conformation parameters and substitution degree. The main experiments were performed on a MOM 3180 ultracentrifuge (Hungary) at $n = 40 \cdot 10^3$ rev/min and a maximum acceleration $a \sim 180 \cdot 10^3 g$ at $R_{max} \sim 6.5$ cm. The refractive-index gradient d_n/d_x was measured using Filpot-Swenson's scheme with a phase-contrast plate. The registration angle was $3 - 15^0$ depending on the solution concentration. Standard wood pulp and cotton NC specimens were examined. Because the sedimentation constants for NC solutions in acetone are shown for $T = 20^0C$ in the literature, our analysis of the overall results was for that temperature. We measured sedimentation constants for NC solutions in ethyl acetate at different temperatures and, for joint analysis, converted the results to 25^0C by $S_{25} = S_T \eta_{oT} (1-V_{25}\rho_{o25})/ \eta_{o25}(1-V_T\rho_{oT})$, where S_T is the sedimentation constant at temperature T, η_{oT} is the solvent dynamic viscosity at T, ρ_{OT} is the solvent density at T, and V_T is the specific partial volume of NC at T.

EXPERIMENTAL RESULTS AND DISCUSSION

For the NC-acetone system at 293 K, we used two groups of specimens with nitrogen contents of $13.8 \pm 0.2\%$[6-10] and $12.2 \pm 0.4\%$.[6,7,11] Absolute M values were calculated from sedimentation diffusion experiments and by Svedberg's equation,[12] $M=RTS / D(1-V\rho)$, where D is the translational diffusion coefficient, R gas constant, T absolute temperature, S sedimentation constant, V partial specific volume occupied by 1 g of dilute substance, and ρ solvent density.

There were seven NC specimens with N = 12.2%. For those specimens, based on the sedimentation constant and $[\eta]$ measurement experiments we calculated mean hydrodynamic molecular masses $M_{s\eta}$ by Flory-Madelkern's invariant:[13]

$$\Phi_\infty^{1/3}P_\infty^{-1} = N_a\eta_0(1-V_{\rho 0})^{-1}M_{s\eta}^{-2/3}[\eta]^{1/3}S_0 \tag{4}$$

where Φ_∞ and P_∞ are Flory's constants for rotational and translational friction, respectively, $1-V_{\rho o}$ is the buoyancy factor measured by a pycnometer, η_0 is the solvent dynamic viscosity, and N_a is Avogadro number. Equation (4) had been obtained for a Gaussian coiled molecule flowing through a capillary. The theory of rotational[14] and translational friction[15] of a persistent chain showed that $F_\infty R_\infty$ could be invariant for a molecular coil flowing through a capillary, even when F_∞ and R_∞ taken separately did not remain constant when M was varied. For cellulose macromolecules and some of intermediates the experimental value of $F_\infty^{1/3} R_\infty^{-1}$ was very close to its theoretical value of $1.29 \cdot 10^7$ mol$^{-1/3}$.[16] Therefore, taking into account the insignificant polydispersity of the examined specimens, ($M_\infty/M_n = 1.2 - 1.4$) the $M_{S\eta}$ values were used along with absolute M_s.

Figure 1 shows the S_0-M relationship obtained for NC with the two degrees of substitution. The relationship for N = 12.2% was linear in the range of M from 0.16 to $2.7 \cdot 10^5$ g/mol, and can be described as:

$$S_o = (0.28\pm0.08) \cdot 10^{-13}M^{(0.33\pm0.03)} \tag{5}$$

For specimens with N = 13.8%, the range for M, $0.13–25 \cdot 10^5$ g/mol, was much broader, and the exponent β in Eq. (1) decreased as M was increased. A deviation from linear behavior led to the arbitrary choice of linear sections yielding the following constants:

$$S_o = (0.50\pm0.16) \cdot 10^{-13}M^{(0.28\pm0.02)} \quad \text{for } M = (0.13\text{-}4.8) \cdot 10^5 \tag{6}$$

$$S_o = (0.10\pm0.09) \cdot 10^{-13}M^{(0.4\pm0.01)} \quad \text{for } M = (4.8\text{-}25) \cdot 10^5 \tag{7}$$

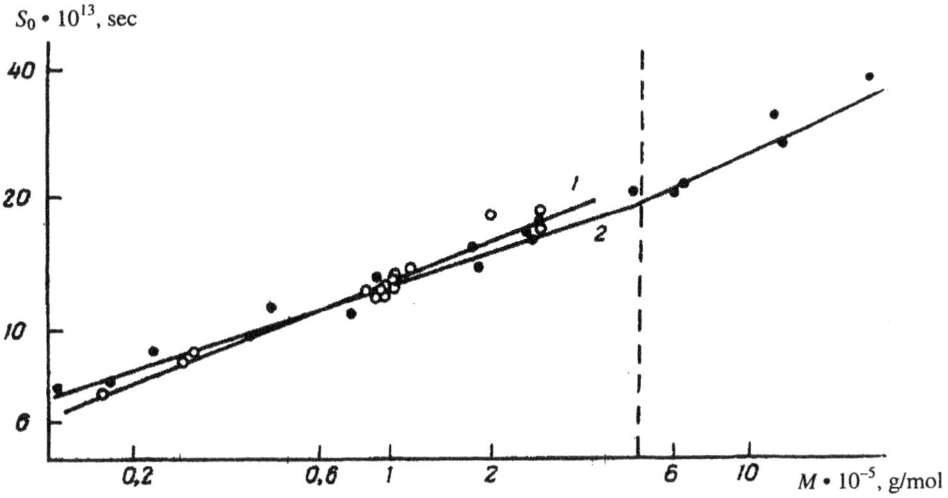

Figure 1. Experimental dependence of sedimentation constant S_o on M for NC solutions in acetone at 293 K. (1) – 12.2% of N; (2) – 13.8 % of N.

While these values of K_S and β are credible for the first section, there is high uncertainty about their values for the second section due to fewer experimental points. This linearity breach in a *log*[η] – *log*(M) plot was reported by Schulz[4] for NC with N = 13.8% over a wide range of M. The linear section boundaries identified by him coincide with those chosen here. Theories[14,15,17,18] unambiguously associate this character of *log*(S_0)–*log*(M) and *log*[η]–*log*(M) plots for rigid polymers with a decreasing flow-through ability of macromolecules as their chains become longer. In addition, the effect of excluded volume can be neglected for NC due to the high equilibrium rigidity of its molecular chains.[19,20]

As mentioned above, the nitrogen content greatly affects the conformation of NC macromolecules and, hence, knowledge of the conformational characteristics would provide information on the degree of substitution. To compute the conformational parameters it is feasible to employ a persistent chain model that describes the hydrodynamic behavior of polymer molecules with conformations varying from a curved rod to Gaussian coil. Translational friction of a persistent chain flowing through a capillary in the absence of the excluded volume effect was examined using a necklace model with closely positioned "beads" (Q = 1.431)[17] and a worm-like cylinder model (Q = 1.0561),[14] to yield the following result:

$$[So] \equiv S_0\eta_0(1\text{-}V\rho_0)^{-1} \equiv M\eta_0 D_0(RT)^{-1}$$
$$\equiv (P_\infty N_a)^{-1}(M_l/A)^{1/2}M^{1/2}+(M_l/3\pi N_{63952*1})[\mathrm{In}(A/d)\text{-}Q] \qquad (8)$$

where $[S_0]$ is the intrinsic sedimentation constant, M_L is the molecular weight M of a chain length unit, A is the length of Kauhn's statistic segment, and d is the hydrodynamic diameter of the chain.

We used Eq. (8) to evaluate the equilibrium rigidity of the NC chain with different substitution degrees. Figure 2 shows the results. The experimental points are positioned well on straight lines with correlation coefficients 0.98 – 0.99. Table 1 compares conformational parameters calculated by Eq. (8) with those from the literature. Figure 3 shows values of A determined for 8 cotton NC specimens dissolved in acetone with

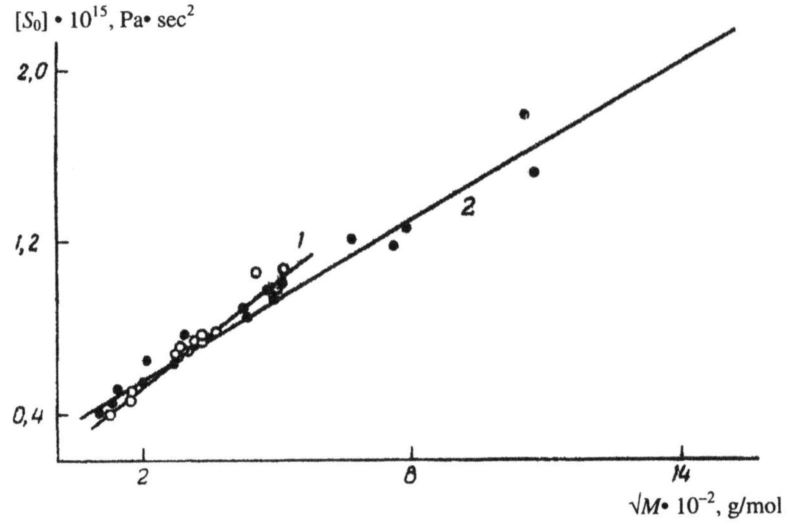

Figure 2. Dependence of intrinsic sedimentation constant $[S_0]$ on M.
(1) – 12.2% of N; (2) – 13.8 % of N.

Table 1. NC conformational parameters calculated from the data on translational friction.

V	A (Å)	Solvent	T (K)	V (10^3 m^3/kg)	d (Å)	Ref
2.870	404 ± 55	Acetone	293	0.57±0.02	5.7±0.9	-
2.322	205 ± 40	Acetone	293	0.57±0.02	5.7±0.9	-
2.724	400 ± 50	Ethyl acetate	294	0.59	12	[1,20]
2.653	300 ± 50	Ethyl acetate	293	0.59	22 ± 7	[21]
2.386	260 ± 40	Ethyl acetate	298	0.59	18 ± 5	[22]
2.291	200 ± 30	Ethyl acetate	298	0.63	7 - 10	[23]
2..039	130 ± 30	Ethyl acetate	298	0.57±0.05	10 ± 2	[24]
1.887	150 ± 30	Methylethylketone	299	0.54	14 ± 5	[25]
1.143	150 ± 30	Dimethyl acetamide+ 6%	298	0.57±0.01	12 ± 5	[22]
0	100 ± 5	Cadoxene + water	293	-	10	[26]

substitution degrees $\gamma = 2.52 - 2.48$ and $\gamma = 2.12 - 2.27$.[29] The specimens had been prepared by treating the starting NC specimens with 0.5 N KOH solution at 42^0C for 0, 30, 60 and 120min, with the initial $M_{S\eta} \approx M_W = 2.57 \cdot 10^5$ and $M_w/M_n = 1.2 - 1.35$.

Table 1 and Fig. 3 show that the present values of A are an order of magnitude higher than those typical for flexible-chain polymers, and increasing with increasing γ as shown before.[1,16] Yet, in the literature we found no discussion of this dependence over a wide range of γ. In addition, there have been widely differing views of the molecular rigidity of cellulose intermediates. Some have associated high values of A with the polyglucoside chain structural details. For instance, the high A values of cellulose nitrates have been attributed exclusively to steric hindrances due to the bulky NO_2 group in the C_2 position of the polyglucoside ring.[27]

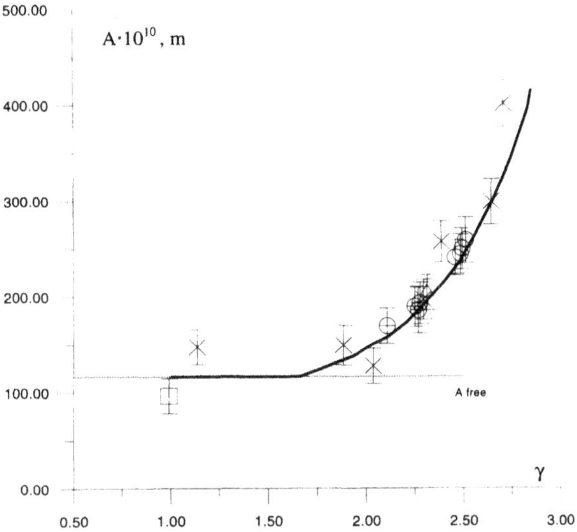

Figure 3. Dependence of Kauhn's segment value A on the substitution degree γ.

The measured molecular size of the same cellulose ether may vary dramatically in different solvents,[28] demonstrating that the effect of intramolecular hydrogen bonds should not be neglected. Evidently, local hydrogen bonds interact inside the molecular chain and thereby enhance its skeletal rigidity. Moreover, NO...O bonds add to OH...O in NC. As seen in Fig. 3, the $A = f(\gamma)$ function describes fairly well the A values obtained in various solvents. With regard to Kauhn's segment values at free rotation (A_{fr} = 120 Å),[2] it is possible to say that A only slightly exceeds A_{fr} before γ reaches 2. Consequently, we may suppose that an acute rise of A at $\gamma > 2$ should result from substitution of the third (last) OH group, which is likely to be in the C_2 position, by a more bulky NO_2 group. The A (γ) dependence in analytical form can be represented by:

$$A = 120+37.5\exp[2.54(\gamma-2)] \tag{9}$$

The value of the hydrodynamic diameter d of the NC molecular chain does not depend (within tolerable accuracy) on the substitution degree when the same solvent is used. The difference in the d values for acetone (d = 5.7Å) and ethyl acetate (d = 12Å) is likely to originate from the different solvating ability of the solvents.

We used the tabulated data and Fig. 3 to determine the dependence of S_0 and D_0 on M through Eq. (8). To analyze the effect of γ on $S_0(M)$ for NC in acetone, we chose γ values from 0 to 0.26. We defined M_L by $M_L = M_0/l_0$, where $M_0 = (162 \pm 45\ \gamma) \cdot 10^{-3}$ kg/mol is the molecular weight of the NC ring and L_0 = 5.15 Å is the contour length of the cellulose ring.[2] The model's parameters also included values computed by Eq. (9) and d = 5.7 Å (Q=1.0561). Analysis of the relationship of S_0, M and γ revealed that it was necessary to take into account γ, except for the section where $\gamma = 1.5 - 2.0$ and the contribution of γ to S_0 was comparable with the 2-5% experimental uncertainty of measured S_0. Such regularities may also apply for the other solvents listed in Table 1. The variation of d leads only to a change in the share of the flow-through effect in S_0 (the second term in the right part of Eq. (8)) and is manifested as a shift of the linear relationship between S_0 and $M^{1/2}$. Therefore, if we assume γ is within 0 – 2.89 for true solutions, it is possible to attain sufficiently accurate relationships between S_0 and D_0 and M with due regard to γ, thereby avoiding laborious experiments.

There are a few publications[1,20,23,24,30] that deal with the hydrodynamics of NC macromolecules in ethyl acetate solution. It seems reasonable to carry out a joint analysis of those results and define constants K_S in Eq. (1) as well, as they would make it possible to calculate the NC molecular mass M in the NC-ethylacetate system. Four non-fractionated cotton NC specimens with N of about 13.5 % were studied.[20] From sedimentation- diffusion measurements, M_{SD} was calculated through Svedberg's equation. The range of molecular mass changes M_{SD} was about $(0.93 - 1.5) \cdot 10^5$. Polydispersity coefficients were not cited, but, as a rule, these do not exceed $M_w/M_n \sim 1.2 - 1.4$. Statistical processing of the results[20] leads to:

$$S_0 = (1.50\,^{+1.02}_{-0.61})M_{SD}^{0.18\pm0.04} \tag{10}$$

Those results do not, however, conform to other papers. For example, the mean molecular mass M_w was determined on six NC fractions with N ~ 13.5% by sedimentation measurements and light scattering to yield $M_w \sim (0.41 - 5.73) \cdot 10^5$.[30] After processing, the results could be described by:

$$S_0 = (0.294\,^{+0.498}_{-0.185})M_W^{0.28\pm0.08} \tag{11}$$

The intrinsic viscosity, sedimentation and diffusion were measured for 23 fractions of low-molecular colloxylins with an N content of 11 – 11.5%,[1, 23, 24] 15 NC fractions with N ~ 13.3%,[1] and 26 fractions with N ~ 12.2%.[23] The molecular mass M_{SD} was calculated by Svedberg's equation and varied from $5.4 \cdot 10^3$ to $9.8 \cdot 10^5$. The results could be described by:

$$S_0 = (0.199\,^{+0.027}_{-0.024})M_{SD}^{0.3158\pm0.011} \tag{12}$$

Figure 4 shows that the results from several publications[1,23,24,30] can be expressed fairly well by:

$$S_0 = (0.201\,^{+0.0263}_{-0.0232})M^{0.3139\pm0.0106} \tag{13}$$

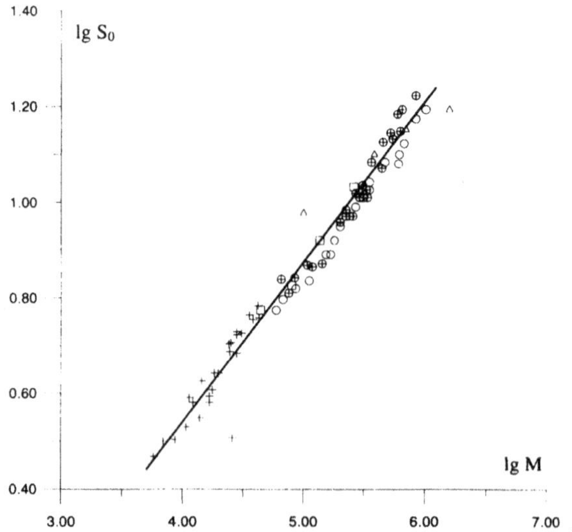

Figure 4. Dependence of intrinsic sedimentation constant S0 on M for NC solutions in ethylacetate at 298 K. △ – [30]; □ – [20]; +, O, ⊕ – [23-25].

An attempt was made to consider the influence of the substitution degree on the S_0 - M relationship in these publications through:

$$S_0 = K_S(1+\alpha N)M_{SD}^{1-b(1+\gamma N)} \tag{14}$$

where N is the nitrogen content of NC and α and γ are constants. No such influence was found for the NC-ethylacetate system. As a whole, the published results are not contradictory. For NC-ethylacetate solutions, Eq. (13) is currently the most accurate amongst all suggested relationships, as it is based on numerous reliable measurements.

Research is needed to elucidate the influence of the substitution degree of NC on the hydrodynamic parameters of its macromolecules in ethylacetate solutions.

REFERENCES

1. G.M. Pavlov, A.N. Kozlov, G.N. Marchenko, and V.N. Tsvetkov, *Visokomolek. Soed.* B. 24: 284 (1982).
2. V.N. Tsvetkov, *Zhestkotsepnie Polimerniew Molekuli*, Leningrad (1986).
3. E.L. Akim and G.L. Akim, *Тр. Ленинград. лесотехн. акад.* им. С.М.Кирова. 102: 167 (1962).
4. M. Marx-Figini and G.V. Schulz, *Makromolek. Chem.* 54: 102 (1962).
5. C. Lindsley and M.B. Frank, *Ind. Eng. Chem.* 45: 2491 (1953).
6. I. Jullander, *J.Polymer Sci.* 2: 329 (1947).
7. I. Jullander, *J.Polymer Sci.* 3: 631 (1948).
8. G. Meyerhoff, *Naturwiss.* 41: 13 (1954).
9. G. Meyerhoff, *J.Polymer Sci.* 29: 399 (1958).
10. G. Meyerhoff, *Makromolek. Chem.* 32: 249 (1959).
11. H. Mosimann, *Helv. Chim. Acta* 26: 61 (1943).
12. Th. Svedberg and K.O. Pederson, *Ultracentrifuges,* Oxford (1940).
13. L. Madelkern and P.J. Flory, *J.Chem. Phys.* 20: 212 (1952).
14. H. Yamakawa and M. Fujii, *Macromolecules* 6: 407 (1973).
15. H. Yamakawa and M. Fujii, *Macromolecules* 7: 128 (1974).
16. Ю.Е. Эйзнер and О.Б. Птицын, *Высокомолек. соед.* 4: 1725 (1962).
17. J.E. Hearst and W.H. Stockmayer, *J. Chem. Phys.* 37: 1425 (1962).
18. J.E. Hearst, *J. Chem.Phys.* 38: 1062 (1963).
19. G.V. Schulz and E. Penzel, *Makromolek. Chem.* 112: 260 (1968).
20. M.L. Hunt, S. Newman, H.A. Scheraga, and P.J. Flory, *J. Phys. Chem.* 60: 1278 (1956).
21. Н.В. Погодина, К.С. Поживилко, А.Б. Мельников, С.А. Диденко, Г.Н. Марченко, and В.Н. Цветков, *Высокомолек. соед. А* 23: 2454 (1981).
22. С.В. Бушин, Е.Б. Лысенко, В.А. Черкасов, К.П. Смирнов, С.А. Диденко, Г.Н. Марченко, and В.Н. Цветков, *Высокомолек. соед. А* 25: 1899 (1983).
23. Н.В. Погодина, П.Н. Лавренко, К.С. Поживилко, А.Б. Мельников, Т.А. Колобова, Г.Н. Марченко, and В.Н. Цветков, *Высокомолек. соед. А* 24: 332 (1982).
24. Н.В. Погодина, А.Б. Мельников, О.И. Микрюкова, С.А. Диденко, and Г.Н. Марченко, *Высокомолек. соед. А* 26: 2515 (1984).
25. Н.В. Погодина, К.С. Поживилко, Н.П. Евлампиева, А.Б. Мельников, С.В. Бушич, С.А. Диденко, Г.Н. Марченко, and В.Н. Цветков, *Высокомолек. соед. А* 23: 1252 (1981).
26. С.Я. Любина, С.И. Кленин, И.А. Стрелина, А.В. Троицкая, А.К. Хрипунов, and Э.У. Уринов, *высокомолекл. соед. А* 19: 244 (1977).
27. H.A. Swenson, C.A. Schmitt, and N.S. Tomson, *J. Polymer Sci. C* 3: 243 (1965).
28. W.R. Moore and R. Shuttleworth, *J. Polymer Sci. A* 1: 733, 1985 (1963).
29. В.А. Мальчевский and Б.А. Лерье, *Высокомолек. соед. А* 42: 1 (2000).
30. S. Newman, L. Loeb, and C.M. Conrad, *J. Polymer Sci.* 10: 463 (1953).

MECHANICAL PROPERTIES OF NITROCELLULOSE COMPOSITE MATERIALS

V. A. Malchevsky

Russian Chemical Technological University, Moscow, Russia

INTRODUCTION

The processing of double-base rocket propellants includes significant thermal and mechanical loads on the material at all steps. The molecular mass distribution (MMD) of the polymeric base in such formulations is to a great extent responsible for their properties. Changes in the MMD convey important information on the influence of the processing steps.

Here, we are concerned with the mechanical properties of nitrocellulose (NC) composite materials, including strain kinetics, fracture mechanics and thermal failure and their relation to the content and activity of the plastisizer, chemical transformations, and destruction of the polymeric base. We focus on the analysis of MMD changes in NC formulations during processing, and the chemistry of destructive transformations in the polymeric base.

EXPERIMENTAL

A model NC-plastisizer system with a ratio of ~1 was selected as a research subject. The plastisizer was a 1:2 NG-TM mixture, where TM is a mixture of three linear nitroamines. The research was performed on an ultracentrifuge MOM-3180 (Hungary) and by the method of dilute solutions viscosimetry. The MMD was determined in NC after its extraction from the formulation. Both cotton (4 batches) and woodpulp (7 batches) NC with standard physical-chemical parameters were employed. The measured stress-strain parameters of formulations based on this NC differed slightly from those predicted by a model that relates mechanical characteristics to the physico-chemical properties of the polymeric base. The scatter in MMD values for NC from cotton was negligible and comparable with the accuracy of their determination: Mn ($71–83 \cdot 10^3$), Mw ($101–116 \cdot 10^3$), Mz ($139–157 \cdot 10^3$), Mw/Mn (1.38–1.42). On the other hand, the different batches of woodpulp-based NC significant variations: Mn ($53–96 \cdot 10^3$), Mw ($83–145 \cdot 10^3$), Mz ($123–222 \cdot 10^3$), Mw/Mn (1.39–1.56).

Processing by Centrifugation, Edited by Regel and Wilcox
Kluwer Academic/Plenum Publishers, New York, 2001

We determined the development of the formulation's mechanical characteristics (σ_p, ε_p and $E_{2\%}$) during mechanical processing on rolls. It has been claimed that the mechanical properties depend directly on the level of microflaws ΣV from evaluation of composite materials by small-angle x-ray diffraction. The optimal σ_p, ε_p and $E_{2\%}$ corresponded to minimal ΣV_i. We chose mechanical loads in processing corresponding to n = 8, 30 and 60 rolling runs, characteristic of the three different states of the formulation. At n = 8 the plastification was not complete and microflaw in the rolled sheet were relatively high ($\Sigma V_I \sim 0.43 \cdot 10^{-4}$ cm^3/cm^3) and suitable physico-mechanical properties had not developed (σ_p=3.1 MPa, ε_p=40%, $E_{2\%}$=35 MPa). At n = 30 the mechanical properties were optimal (σ_p = 11.2 MPa, ε_p = 64%, $E_{2\%}$ = 14 MPa) and the microflaw concentration was minimal ($\Sigma V_I \sim 0.34 \cdot 10^{-4}$ cm^3/cm^3). At n = 60, destruction of the formulation had started, the microflaws increased ($\Sigma V_I \sim 0.52 \cdot 10^{-4}$ cm^3/cm^3) and the strength declined (σ_p=9.8 MPa, ε_p=65%, $E_{2\%}$=11.5 MPa).

Dilute-solution viscosimetry of the polymeric binder in acetone showed that η_{sp}/C was a linear function of the NC concentration C (Fig. 1), with a correlation coefficient of 0.97–0.99 and the same slope for all specimens. When the experimental results were described by Huggins's equation, $\eta_{sp}/C = [\eta] + K_x [\eta]^2 C$, the value of K_x was 0.48–0.51. During the formulation processing, the NC intrinsic viscosity $[\eta]$ decreased by ~24% from the initial value. Figure 2 shows the variation of $[\eta]$ with n, and has not been reported for other polymers.

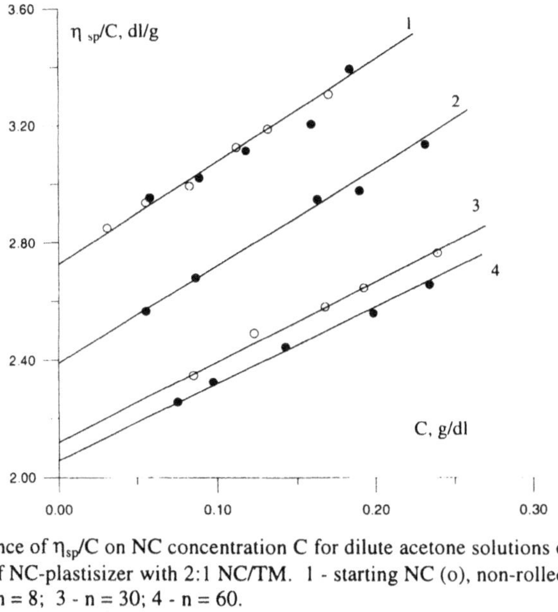

Figure 1. Dependence of η_{sp}/C on NC concentration C for dilute acetone solutions of NC extracted from rolled sheets of NC-plastisizer with 2:1 NC/TM. 1 - starting NC (o), non-rolled material (•); 2 – after 8 rollings, n = 8; 3 - n = 30; 4 - n = 60.

Figure 3 and Table 1 show the molecular weight distribution MMD results. Note that the MMD of NC extracted from unrolled material coincides with that of the starting NC. Mechanical rolling changed the MMD dramatically, and at n = 8 the MMD curve even changed its shape. Table 1 lists the key quantitative parameters of MMD of the studied specimens. The Mn value of the low molecular weight part increased as the processing time was increased. The evolution of Mw and Mn values for mean and high molecular weight parts, respectively, was ambiguous. The polydispersity index dropped at the beginning of the processing from Mw/Mn ~1.5 for starting NC to ~1.24 at n = 8, rose to ~1.4 at n = 30, and decreased again to ~1.1 at n = 60. Those changes are beyond measurement accuracy limits. They testify to the complexity of the processes evolving in the polymer during processing.

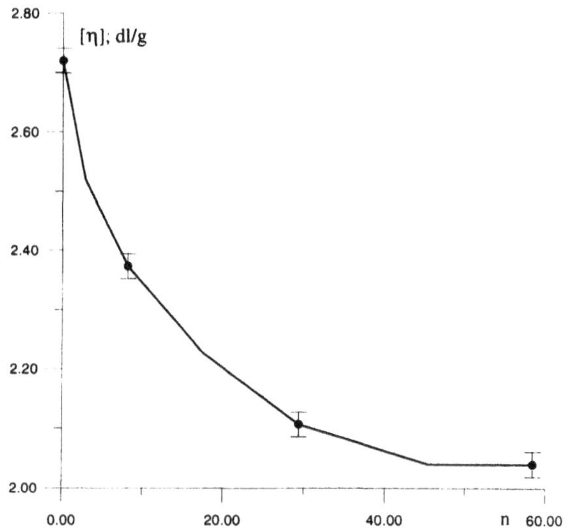

Figure 2. Decreasing [η] for dilute solutions of NC extracted from sheets rolled n times.

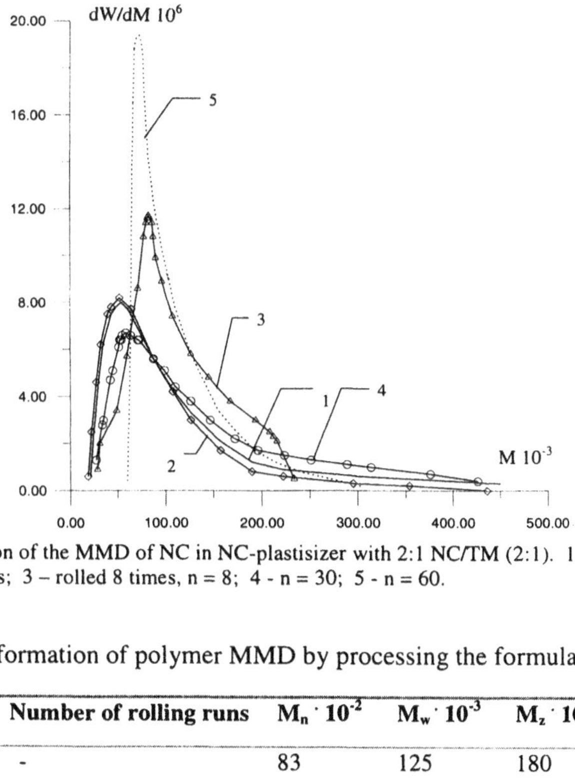

Figure 3. Evolution of the MMD of NC in NC-plastisizer with 2:1 NC/TM (2:1). 1 - starting NC; 2 – non-rolled mass; 3 – rolled 8 times, n = 8; 4 - n = 30; 5 - n = 60.

Table 1. Transformation of polymer MMD by processing the formulation.

Specimen	Number of rolling runs	$M_n \cdot 10^{-2}$	$M_w \cdot 10^{-3}$	$M_z \cdot 10^{-3}$	M_w/M_n
NC	-	83	125	180	1.51
Non-rolled mix	-	80	123	176	1.53
Rolled sheet	8	96	119	139	1.24
Rolled sheet	30	108	152	200	1.40
Rolled sheet	60	114	120	151	1.06

DISCUSSION

In a polymer binder, two types of mechanical-chemical transformations are likely to occur simultaneously - destruction and synthesis. We may assume here that the mechanical synthesis is associated with building specific aspherical macromolecular structures of stiff-chain polymers. Such structures are characterized not only by long stick-like molecules, but also by branched structures. It seems acceptable to suppose that some of those macromolecules may have an elevated ability to flow through a capillary, despite a build-up of their size.

When destruction involves a stochastic process, the transformation of a polymer MMD can be forecast.[1] A broad MMD is predicted to become narrower (Mw/Mn>2) and tend to the most probable limit Mw/Mn ~ 2. For polymers with narrow starting MMD (Mw/Mn<2), the broadening of the MMD does not result from a destruction mechanism. Yet, experimental facts seldom confirm these ideas, as in the present research. the NC MMD transformation to a feasible formulation confirms previous conclusions on mechanical-chemical transformations in a material and, moreover, enables observation of some other specific features of those transformations.

The present results indicate that the polymeric base of the formulation is responsible for both the availability and the course of all chemical transformations occurring under mechanical deformation of the propellant mass. The totality of the accumulated data, including the above, affords an opportunity to judge the probable chemical transformations during the initial steps of nitrocellulose mechanical destruction.

The ability of OH groups in the cellulose macromolecule to be nitrated tends to follow the sequence $C_6 > C_3 > C_2$.[2] Since the substitution degree in our colloxylin specimen was $\gamma = 2.26$, it is possible to believe that three out of four rings contained one unsubstituted OH group at C_2. Hence, the formula of colloxylin may be presented as:

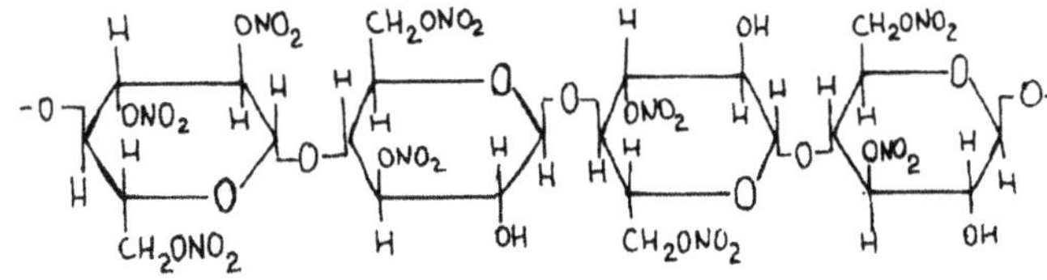

Based on data for glycopyranose ring electronegativity[3] and its influence on the strength of the glycoside bond 1-4, we may consider the bond 1-4 between rings I and II of the initial colloxylin fragment to be the weakest:

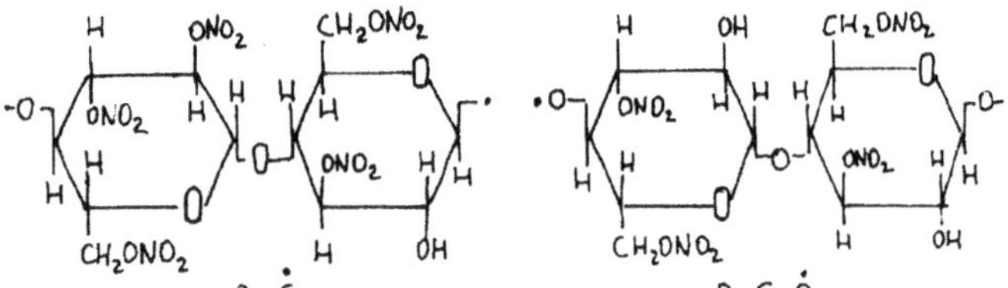

R-$\dot{C}_1$ R-$C_4\dot{O}$

The resulting radicals, R-C_1 and R-C_4O, having excess energy, undergo further transformations accompanied by energy release. R-C_1 can partially transform to R-C_5, and R-C_4O fully transforms to the alkyl-substituted radical R-C_5:

$$m\,R\text{-}\dot{C}_1 \rightarrow (m-n)R\text{-}\dot{C}_1 + n\,R\text{-}\dot{C}_5$$

$$R\text{-}C_4\dot{O} \rightarrow R\text{-}\dot{C}_5$$

The relative stability of the radicals at C_5 results from the screening action of the $-C_6H_2ONO_2$ fragment. There are several ways to form alkyl-substituted radicals at C_5, as discussed publications devoted to cellulose-dispersion products in polysaccharides. We examined the dispersion products characteristic of two types of alkyl-substituted radicals. They may be assigned to radicals R-C_I and R-C_5.

Among the potential transformations related to the death of the radicals R-C_I and RC_5, most probable are reactions of hydrogen atoms broken off from the neighboring NC macromolecules. A negative effect of the glycopyranose ring[3,4] actuates hydrogen mobility at C_I, where the reaction center is formed, due to the available quasidouble bond: 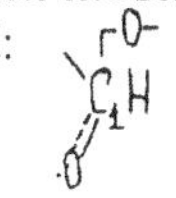

The relatively positive electronic density of C_2 and C_3 measured on pentapyrans is about half that of C_3. That accounts for the high hydrogen mobility in hydroxyls and partially substituted celluloses. Two reaction centers in the NC macromolecule are likely to be most active. The first is connected with hydrogen mobility at C_I, and the second with the mobility of hydrogen in the unsubstituted hydroxyl group, which theoretically may be located at C_2.

Therefore, the interaction of radicals R-C_I and R-C_5 with the NC macromolecule yields central macroradicals R-C_2O and R-C_1O-:

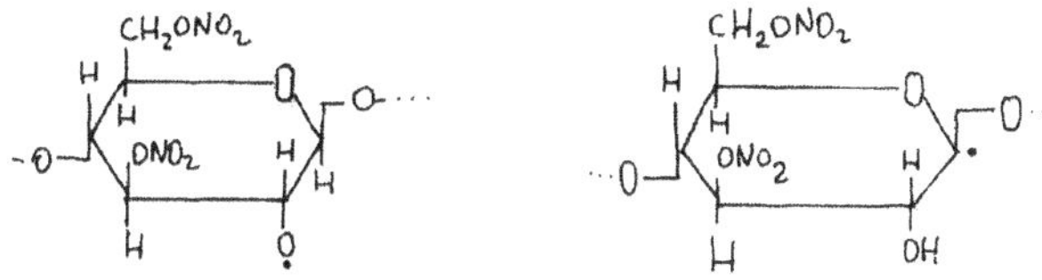

Those radicals are unstable and decompose releasing NO_2:[4]

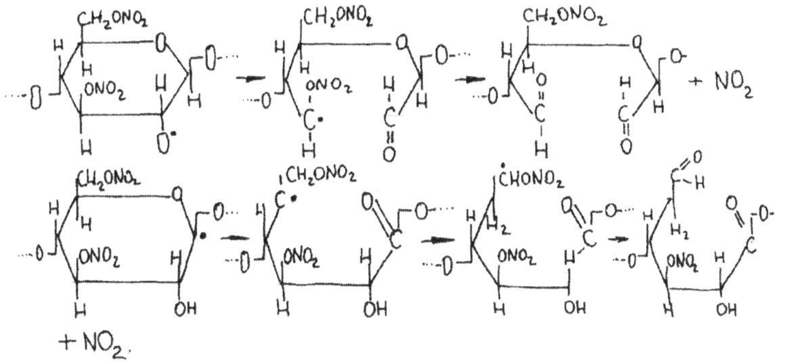

These same radicals are capable of building branched structures.

The transformations above proceed simultaneously with ongoing mechanical dispersion of colloxylin. As shown experimentally, the system acquires stability when a certain ratio of alkyl-substituted radicals R-C_I, R-C_5 and NO_2 is reached in the destruction products. For colloxylin, alkyl-substituted radicals obey the relationship [R-C_I]/[R-C_5] $\approx$ 2. The ratio of alkyl-substituted radicals ΣR-C_I to NO_2 is about 0.43 in dispersion products, i.e. [ΣR-C_I]/[NO_2] $\approx$ 0.43. Only alkyl-substituted radicals and NO_2 are in the dispersion products of the model systems with solid rocket-propellants (SRP). Their relationship varies with the composition, as illustrated by the data in Table 1. The more secondary transformations there are, the less amount of alkyl-substituted radicals and higher content of NO_2 are needed to stabilize the system of dispersion products. Alternatively, the sooner the product system becomes stabilized, the more alkyl-substituted radicals and the less NO_2 it contains. Hence, the ratio ΣR-C_i/NO_2 = K may be understood as a criterion of SRP mechanical-chemical stability. Therefore, mechanical-chemical stability is a controllable characteristic of SRP, provided that stabilization of primary alkyl-substituted macroradicals is achieved by hampering their participation in the secondary reactions that yield NO_2.

REFERENCES

1. A. Casale and R. Porter, *Polymer Stress Reactions*, Academic Press, New York (1978) p. 441.
2. А.Ф. Бочков, А.А. Афанасьев, Г.Е. Заиков, *Образование и расщепление гликозидных связей*, Наука (1978) с 179.
3. В.П. Панов, Р.Г. Жбанков, В.П. Евдаков, Конформационный анализ тринитрата целлюлозы, *Высокомолек. соед. А* 15: 1468 (1973).
4. N. Bikales and L. Segal, *Cellulose and Cellulose Derivatives*, New York (1971) pp 1-5.

CENTRIFUGAL CASTING OF Al-25%w Cu ALLOY WITH ELECTROMAGNETIC STIRRING AND WATER COOLING

Weiqiang Zhang,[1] Yuansheng Yang,[2] and Zhuangqi Hu[2]

[1]Liaoning University of Engineering and Technology, Fuxin 123000, China
[2]Institute of Metal Research, Shenyang 110015, China

ABSTRACT

Electromagnetic stirring and water cooling were applied to centrifugal casting of Al-25% Cu alloy. A full eutectic zone formed due to macrosegregation of Cu caused by the centrifugal force. Both electromagnetic stirring and water cooling refined the primary grains. Moreover, the equiaxed dendrites of α-Al were transformed into globular grains and the eutectic fraction in the alloy increased. These effects are attributed to increased undercooling for growth of the primary phase.

INTRODUCTION

The three methods usually applied during solidification to refine as-cast structures of metals are: (1) mechanically or electromagnetically stirring during freezing; (2) increasing the cooling rate or the undercooling of the melt; (3) adding chemicals to the alloy to increase the nucleation rate. Electromagnetic centrifugal casting is a novel method of metals processing in which metals or alloys are solidified with both centrifugal and electromagnetic forces applied concurrently. The centrifugal force is responsible for melt forming, while stirring or forced convection in the melt is induced by the electromagnetic force. This convection affects heat, mass and momentum transfer during solidification, and determines the as-cast structures and segregation. In recent years, work has been performed on solidification of Al-Cu, Al-Si eutectic and Fe-Cr-Ni hypoeutectic alloys by this method.[1-3] In addition, Yeh *et al.*[4] employed water cooling during centrifugal casting of Al alloys to obtain a refined structure. In the present work, both electromagnetic stirring and water cooling were applied during centrifugal casting. The Al - 25%Cu hypoeutectic alloy was solidified under these conditions.

Processing by Centrifugation, Edited by Regel and Wilcox
Kluwer Academic/Plenum Publishers, New York, 2001

EXPERIMENTAL PROCEDURE

An Al - 25%w Cu alloy was prepared from pure aluminum and copper and cast into ingots. This alloy was remelted by heating up to 730°C in a resistance furnace, and poured into a brass casting mold (80-mm diameter and 100-mm long) using the setup schematically shown in Fig. 1. The mold was located between two magnetic poles and was rotated at 2000 rpm. A stationary d.c. magnetic field was produced between the two poles, with its strength controlled by the current passed through the coils. A water-cooling system was installed outside the mold, whereby cooling water was sprayed directly onto the mold.

Solidification occurred as the alloy was either air cooled or water cooled while a magnetic field of 0-0.17 tesla was applied. The microstructure of cross sections of cast specimens was examined, and the size of the pro-eutectic phase and the volume fraction of eutectic structure were quantitatively measured via an optical microscope. The radial macrosegregation of Cu was measured by an electronic microprobe apparatus with a scanning spot 200 μm in diameter.

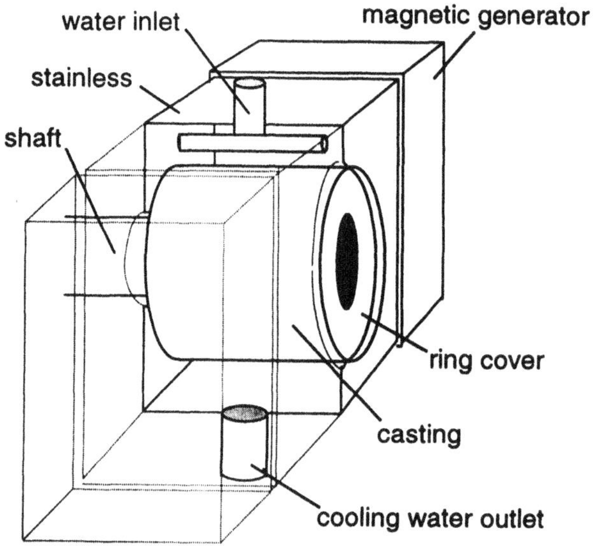

Figure 1. Experimental setup for casting.

RESULTS AND DISCUSSION

Forces

As the melt rotates along with the mold during casting, three body forces are imposed: gravitational, centrifugal and electromagnetic. The parameters are shown in Fig. 2. The volumetric gravitational force is $\mathbf{f}_g = \rho\mathbf{g}$, where ρ is the density of the melt and **g** is the acceleration due to gravity. This force is in a constant direction and has no effect on the convective velocity of the melt.[5] The centrifugal force $\mathbf{f}_c = \rho\Omega^2\hat{r}/r$ is always in the radial direction and is responsible for forming the tube down the center. The rotation of the melt in a magnetic field produces an induction current. An electromagnetic force is consequently generated by the interaction of the current and the magnetic field. The electromagnetic force per volume of melt is $\mathbf{f}_e = \mathbf{j}\times\mathbf{B} = \sigma_e(-\nabla\Phi_e+\mathbf{V}\times\mathbf{B})\times\mathbf{B}$, where **V** and v are, respectively, the vector and scalar forms of the rotating velocity of the liquid metal, σ_e is the conductivity of the melt, and Φ_e is the static potential. The electromagnetic force has one tangential component that is in the opposite direction to the melt movement, leading to a flow opposite to the mold revolution. Electromagnetic stirring of the melt is produced thereby.

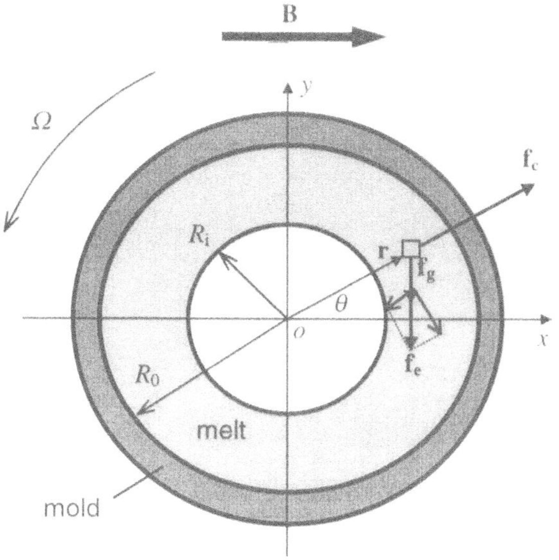

Figure 2. Body forces imposed on the melt.

Macrostructures

Figure 3 shows the macrostructure of an air-cooled specimen containing three distinguishable regions. Region I is characterized by fine equiaxed grains, and formed where the melt contacted the wall of mold and so solidified rapidly. This region should be thickened by water cooling. Region II consisted of a columnar structure and should be diminished by electromagnetic stirring, which promotes the formation of equiaxed grains. Coarse equiaxed grains could be observed in region III where electromagnetic stirring or water-cooling refined the grains.

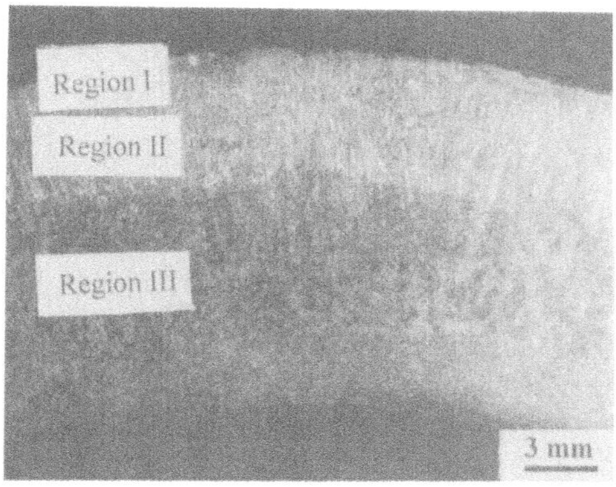

Figure 3. Macrostructure of Al-25%Cu alloy centrifugally cast with air cooling.

Microstructures

Solidified microstructures are shown in Fig. 4. In region I, the chilled microstructure includes a fine primary α-Al plus lamellar Al-$CuAl_2$ eutectic. Fig. 4(a) shows the microstructure in the columnar region (region II) without electromagnetic stirring.

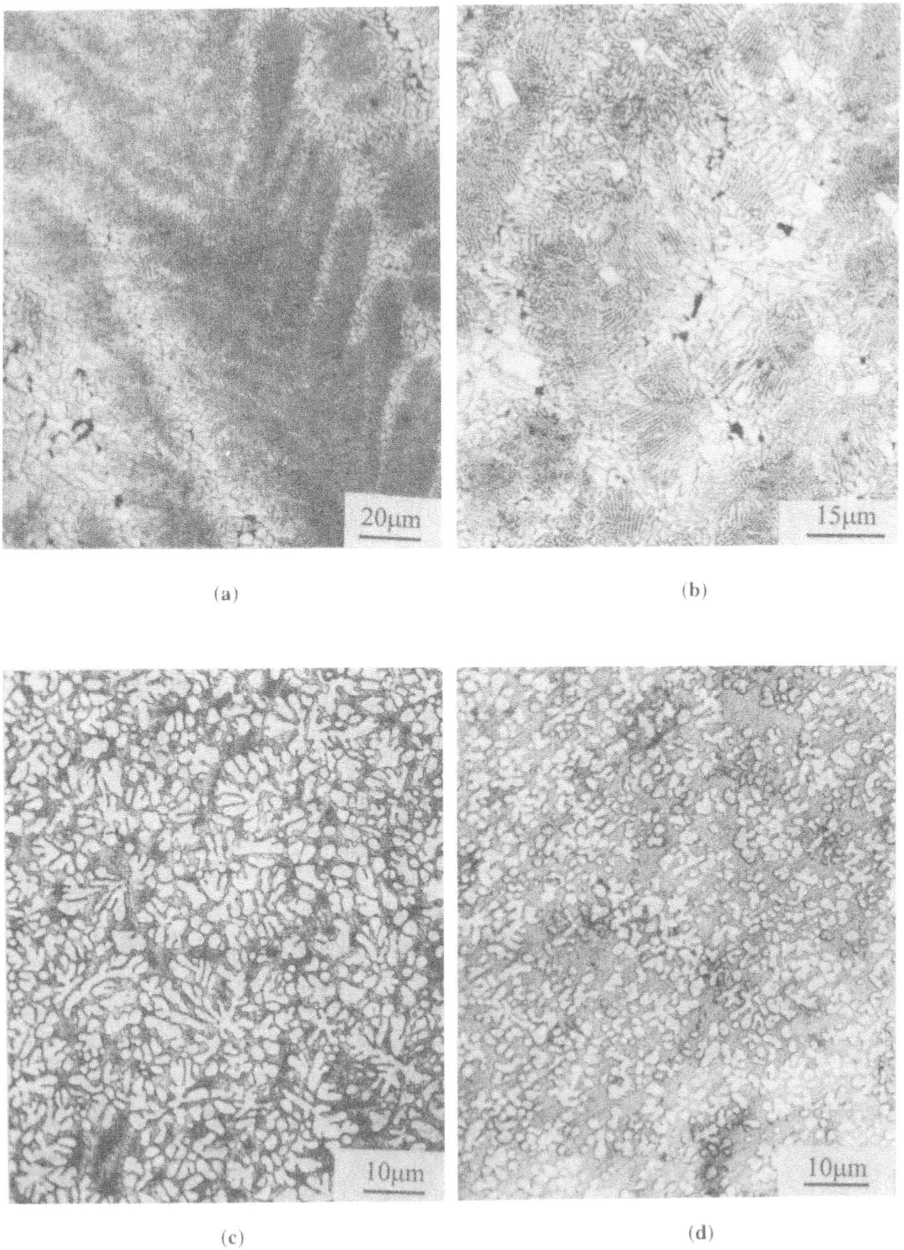

Figure 4. Microstructures of air-cooled specimens: (a) region II of sample prepared without electromagnetic stirring; (b) region II of electromagnetically stirred sample; (c) region III in sample prepared without electromagnetic stirring; (d) region III of electromagnetically stirred sample.

Although the alloy composition was hypoeutectic, little primary α-Al formed in this region, indicating that the local composition was approximately eutectic because segregation of copper occured, which is induced by centrifugal force. It was also observed that the structure was a lamellar eutectic of Al-$CuAl_2$ forming dendritic grains, and these dendritic eutectic grains were changed by electromagnetic stirring. As shown in Fig. 4(b), eutectic colonies were transformed into equiaxed morphology when electromagnetic stirring was applied. Moreover, proeutectic phase, determined to be $CuAl_2$ via electron microprobe, began to form; this phenomenon has been reported by Zhang *et al.*[1] Fig. 4(c) and (d) indicate the effects of electromagnetic stirring on the microstructure in region III. The equiaxed zone contained a typical hypoeutectic structure, i.e., the proeutectic α-Al phase plus lamellar Al-$CuAl_2$ eutectic.

As shown in Fig. 4(c), the proeutectic α-Al phase formed in coarse-grained 'rosette' patterns. Electromagnetic stirring caused the proeutectic α-Al phase to change into globules, with the α-Al grains remarkably refined. When the magnetic flux density was increased up to 0.17 tesla, the rosette pattern was absent, as shown in Fig. 4(d). Water cooling, which increases the freezing rate and the undercooling of the melt, remarkably refined the proeutectic grains. The grain sizes can be quantitatively compared in Fig. 5 (a) and (b). With air-cooling and without stirring, the maximum grain size reached 25.6 μm, and was reduced to less than 20 μm when a field of 0.17 tesla was applied. When the alloy was frozen with water cooling, the grain size near the inner surface was 16 μm without electromagnetic stirring, and became 12.8 μm when the alloy was stirred at 0.17 tesla. Although the inner grains were larger than the outer ones, the refinement of α-Al grains by electromagnetic stirring and the additional refining caused by water-cooling were readily observed.

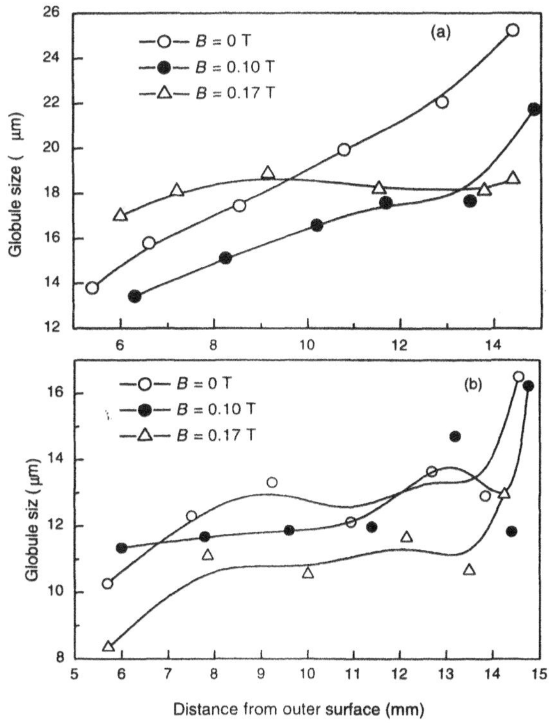

Figure 5. Distribution of primary Al grain size in Al-25 %Cu alloy solidified by centrifugal casting with electromagnetic stirring. Top (a): with air-cooling. Bottom (b): with water-cooling.

Refined and globular proeutectic grains have also been observed in rheocast or semi-solid processed alloys stirred mechanically during solidification.[6,7] Flemings[8] stated that this kind of microstructure formed possibly because the dendritic arms were mechanically broken or remelted off due to a change of solute distribution.

Another microstructural change caused by electromagnetic stirring and water-cooling was an increase of eutectic volume fraction, which can be seen by comparing Fig. 4(c) and (d) and from the quantitative plots in Fig. 6. In regions I and II of the specimens, the eutectic fraction, measured by quantitative metallography, varied between 70% and 95% depending on electromagnetic stirring and cooling method. The eutectic fraction was a maximum in region II. As the distance from the outer surface increased, the eutectic fraction decreased, implying that macrosegregation was produced by the centrifugal force.

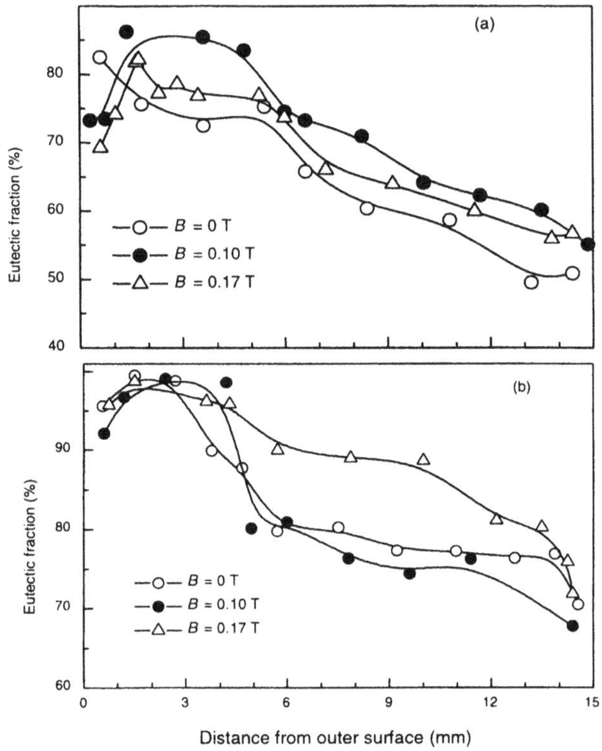

Figure 6. Distribution of eutectic fraction in Al-25% Cu alloy centrifugally solidified with electromagnetic stirring. Top (a): with air-cooling. Bottom (b): with water-cooling.

The cooling method influenced the dependence of eutectic fraction on electromagnetic stirring. With both air and water cooling, the more intense the electromagnetic stirring, the more the eutectic structure. More eutectic structure was present in specimens prepared with water cooling than with air cooling. An increase in the amount of eutectic structure implies an increase of Cu content in the alloy, except that high undercooling is generated to enhance coupled growth in hypoeutectic alloys. Figure 7 shows the segregation ratio of Cu in air-cooled samples. Here, the segregation ratio is defined as the ratio of measured copper concentration to the average of the alloy. It can be observed that the Cu concentration was independent of electromagnetic stirring, indicating that the increase in amount of eutectic structure in electromagnetically stirred samples was not caused by an increase in copper content. Therefore, the eutectic structure formed with electromagnetic stirring must contain less Cu than the eutectic composition on the phase diagram (33.2% Cu).

In nonequilibrium solidification where the alloy is cooled at a high rate (e.g., by water cooling), the freezing temperature of the alloys may be lowered into the coupled zone, as shown in Fig. 8. Then the two phases tend to grow cooperatively and form a pseudo-eutectic structure that is similar to real eutectic in morphology but different in composition. It seems that electromagnetic stirring has an effect similar to water-cooling on promoting cooperative growth.

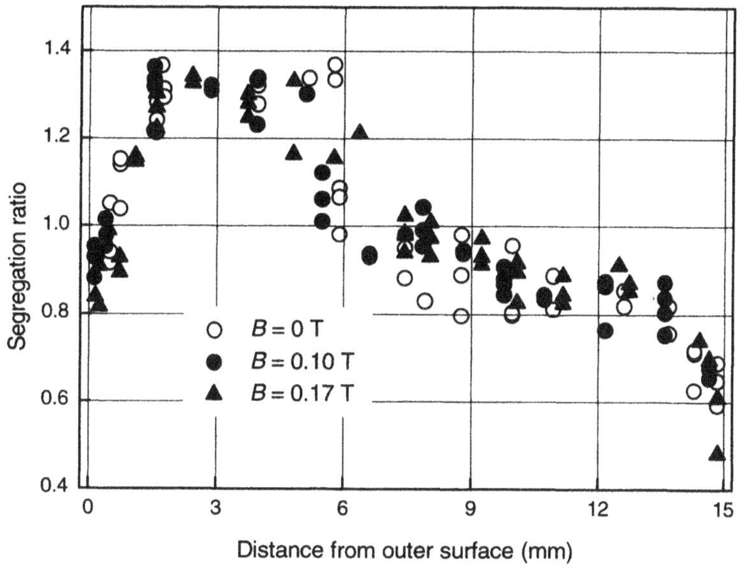

Figure 7. Distribution of copper in specimens prepared with air cooling and different magnetic fields.

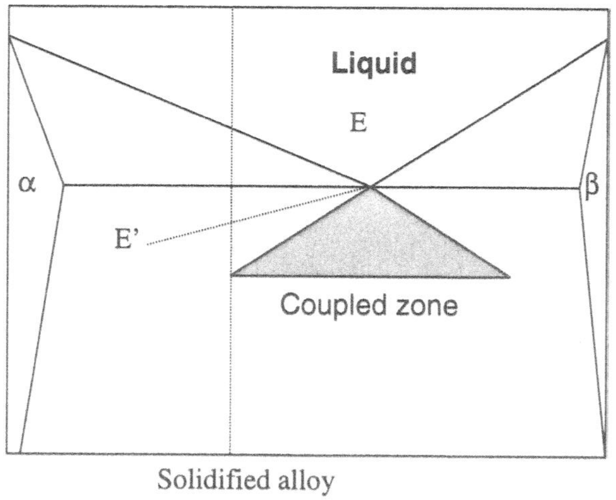

Figure 8. Schematic diagram of a eutectic phase diagram showing the coupled zone.

Kurz and Fisher[9] claimed that the size of the coupled zone depends on the relative undercooling of both primary phases and the eutectic structure. If a growing primary phase requires less undercooling than the eutectic structure, then primary phase plus eutectic structure will be observed; otherwise, only eutectic structure is found. However, the effects of electromagnetic stirring on undercooling are complicated. As for the growth of regular eutectic (e.g. Al - $CuAl_2$ eutectic), the total undercooling (ΔT_e) determining the eutectic morphology consists of two dominant parts: constitutional and capillarity undercooling. Convection induced by electromagnetic stirring decreases the constitutional undercooling of the freezing interface of the eutectic structure. The total undercooling may be expressed by $\Delta T_d = GD/R + K_d R^w$ for dendrites and by $\Delta T_e = K_e R^{1/2}$ for coupled growth, where G is the interfacial temperature gradient, R is the growth rate, D is the diffusion coefficient in the melt, and K_d and K_e are constants. Although they are for convection-free solidification, both of these equations indicate that enhancing mass transport (or diffusion) tends to increase the total undercooling for dendritic growth. Forced convection increases both mass and heat transport, and should decrease ΔT_e more than ΔT_d so that coupled growth becomes favored. The location of the boundary for coupled zone on the hypoeutectic side should be altered as shown by line EE' in Fig. 8.

CONCLUSIONS

- Centrifugal force enhanced macrosegregation of Cu and the formation of eutectic structure (region II) consisting of columnar grains in an Al-25% Cu alloy.
- The primary α-Al phase forming coarse-grained equiaxed dendrites was refined and transformed into globular grains by application of electromagnetic stirring and water-cooling during freezing.
- The amount of eutectic structure was increased by applying electromagnetic stirring and water cooling during freezing. This behavior is attributed to changes in the undercooling of primary phase and eutectic structure.

REFERENCES

1. W.Q. Zhang *et al.*, *Metall. Mater. Trans. A* 29A: 404-408 (1998).
2. W.Q. Zhang *et al.*, *Mat. Sci. Technology* 14: 306-311 (1998).
3. Y. Yang *et al.*, *ISIJ Int.* 35: 389-392 (1995).
4. J.W. Yeh and S.H. Jong, *Metall. Mater. Trans. A* 25A: 643-650 (1994).
5. W.Q. Zhang *et al.*, *Model. Simul. Mater. Sci. Eng.* 4: 421-432 (1996).
6. D.B. Spencer *et al.*, *Metall. Trans.* 3: 1925-32 (1972).
7. D.H. Kirkwood, *Int. Mater. Rev.* 39: 173-190 (1994).
8. M.C. Flemings, *Metall. Trans. A* 22A: 957-981 (1991).
9. W. Kurz and D.J. Fisher, *Int. Mater. Rev.* 5&6: 177-204 (1979).

THE INFLUENCE OF HIGH ARTIFICIAL GRAVITY ON SHS PROCESSES

V.I. Yukhvid, V.N. Sanin and A.G. Merzhanov

Institute of Structural Microkinetics and Materials Science,
Russian Academy of Sciences, Chernogolovka, 142432 Russia

INTRODUCTION

The influence of centrifugation on combustion of gasless systems was first studied by Serkov, Maksimov and Merzhanov in 1968.[1] Centrifugal acceleration up to 900 *g* could cause a six-fold increase in the combustion velocity in a thermit system. Further experimental[2-16] and theoretical[17-22] studies by the present authors and their colleagues showed that gravity exerts a significant influence on chemical conversion in combustion waves, the product-phase composition, and macro- and micro-structures in a wide range of thermit and elemental systems. Use of centrifugal SHS helped to solve such practical problems as production of protective coatings, tubular articles, gradient materials and the like from refractory inorganic compounds and hard alloys.[22-29] The influence of gravity on the SHS processes is being intensely studied in Japan, USA, Italy, China, and others.[28-34]

The present paper reports on results obtained during a long-term investigation by the authors and their co-workers at the Institute of Structural Macrokinetics and Materials Science.

EXPERIMENTAL

The experiments were carried out in radial (Fig. 1a) and axial (Fig. 1b) centrifugal installations with an acceleration from 2 to 1000*g*. In experiments in the radial centrifuge, the directions of the vectors of combustion velocity and centrifugal acceleration were the same, while in the axial centrifuge they were perpendicular to each other.

Synthesis was performed according to the following elemental and thermit schemes: $Ti\text{-}xC \rightarrow TiC_x$, $Ni\text{-}xAl \rightarrow NiAl_x$, $CrO_3 + (x{+}2)Al \rightarrow CrAl_x + Al_2O_3$, $NiO + (2{+}x)Al \rightarrow NiAl_x + Al_2O_3$, $WO_3 + 2Al + C \rightarrow WC + Al_2O_3$, *etc.* The green mixtures were burnt either in quartz glass or graphite vessels. In some experiments in the axial centrifuge, the green charge was drilled through to have a 5 mm wide channel along the symmetry (rotation) axis.

Processing by Centrifugation, Edited by Regel and Wilcox
Kluwer Academic/Plenum Publishers, New York, 2001

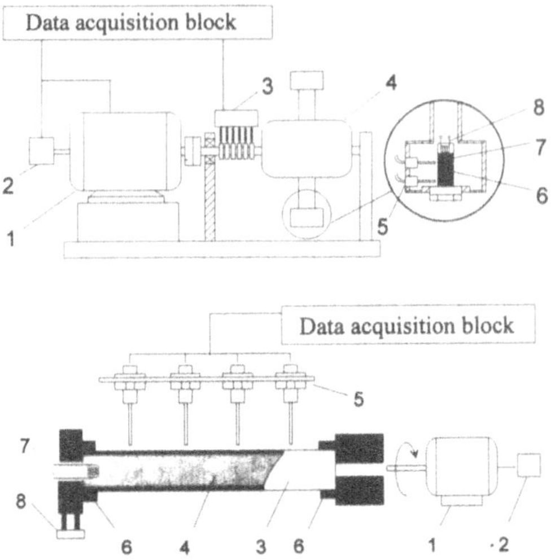

Figure 1. Schematic diagrams of radial (top) and axial (bottom) centrifuges:
Top: 1 - electrical motor, 2 - tachometer, 3 - collector, 4 - rotor, 5 - photodiodes, 6 - quartz matrix, 7 - exothermic mixture, 8 - electrical spiral.
Bottom: 1 - electrical motor, 2 - tachometer, 3 - rotor, 4 - exothermic mixture, 5 - photodiodes, 6 - checks, 7 - electrical spiral, 8 - collector.

For the elemental and thermit systems we determined the average linear combustion velocity u_c, the product chemical and phase compositions, their macro- and micro-structures, the sample relative elongation $\varepsilon = L_f/L_0$ (L_f and L_0 are the final and initial lengths of the sample), and degree of phase separation $\eta = (m/m_{calc}) \cdot 100$ (m and m_{calc} are the experimental and calculated masses of the metal layer).

The Influence of Gravity on SHS processes in Elemental Systems

Video records of the combustion process in the Ti+C and Ni+Al systems under natural gravity and in the centrifuge showed that upon ignition a flat combustion front propagated at a constant velocity through the mixture. At 1 *g*, combustion of Ti + C samples was accompanied by heat release, resulting in sample expansion (elongation), while the Ni + Al sample height did not change since combustion proceeded without gas release. Under the compression caused by centrifugation the sample height was significantly reduced and the combustion products were much denser than the green charge.

Increasing the centrifugation from 1 to 1000 *g* noticeably increased the combustion velocity and the chemical conversion depth in both Ti + C and Ni + Al compositions (Fig. 2). The most pronounced increase in velocity (~ 4 times) was observed in the latter system. At about 250 *g* the initial and final heights of the Ti + C sample were the same ($\varepsilon = 1$).

The composition of the combustion product layer in both Ti + C and Ni + Al systems was not changed much by increasing the acceleration from 1 to 1000 *g*. According to X-ray data, this was practically single-phase TiC or NiAl. Centrifugation caused a reduction of the final height (Fig. 3) and changed the microstructure (Fig. 4). The largest deformation was exhibited by samples with a high content of the liquid phase in the combustion wave.

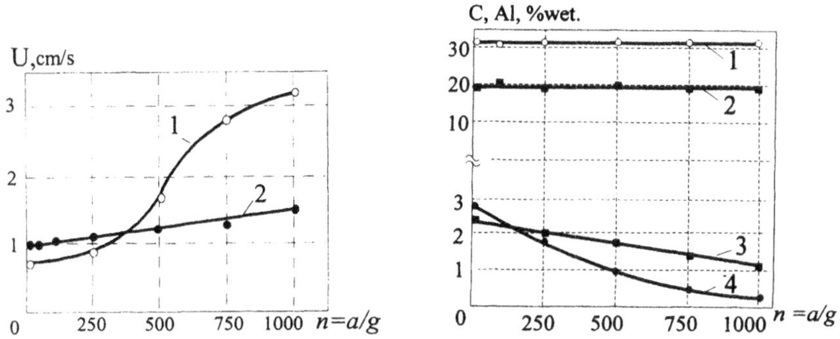

Figure 2. Acceleration dependence of the combustion velocity (left) and the degree of chemical conversion (right) in the Ni + Al and Ti + C systems. left: 1 is Ni + Al (for $\rho = 3.1$ g/cm^3), 2 is Ti + C (for $\rho = 0.8$ g/cm^3). right: 1 is total Al, 2 is total C, 3 is free Al, 4 is free C.

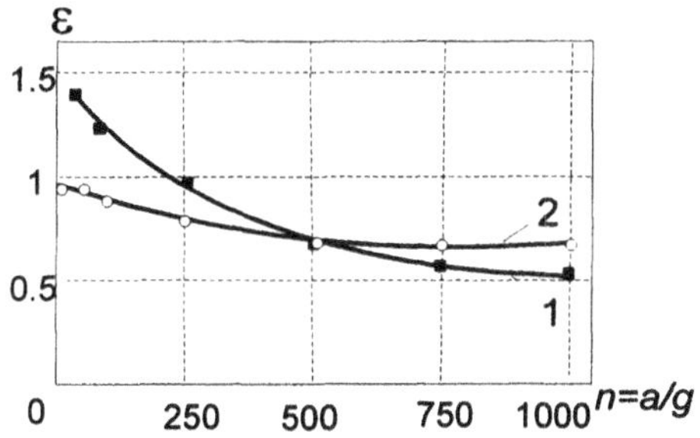

Figure 3. Acceleration dependence of the relative sample height in the Ti+C (1) and Ni+Al (2) systems.

According to thermodynamic calculations, liquid constituted 40% of the combustion products in the Ni-Al system. As the acceleration was increased from 1 to 1000*g* the average porosity of the final product NiAl was reduced to 5-15%. The porosity was nonuniformly distributed along the vertical cross-section (Fig. 5a-c) of the sample, whereas the pores were arranged into circular structures over the horizontal cross-section (Fig. 5d).

The Influence of Centrifugation on SHS Processes in Thermit Systems

When performed in thermit mixtures yielding liquid-phase products, SHS is a multi-stage process. The general case consists of three stages: (1) combustion and formation of the chemical composition of the product melt; (2) lamination of mutually insoluble metallic and oxide phases; (3) crystallization and formation of the phase composition and structure of the SHS products. Each stage involves a number of chemical and physical processes. Gravity can significantly affect all these processes.

Figure 6 shows that the qualitative and quantitative influence of centrifugation on the mean linear combustion velocity depended on the composition of the green mixture (through the combustion temperature) and on the angle between the vectors of acceleration a and combustion velocity. With orthogonal vectors ($u \perp a$) (Fig.7) the dependence of velocity u on acceleration a was affected by a channel through the green charge and the mutual orientation of the vectors of the combustion velocity (u) and the two-phase (the gas + the dispersion products) flow (v) through the channel.

Comparison of data for combustion velocity obtained on elemental and thermit

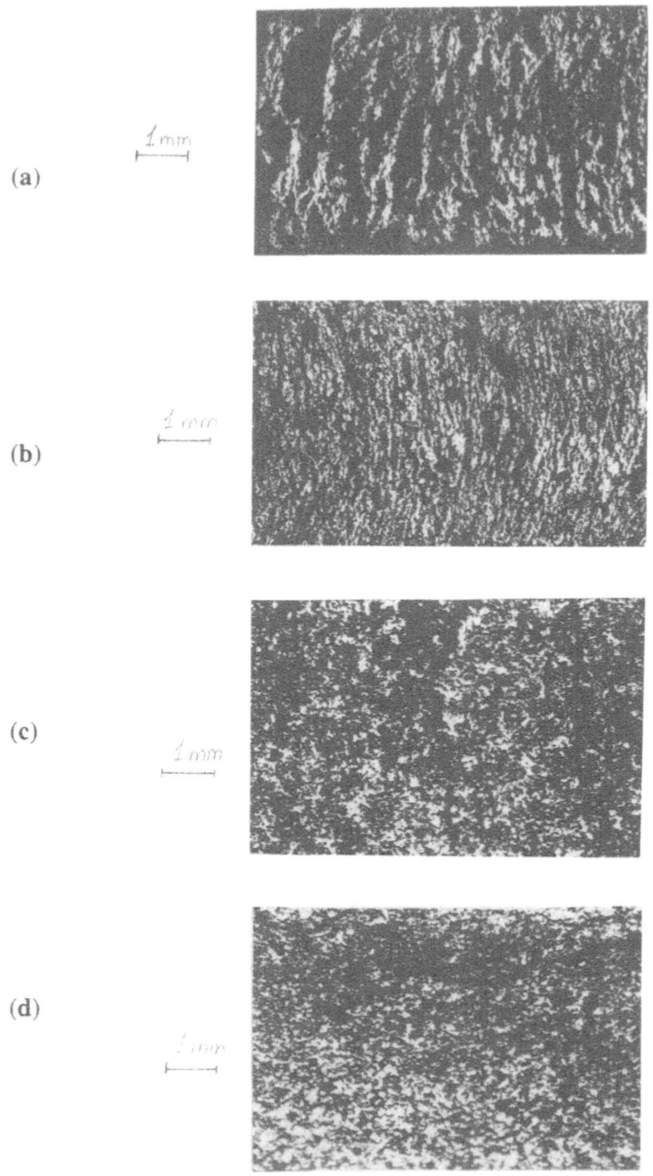

Figure 4. The dependence of the product microstructure on centrifugation for 1g (a), 15g (b), 500g (c), 1000g (d) in the Ti + C system. Each sample shown was approximately 7 mm long.

showed that the thermit systems were much more sensitive to acceleration than those with elemental compositions. For example, in the NiO + Al system with its completely fusible initial components and final products, the combustion velocity was increased by 16 times (see curve 1 in Fig. 6).

In the majority of cases of combustion of highly exothermic thermit compositions, the degree of the oxide reduction and transition of the reduced metals into an ingot achieved 100%. On the contrary, the content of non-metals in the ingot was much less than shown by calculations. In some cases, the degree of non-metal conversion could be regulated by varying its dispersion in the green mixture (Fig. 8).

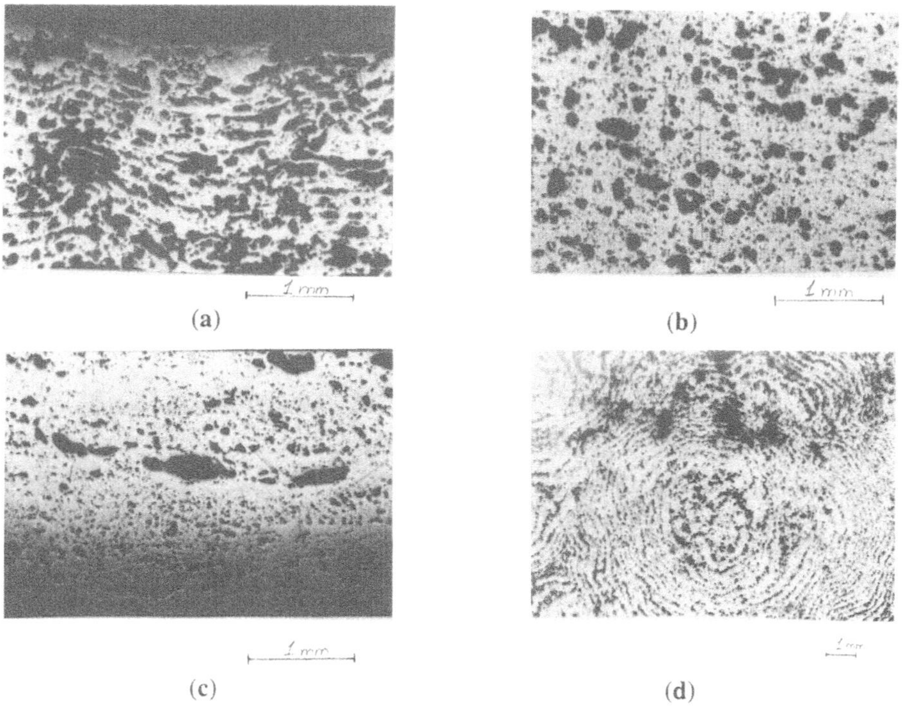

Figure 5. Anisotropy of the combustion product microstructure in the Ni + Al system: (a,b,c) vertical cross sections (~ 3.8 mm long); (d) horizontal cross-section (~ 11.7 mm long).

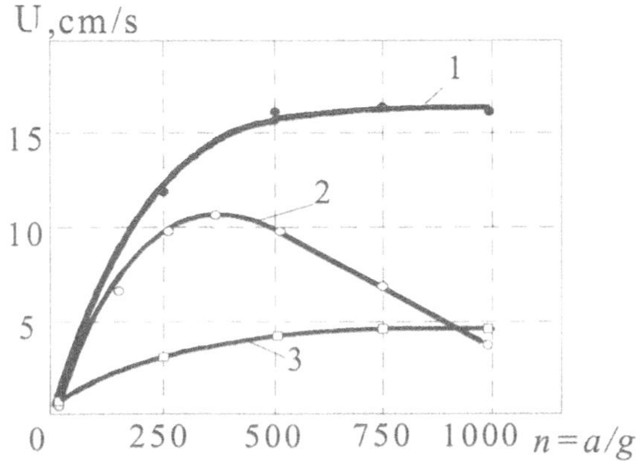

Figure 6. The influence of centrifugation on the combustion velocity for $u \parallel a$ in the thermit systems: 1 - NiO + Al; 2 - FeO + TiO_2 + Cr_2O_3 + Al + C; 3 - CrO_3 + TiO_2 + NiO + Al.

Figure 9 shows that gravity exerted a strong effect on the formation of the macro- and micro-structures. Thus, increasing the acceleration increased the degree of separation of the metallic and oxide phases in the combustion products of the thermit systems. The completeness of the participation of the 'metallic' phase (carbides, borides, hard alloys, *etc.*) in the formation of an ingot could be regulated from the calculated value down to zero by variation of green composition and acceleration. This approach allowed us to synthesize products of three types: separated, partial separated (gradient), and cermet materials. These have all found practical applications.

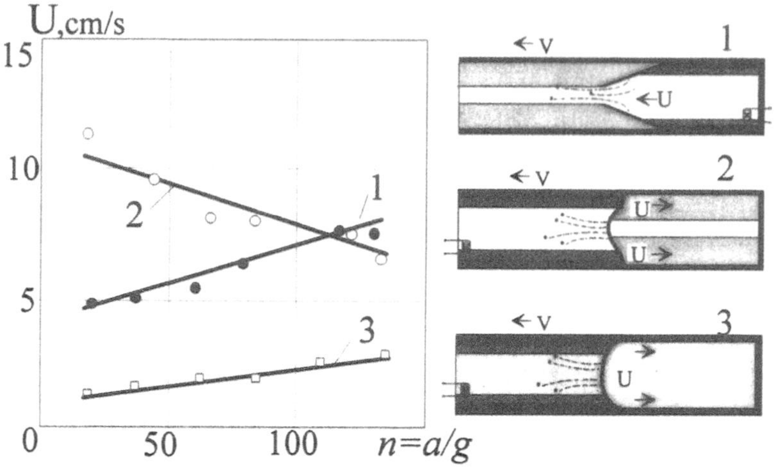

Figure 7. The influence of centrifugation and the direction of gas flow on the combustion velocity: (a) $Fe_2O_3 + Cr_2O_3 + Al + C$ with $u \perp a$; (b) experimental scheme. There were through channels in green mixtures 1 and 2. The direction of *u* and *v* vectors were opposite in 2 & 3 and parallel in 2.

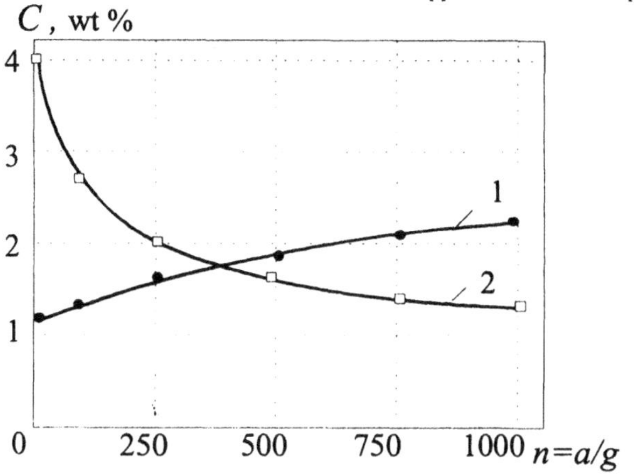

Figure 8. The influence of centrifugation on the dependence of the C content in W-C-Ni ingots. The green mixture was $WO_3 + Al + Ni + C$. Carbon source: 1 - carbon black, 2 - graphite.

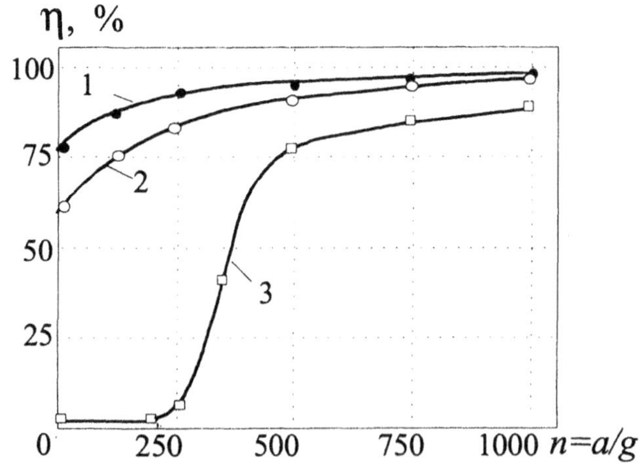

Figure 9. The influence of centrifugation on the degree of phase separation in systems of the thermit type: 1 - $NiO + Al$; 2 - $WO_3 + NiO + Al + C$; 3 - $WO_3 + Al + C$.

The direction of phase separation in centrifugal SHS can be controlled by varying the densities of the metallic and oxide phases. Thus, diminishing the metallic phase by addition of excess aluminum in CrO_3-Al mixtures resulted in inverse phase separation of the liquid-phase combustion products, Cr-Al and Al_2O_3.

Application of high acceleration caused reduction of the grain size of the carbide phase contained in hard alloys and composite materials. Centrifugation during SHS of the TiC-Cr_3C_2-Ni-Mo hard alloy from a multi-component mixture of the thermit type caused a 5 to 6-fold decrease in the carbide grain size.[3,5]

Formation of layered compositions by combustion of highly exothermic thermit compositions on steel substrates. Experimental studies showed that maintaining certain conditions could help to achieve uniform spreading of the metallic phase of the combustion product over a steel surface, leading to formation of a cast coating that tightly adhered to the substrate.[17,18,21,26] From local chemical analyses, the coating formed from the elements of the green mixture and the surface layer of the substrate.

Centrifugal force exerted a strong influence on all stages of SHS -- phase separation and metallic melt spreading over the substrate surface. In turn, these factors determined the chemical composition, phases and microstructure of the coatings. First of all, it should be noted that fine (down to 1 mm), uniform coatings were obtained with centrifugation. In addition, by increasing the acceleration the Fe content in the cast coating was reduced from ~35 to 1%w (curve 2 in Fig. 10) and the content of the carbide components was increased (curves 1, 4 and 5 in Fig. 10).

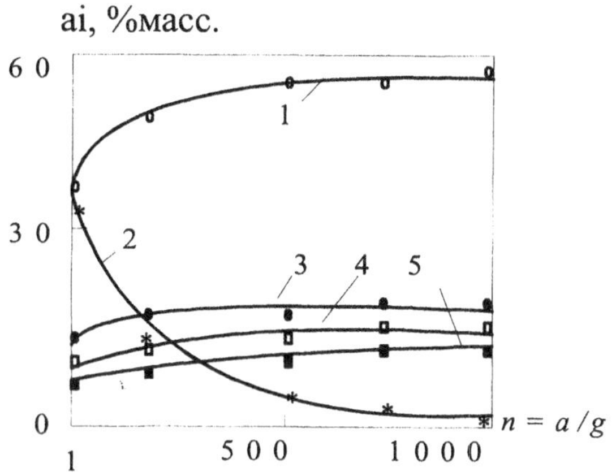

Figure 10. The influence of centrifugation on the chemical composition of cast coatings: 1 - Cr, 2 - Fe, 3 - Ni, 4 - C, 5 - Ti.

MODELING OF SHS IN A GRAVITATIONAL FIELD

Deformation and Convective Processes during Combustion of Elemental Systems

The following most general conclusions can be drawn from analysis of the experimental data:

- Centrifugation increases the combustion velocity in elemental systems.
- Combustion in elemental systems is accompanied by strong deformation of the reacting medium, i.e., expansion or contraction.
- Gravity causes infiltration of the high-temperature liquid phases in the combustion wave.

Let us consider the mechanisms of these processes and their inter-relationships. Heating in the combustion wave is accompanied by release of the component (Ti, Ni, Al) that melts initially and the gases (H_2, N_2, suboxides) released inside the pores. Formation of liquid phases results in a decrease in the sample strength. The action of the gas pressure in the pores P_1 and the hydrostatic pressure P_2 on the hot zone is the cause of the medium expansion or contraction. Deformation begins when the total resultant pressure, $P_\Sigma = -P_1 \pm P_2$, becomes higher than the material's stretching or compression strength σ and when the time required for sample deformation t_1 exceeds the lifetime of the liquid phases t_2. These necessary conditions for deformation can be written as $|P_\Sigma| > |\sigma|$ and $t_1 \leq t_2$. For $P_\Sigma < 0$ and $\sigma < 0$, the material is stretched, while for $P_\Sigma > 0$ and $\sigma > 0$ it is compressed. The values of t_1, t_2, and P_Σ can be estimated from the simple ratios: $t_1 = \mu/P_\Sigma$, $t_2 = d_2/\chi$, $P_1 = (n - n_1)RT_c/V_p$, and $P_\Sigma = \rho ah$, where μ is viscosity, d and χ are the sample diameter and heat diffusivity, respectively, n and n_1 are the respective numbers of moles of gas released in the deformation zone and withdrawn therefrom for t_2 and referred to a unit volume of the mixture, and V_p is the pore volume in the combustion wave.

Because of the density gradient, different parts of the combustion wave propagate at different linear velocities. However, for a steady-state combustion mode, their volume velocities are the same, so that the relationship between the linear velocity of various points of the temperature profile before deformation u_0 and the linear velocity of the combustion front u_* can be written as $u_* = u_0\rho_0/\rho_*$, where ρ_0 and ρ_* are the initial density and the average density in the reaction zone. Using this ratio we can derive expressions for the front velocity in the limiting cases of 'fast' and 'slow' deformation:

- $\rho_* = \rho_f$ and $u_* = u_0r_0/r_f$ for $t_1 < \chi/u_0^2$ (fast deformation)
- $\rho_* = \rho_0$ and $u_* = u_0$ for $t_1 >> \chi/u_0^2$ (slow deformation)

Here, $u_0 = (L_f - L_0)/t_c$. With fast deformation, the density is completely changed in front of the combustion front, which therefore propagates along material with a high density. Slow deformation is characterized by deformation behind the combustion front, which thus propagates along material with the initial density.

In the general case, the parameters of combustion and deformation are derived by concurrent solution of the equations of combustion and deformation. The conditions of densification, the medium viscosity, and the gas pressure in the pores of the combustion front can be evaluated from the above-mentioned expressions. Evaluation shows that for large positive accelerations with the Ti + C system, deformation proceeds in the 'intermediate' regime, the viscosity in the combustion wave is equal to 10^2 Pa.s, and the pressure excess in the deformation zone constitutes 1 atm.

As mentioned above, the combustion velocity is strongly affected by centrifugation. The combustion velocity in the Ni + Al system is increased by four-fold by increasing the acceleration from 1 to 1000 *g*. Such a strong effect cannot be explained by deformation of the charge alone. The most probable cause of so large an increase is intensification of the forced infiltration under the action of centrifugation. The infiltration rate determines[4] the effective coefficient of thermal diffusivity χ, which enters the formula for the combustion velocity: $u_f^2 = \chi\exp(-E/RT_c)$.

Modeling of SHS Processes in Systems of the Thermit Type

SHS of cast refractory materials is a complicated multi-stage process. It can be divided into three main stages: combustion, phase separation and precipitation of the combustion products. Below we consider the mechanisms of the processes involved in each stage. The two latter stages are described by simple qualitative mathematical models. The major attention is paid to three-component systems composed of a metal oxide, aluminum, and carbon.

In the combustion wave, the green mixture undergoes a number of physical and chemical transformations. Simultaneously, in the heating zone a continuous medium with Al drops and C particles distributed through it is formed upon melting the initial metal oxide MO_2 (Fig. 11).

At the boundary of the zone of chemical conversion, the drops and the particles undergo independent reactions with the initial oxide:

$$MO_2 + Al \rightarrow M + Al_2O_3 \quad (1)$$

$$MO_2 + C \rightarrow M\text{-}C + CO \quad (2)$$

Then the formation of an Al_2O_3 'bridge' is followed by exchange reactions and carbonization of the metal reduced via reaction (1):

$$M + C \rightarrow M\text{-}C \quad (3)$$

Thus, two condensed products, M-C and Al_2O_3, and CO are formed within the proposed reaction scheme. According to thermodynamic data, other gases, *e.g.* the metal vapors and suboxides, should also be present among the reaction products.

Evolution of the gas formed in reaction (2), as well as the vapor and the suboxides from the melt during combustion, results in the melt dispersing. The character of this dispersion depends on the ratio of the gas volume V_g to the volume of the condensed phase V_c. For $V_g >> V_c$, the mixture is completely dispersed to form a suspension. For $V_g < V_c$, the gas is evolved from the melt as separate bubbles, which slightly disperse the melt surface. The condition for the transition of one regime to the other can be conventionally represented by:

$$V_g/V_c \sim 1 \quad (4)$$

A mass balance gives the dependence of V_g on the experimental parameters:

$$V_g = (AP)[u/(u + V_p)] \quad (5)$$

where A is a constant, P the pressure, u the combustion velocity, and $V_p = f(a/g)$ the rate of Stokes motion of the gas in the melt. As seen from Eqs. (4) and (5), suppression of melt dispersion induced by the elevated gas pressure and centrifugation can be caused by the reduction of the gas volume in the combustion product melt.

The dependence of the linear velocity u of stationary combustion in the kinetic regime on the Al and C particle sizes (d_{Al} and d_C) can be derived by solving the set of equations describing the combustion process and the equation of heat exchange between the carbon particles and the melt:

$$u_0^2 \sim 1/d_{Al}[\exp(-E/RT_C)] \quad (6)$$

where E is the activation energy of chemical reaction (1), T_C is the d_C-dependent combustion temperature, and R is the gas constant. The qualitative dependence of T_C on d_C (Fig. 12) can be divided into two parts. In region I, through endothermic reaction (2) the rather fine carbon particles compete intensely with aluminum in reduction of the initial oxide. In region II, the large particles of carbon do not participate in the initial oxide reduction, being practically inert in the leading zone where they only participate in by consuming heat as they warm up.

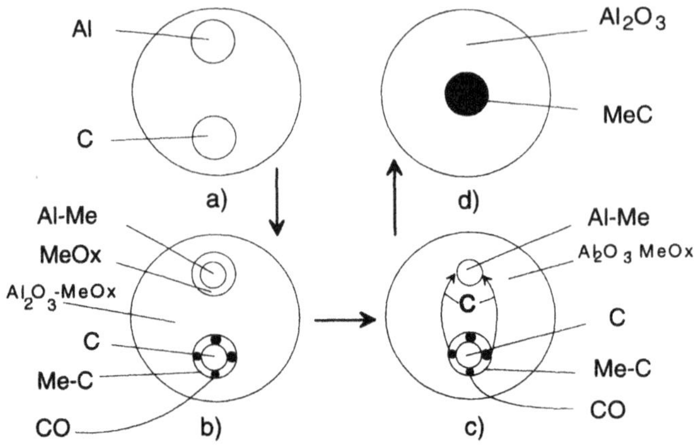

Figure 11. Behavior in the combustion wave.

The formulas for the minimum T_1 and maximum T_2 combustion temperatures and the T_c dependence on the characteristics of combustion in region II were obtained by solution of the equations of combustion and heat exchange between the C particles and the melt.[19]

After passage of the combustion front, there is a two-phase liquid consisting of a continuous phase of the oxide of the metal-reductant with drops of the 'metallic' phase distributed through it. Because of the density gradient, the phases are rearranged under the action of gravity or centrifugal force. The phase motion proceeds in the cooling melt under the condition of variable viscosity of the oxide melt. This phase separation ceases either at the moment of complete removal of the drops from the melt into a 'metallic' ingot or at the moment of oxide phase crystallization (incomplete phase separation). From the set of equations[17] describing the incorporation of the metallic phase in the ingot the dependence of the degree of phase separation η_{ph} on the major parameters can be represented by:

$$\eta_{ph} = A\frac{d_{cr}a}{\alpha\mu_0 r + L_0}\exp(-E_B/RT_c)\{1-\exp[-E_B(T_c - T_{ph})/RT_c^2]\} \qquad (7)$$

where a is the total acceleration and T_{ph} is the temperature of the end of the phase separation (for $\eta_{ph} < 1$, $T_{ph} = T_{m.p}(Al_2O_3)$). It follows from Eq. (7) that the parameter that allows regulation of the completeness of phase separation is the combustion temperature T_c. Among other parameters that can be easily varied during the experiment are acceleration *a*, the coefficient of heat withdrawal, the system geometry (r and L_0), and the viscosity pre-exponent (which can be varied by adding fluxes). With intense cooling, considerable overheating of the melt above T_c is required for complete phase separation. For the parameters typical for the oxide-aluminum-carbon system, L_0 = r = 1 m and E_B = 168 kJ/mol, Eq. (7) gives T_c (min) = 2660 K for $T_{m.p.}$ (Al_2O_3) = 2300 K.

Modeling of SHS Surfacing

A simplified mathematical model is proposed for the study of gravity-induced separation of molten combustion products under non-isothermal conditions. Numerous attempts to theoretically analyze the melt interaction with the parent metal have been undertaken.[17,18] However, simplification of these models could only be achieved by such inflexible assumptions that application to experimental data was severely restricted. In addition, the process was considered only up to the stage of phase crystallization, so that the further dynamics of the process remained obscure.

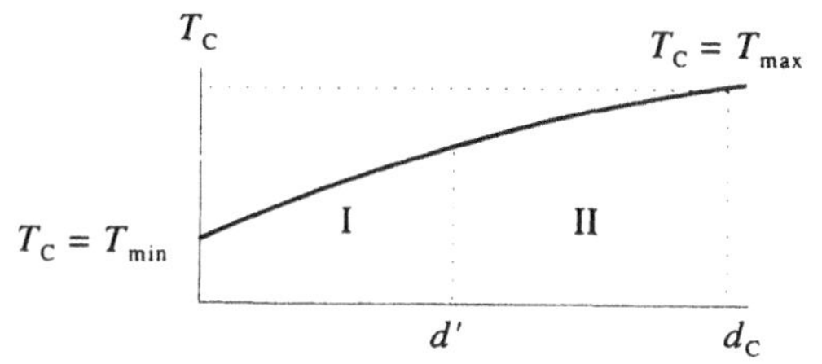

Figure 12. The qualitative influence of carbon particle size d_C on the combustion temperature T_C in the leading zone.

In the present model, the significant non-isothermicity of the process and the dependence of the viscosity and thermal conductivity on temperature and composition of the green mixture are taken into account using a numerical solution of our mathematical model.[20] A model is considered for the interaction of the combustion product layer (L_f thick and T_c hot) and the cold metal substrate (L_c thick and T_0 hot). This interaction is accompanied by two concurrent processes:

1. Phase motion under the action of Archimedean force (phase separation).
2. Heat transfer from the 'hot' melt to the 'cold' substrate resulting in cooling of the melt and heating of the substrate.

Phase motion proceeds until the temperature of the oxide phase crystallization is reached. The second process above stops when the melt and substrate temperatures are the same.

The model requires solution of the following equations (indexes 1, 2, 3 belong to the metal, oxide, and the substrate, respectively). The equation for motion of a metal particle, with the particle being treated as a solid sphere, is:

$$\frac{4}{3}\pi R_1^3 \rho_1^0 \frac{dV_1}{dt} = -\frac{4}{3}\pi R_1^3 \rho_1^0 g + \frac{4}{3}\pi R_1^3 (\rho_1 + \rho_2) g - k\mu_0 R_1 (V_1 - V_2) \exp(E/RT) \quad (8)$$

The equation for conservation of energy with conductive heat transfer is:

$$\frac{\partial}{\partial t}[(c_1\rho_1 + c_2\rho_2 + c_3\rho_3)T] + \frac{\partial}{\partial x}[(c_1\rho_1 V_1 + c_2\rho_2 V_2)T] = \frac{\partial}{\partial x}\left(\lambda \frac{\partial T}{\partial x}\right) - q_1\rho_1 f_1 - q_2\rho_2 f_2 - q_3\rho_3 f_3 \quad (9)$$

where the dependence of thermal conductivity λ on composition is given by:

$$\lambda = \lambda_{3c}(1-\eta_3) + \lambda_{3f}\eta_3 \quad (10)$$

in the vicinity of the substrate and by:

$$\lambda = \frac{\rho_1}{\rho_1^0}[\lambda_{1c}(1-\eta_1) + \lambda_{1f}\eta_1] + \frac{\rho_2}{\rho_2^0}[\lambda_{2c}(1-\eta_2) + \lambda_{2f}\eta_2] \quad (11)$$

in the zone of melting. Here, indices c and f denote the condensed and liquid states, h_1, h_2 and h_3 are the values characterizing the volume portion of the respective components. Conservation of the mass of the final metal is expressed by:

$$\frac{\partial \rho_1}{\partial t} + \frac{\partial}{\partial x}(\rho_1 V_1) = 0 \tag{12}$$

Conservation of the mass of the oxide is given by:

$$\frac{\partial \rho_2}{\partial t} + \frac{\partial}{\partial x}(\rho_2 V_2) = 0 \tag{13}$$

The law of conservation of volume is:

$$\frac{\rho_1 V_1}{\rho_1^0} + \frac{\rho_2 V_2}{\rho_2^0} = 0. \tag{14}$$

Phase conversion is described by a specially derived equation of macrokinetics. It is determined by the governing parameters that regulate the width of the zone of phase transition on temperature and the character of the process (melting or solidification).

Taking into account the foregoing, the law of mass conservation for the molten metal becomes:

$$\frac{\partial}{\partial t}(\rho_1 \eta_1) + \frac{\partial}{\partial x}(\rho_1 V_1 \eta_1) = \rho_1 f_1 \tag{15}$$

where $f_1 = \begin{cases} \operatorname{sign}\left(\frac{\partial T}{\partial x}\right) k_{f1} \exp\left[\operatorname{sign}\left(\frac{\partial T}{\partial x}\right)\frac{T - T_{f1}}{\Delta_1 T_{f1}}\right] \\ 0, \text{if } \eta_1 = 0 \text{ or } \eta_1 = 1 \end{cases}$

Mass conservation for the molten oxide is expressed by:

$$\frac{\partial}{\partial t}(\rho_2 \eta_2) + \frac{\partial}{\partial x}(\rho_2 V_2 \eta_2) = \rho_2 f_2 \tag{16}$$

where $f_2 = \begin{cases} \operatorname{sign}\left(\frac{\partial T}{\partial x}\right) k_{f2} \exp\left[\operatorname{sign}\left(\frac{\partial T}{\partial x}\right)\frac{T - T_{f2}}{\Delta_2 T_{f2}}\right]; \\ 0, \text{ if } \eta_2 = 0 \text{ or } \eta_2 = 1 \end{cases}$

For the molten substrate:

$$\frac{\partial \eta_3}{\partial t} f_3 = \begin{cases} \operatorname{sign}\left(\frac{\partial T}{\partial x}\right) k_{f3} \exp\left[\operatorname{sign}\left(\frac{\partial T}{\partial x}\right)\frac{T - T_{f3}}{\Delta_3 T_{f3}}\right] \\ 0, \text{if } \eta_3 = 0 \text{ or } \eta_3 = 1 \end{cases} \tag{17}$$

The boundary conditions are determined by heat exchange with the environment. The following cases are considered:

- the absence of heat exchange (adiabatic): $\partial T/\partial x = 0$

- heat loss from below: $\lambda_3 \frac{\partial T}{\partial x} = -\alpha(T - T_0)$
- radiation-induced heat loss from above: $\lambda_2 \frac{\partial T}{\partial x} = -\beta(T_4 - T_0^4)$.

Here, t is time, T is temperature, R_1 the radius of the reduced metal drop in the melt, V_1 and V_2 the velocities of the metal and the oxide motion; r_1, r_2 and r_3 the densities; T_{f1}, T_{f2} and T_{f3} the temperatures of the phase transitions; D_1, D_2 and D_3 the parameters responsible for the width of the zone of the phase transition and depend on temperature; c_1, c_2 and c_3 the heat capacities; q_1, q_2 and q_3 the heats of fusion; λ_{1f}, λ_{2f}, λ_{3f}, λ_{1c}, λ_{2c} and λ_{3c} the thermal conductivity coefficients; *g* is the acceleration due to earth's gravity; and k, m_0, e and R are parameters.

This model takes into account gravity-induced separation of the components with a non-isothermal process accompanied by melting and crystallization, the dependence of the thermal conductivities on composition, and dependence of the viscosity of the melt on temperature. At the same time, local thermal uniformity is assumed. The model does not account for polydispersivity of drop size, coalescence and disintegration of drops, convection, and the complicated hydrodynamics of mutual motion. The model was solved numerically for the following parameters: E = 71200 J/mol, $m_0 = 10^{-3}$ Pa.s, $l_{c1} = l_{c3} = 75$ Wt/mK, $l_{c2} = 2$ Wt/mK, $l_{f1} = l_{f2} = l_{f3} = 750$ Wt/mK, $c_1 = c_3 = 5 \cdot 10^2$ J/kgK, $c_2 = 10^3$ J/kgK, $r^0_1 = r^0_3 = 6 \cdot 10^3$ kg/m^3, $r^0_2 = 2.6 \cdot 10^3$ kg/m^3, $T_{f1} = T_{f3} = 1800$ K, $T_{f2} = 2300$ K, $g = 9.8$ m/s^2, $q_1 = q_3 = 2.7 \cdot 10^5$ J/kg, $q_2 = 10^5$ J/kg, e = 0.32, $k_{f1} = k_{f2} = k_{f3} = 10^{-3}$ s^{-1}, $D_1 = D_2 = D_3 = 10^{-2}$.

In the adiabatic case (a = b = 0) at $L_f = L_c = 0.1$ m, heat transfer from the hot melt to the cooler substrate causes the melt to cool and the substrate to warm up (Fig. 13). This heat exchange is accompanied by phase formation and the motion of the more-dense metal drops in the oxide melt, resulting in the formation of a metallic layer on the substrate surface, and squeezing of the less-dense oxide into the upper layer of the melt. Heating of the metal substrate leads to the appearance of a melting front. The dynamics of this melting front $X_{III}(t)$ is shown in Fig. 13a. The melting front propagation comes to a stop at $t = t_{III} = 1000$ s, when the substrate has been completely fused ($X_{III} = L_c = 0.1$ m). Growth of the metallic layer on the surface of the substrate due to phase separation is also of a frontal character (Fig. 13a, $X_I(t)$), and is completed when all metal drops have moved into the layer, long prior to substrate fusion ($t_I << t_{III}$). By the beginning of oxide layer crystallization, the metallic and oxide phases are separated into two distinct layers. The oxide layer crystallization is of a frontal character (Fig. 13a, *curve 2*) and is completed before metal substrate fusion, $t_{II} < t_{III}$. For a = b = 0 and $L_f = L_c$, thermal equilibrium is established at a temperature that is much higher than the substrate melting temperature (Fig. 13a).

In contrast to the adiabatic case above, for a = 10^2 Wt/m^{2}K and b = 10^2 Wt/m^{2}K^4 the times for phase separation, substrate melting, and oxide layer crystallization are comparable (Fig.13b). Points I, III, IV, V and II in Fig. 13 are for completion of phase separation, metal substrate fusion, metal substrate solidification, and metal layer solidification, and oxide layer crystallization, respectively. (The metal layer consists of the molten portion of the metal substrate plus that sedimented onto it during phase separation). There are two fronts of crystallization in the oxide layer that propagate towards each other (upward and downward). The T(x) dependence has a maximum (Fig. 13b). In this case, some portion of the metal drops remains entrapped by the oxide phase at the moment $t = t_{II}$. The crystals of the oxide phase are squeezed to the undersurface of the metallic layer until $t = t_V$. As a consequence, the oxide phase may appear near the region of $X = L_f + L_c$ and $X = X_{II}$. In this example, the metal layer height is determined from the condition that $r_1 = (5.6\text{-}6) \cdot 10^3$

kg/m^3, *i.e.* the portion of the oxide phase does not exceed 10 wt%. Metal substrate melting is followed by solidification of the total metal layer. It should be noted that in the case under consideration the thickness of the fused substrate layer is an order-of-magnitude less than in the adiabatic case, while the maximum heating of the substrate is much lower. The influence of the initial parameters on the dynamics and the final characteristics of the process are considered elsewhere.[20]

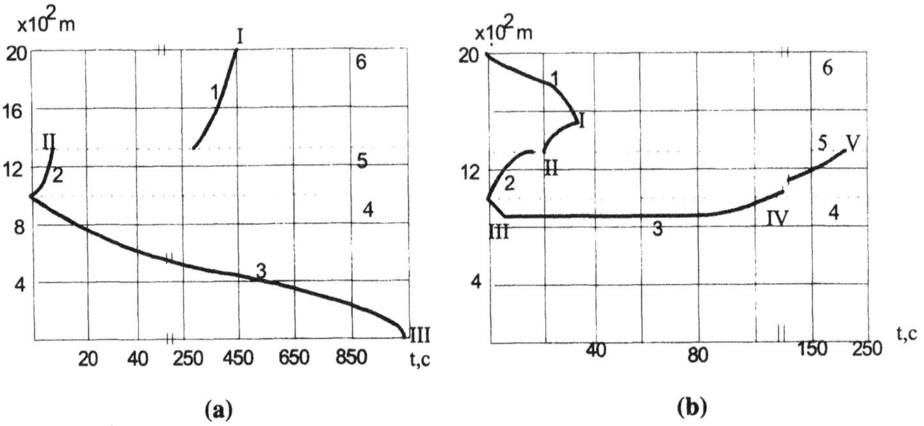

Figure 13. Dynamics of SHS in the melt-substrate system: 1 - crystallization of the oxide phase, 2 - phase separation, 3 – melting and re-solidification of the metallic substrate. Completion of: I - crystallization, II - phase separation, III - melting of the metallic substrate, IV - re-solidification of the molten layer of the metallic substrate, V - solidification of the entire metallic layer. **(a)** $\alpha = \beta = 0$; **(b)** α = 100 Wt/m^{2}K and β = 100 Wt/m^2.K^{-4}.

PRACTICAL APPLICATIONS

Experiments performed in horizontal centrifuges show that centrifugal force enhances the casting of tubes using high-temperature SHS melts.[23-25, 28-31] The degree of separation of the oxide and metallic phases can be controlled by selecting the green-mixture composition and the acceleration. As a result, two-layer, graded composition, and cermet tubes can be fabricated. Study of the properties of the cermet tubes showed that, in contrast to single-phase oxide tubes of Al_2O_3, they exhibited high resistance not only to high temperatures, oxides, alkalies, metal melts but to thermal shock as well. The highest stability cermets were those with a high volume-fraction of the metallic phase containing Cr and Ni. The cermet Al_2O_3-Ti-C-Fe has been used to manufacture metal channels for casting furnaces. These have been tested for casting of Silumin and copper in addition to pig-iron. Two-layer and graded-composition tubes can be also employed in the chemical and metallurgical industries.

From centrifuge experiments, the coating thickness could be varied from 1-2 to 2-3 mm dependent on the green-mixture weight and the under-surfacing area. At present, the SHS technology has been developed for surfacing with a wide range of hard-alloy coatings: Cr-C-Fe, Cr-Ti-C-Fe, Cr-Ti-C-B-Ni-Fe, Cr-Ti-C-Ni-Fe-Mo, *etc.* Their wear-resistance is much higher than that of the industrial coatings. From metallographic analysis, this result can be attributed to a high concentration of the carbide and boride phases with their higher microhardness. An industrial test promises an increase in the coating efficiency from 2 to 20 times.[5,28,31]

REFERENCES

1. B.B. Serkov, E.I. Maksimov, and A.G. Merzhanov, Combustion of condensed systems in the field of the mass force, *Fiz. Gor. Vzryva.* 4: 600 (1968).
2. A.G. Merzhanov, V.I. Yukhvid, and I.P. Borovinskaya, Self-propagating high-temperature synthesis of cast refractory inorganic compounds, *Dokl. Akad. Nauk SSSR* 255: 120 (1980).
3. A.R. Kachin, V.I. Yukhvid, and G.A. Vishnyakova, The regularities of formation of the composition and structure of a cast hard alloy based on complex titanium-chromium carbide in the SHS processes, *The Problems of Technological Combustion*, OIKhF, Chernogolovka, 23: 22 (1981).
4. S.A. Karataskov, V.I. Yukhvid, and A.G. Merzhanov, The regularities and mechanism of combustion of fusible heterogeneous systems in the field of the mass force, *Fiz. Gor. Vzryva.* 6: 41 (1985).
5. A.G. Merzhanov and V.I. Yukhvid, The self-propagating high-temperature synthesis in the field of centrifugal forces, in: *Proceedings of the 1st US-Japanese Workshop on Combustion Synthesis*, National Research Institute for Metals, Tokyo, Japan. 1 (1990).
6. V.I. Yukhvid, Combustion processes forming high-temperature melts, in: *Joint Meeting of the Combustion Institute*, Tacchi-Editore, Pisa, Italy. 431 (1990).
7. V.I. Yukhvid, The structural dynamics of the systems of metal oxide-aluminum-carbon in the processes of combustion and chemical conversion, in: *The Problems of Structural Macrokinetics*, ISMAN, Chernogolovka. 108 (1991).
8. V.N. Sanin, S.L. Silyakov, and V.I. Yukhvid, The combustion front propagation along a long-size channel, *Fiz. Gor. Vzryva*, no. 6:29 (1991).
9. V.N. Sanin, S.L. Silyakov, and V.I. Yukhvid, Models of combustion of heterogeneous melting systems in a long channel, *Intl. J. SHS* 1:191 (1992).
10. V.I. Yukhvid, G.A. Vishnyakova, S.L. Silyakov, V.N. Sanin, and A.R. Kachin, Structural macrokinetics of alumothermic SHS processes, *Intl. J. SHS* 5:93 (1996).
11. V.I. Yukhvid, S.L. Silyakov, V.N. Sanin, and A.G. Merzhanov, The effect of gravity on SHS of foam materials, in: *Proceedings of Joint 10th European and 6th Russian Symposium on Physical Sciences in Microgravity*, St. Petersburg, Russia. 196 (1997).
12. V.I. Yukhvid, S.V. Maklakov, P.V. Zhirkov, V.A. Gorshkov, N.N. Timokhin, and A.Yu. Dovzhenko, Combustion synthesis and structure formation in a model CrO_3-Cr SHS system, *J. Mater. Sci.* 32: 1915 (1997).
13. V.I. Yukhvid, The effect of gravity on the self-propagating high-temperature synthesis, in: *Proceedings of the 1st Russian-Japanese Workshop on SHS*, Karlovy Vary, Czech Republic. 70 (1998).
14. V.N. Sanin and V.I. Yukhvid, Effect of artificial gravity on SHS of foam materials based on Ti-C, *J. Mater. Proc. Manuf. Sci.* 7:115 (1998).
15. V.I. Yukhvid, V.N. Sanin, S.L. Silyakov, and T.I. Ignatieva, Effect of mass force on combustion of the Ni-Al system, *Combustion, Explosion, and Shock Waves* 34:29 (1998).
16. V.N. Sanin, S.L. Silyakov, and V.I. Yukhvid, Combustion of thermit systems with orthogonal orientation of overload and combustion-velocity vectors, *Combustion, Explosion, and Shock Waves* 34:51 (1998).
17. V.I. Yukhvid, The regularities of phase separation in the metallothermic processes, *Izv. Akad. Nauk SSSR, ser. Metally*, 6:61 (1980).
18. V.I. Yukhvid, The dynamic interaction of the high-temperature multi-phase melt and the metallic substrate, *Izv. Akad. Nauk SSSR, ser. Metally*, 6: 127 (1988).
19. K.G. Shkadinsky, V.V. Chernetsova, and V.I. Yukhvid, Mathematical modeling of combustion of three-component SHS systems, *Inzh.-Fiz. Zhl.* 65: 455 (1993).
20. I.S. Gordopolova, T.P. Ivleva, K.G. Shkadinskii, and V.I. Yukhvid, Formation of the composition structure under gravity-induced phase separation and heat transfer in the system of high-temperature melt-metal substrate, *Int. J. SHS*, 8:137 (1999).
21. A.G. Merzhanov, V.I. Yukhvid, I.P. Borovinskaya, and F.I. Dubovitsky, The method of production of cast refractory inorganic materials, *Auth. Certif.* 617485 (1975); *GB Pat.* 1497025 (1978); *Fr. Pat.* 2317253 (1978); *Can. Pat.* 1058841 (1979); *BRD Pat.* 2628578 (1980); *Jpn. Pat.* 1294928 (1986).
22. A.G. Merzhanov, A.R. Kachin, V.I. Yukhvid, I.P. Borovinskaya, and G.A. Vishnyakova, The method of fabrication of cast two-layer tubes, *Auth. Certif.* 725326 (1977); *US Pat.* 4217948 (1980); *BRD Pat.* 2837688 (1982); *FR. Pat.* 2401771 (1982).

23. A.G. Merzhanov, V.I. Yukhvid, V.I. Kukushkin, V.N. Tregubov, and I.P. Borovinskaya, An apparatus for centrifugal casting, *Auth. Certif.* 854569 (1979).
24. A.R. Kachin and V.I. Yukhvid, SHS of cast composite materials and pipes in the field of centrifugal forces, *Int. J. SHS* 1:168 (1992).
25. V.I. Yukhvid, Modification of SHS processes, *Pure Appl. Chem.* 64:977 (1992).
26. V.I. Yukhvid, A.R. Kachin, and G.V. Zakharov, Centrifugal SHS surfacing of refractory inorganic materials, *Int. J. SHS* 3:321 (1994).
27. V.I. Yukhvid, V.N. Sanin, and A.G. Merzhanov, SHS of gradient materials in the metal-oxide systems, in: *Proceedings of the 5th International Symposium on Functionally Graded Materials,* Dresden, Germany, 128 (1999)
28. O. Odawara and J. Ikeuchi, Composite materials produced by a centrifugal-thermit process, *Trans. Iron Steel Inst. Japan* 22: B-19 (1982).
29. O. Odawara, Long ceramic-lined pipes produced by a centrifugal thermit process, *J. Amer. Ceram. Soc.* 73:629 (1990).
30. S. Wojcicki, Application of combustion synthesis of cutting tools materials production, in: *Proceedings of the First US-Japanese Workshop on Combustion Synthesis,* Y. Kaieda and J.B. Holt, eds., National Research Institute of Metals, Tokyo, 181 (1990)
31. S. Yin. M. Liu, H.Y. Lai, *et al.*, Stainless steel composite pipes produced by SHS centrifugal technique, in: *Self-propagating High-temperature Synthesis Technology and Materials,* S. Yin and H.Y. Lai, eds., Metal Industry Press, Beijing, 150 (1995)
32. R. Orru', B. Simoncini, P. Virdis and G. Cao, Further studies on a centrifugal SHS process for coating preparation and structure formation in thermit reactions, *Int. J. Self-Prop. High-Temp. Synth.*, 4:137 (1995).
33. R. Orru, B. Simoncini, and G. Cao, Macrokinetic studies of SHS Fe_2O_3 reduction by aluminum in the presence of additives, in: *Book of Abstracts of the 4th International Symposium on SHS*, Toledo, 44 (1997)
34. B.J. Matkowsky, A.P. Aldushin, K.G. Shkadinsky and G.V. Skadinskaya, Further studies on a centrifugal SHS process for coating preparation and structure formation in thermit reactions, *Int. J. Self-Prop. High-Temp. Synth.* 6:243 (1997).

SELF-PROPAGATING HIGH-TEMPERATURE CENTRIFUGAL PROCESSING OF Cu_2O–Cu–Al AND Cu_2O–Al SYSTEMS

A.S. Rogachev,[1] V.N. Sanin,[1] A.E. Sytschev,[1] V.I. Yukhvid,[1]
E. Medda,[2] R. Orrù,[2] and G. Cao[2]

[1]Institute of Structural Macrokinetics and Materials Science, Russian Academy of Sciences, Chernogolovka, Moscow Region, 142432 Russia
[2]Dipartimento di Ingegneria Chimica e Materiali, Centro Studi sulle Reazioni Autopropaganti (CESRA) and Unità di Ricerca del Consorzio Interuniversitario Nazionale di Scienza e Tecnologie dei Materiali, Università degli Studi di Cagliari, Piazza d'Armi, 09123 Cagliari, Italy

INTRODUCTION

Solid-solid and gas-solid chemical reactions are being exploited for interesting and relatively new technological applications based on the so-called Self-propagating High-temperature Synthesis (SHS) technique. As schematically shown in Figure 1, this technique, which belongs to the more general category of Combustion Synthesis (CS), is characterized by the fact that once ignited, a relatively strong exothermic reaction is able to propagate as a combustion wave through the entire reacting mixture, without requiring any other energy supply. The main characteristics of the SHS technique are its simplicity, relatively low power requirements, high combustion temperature (up to 4000 K) and front propagation velocities up to about 25 cm per second, often permitting one to obtain final products with purity and mechanical properties better than those prepared by conventional methods.[e.g.,1-7] This is because SHS processes are characterized by very high temperature gradients (about 10^5 K/cm) and large reaction rates, so that volatile impurities adsorbed on the reactants are eliminated ("self-cleaning") during the reaction, thus leading to materials with high purity. Furthermore, the temperature gradients combined with rapid cooling may form metastable phases and unique structures not possible by conventional methods of furnace synthesis.

Several reactions between the elements that result in the synthesis of refractory borides, carbides, nitrides, silicides, etc., as well as Mg- and Al-thermite reactions (e.g., Table 1) possess a sufficiently high enthalpy of reaction to propagate in a self-sustained manner. Therefore, for all the advantages outlined above, since its discovery in the early '70s SHS has received increasing attention for the synthesis of a variety of advanced materials such as ceramics, composites, solid solutions, functionally graded materials, etc., as shown in Table 2. Applications include:

Powders of refractory borides, carbides, nitrides, oxides, etc.

- High-temperature industrial heating elements
- Cutting tools
- Ceramic castings (e.g., internal pipe bi-layer coatings)
- Surface-hardened pieces with coatings (e.g., for turbine blades)

Self-propagating reactions have also been proposed recently for environmental applications such as the inertization of hydro-metallurgical wastes,[8] the fixation of high level radioactive wastes,[9] and the degradation of chloro-organic compounds.[10]

An empirical criterion for determining the feasibility of an SHS process is that the adiabatic temperature T_{ad} (the maximum temperature that would be reached with complete conversion of reactants into final products under adiabatic conditions) should exceed 1800-2000 K.

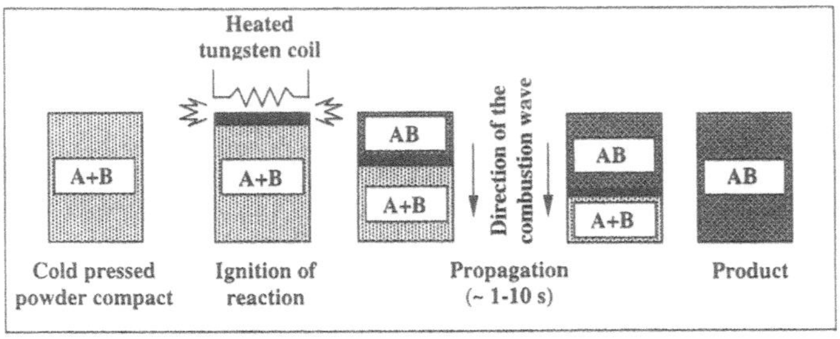

Figure 1. Schematic representation of the SHS process.

Table 1. Some types of reactions utilized in the SHS process.

Type of reactions	Examples
Reactions between elements	$Me(s) + C(s) \rightarrow MeC(s)$ $2 Me(s) + N_2(g) \rightarrow 2MeN(s)$
(Mg, Al) – Thermite reactions	$MeO(s) + Mg(s) + C(s) \rightarrow MeC(s) + MgO(s)$
Reactions between peroxides and lower oxides or elements	$BaO_2 + TiO \rightarrow BaTiO_3$

Table 2. Examples of SHS products.

Borides	Co, Cr, Fe, Hf, La, Mo, Nb, Ni, Ta, Ti, V, Zr, etc.
Carbides	B, Cr, Hf, Mo, Nb, Sc, Si, Ta, Ti, V, W, Zr, etc.
Chalcogenides	Mg, Mo, Nb, Ta, W, etc.
Hydrides	Nb, Ti, Zr, etc.
Nitrides	Al, B, Hf, Nb, Si, Ta, Ti, V, Zr
Silicides	B, Mg, Mo, Nb, Ta, Ti, Ti-C, V, Zr
Binary and ternary oxides	Titanates and zirconates of Ba, Pb; Niobates of Ba, K, Li, Na; Y-Al, Y-Fe, Gd-Ga garnets; Ni-Zn, Co, Ba, Li-ferrites; Ba, Mn metasilicates; Y-123 superconductors
Composites	$MoB-Al_2O_3$, NbC-NbN, $SiC-Al_2O_3$, $SiC-B_4C$, TaC-TaN, TiC-Ni, TiB_2-MoB_2, TiC-TiN, $TiC-TiB_2$, $TiC-Al_2O_3$, TiC-WC, $WC-Al_2O_3$, ZrC-ZrN, etc.

Many systems, e.g., the majority of intermetallics, are exothermic but with $T_{ad} < 1800$ K.

Consequently, the synthesis of these phases has been mainly confined to simultaneous combustion or thermal explosion, which involves the gradual heating of the reacting mixture up to a temperature where the reaction takes place simultaneously over the entire sample. Alternatively, the reactants can be preheated up to a temperature at which a self-propagating reaction wave can be initiated and sustained. Recently, an innovative method based on the use of an electric field to activate SHS reactions in less exothermic systems was developed.[11]

Although the many advantages mentioned above make the SHS technique attractive for industrial applications, process scale-up and control are not straightforward due to the extreme processing conditions. In addition, combustion synthesis processes involve complex physico-chemical phenomena, such as melting and diffusion of reactants, chemical reactions with formation of intermediate phases, nucleation and grain growth, all of which influence the microstructure of the product. Several techniques have been proposed to investigate the mechanisms of SHS reactions.[e.g.,12] In particular, the Combustion Front Quenching (CFQ) technique, based on the rapid extinction of the reaction front during its progress, was demonstrated to be very useful since it causes the simultaneous freezing of intermediate phases and final products. Consequently, as schematically shown in Figure 2, the reaction evolution may be followed during preheating, combustion and post-combustion, by studying the microstructure and composition at different areas relative to the location where the combustion front was extinguished.

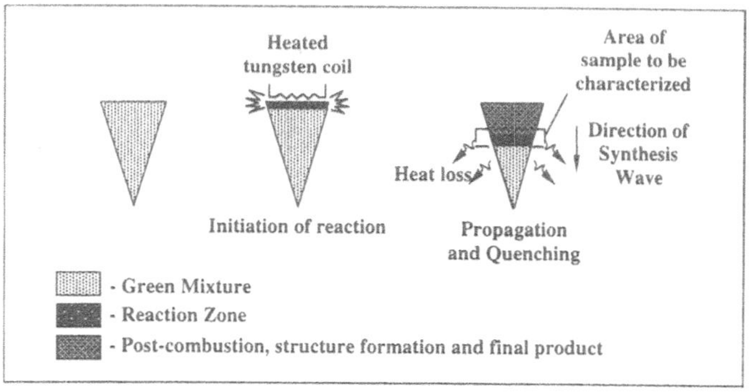

Figure 2. Principle of the combustion-front quenching technique.

The structure and properties of materials produced by SHS strongly depend upon process conditions such as reactants composition, green density, and particle size, as well as on external fields. In particular, the effect of the gravitational field on self-propagating reactions has recently been receiving attention. The driving force for such studies is that self-propagating reactions involve a complex series of stages, most of which are significantly affected by gravity (e.g., melting of reactants and products, spreading of the melt, drop coalescence, convection, buoyancy of solid particles, densification of liquid products).[13,14] Thus it is important to investigate the effect of gravity in order to identify the detailed mechanisms of reaction evolution and structure formation. Along these lines, interesting results have been recently obtained in the USA, Japan, Canada and Russia, using the low gravity environment (see Mukasyan *et al.*[15] and references therein). These results show that gravity may significantly influence structure formation. In particular, products with finer and more uniform microstructure are obtained under microgravity conditions. A summary of the relevant research in combustion synthesis in microgravity is given in Table 3.

Table 3. Relevant research on combustion synthesis in microgravity.

System	Accel (a) in *g*	Observed effects in microgravity	Country	Ref
Ti + xC + additives	10^{-2}	Higher porosity and more uniform pore distribution	Russia	16
B_2O_3-C-Al	10^{-2}	Higher porosity	USA	17
Ti-Al-B	10^{-2}-10^{-5}	Finer and more uniform microstructure	Japan	18
Zr-Al-Fe_2O_3	10^{-2}	Difference in phase composition and finer particles	Japan	18
TiO_2-C-Al, ZrO_2-B_2O_3-Al, B_2O_3-C-Al	10^{-2}	Finer and more uniform microstructure	USA	19
Zn-S	10^{-2}	Larger crystals, closer to the ideal wurzite structure	Canada	20
Ti-B	10^{-2}	Higher combustion temperature	USA	21
GaP, InP, (Ga,In)P	10^{-4}	Difference in phase composition, finer and more uniform microstructure	Japan	22
(3Ni+Al)+(Ti+2B), Ti+0.8C, Ti+C+xNi, (Ti+C)+(Ni+Al), Ni+Al	10^{-4}	Finer size of refractory phase (TiB_2), more uniform microstructure, higher porosity (Ti-C system)	USA	15, 23
Ti+C, Ti+C+Ni, Ti+SiC, Ni+Al; NiO+Ni+Al	10^{-2}	"Contactless" SHS, finer and more uniform microstructure, unusual form and structure of products	Russia	24-26

On the other hand, the combination of centrifugation and self-propagating reactions has allowed investigators to demonstrate the possibility of obtaining layered and functionally graded materials, ceramic-lined pipes, compact materials, metal-matrix composites, alloys, etc.[14, 27-30] The microstructure of the products strongly depended on the acceleration, with the effect of gravity mainly related to the volume fraction of liquid formed and its corresponding lifetime.

In this work, the effect of centrifugal processing on self-propagating reactions taking place in the Cu_2O-Al and Cu_2O-Cu-Al systems was determined. The Cu-Al system is particularly important since, due to the relatively low exothermic character when starting from elemental powders, the corresponding intermetallic phases can be synthesized by combustion only in the thermal explosion mode.[e.g.,31] The effect of centrifugation on the mechanism of microstructure formation was also examined by taking advantage of the combustion front quenching technique.

EXPERIMENTAL SET-UP AND PROCEDURES

The reactant powders used in this work are listed in Table 4. The green mixtures were prepared by mixing the reactants in the required stoichiometry according to the following reactions:

$$3Cu_2O + 6Al \rightarrow 2Cu_3Al_2 + Al_2O_3 \quad (1)$$

$$3Cu_2O + 8Al \rightarrow 6CuAl + Al_2O_3 \quad (2)$$

In order to increase the amount of the intermetallic compound formed during reaction (2), Cu and Al were added to the starting powders in a stoichiometric ratio according to:

$$Cu + Al \rightarrow CuAl \tag{3}$$

Hereafter, the systems corresponding to reaction (1) and reaction (2) combined with (3) are indicated by S_1 and S_2, respectively.

Table 4. Characteristics of the reactant powders.

Powder	Size, μm	Purity, %	Vendor
Cu_2O	< 5	97	Aldrich
Cu	3	99.7	Aldrich
Al	<75	+99	Aldrich

Two different methods of mixing, i.e. dry mixing in a china mortar for about 10 min and wet mixing by mechanically blending with acetone as a dispersing agent for 6 hr, were tested in order to see if that may cause differences in the combustion products. The starting mixtures were placed into cylindrical quartz tubes (50 mm long and 16 mm internal diameter) and ignited by an electrically heated tungsten coil (d_{coil} = 3.5 mm, d_{wire} = 0.5 mm). Experiments were carried out under normal and artificial gravity in air or argon atmosphere (P = 0.1 MPa) using the apparatuses described below.

While the experiments under normal gravity made use of a typical SHS reaction chamber, centrifugal SHS experiments were performed using the apparatus shown schematically in Fig. 3. Basically, it consisted of a rotating system to which the reaction chambers used for SHS experiments were connected. Two sections, indicated by A and B in Fig. 3, were available and corresponded to two different acceleration ranges. An acceleration in the range of 10 to 1000 *g* could be applied using the apparatus depicted in section A. On the other hand, the acceleration that could be attained in the apparatus shown in section B was 1-15*g*. For the experiments described here, *a* = 1-100*g* (0-680rpm).

A wedge copper block was used to perform quenching experiments to investigate the microstructure formation mechanism at different accelerations from 1 to 15*g* using section B. The reacting mixture was first pressed inside the wedge-like cut of the massive copper block and then ignited at the base of the wedge. As the combustion front traveled through the mixture towards the apex of the wedge, it was quenched due to increasing heat loss. The wedge copper block was placed in the centrifugal apparatus in order to have the acceleration vector parallel to the direction of the combustion wave propagation, as shown schematically in Fig. 4.

After reaction, the phase composition of the combustion products was identified by X-ray diffraction and the microstructure was investigated by scanning electron microscopy. The process parameters examined were the average velocity U of the combustion wave propagation, the degree of expulsion η_s, and the degree of sample shrinkage ΔH, all defined as follows:

$$U = \frac{L}{t_{comb}} \tag{4}$$

$$\eta_s = \frac{(M_0 - M_{end})}{M_0} 100\% \tag{5}$$

$$\Delta H = \frac{(L_0 - L_{end})}{L_0} 100\% \tag{6}$$

where L is the sample length, t_{comb} is the duration of combustion, M_0 and M_{end} are the initial and final samples mass, respectively, and L_0 and L_{end} are the initial and final samples length, respectively.

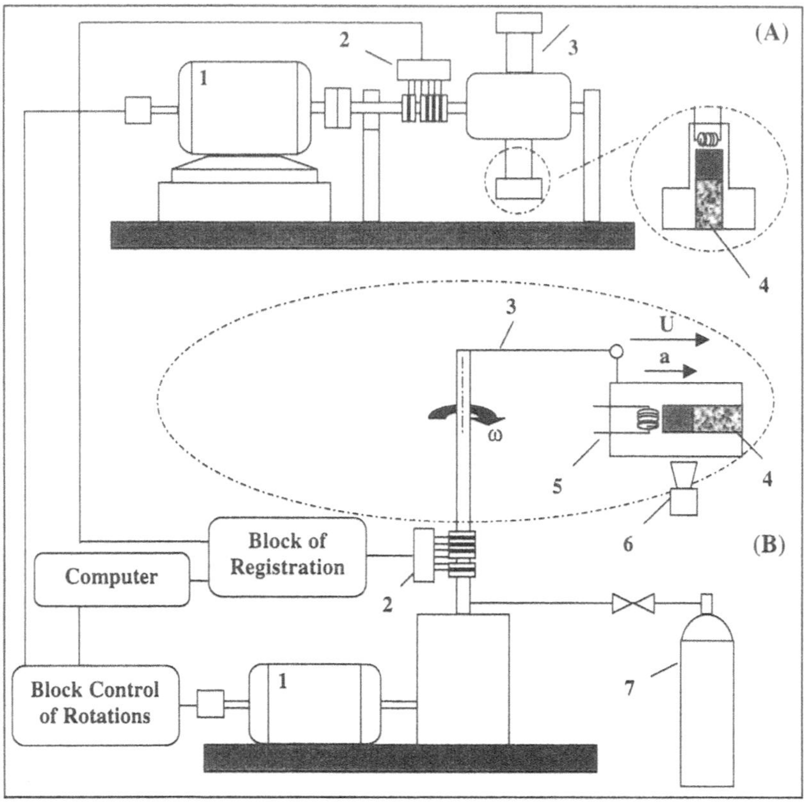

Figure 3. Schematic representation of the experimental set-up used for centrifugal SHS: (1) electric motor, (2) collector, (3) rotor, (4) reacting sample, (5) igniter, (6) video camera, (7) gas tank.

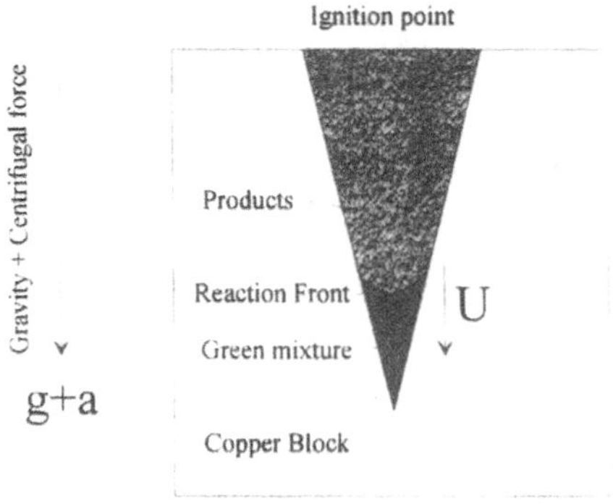

Figure 4. Schematic diagram of the wedge used to contain the reaction mixture during quenching experiments.

EXPERIMENTAL RESULTS AND DISCUSSION

Experiments at 1g

Preliminary experiments performed under normal gravity revealed that the reaction atmosphere, i.e. air or argon, as well as the method of mixing, i.e. dry or wet, had no influence on the phase composition of the product. The reaction products consisted of Cu_9Al_4 and Al_2O_3 for system S_1 (Cu_2O-Al), and CuAl, $CuAl_2$, and Al_2O_3 for system S_2 (Cu_2O-Cu-Al). The computed adiabatic temperature for S_1 was 2397 K. The effect of the (Cu+Al) content in system S_2 on wave velocity U and degree of expulsion η_s are shown in Figure 5, where α represents the weight percentage of (Cu+Al) in the starting mixture of system S_2. It can be seen that the reaction propagated in a self-sustaining manner in the range of α = 0-30%. Within this range, both wave velocity and degree of expulsion were decreased by increasing α. While U decreased almost linearly with α, the degree of expulsion rapidly decreased with increasing contribution of reaction (3), becoming negligible when $\alpha \geq 20$ %. These results are consistent with the relatively low exothermic system (Cu+Al) reducing the high reactivity of the thermite one (Cu_2O+Al), as shown in Fig. 6, which shows that when $\alpha \geq 40\%$ the adiabatic temperature is lower than 1800 K so that the reaction cannot propagate in a self-sustaining mode.

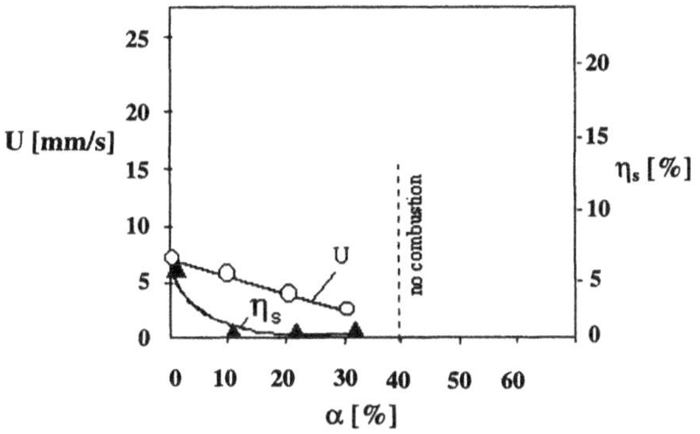

Figure 5. Effect of (Cu+Al) content in the starting mixture in the system S_2 (Cu_2O-Cu-Al)on combustion velocity U and degree of expulsion η_s.

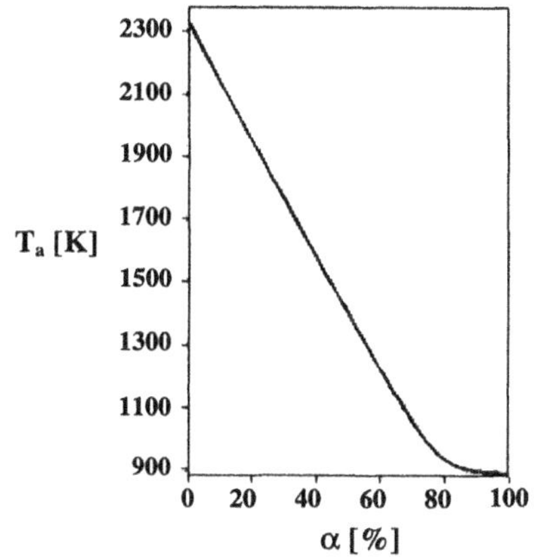

Figure 6. Effect of (Cu+Al) content in the starting mixture in the system S_2 (Cu_2O-Cu-Al) on calculated adiabatic temperature.

Centrifuge experiments

Based on the results described above, α was chosen as 20%w for the centrifuge experiments in order to minimize the degree of expulsion while keeping the reaction propagating in an SHS manner. The adiabatic temperature of this system is 1952 K.

Figure 7 shows the influence of centrifugation on the combustion wave velocity and the degree of expulsion for thermite system S_1. Figure 8 shows the results for system S_2 in terms of combustion wave velocity and degree of shrinkage as a function of acceleration. In the range of 1-20*g*, η_s decreased markedly when the acceleration was increased, while both wave velocity and the degree of shrinkage were not significantly affected. On the other hand, these parameters were very sensitive to acceleration in the range of 20-100*g*. Apparently, centrifugation caused a compression effect that tended to limit the expulsion of gases and solid materials even at relatively low accelerations. As the acceleration was increased, the reacting sample was more compressed, thereby increasing continuity among particles and consequently an increase of the combustion wave velocity. The same centrifugal compression shrank the sample as the reaction proceeded, showing the possibility of pellet densification using centrifugation.

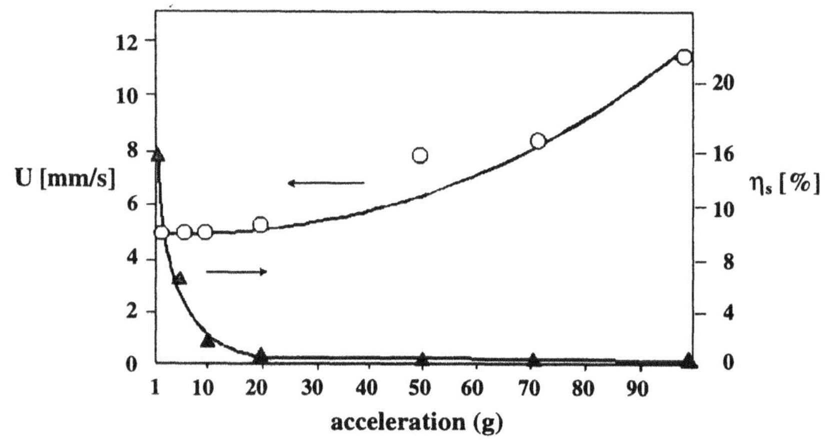

Figure 7. Effect of gravity on combustion wave velocity and degree of expulsion for system S_1.

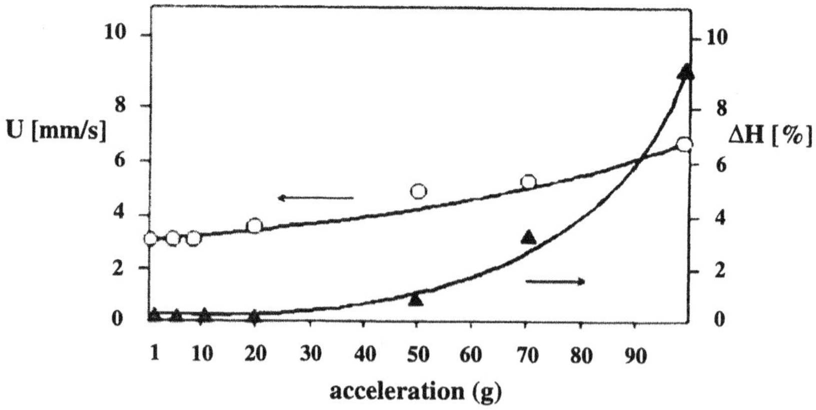

Figure 8. Effect of gravity on combustion wave velocity and sample shrinkage for system S_2 with $\alpha = 20\%$.

The combustion products obtained with system S_1 were two-layered. The upper less-dense portion consisted of corundum (Al_2O_3) and the lower portion of Cu_9Al_4. These compositions did not change with acceleration up to 100*g*.

The final products obtained with system S_2 consisted of an intermetallic matrix, inside which oxide inclusions were dispersed. X-ray diffraction patterns displayed the presence of Al_2O_3 (corundum) and two intermetallics, CuAl and $CuAl_2$, independent of acceleration up to 100g. The effect of acceleration on combustion product microstructure is shown in Figure 9 for the case of the system S_2. The main effect was coarsening of the intermetallic. This may have been caused by infiltration of the liquid product followed by agglomeration.

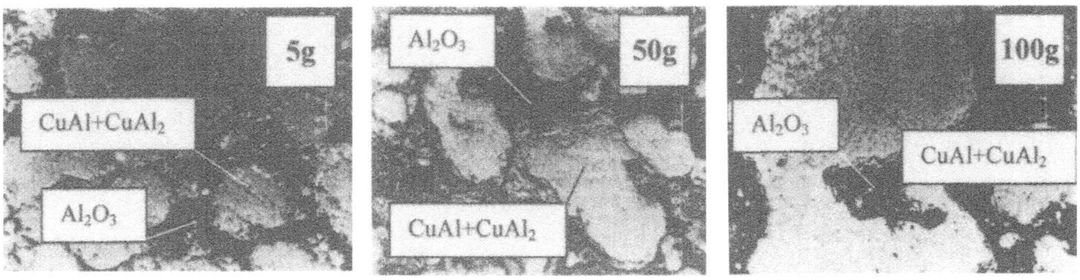

Figure 9. SEM micrographs (40x) of combustion products for system S_2.

Quenching experiments

Centrifugation did not have a significant influence on the microstructure for system S_2 at 1*g* and 15*g*. Figure 10 shows the microstructure of four different regions where the reaction front stopped. The microstructure in the wider end of the wedge-like samples shown in Figs. 10(h) and 10(i) were identical to the microstructure of the final products when the reaction went to completion at the same accelerations. Three different phases can be distinguished in Zone IV of Fig. 10. The white phase p_1 in Figs. 10(h) and 10(i) consists of dendrites with a relatively high Cu content. The dark-gray phase p_2 is a matrix rich in Al. The light-gray phase p_3 is located at the boundary between the p_1 and p_2 phases. Quantitative EPMA (microanalysis with ZAF correction) for these three phases and for the corresponding phases from the Cu-Al phase diagram are reported in Table 5. The sequence of transformations involved in the four regions shown in Fig. 10 are summarized in Table 6.

The main differences observed under the two gravity conditions are summarized as follows. Zone II was narrower and less pronounced for 15*g*. In Zone III, where coalescence phenomena are in evidence, the coalescence process proceeded faster at 15*g*. However, the differences seemed to vanish when reaching Zone IV. In fact, in this latter zone, globules sizes became approximately identical, i.e. about 300-400 μm, under both gravity conditions investigated.

Taking into account the Cu-Al phase-diagram and the combustion temperatures, we may assume that the microstructures of the final products formed as a consequence of crystallization from the metal melt. Coalescence of the latter proceeded faster at 15*g* than at normal gravity, but the microstructure of the globules in the wider end of the wedge-like samples (e.g., Figs 10(h) and 10(i)) were identical. These observations lead us to the conclusion that the post-reaction stage, i.e. crystallization from the melt, caused the product formation to be independent of the previous process steps.

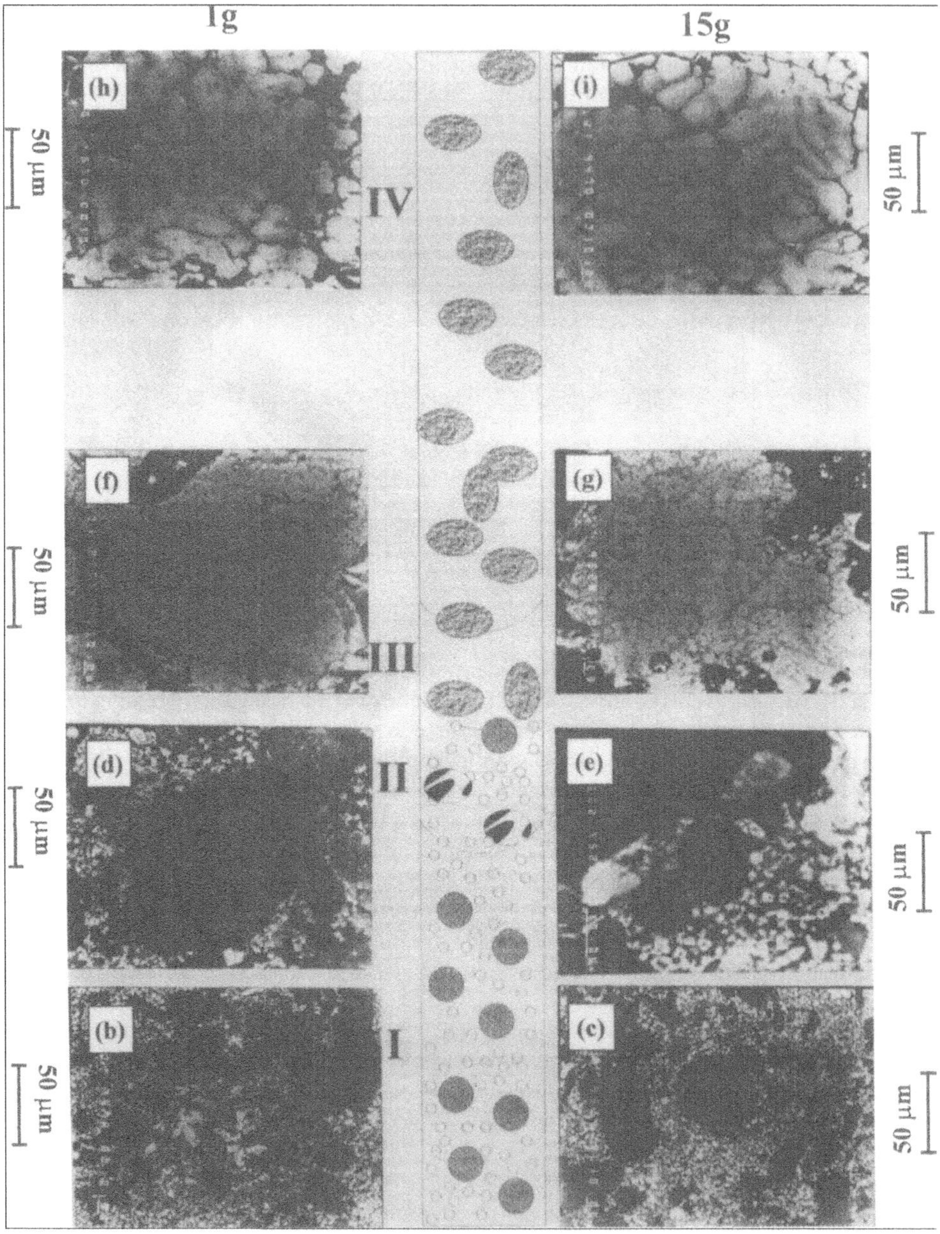

Figure 10. SEM micrographs of different regions where the reaction front had been stopped.

Table 6. Quantitative EPMA analyses of the three phases p_1, p_2 and p_3 in Zone IV.[§]

Accel. *(a)*	Phases	Cu, mass%	Al, mass%	Sum %m	Corresponding phases in Cu-Al phase diagram
	p_1	74.07	24.47	98.54	Cu_3Al_2 (ς_2 phase)
1*g*	p_2	54.35	45.63	99.98	$CuAl_2$ (θ phase)
	p_3	71.71	27.38	99.09	CuAl (η_2 phase)
	p_1	74.05	24.47	98.52	Cu_3Al_2 (ς_2 phase)
15*g*	p_2	54.48	45.13	99.61	$CuAl_2$ (θ phase)
	p_3	73.20	27.00	100.00	CuAl (η_2 phase)

[§]Experimental error is about ±0.5%m.

Table 7. Transformation zones and their characteristics for the system Cu_2O-Cu-Al.

Zone	Nature of structural transformation	Result of chemical interaction
(I)	No transformation	No chemical interaction
(II)	Melting of Al	Formation of Al-based melt
	Dissolution of Cu and Cu_2O into molten Al	
	Chemical reaction	Formation of Al_2O_3
(III)	Melting of Cu and Cu_2O	Formation and separation of liquid (Al-Cu) and Al_2O_3
	Coalescence of liquid reactants	
(IV)	Crystallization from the melt	Formation of $CuAl_2$, CuAl and Al_2O_3

However, this conclusion may be valid only in the gravitational range examined in this work. The situation may change at higher accelerations. In fact, based on the results discussed previously on SHS products obtained from 5*g* to 100*g* centrifugation influences the combustion product microstructure as shown in Fig. 9. Therefore, more CFQ investigations at higher accelerations, e.g. 100*g* or higher, are needed to fully elucidate the influence of gravity on microstructure formation during SHS processes for system S_2.

Acknowledgement

The financial suppcrt of Agenzia Spaziale Italiana is gratefully acknowledged.

REFERENCES

1. A.G. Merzhanov and I.P. Borovinskaya, Self- propagating high-temperature synthesis of refractory inorganic connections, *Dokl. Akad. Nauk* 204: 366-369 (1972).
2. J.B. Holt, Combustion synthesis: a new area of research in materials science, in: *Energy and Technology Review LLNL* (January 1984) pp 11-20.
3. Z.A. Munir and U. Anselmi-Tamburini, Self-propagating exothermic reactions: the synthesis of high-temperature materials by combustion, *Mater. Sci. Rept.* 3: 277-365 (1989).
4. A. Varma and J-P. Lebrat, Combustion synthesis of advanced materials, *Chem. Engng. Sci.* 47: 2179-2194 (1992).
5. A.G. Merzhanov, History and recent developments in SHS, *Ceramics International* 21: 371-379 (1995).

6. V. Hlavacek and J.A. Puszynski, Chemical engineering aspects of advanced materials, *Ind. Eng. Chem. Res.* 35: 349-377 (1996).
7. A. Varma, A.S. Rogachev, A.S. Mukasyan, and S. Hwang, Combustion synthesis of advanced materials: principles and applications, *Adv. Chem. Eng.* 24: 79-226 (1998).
8. R. Orrù, M. Sannia, A. Cincotti, and G. Cao, Treatment and recycling of zinc hydrometallurgical wastes, *Chem. Engin. Sci.* 54: 3053-3061 (1999).
9. I.P. Borovinskaya, T.V. Barinova, V.I. Ratnikov, V.V. Zakorzhevsky, and T.I. Ignatjeva, Consolidation of radioactive wastes into mineral-like materials by the SHS method, *Int. J. SHS* 7: 129-135 (1998).
10. G. Cao, A. Doppiu, M. Monagheddu, R. Orrù, M. Sannia, and G. Cocco, The thermal and mechanochemical self-propagating degradation of chloro-organic compounds: the case of hexachlorobenzene over calcium hydride, *Ind. Eng. Chem. Res.* 38: 3218-3224 (1999).
11. A. Feng and Z.A. Munir, The effect of an electric field on self-sustaining combustion synthesis: part ii. field-assisted synthesis of β-SiC, *Metall. Mater. Trans. B* 26B: 587-593 (1995).
12. A.S. Mukasyan and I.P. Borovinskaya, Structure formation in SHS nitrides, *Int. J. SHS.* 1: 55 (1992).
13. S.A. Karataskov, V.I. Yukhvid and A.G. .Merzhanov, Regularities and mechanism of combustion of melting heterogeneous systems in a field of mass forces, *Fizika gorenia i vzriva*, no.6: 41-43 (1985)
14. A.G. Merzhanov and V.I. Yukhvid, The self- propagating high temperature synthesis in the field of centrifugal forces, in: *Proceedings of the First US-Japanese Workshop on Combustion Synthesis*, Tokyo, Japan (1990) pp 1-22.
15. A.S. Mukasyan, A. Pelekh, A. Varma, and A. Rogachev, Effects of gravity on combustion synthesis in heterogeneous gasless systems, *AIAA Journal* 35: 1821-1828 (1997).
16. A.S. Shteinberg, V.A. Scherbakov, V.V. Martynov, M.A. Mukhoyan, and A.G. Merzhanov, Self-propagating high-temperature synthesis of high-porosity materials under zero-g conditions, *Soviet Physics Doklady* 36: 385-387 (1991) .
17. J.J. Moore, H.J. Feng, K.R. Hunter, and D.G. Wirth, Combustion synthesis of ceramic and metal-matrix composites, in: *Proceedings of the 2nd International Microgravity Combustion Workshop*, NASA CP 10113, Cleveland, OH (1992) pp 157-162.
18. O. Odawara, K. Mori, A. Tanji, and S. Yoda, Thermite reaction in a short microgravity environment, *J. Mater. Synth. Proc.* 1: 203-207 (1993).
19. K.R. Hunter and J.J. Moore, The effect of gravity on the combustion synthesis of ceramic and ceramic-metal composites, *J. Mater. Synth. Proc.* 2: 355-365 (1994).
20. S. Goroshin, J.H.S. Lee, and D.L. Frost, D.L., Combustion synthesis of ZnS in microgravity, in: *25th International Symposium on Combustion*, Combustion Institute, Pittsburgh, PA (1994) 1651-1657.
21. C.C. Lantz, P.A. Tefft, J..J. Moore, and D.W. Readey, Self propagating synthesis of ceramics in a microgravity environment, in: *7th International Symposium on Experimental Methods for Microgravity Materials Science*, R.A. Schiffman, ed., TMS Publication, Warrandale, PA (1995) 41-44.
22. O. Odawara, N. Kanamaru, T. Okutani, H. Nagai, Y. Nakata, and M. Suzuki, Combustion synthesis of GaP, InP, and (Ga,In)P under microgravity environment, *Int. J. SHS* 4: 117-122 (1995).
23. A. Mukasyan, A. Pelekh, and A. Varma, Combustion synthesis in gasless systems under microgravity conditions, *J. Mater. Synth. Proc.* 5: 391-400 (1997).
24. A.G. Merzhanov, A.S. Rogachev, V.N. Sanin, V.A. Scherbakov, A.E. Sytshev, and V.I. Yukhvid, Self-propagation high-temperature synthesis (SHS) under microgravity, *J. Japan Society Microgravity Appl.* 15: 550-555 Supplement II (1998).
25. A.G. Merzhanov, A.S. Rogachev, and A.E. Sytschev, SHS in space. First experiments, *Dokl. Russ. Akad. Nauk* 362: 217-221 (1998).
26. A.G. Merzhanov, V.N. Sanin, and V.I. Yukhvid, Peculiarities of structure formation of high caloric systems during combustion under microgravity, *Dokl. Russ. Akad. Nauk* 371: 94-97 (2000).
27. O. Odawara, Long ceramic-lined pipes produced by a centrifugal-thermite process, *J. Amer. Ceram. Soc.* 73: 629-638 (1990).
28. A.R. Kachin and V.I. Yukhvid, SHS of cast composite materials and pipes in the field of centrifugal forces, *Int. J. SHS* 1: 168-171 (1992).
29. R. Orru', B. Simoncini, P. Virdis, and G. Cao, Further studies on a centrifugal SHS process for coating preparation and structure formation in thermite reactions, *Inter. J. Self-Prop. High-Temp. Synth.* N4: 137-147 (1995).
30. V.I. Yukhvid, S.L. Silyakov, V.N. Sanin, and A.G. Merzhanov, The effect of gravity on SHS of foam materials, in: *Proceedings of the Joint Xth European and Vith Russian Symposium on Physical Sciences in Microgravity* (1997) pp 397-400.
31. L.L. Wang, Z.A. Munir, and J.B. Holt, The combustion synthesis of copper aluminides, *Met. Trans.* 21B: 567-577 (1990).

CENTRIFUGAL-THERMIT PROCESS FOR PRODUCTION OF COMPOSITE PIPES OF VARIOUS SIZES

Eiji Miyazaki and Osamu Odawara

Department of Materials Science and Engineering
Tokyo Institute of Technology, Yokohama 226-8502, Japan

INTRODUCTION

The Centrifugal-Thermit (C-T) process is a unique technique for producing metal-ceramic composite pipes.[1] It has been proposed and investigated as a simple, rapid and economical method for forming a ceramic-lining layer on the inside surface of pipes. The C-T process is characterized by reaction propagation with centrifugal force, a large amount of reaction heat, and centrifugal force applied to the reaction products. In our previous work, we investigated the effect of thermal insulation,[2] centrifugal force,[3] environmental gas content and pressure,[4] additives to the thermit powders,[5] etc. The C-T process has successfully produced long (>3m) metal-ceramic composite pipes of more than 80 mm diameter.[6]

It is very important in improving the quality of the composite pipes to find the proper conditions for the reactants and the centrifugal force. The former mainly determines the product's ceramic layer quality, and the latter the achievement of layer separation and the mechanical properties of the pipes. Reaction propagation in the C-T process influences the mechanical properties of the resulting composite pipes. The innermost ceramic layer is compressed by the intermediate metal, so that the residual stress in the outer steel pipe changes from tension to compression as one moves outward radially. In the intermediate metal layer mainly tensile stress remains. The metal-ceramic composite pipes produced by the C-T process have great potential compared to those obtained by conventional techniques.[6]

As a result of our previous work on the influence of centrifugation and reaction heat on large-diameter pipes, we concluded that the optimum centrifugal acceleration is between $50g$ and $200g$[7] and the heat liberated by the reaction should be between 1kJ and 3 kJ per gram of outer metal pipe mass.

Because of the recent demand for composite pipes of various sizes, especially small diameter, it became desirable to investigate the optimum conditions for small-diameter pipe production. In order to assess the potential of the C-T process for producing small-diameter pipes (less than 20 mm), the characteristics of thermit reaction propagation were investigated in various packing densities of thermit powder and for various inside pipe diameters. The properties of the small-diameter pipes were also determined versus reaction heat and g.

PRINCIPLES OF THE C-T PROCESS

The procedure for the C-T process is as follows: (1) a powdered thermit mixture composed of a strong reducing element and a reducible metal oxide is placed on the inner surface of a metal pipe; (2) the metal pipe is rotated around its axis so that the thermit mixture is pressed against the inner surface of the pipe by the centrifugal force to form a layer; (3) the thermit mixture is ignited at one point , so that the thermit reaction takes place and the reducible metal oxide is reduced to the molten metal, while the strong reducing element is oxidized to the molten oxide; and (4) the resulting pipe, which now has stratified layers of metal and ceramic oxide (innermost layer), is cooled down.

In a typical thermit reaction, aluminum oxide and metal (Me) are produced by the reaction of $3MeO+2Al=Al_2O_3+3Me+Q$, where Q is the heat of reaction. The temperature of the system immediately increases to reach levels above the melting points of the products formed by the reaction. The inner surface of the pipe melts and forms a composite by reacting with the products from the thermit reaction (see Fig. 1). Since the density of aluminum oxide is smaller then that of the metal, a thick ceramic layer is formed on the inner side of the pipe and a new metal product is formed in the intermediate region between the pipe and the ceramic layer. The intermediate metal layer provides the ceramic-lined pipe with high strength, toughness and high resistance to thermal and mechanical shock. High resistance to corrosion and abrasion is provided by the ceramic.

Two of the main characteristics of the C-T process are that a large amount of reaction heat is generated and a centrifugal force is applied to the pipe, which together determine

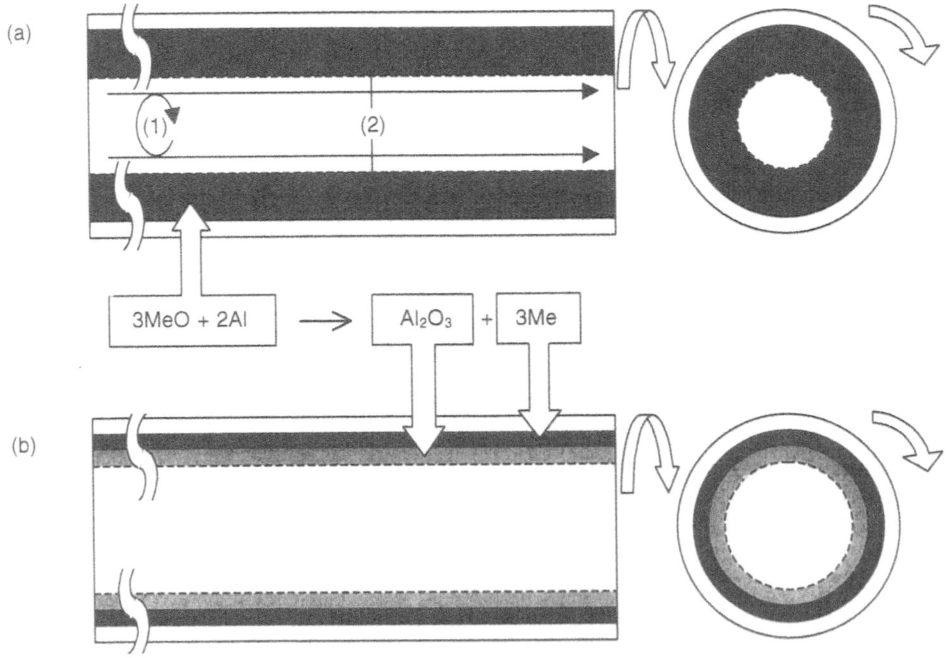

Figure 1. Principles of the C-T process: (a) ignition and reaction propagation (direction of reaction propagation is (1) → (2)); (b) after reaction.

the behavior of reaction propagation. First, the thermit reaction proceeds rapidly along the surface of the reactant layer formed by centrifugal force, and then into the reactant layer simultaneously in the radial and longitudinal directions, resulting in a homogeneous layer in the direction of pipe length, provided that the proper amount of thermit powders and centrifugal force are used. Out of those two parameters, it is the centrifugal force that determines the unique mechanical properties of the ceramic-lined pipes. Due to the tensile stress that remains in the metal layer, the reaction stays under compression stress, providing the C-T process with a high potential for success compared to conventional ceramic-coating techniques.

The quality of the resulting ceramic lining is mainly determined by the heat and the stoichiometry of the reactants, under centrifugal force. The effect of centrifugal force is considered to be the main factor for producing uniform layers and enhancing properties such as low porosity and high strength. As an example, the reaction products in a ceramic lining produced by the reaction of the thermit powders Al and Fe_2O_3, are corundum, which grows dendritically in the pipe radial direction, and hercynite (FeO-Al_2O_3 spinel). The hercynite represented 12%w of the total product and was distributed at the boundary region between the ceramic and metal layers. Although the hercynite plays a role in improving the mechanical properties of the ceramic layer, such as toughness and thermal shock resistance, it should be reduced in content in order to improve corrosion resistance. This is because the FeO component of hercynite is attacked in highly acidic conditions.

EXPERIMENTAL PROCEDURES

Measurement of Thermit Reaction Propagation Rate

A thermit powder was prepared by blending Fe_2O_3 powder and Al powder with stoichiometric ratio for $Fe_2O_3 + 2Al \longrightarrow Al_2O_3 + 2Fe$. The thermit powder was filled inside carbon-steel pipes of 16-mm inner diameter and 2-mm wall thickness. Various diameters of core that has not been filled by thermit powder, lengths, and packing densities were used, as shown in Table 1.

A tungsten wire with 0.5 mm in diameter was electrically heated at one end of the sample powders to start the thermit reaction. For measuring the temperature during reaction propagation, thermocouples were attached to the samples as shown in Figs. 2 and 3. These were K-type thermocouples (alumel-chromel) with 0.254-mm diameter. The temperature data were recorded by a data logger with high speed sampling for voltage input (L840 by UNIPULSE).

Table 1. Sample configurations for measuring the thermit reaction rate.

Diameter of core	Length	Packing density	Reaction heat*	Run	Experimental condition
0	110 mm	25 %	0.92 kJ/g	0 – 1	A
		30 %	1.10 kJ/g	0 – 2	A
		40 %	1.47 kJ/g	0 – 3	A
3 mm	200 mm	31 %	1.10 kJ/g	3 – 1	C
		42 %	1.47 kJ/g	3 – 2	B
		42 %	1.47 kJ/g	3 – 3	C
6 mm	300 mm	29 %	0.92 kJ/g	6 – 1	C
		35 %	1.10 kJ/g	6 – 2	C
		47 %	1.47 kJ/g	6 – 3	B
		47 %	1.47 kJ/g	6 – 4	C

* Reaction heat per gram of outer metal pipe.

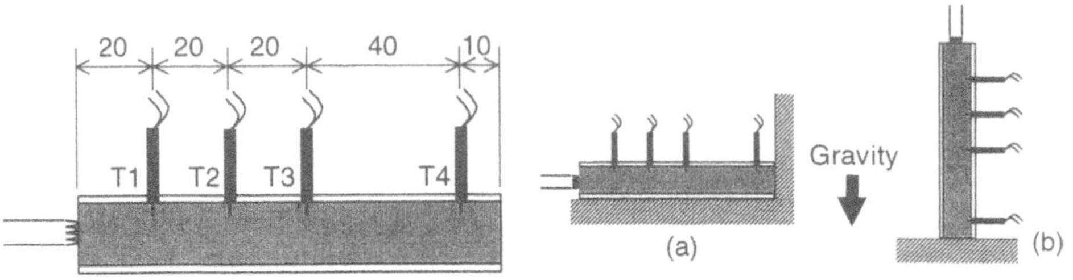

Figure 2. Experimental condition A (dimensions in mm, T1 – T4 are thermocouples): (a) horizontal; (b) vertical.

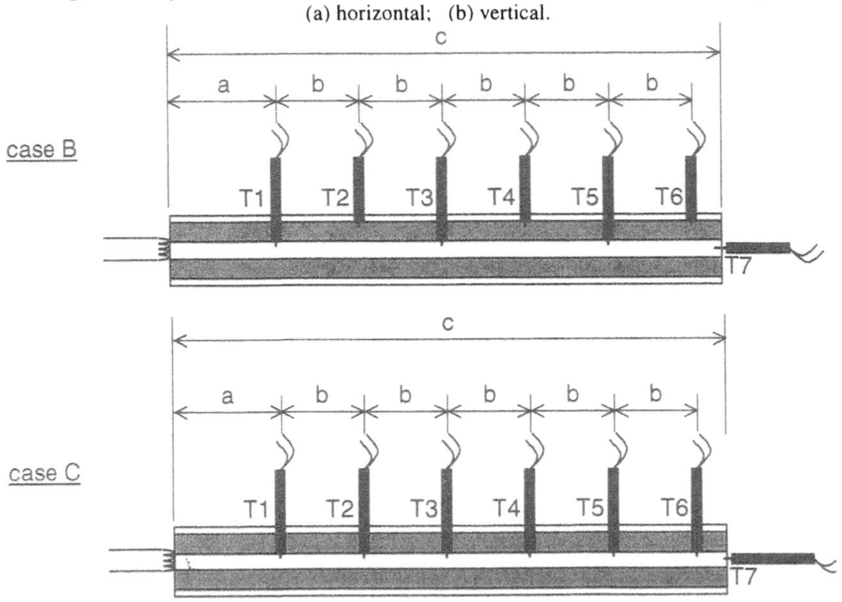

Figure 3. Thermocouple locations for experimental conditions B and C: dimensions are shown in Table 2.

Table 2. Dimensions of samples for experimental conditions shown in Fig. 3.

Experimental condition	Run	Dimensions		
		a mm	b mm	c mm
B	3 - 2	40	30	200
	6 - 3	70	40	300
C	3 - 1	40	30	200
	3 - 3			
	6 - 1	70	40	300
	6 - 2			
	6 - 4			

The resulting data were analyzed to calculate the reaction propagation rate.

Three sets of experiments were performed to measure the reaction propagation rate in each region: across the compact body for two directions of gravity (experiment A); across the compact body and on the surface of the core (experiment B); and on the surface of the core (experiment C). The reaction propagation rate was calculated from the distance between two thermocouples and the differences in time to reach 1073K (0.016V). Experiments B and C were performed with one end of the pipe open and the end far from ignition closed.

Small Pipe Production by the C-T Process

The thermit mixture prepared by blending Fe_2O_3 and Al powders was the same as for the first experiments. The outer metal pipe was the same in diameter and wall thickness: 16mm inner diameter and 2mm wall thickness. The pipe length was 110mm. Sufficient thermit powder was placed in the pipe to provide a reaction heat of 1.0 kJ or 1.5 kJ per gram of outer metal pipe. The core was the same as the first experiment. A centrifugal acceleration between 0 and 150*g* at the inner surface of the outer metal pipe was generated by rotating the pipe in the machine shown in Fig. 4. The experimental conditions are summarized in Table 3. The thermit reaction was ignited by the 0.5 mm in diameter tungsten wire heated electrically while the centrifuge machine was rotating the sample. After the reaction finished, the samples were unloaded from the centrifuge machine and the microstructure of cross sections of the product was determined.

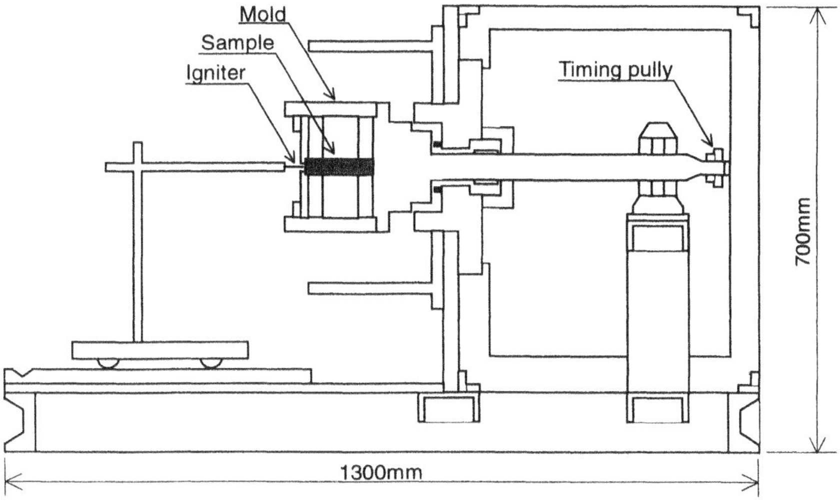

Figure 4. Schematic illustration of the centrifuge machine.

Table 3. Experimental configurations for investigating production of small pipes by the C-T process.

Diameter of core mm	Length mm	Packing density %	Reaction heat* kJ/g	Run	Experimental condition
0	110	27	1.0	0 – 3	D
		41	1.5	0 – 4	
3	110	28	1.0	3 – 4	
		42	1.5	3 – 5	
6	110	32	1.0	6 – 5	
		47	1.5	6 – 6	

* Reaction heat per gram of outer metal pipe.

RESULTS AND DISCUSSION

Thermit Reaction Propagation Rate of the Compact Body with a Core

For runs with experimental condition A, the pipe was placed either horizontally or vertically to have a different direction of gravity during the thermit reaction. Although horizontal samples 0-1 and 0-2 were ignited, the reaction didn't propagate. Horizontal run 0-3 ignited and the reaction propagated successfully with a reaction propagation rate of about 12mm/s. All runs with a vertical pipe ignited and the reaction propagated at about 10mm/s (see Fig. 5). For vertical runs, the rate decreased with increasing packing density. The reaction propagation rate between ignition and the end of the sample may have been almost constant because the data have ~20% scatter. These results show that heat conduction mainly occurred for horizontal runs, and that the diffusion of high temperature vapor among the powder mainly occurred. The same results were obtained in previous research as here for vertical experiments.

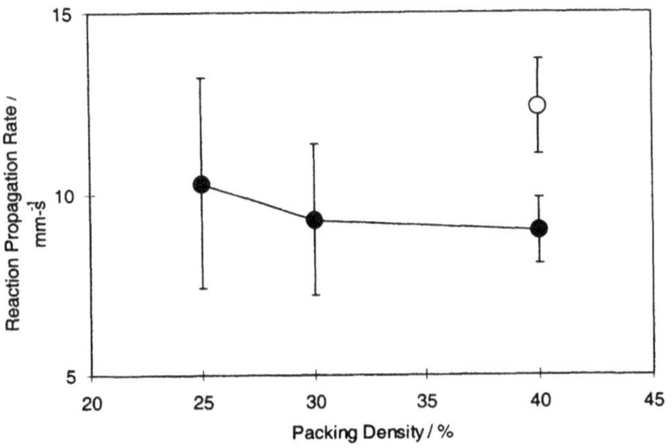

Figure 5. Typical results with experimental condition A: O horizontal; ● vertical.

For experimental condition B with all pipes horizontal, the end of the pipe far from ignition was either open or closed, to determine the influence of the core on reaction propagation. When sample 3-2 was ignited, the reaction propagated along the inside surface first and then proceeded to the outer pipe wall (see Fig. 6 and Table 4). This result shows that the high temperature vapor generated by the thermit reaction moved very fast along the inside tube. After this, the reaction propagated from the inner surface of the thermit mixture to its outer surface at the metal pipe wall. The difference in results was small between having both ends open and one end closed. Only the time between T1 rising and T6 rising was longer for the closed sample than that of open sample. When 6-3 was ignited, the reaction propagated the same as for 3-2. However, the time between T1 rising and T6 rising was much longer for a closed pipe than for an open pipe. This result shows that the high temperature vapor's movement occurred the same as in our previous work. The closure of one end can inhibit the movement of vapor in the central open tube.

To measure the reaction propagation rate, experiments were performed under condition C with all pipes horizontal. The end of the pipe far from ignition was either open or closed, to determine the influence of the central tube on reaction propagation. For convenience, an interval INn is defined that is between Tn and Tn+1, e.g. IN2 is between T2 and T3.

In runs 3-1 and 3-3, when both ends of the pipe were open, the reaction propagation rate between IN1 and IN4 was about 300 mm/s, and accelerated to 1000 mm/s in IN5. However, when one end was closed, the rate had a peak around IN3 or IN4. In run 3-1, the rate from start to finish was about 200 mm/s with a 700 mm/s peak at IN3.

In run 3-3, the rate from start to finish was about 200 mm/s with a 450 mm/s peak (see Fig. 7). Product blew out of both open ends of the pipe. The cases of samples 6-1, 6-2 and 6-4 are discussed as follows: when both ends are open, the rate is accelerated: samples 6-1 and 6-2 are from 600 mm/s to 1500 mm/s; sample 6-4 is from 1000 mm/s to 1500 mm/s.

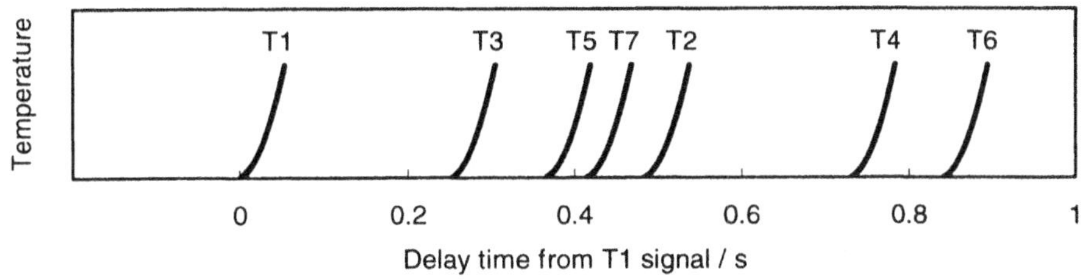

Figure 6. Typical results for experimental condition B (see Table 4).

Table 4. Results for experimental condition B (see Table 2).

Run	End*	Delay time from T1[#]					
		Inner			Outer		
		T3	T5	T7	T2	T4	T6
		ms	ms	ms	ms	ms	ms
3 - 2	open	253	366	415	484	730	842
	closed	283	404	444	462	775	887
6 - 3	open	124	143	136	250	587	779
	closed	226	250	260	364	470	777

* open: both ends of the pipe were open; closed: the end far from ignition was closed.
[#] Calculated from the difference in time that the temperature rose to 1073K (0.016V).

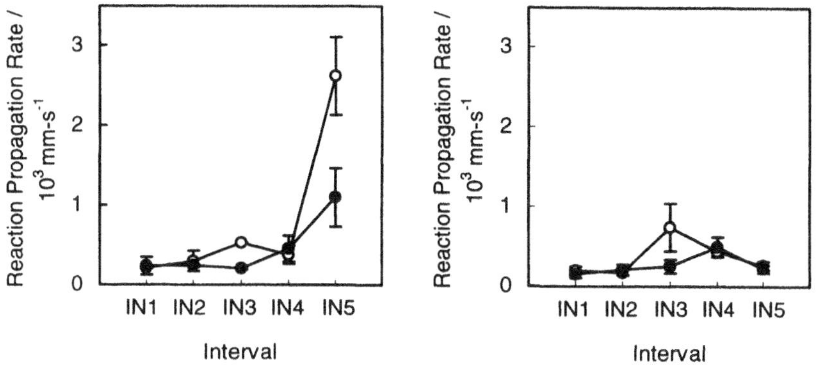

Figure 7. Typical results with experimental condition C. O run 3-1. ● run 3-3.
Left: both ends of the pipe open. Right: one end closed.

Blowing off of product occurred for all samples. When one end was closed, the rate was increased in two samples and decreased in one. Samples 6-1 and 6-2 were from 200 mm/s to 500 mm/s; sample 6-4 decreased from 700 mm/s to 100 mm/s (see Fig. 8). These results show that the reaction propagated rapidly down the interior surface first because of the rapid movement of the high temperature vapor down the open tube. This was the same result as in our previous work. The closure of one end of the pipe caused the high temperature vapor to move less rapidly, decreasing the rate of propagation of the reaction.

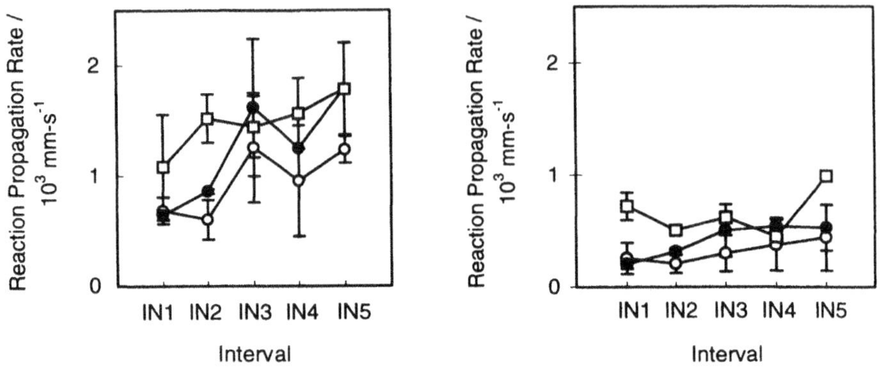

Figure 8. Typical results for experimental condition C. O - run 6-1. ● - run 6-2. □- run 6-4. Left: both ends open. Right: one end closed.

Small Pipes Produced by the C-T Process

Experiments were performed under condition D in order to characterize the production of small pipes by the C-T process. The samples were C-T processed, and then cross sections of each sample were examined to determine parameters such as the thermit reaction propagation and the separation of metal layer and ceramic layer. Table 5 shows the results.

Table 5. Experimental results using condition D (see Table 3) versus centrifugal acceleration.

Run	Reaction heat* kJ / g	Degree of completion of the C-T process#					
		0*g*	50*g*	80*g*	100*g*	120*g*	150*g*
0 - 3	1.0	N	N	--	--	--	--
3 - 4		CP	CP	N	--	--	--
6 - 5		CP	CP	CPL	CPL	CPL	CPL
0 - 4	1.5	CP	--	N	--	--	N
3 - 5		CP	--	CP	--	--	P
6 - 6		CP	--	CPL	--	--	CPL

* Reaction heat per gram of outer metal pipe.

CPL: complete propagation with layer separation. CP: complete propagation without layer separation. P: self-propagation with non-reacted removal. N: non-propagation. --: no experiment performed.

For runs performed without a central tube open (0-3 and 0-4), reaction propagation without centrifugation (0*g*) occurred only for 0-4. For runs with a 3-mm diameter open tube (3-4 and 3-5), although the reaction propagated completely with low centrifugation, at high acceleration the reaction propagated incompletely or not at all. However, 6-5 and 6-6 had complete propagation at all accelerations. This result shows that the balance between liberated heat of reaction and removed heat depended on acceleration, with a larger rotation rate giving more rapid heat removal. For runs with 0-mm and 3-mm open central cores, the reaction ceased because the rate of liberation of heat of reaction was less than the rate of heat loss. For runs with a 6-mm open central core, because the reaction propagated rapidly along the inner surface, the reaction heat was liberated rapidly. Then the reaction continued to completion.

When the reaction was complete, the resulting structure depended on the centrifugal acceleration and a separated metal/ceramic layer was apparent. For 6-5 and 6-6, the layer separation occurred at high rotation rate. This result agrees with our previous results for larger diameter pipes.

CONCLUSIONS

We confirmed that the C-T process can be used to coat the interior of small-diameter pipes. We investigated the characteristics of thermit reaction propagation in small-diameter pipes as a function of the packing density of the thermit powder, the diameter of the inner open core, vertical or horizontal orientation, and centrifugal acceleration. For vertical pipes with no open core (completely filled pipe), radial heat conduction and movement of high temperature vapor through the thermite powder predominated. This was the same as our previous results for vertical experiments with larger diameter pipes. For pipes with a central open core, the high temperature vapor generated by the thermit reaction moved rapidly through the core, and subsequently the reaction propagated from the inner surface to the metal pipe wall. When one end of the pipe was closed, the speed of reaction propagation along the inner surface was slower than when both ends were open.

The characteristics of small-diameter pipes produced by the C-T process were investigated. The reaction propagation was controlled when the central core inside the compacted powder was properly formed. Whether reaction propagation occurred or not depended on the relative rates of release of the heat of reaction and the rate of heat loss. Because the reaction propagation speed was small for pipes with a 0 and 3-mm central core, the rate of liberation of heat of reaction heat was not sufficient for continued reaction propagation. Therefore, with a 6-mm core the reaction propagated to completion, while it did not for 0 and 3-mm core samples. For the 6-mm core runs with complete reaction propagation, the larger the rotation rate, the better was the structure produced. Separation of the metal layer and the ceramic layer was complete for a run at 80*g* for a 6-mm core. This result also agrees with our previous results for larger diameter pipes.

REFERENCES

1. O. Odawara, *U.S. Patent* 4,363,832 (1982); *Japan Patent.* 1,144,098 (1983).
2. O. Odawara, J. Ikeuchi, Y. Ishii, H. Yamasaki, and M. Sato, Effect of thermal insulation in the production of metal-ceramic composite pipes by a centrifugal-thermit process, *J. Japan Inst. Metals* 50:763 (1986).
3. O. Odawara, K. Nagata, K. S. Goto, Y. Ishii, H. Yamasaki, and M. Sato, Ceramic layer composition of metal-ceramic composite pipes produced by a centrifugal-thermit process, *J. Japan Inst. Metals* 52:116 (1988).
4. O. Odawara, Ceramic lined pipes produced by a centrifugal-thermit process, *Trans JIM* 26:578 (1985).
5. O. Odawara and J. Ikeuchi, Alumina and zirconia ceramic lined pipes produced by a centrifugal-thermit process, *Trans JIM* 27:702 (1986).
6. O. Odawara, Long ceramic-lined pipes produced by a centrifugal-thermit process, *J. Am. Ceram. Soc.* 73:629 (1990).
7. O. Odawara and J. Ikeuchi, Effect of centrifugal force in the production of composite pipes by a centrifugal-thermit process, *J. Japan Inst. Metals* 49:801 (1985).

CERAMIC-LINED COMPOSITE PIPES PRODUCED BY THE GS-T SHS PROCESS WITH MECHANICAL VIBRATION

Jianjiang Wang, Zhongmin Zhao, Minghui Ye, Xinkang Du, Jinhua Wen

Institute of Advanced Materials, Shi Jia Zhuang Mechanical Engineering College, Shi Jia Zhuang, P.R. China, 050003

ABSTRACT

Self-propagating High-temperature Synthesis (SHS) with gravitational sedimentation and mechanical vibration was used to produce ceramic-lined composite pipes. Increasing the frequency of mechanical vibration increased the combustion temperature, propagation rate, SHS transformation rate, Al_2O_3-Fe liquid-liquid phase separation, and the ceramic solidification rate. Increasing the frequency also changed the size, growth orientation and morphology of the Al_2O_3 dendrites. The mechanical properties of the pipes and the surface quality of lined ceramics were improved to a great extent following mechanical vibration with increasing frequency.

INTRODUCTION

As described elsewhere in these proceedings, Odawara has developed the centrifugal-thermit process for coating the interior of large-diameter straight pipes [1]. With the traditional Al-Fe_2O_3 mixture, rapid rotation during SHS combustion causes an iron layer to bond with the initial pipe, followed by a layer of Al_2O_3. In 1996 we invented a new method to produce ceramic-lined composite pipes, the SHS Gravitational-Separation Thermit (GS-T) process.[2] It is especially suitable for production of ceramic-lined pipes with small inside diameter and bends. The initial pipe is held vertically and filled with the thermit powder mixture. The top of the thermit is ignited. As the combustion wave travels downward, gravity causes the metal and ceramic phases to separate such that the ceramic becomes bonded directly to the initial pipe, followed by a layer of metal. Because of their resistance to corrosion, wear, thermal shock and mechanical impact, these ceramic-lined pipes have broad prospects for applications. For example, they have been used successfully for fine coal delivery systems in large iron and steel enterprises.[3,4] As a result, the process is attracting more and more attention.

Thermit combustion and ceramic solidification are basic and important processes in

Processing by Centrifugation, Edited by Regel and Wilcox
Kluwer Academic/Plenum Publishers, New York, 2001

use of the SHS gravitational-separation thermit process for production of ceramic-lined composite pipes.[3] Their control is key for improvement of structure and properties, decreasing the qualification time and reducing the cost.

Here, we used the GS-T SHS process to produce ceramic-lined composite pipes with mechanical vibration using different amplitudes and frequencies of up-and-down motion with one degree of freedom. We determined the influence of mechanical vibration on thermit combustion, ceramic solidification, and the resulting structure and properties of pipes in order to search for a path for quality control of ceramic-lined composite pipes.

EXPERIMENTAL MATERIALS AND PROCEDURES

Figure 1 shows a sketch of the mechanical vibration device. For the experiments we used stainless steel pipes of 1Cr 18Ni 9Ti with an outer diameter of 24 mm and a wall thickness of 2.5 mm cut to a length of 300 mm. The thermit mixtures were Al powder (100~+200 mesh, purity 99.5%) and Fe_2O_3 powder (-200~+325 mesh, purity 97.5%), with added SiO_2 powder (-100~+200 mesh, purity 99%) and CrO_3 powder (-100~+200 mesh, purity 99%). After drying and screening, the mixture was blended according to the molar proportions Fe_2O_3 + Al +2%SiO_2 +10%CrO_3. After blending with a mixing machine, the mixture was poured into a vertical pipe, which was sealed with an aluminum cap at the bottom end under a certain pressure. The mixture was ignited by electrically heating the top with a tungsten wire, 0.5 mm in diameter. After combustion was started, the mechanical vibration device was switched on to exert mechanical vibration with the chosen amplitude and frequency in the form of an up-and-down motion with one degree of freedom. The parameters of mechanical vibration for each sample are shown in Table 1, in which each sample is identified with a corresponding label. The combustion time was recorded with an electronic stopwatch. The temperature at the combustion front was examined with an WFF-202 full radiator thermometer.

The samples cut from ceramic-lined pipes were used for property tests and structure analysis after polishing. The phases constituting the ceramic layers were determined with x-ray diffraction. The porosity and hardness of the lined ceramics were obtained by Archimedes method and microhardness tests, respectively. The compressive strength $\bullet_p$ and compressive shearing strength $\bullet_s$ were measured.

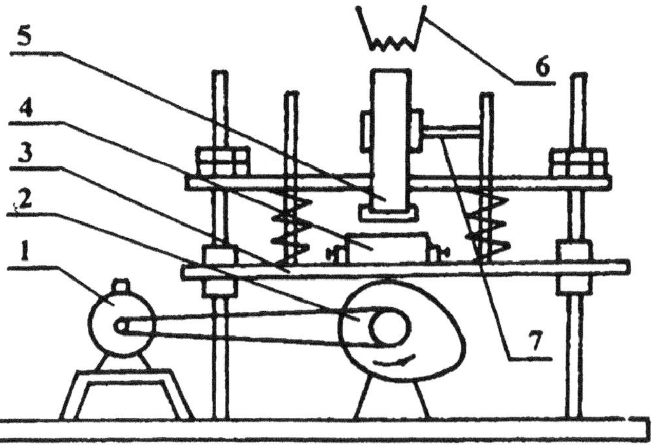

Figure 1. Schematic diagram of mechanical vibration device. 1 - electrical motor with stepless speed regulation, 2 - cam, 3 - vibration platform, 4 - graphite crucible, 5 - steel pipes, 6- W ignition wire, 7 – fixture.

Table 1. Parameters of mechanical vibration for each sample

Sample identification	A1	A2	A3	A4
Frequency (Hz)	0	2	5	5
Amplitude (mm)	0	5	5	10

RESULTS AND DISCUSSION

Influence of Mechanical Vibration on Thermit Combustion and Phase Structure

Table 2 shows the experimental combustion temperature and propagation rate for each sample. Increasing the frequency of mechanical vibration greatly promoted the SHS combustion process. Increasing the amplitude decreased the combustion temperature and increased the propagation rate. When the GS-T SHS process was used to produce ceramic-lined pipes, thermit combustion started after the Al powder had melted, and proceeded mainly by a diffusion mechanism with movement by a combustion wave with layered advance.[3] As learned from a combustion model for layered reactants based on the geometry of a layered combustion front, the combustion rate of layered reactants is directly proportional to the effective thermal conductivity K_{eff}, diffusion coefficient D_{eff} and combustion temperature T_c, and is inversely proportional to the square of the total thickness L of the layer zones in the reaction area and to the activation energy E^*. Mechanical vibration reduces the internal friction and shearing strength among particles in thermit,[7] causing a large quantity of gas in the preheating zone to escape upward, in turn leading to an increase in the mixture density and contact between particles in the zone. Hence, vibration increased the effective thermal conductivity K_{eff} on the layered combustion front. Meanwhile, under the inertia force caused by mechanical vibration, the flow of molten Al around nearby Fe_2O_3 particles was increased, thereby increasing the contact efficiency and diffusion area between Fe_2O_3 particles and molten Al. As a result, mass transport between Al and Fe_2O_3 particles was enhanced, resulting in an increase in effective diffusion coefficient D_{eff}. Mechanical vibration also increased convective heat transfer between the combustion front zone and the synthesis zone, decreasing the temperature gradient in the entire combustion area, further increasing the combustion temperature T_c. In brief, mechanical vibration effectively promoted the combustion process through increasing K_{eff}, D_{eff} and T_c.

Table 2. Influence of mechanical vibration on combustion temperature and rate.

Sample	A1	A2	A3	A4
Combustion temperature (K)	2950	3050	3150	2900
Propagation rate (mm/s)	9.8	11.7	15.3	18.1

The results for A4 made with increased vibration amplitude can be attributed to throwing out partially melted mixtures from the pipe, causing the distance to the combustion wave to move downward and be shortened, thus increasing the subsequent propagation rate. But because the heat-loss was increased following heat production in the synthesis zone, the combustion temperature was subsequently decreased.

It can be seen from the x-ray diffraction patterns in Fig. 2 that the ceramic layers in A1 and A3 were both composed of an •-Al_2O_3 matrix and $FeO \cdot Al_2O_3$ spinel. Some SiO_2

quartz phase and CrO_3 were found, but did not change the phase compositions in the ceramic liner, indicating that the constituents of the ceramic layers were not changed by mechanical vibration. As observed from the microstructure of the ceramic layers in A1 and A3 shown in Fig. 3, the amount of spinel phases in A3 was less than in A1. The volume fraction of spinel phase in the ceramic layers of A1, A2, A3 and A4 was 21.6%, 19.2%, 15.9% and 24.7%, respectively, via quantitative metallographic analysis.

The above results confirm that mechanical vibration with increasing frequency greatly accelerated the thermit combustion process, which was mainly controlled by a diffusion mechanism. From the theory for SHS transformation rate,[6] increasing the frequency of mechanical vibration caused the propagation rate V rate to increase, leading to reductions in the mixture density ρ the preheating zone, the effective thermal conductivity K_{eff}, the combustion temperature T_c, and the temperature gradient dT/dx in the whole combustion volume. This increased the SHS transformation rate $\eta(x)$ with increasing frequency of mechanical vibration. The vibration reduced the amount of metastable FeO in the molten synthesis region and the content of $FeO \cdot Al_2O_3$ spinel phase in the resulting ceramic layers. The SHS transformation rate was reduced owing to the reduction in • , T_c and dT/dx.

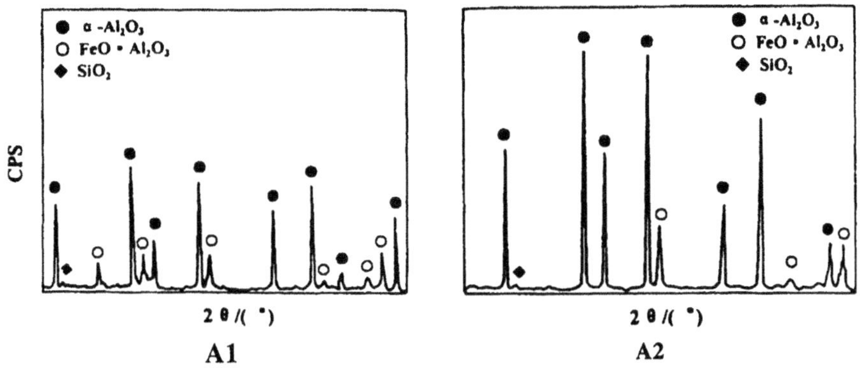

Figure 2. X-ray diffraction patterns for ceramic layers

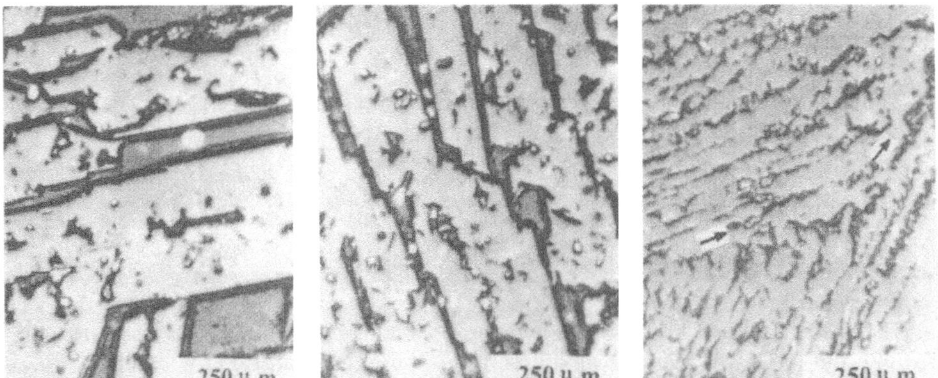

Figure 3. Microstructure of ceramic layers

Influence of Mechanical Vibration on Solidification Structure of Ceramic Layers and Properties of Composite Pipes

Table 3 shows the volume fraction of Fe phase, the average diameter of Fe particles, the porosity and the average thickness of the ceramic layers. It can be seen that the Al_2O_3-Fe liquid gravitational separation was accelerated by mechanical vibration, and the gas-phase content of the ceramic layers was greatly reduced. Combined with a study of the separation mechanism[8] for two immiscible liquids, it was concluded that increasing the frequency of vibration increased the temperature of multiphase melts by increasing the combustion temperature, thereby decreasing the viscosity of the melts. As a result, the migration rate of Fe drops in molten Al_2O_3 was accelerated, increasing the probability of Fe droplets impacting one another and ripening to larger sizes. Hence, the size of Fe drops in molten Al_2O_3 was increased and the volume fraction of Fe drops was reduced, increasing the gravitational separation of molten Al_2O_3-Fe via Stokes migration, as shown in Fig. 4. At the same time, the nucleation rate of bubbles also increased, and the merger of bubbles was enhanced. Hence, vibration increased the removal of gas from the melts, resulting in a reduction in porosity of the ceramic layers and an increase in the relative density of the ceramic liner. Moreover, owing to the improved fluidity of the melts, vibration decreased the volume of molten ceramic adhering to the inside surface of the initial pipe, decreasing the average thickness of the ceramic layer and making the final surface of the inside tube more even. For experiment A4, the temperature of the melt was reduced by the decrease in combustion temperature caused by increasing the vibration amplitude, resulting in a sharp increase in the viscosity of the melt and thereby retarding the liquid-liquid phase separation and ceramic densification. Compared with experiment A1, the average distance between dendrites in ceramic layers of experiments A2 and A3 was reduced gradually, dendrites were refined gradually, the growth orientation of dendrites became more evident, and branching was reduced, as shown in Fig.3.

Table 3. Mechanical properties of ceramic layers on composite pipes.

Exp	Fe volume %	Fe particle diameter •m	Ceramic porosity %	Ceramic hardness Kg_f/mm^2	Ceramic thickness mm	Compressive strength MPa	Shear strength MPa
A1	5.2	1.2	8.8	1503±203	1.2	387	16.5
A2	4.0	1.0	6.6	1630±216	0.8	408	17.8
A3	2.4	0.8	3.5	1721±105	0.6	430	20.5
A4	10.8	5.6	12.7	1320±305	1.0	352	10.3

In addition to vibration decreasing the temperature of the melt via acceleration of the combustion process, it also increased convection in the melt during ceramic solidification. As a result, heat exchange of the melt with the outside and the supercooling of liquid ceramic film adhering to the surface of the pipe were both increased, resulting in an increase in the nucleation rate of Al_2O_3 crystals in the melt and refinement of the dendrites in the ceramic layers during solidification. Moreover, because the violent convection caused by vibration severely disturbed the temperature along the liquid-solid interface, the tips of the dendrites were uneven in different areas. Thus, those dendrites deep inside the liquid grew more quickly because solute-enriched melt near their tips was more rapidly displaced and heat transport in those areas was more intense. Those dendrites that lagged behind were slowed down and eventually stopped. Therefore, a preferred orientation of dendrites in the ceramic layers of experiments A2 and A3 gradually became more pronounced and branching gradually weakened.

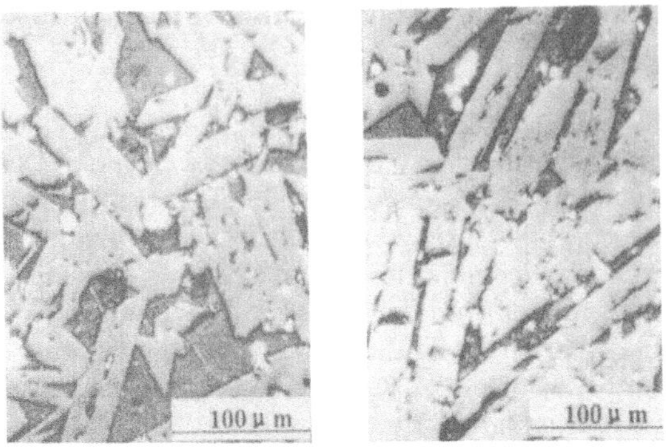

Figure 4. Iron particles in ceramic layers

When the convection became turbulent, the dendrite arms of the Al_2O_3 primary crystals were continually removed by flowing melt. When dendritic arms broke off they served to increase the number of nuclei in the melt and caused an equiaxed microstructure, as shown in Fig. 5 for experiment A3. In addition, the growth orientation of • -Al_2O_3 dendrites was changed in some areas, as indicated by the arrow in Fig. 3(c). That behavior may be attributed to enhanced flow of melts around Al_2O_3 primary dendrites, altering the temperature and composition of the melt at the dendrite tips in a manner depending on the direction of the flow relative to the dendrite orientation. Those dendrites growing in the direction opposite to the flow grew faster, while those growing in the same direction grew slower. Hence, the increased frequency of vibration could cause the growth orientation of some Al_2O_3 dendrites to change.

As could be seen in Table 3, comparing the results of experiments A1 and A3 shows that vibration with amplitude 5 mm and frequency 5 Hz increased the hardness of the ceramic layers by 14.5%, the compressive strength by 11.1%, and the shear strength of the pipes by 24.2%. Hence, the hardness depended on the volume fraction of spinel in the ceramic layer. The compressive strength depended on the porosity of the ceramic liner and the refinement of the • -Al_2O_3 dendrites. The compressive shear strength depended on the heat produced by the combustion and by the total amount of thermal expansion of the steel pipes.

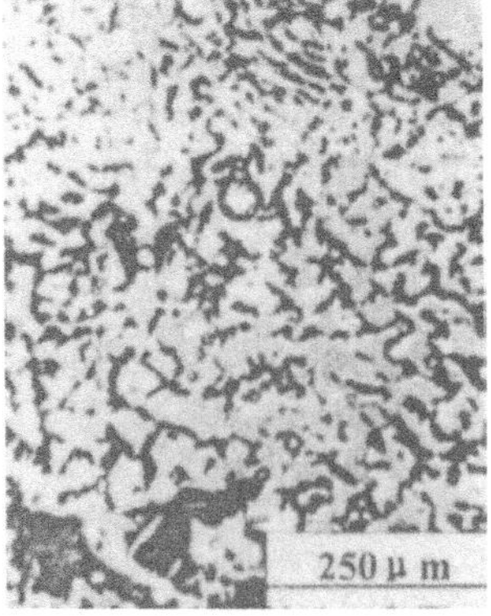

Figure 5. Broken dendrites in A3.

CONCLUSIONS

1. Increasing the frequency of mechanical vibration effectively enhanced the thermit combustion process by increasing the combustion temperature, the propagation rate and the SHS transformation rate.
2. Increasing the vibration frequency effectively accelerated the Al_2O_3-Fe liquid-liquid phase gravitational separation and ceramic densification, caused the dendrites in ceramic liners to be refined, and yielded equiaxed grains in some regions of the ceramic layers.
3. Increasing the vibration frequency effectively improved the microstructure, the properties of composite pipes, and the surface quality of ceramic liners.

Acknowledgement

This work was supported by the National Natural Science Foundation of the Peoples Republic of China via grant number 59972043.

REFERENCES

1. O. Odawara and J. Ikeuchi Jun, Study on composite materials with a centrifugal-Thermit process, *Trans. Japan Inst. Materials* 45: 316-321 (1981).
2. Z. Zhao, J. Wang and M. Ye, Structure and properties of the ceramic-lined pipes produced by SHS gravitational separation-Thermit process, *Mechanical Engineering Materials* 22: 34-37 (1998).
3. L. Zhang, Z. Zhao and J. Wang , Study on synthesis process of ceramic-lined composite pipes using gravitational separation SHS method, *Foundry* 6: 4-10 (1998).
4. J. Wang and Z. Zhao Zhongmin, Production of a novel ceramic-lined coal-dust spurting gun, *Ironmaking* 17: 35-37 (1998).
5. P. Huang, *Powder Metallurgy Principles*, Metallurgical Industry Press, Beijing (1982) pp 373-374.
6. Z.A. Munir, Self-propagation exothermic reaction: the synthesis of high-temperature materials by combustion, *Materials Science Reports* 3: 277-365 (1989).
7. W. Zhang, *Mechanical Vibration Principles,* Mechanical Industry Press, Beijing (1983) pp 1-2.
8. J. Zhao, *Liquid-liquid phase separation mechanism and solidification behavior of Zn-Pb alloy*, Ph.D. Thesis, Harbin Institute of Industry, Harbin, China (1994).

DIRECTIONAL SOLIDIFICATION OF ALUMINUM AND LEAD IN A CENTRIFUGE

Sh. Mavlonov and H. Shodiev

Physics Department, Samarkand State University
703004 Samarkand, Uzbekistan

ABSTRACT

Aluminum and lead were solidified in a centrifuge at an acceleration of up to 1000*g* at a temperature of 1050K. With increasing acceleration, the segregation coefficient decreased by one order of magnitude and the purity of aluminum was increased up to 99.9999% and lead up to 99.999%.

INTRODUCTION

Several mechanisms influence compositional segregation during solidification.[1] A Centrifugal Force Field (CFF) has an effect on each of these mechanisms. Studies on the influence of acceleration are traditionally divided into two ranges:[2]

- Microgravitational segregation studied in space conditions where the average acceleration is small (down to $10^{-6}g$).
- Macrogravitational segregation studied at higher than 1 *g* acceleration.

Here, we describe directional solidification of Al and Pb in CFF. The correlation function of regular solution theory was utilized to interpret the experimental results.

DIRECTIONAL SOLIDIFICATION OF ALUMINUM

Solidification of Al was carried out in a centrifuge that allowed us to obtain an acceleration up to 1000*g* at 1050 K. The freezing rate was about 2 cm/hour, the length of the melt was 1 cm, and the sample's length was 10 cm.[3] Neutron-activation analysis was used to measure impurity concentrations. Figure 1 shows the resulting concentration profiles for experiments performed without centrifugation (1 *g*). Curve 1 is what one would expect for a segregation coefficient of 0.1 without convection in the melt, while curve 2 is with convection. By fitting such concentration profiles to theoretical relationships[4] we obtained the segregation coefficients shown in Table 1 for $N = a/g$.

Processing by Centrifugation, Edited by Regel and Wilcox
Kluwer Academic/Plenum Publishers, New York, 2001

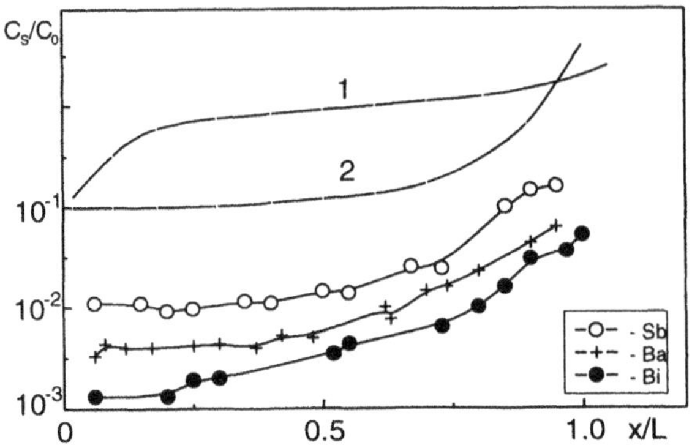

Figure 1. Concentration profiles for Sb, Ba and Bi in Al from solidification at 1*g*.

Here, C_o is the initial impurity concentration in the melt, C_s is the impurity concentration at distance x down a sample solidified for length L and $C_s(0)$ is the impurity concentration in the solid first frozen. .

Table 1. Segregation coefficients in Al by solidification at 1*g* (K_g) and at 3*g* (K_a).

Impurity	K_g	C_o	$C_s(0)$	N	K_a
Mn	0.8	$1.1*10^{-4}$	$1.6*10^{-5}$	3	0.14
Cu	0.17	$5*10^{-4}$	$6*10^{-5}$	3	0.05
Zn	0.4	$3*10^{-4}$	$4*10^{-5}$	3	0.13
La	0.09	$4*10^{-5}$	$2*10^{-6}$	3	0.01
Ag	0.07	$9*10^{-5}$	$2*10^{-6}$	3	0.02

Figure 2 shows the dependence of segregation coefficient on acceleration for Zn, Mn, Cu, La and Ag in Al. This shows that acceleration up to about 15 g decreased the segregation coefficient, leading to increased purity of the Al. Beyond 15 g, there was little change in segregation behavior. Because the segregation coefficient deviates increasingly from unity as convection is increased, these data imply that centrifugation increased convection in the melt up to about 15 g, but not appreciably for higher accelerations.

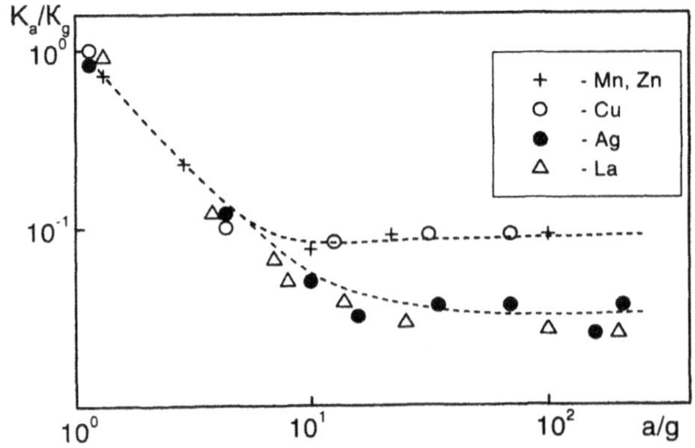

Figure 2. Dependence of effective segregation coefficient in Al on acceleration.

DIRECTIONAL SOLIDIFICATION OF LEAD AT HIGH GRAVITY

Directional solidification is often used to prepare single crystals for use in studying materials and for applications. In addition, high-purity materials are needed for the electrotechnology industry, semiconductor technology and nuclear energy. Directional solidification has been shown to be useful for purification of lead.[4] An impurity can easily be removed by solidification if the segregation coefficient K is much less than 1, while purification is difficult if K is near 1. Table 2 and Fig. 3 show experimental results for solidification of Pb during centrifugation. As with aluminum, centrifugation decreased K, presumably because of increased convection in the melt.

Table 2. Effective segregation coefficients for impurities in Pb.

	%		$K = C_s(0)/C_o$	
Impurity	C_o	$C_s(0)$	**a = 1g**	**a = 1000g**
Sb	0.01	0.003	0.3	0.03
Mg	0.01	0.004	0.4	0.02
Bi	0.01	0.004	0.4	0.01
Ba	0.03	0.003	0.1	0.03
K	0.01	0.004	0.4	0.02

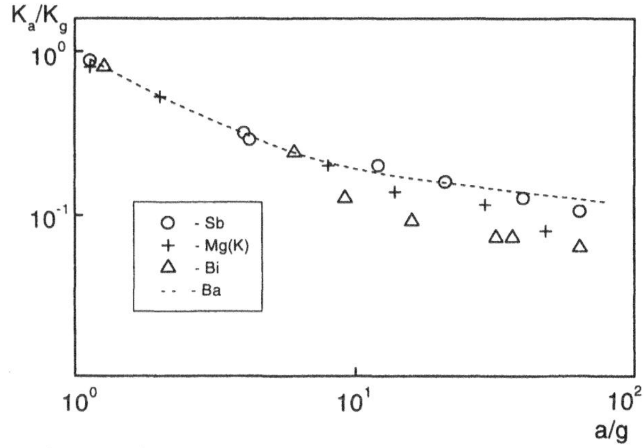

Figure 3. Dependence of effective segregation coefficient in Pb on acceleration.

REFERENCES

1. M. Flemings, *Solidification Processing*, McGraw-Hill, New York (1974).
2. L. Regel, *Cosmic Object Transaction*, sc. v. 35, VINITI, Moscow (1993).
3. Sh. Mavlonov, Equipment for division and cleaning of substance, *Akad. Nauk USSR* 1: 203 (1984); A.S. USSR, N693575.
4. W. Vigdorovoch, Cleaning of metals and semiconductors by crystallization methods, *M. Metallurgia* (1969).
5. Sh. Mavlonov, The phenomena of crystallization in CFF and dynamo effect, *J. Cryst. Growth* 119: 167-175 (1992).
6. D. Shabanov, Sh. Mavlonov, and B. Makhmudov, Crystallization phenomenon conditions of high gravity, *Abstracts Symposium on Experimental Gravitation,* Samarkand, Uzbekistan (1999).
7. Sh. Mavlonov, B. Makhmudov, and T. Alimov, *Crystallization Purification of Al in Condition High Gravity,* Vestnik, St. Petersburg. University Series 4 (1999).
8. H. Endo, K. Tokimura, and M. Lao, Liquid metals and semiconductors under pressure, *Can. J. Phys.* 65: 266-285 (1987).

PHYSICO-CHEMICAL ANALYSIS UNDER CONDITIONS OF HIGH GRAVITY

Sh. Mavlonov and H. Shodiev

Samarkand State University, 703004 Samarkand, Uzbekistan

INTRODUCTION

Many experiments on directional solidification during centrifugation have yielded impurity segregation behavior that depends on the acceleration *a*. Traditionally, these results have been explained by changes in convection due to increased *a*, the Coriolis effect, and the dependence of the acceleration vector on position in the melt. However, centrifugation also causes a change in the pressure at the solid-liquid interface that can alter the equilibrium segregation coefficient. It is this latter effect that is examined here.

All physico-chemical and thermal properties of matter have been obtained in Earth's gravity *g*. With centrifugation, when the acceleration *a* becomes much larger than *g* ($a >> g$), it is not possible to avoid strong gravitational influences on physico-chemical and thermal properties. The goals of this paper are to:

a. Utilize data on segregation coefficient measured experimentally.
b. Calculate the gravitational changes in some phase diagrams of Al and Pb containing impurities.
c. Determine changes in the melting points of lead and aluminum caused by centrifugation.

PHASE DIAGRAMS OF ALUMINUM WITH IMPURITIES

Aluminum containing impurities has been directionally solidified in a centrifuge.[1] We used our same techniques[2,3] to produce an acceleration of 1000*g* and thereby a pressure in the melt of 4.1MPa.

We calculated the influence of centrifugation on the phase diagrams of Al containing some impurities (Zn, Mn, Cu, La, Ag), including determination of the melting temperature of Al when the impurity concentration is less than 1%. The following equations were used to calculate the influence of pressure p and impurity concentration C on the melting point T of material A:

$$\Delta T \equiv T_A - T = RT^2_A C_s / Q_B \qquad (1)$$

Processing by Centrifugation, Edited by Regel and Wilcox
Kluwer Academic/Plenum Publishers, New York, 2001

$$\frac{dT}{dp}=\frac{V_l-V_s}{S_A}=\frac{T_A}{Q_A}\left(\frac{1}{\rho_l}-\frac{1}{\rho_s}\right)=\frac{1}{h\rho_l}\cdot\frac{dT}{dQ_A} \quad (2)$$

Here, $p = \rho_l ah$ is the pressure at the freezing interface, a is the acceleration in the liquid at the freezing interface (8.5×10^3 g in our centrifuge), h is the depth of the melt above the freezing interface (here, 2 cm), C_s is the concentration of impurity B in the solid in equilibrium with C_l in the melt, S_A is the entropy of melting, Q is the latent heat of melting, T is the absolute melting temperature of the mixture of A and B, R is the gas constant; V_l and V_s are the specific volumes of the liquid and the solid, respectively, and ρ is density.

Table 1 shows the calculated results for equilibrium lanthanum concentrations in solid and liquid aluminum at normal and high gravity using a = 8,500 g, h = 2 cm, p = 4.1 MPa, T_A = 933.1 K, ρ_s = 2.700 g/cm³, ρ_l = 2.380 g/cm³, Q_A = 400 kJ/kg. Also shown are equilibrium segregation coefficients $K_0 = C_s/C_l$ measured at 1 g and calculated for 8,500 g both for Al and for other impurities in aluminum. Figures 1 to 4 show the calculated phase diagrams for Ag, La, Mn and Cu impurities in Al. We were able to draw phase diagrams when the concentration of impurity was less than 10^{-4}%. It can be seen that the change in melting point of pure aluminum is 0.5 K for a pressure of 4.1 MPa. There are predicted to be changes in the liquidus temperature and in the solubility of Zn, Mn, Cu, La and Ag in Al. at normal and high gravitation. To avoid repetition, in the table we present detailed data only for the Al-La system. For the other systems we show only the values of K_0.

Table 1. Influence of lanthanum concentration and acceleration on the equilibrium temperature T, solubility in liquid and solid, and equilibrium segregation coefficient. Below are estimated equilibrium segregation coefficients for Ag, Mn, Cu and Zn at normal and high gravity.

La C_l, cm^{-3}	T, K		K_0	La C_s, cm^{-3}	K_0	La C_s, cm^{-3}
	a=1g	8,500g	a=1g		a=8,500g	
1.3*10^{16}	934.00	933.50	0.09 for La	1.2*20^{15}	0.01 for La	1.3*10^{14}
1.3*10^{17}	933.90	933.40	0.09 for La	1.2*10^{16}	0.01 for La	1.3*10^{15}
1.3*10^{18}	933.80	933.30	0.09 for La	1.2*10^{17}	0.01 for La	1.3*10^{16}
1.3*10^{19}	933.65	933.20	0.09 for La	1.2*10^{18}	0.01 for La	1.3*10^{17}
			0.07 for Ag		0.02 for Ag	
			0.8 for Mn		0.14 for Mn	
			0.3 for Cu		0.05 for Cu	
			0.4 for Zn		0.13 for Zn	

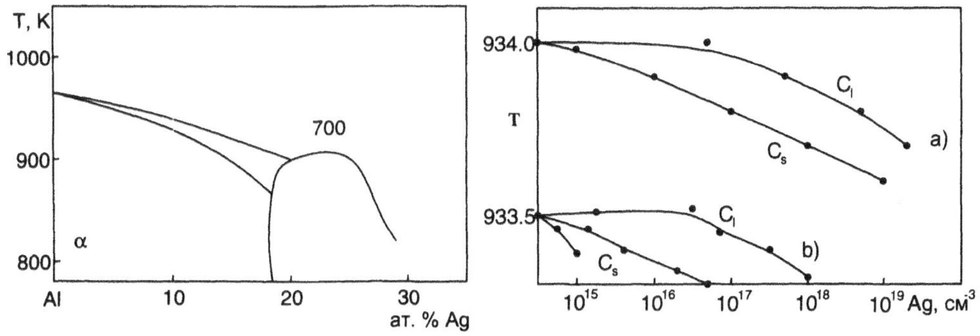

Figure 1. Phase diagrams for Al-Ag at a) a=1g, b) a=8,500g.

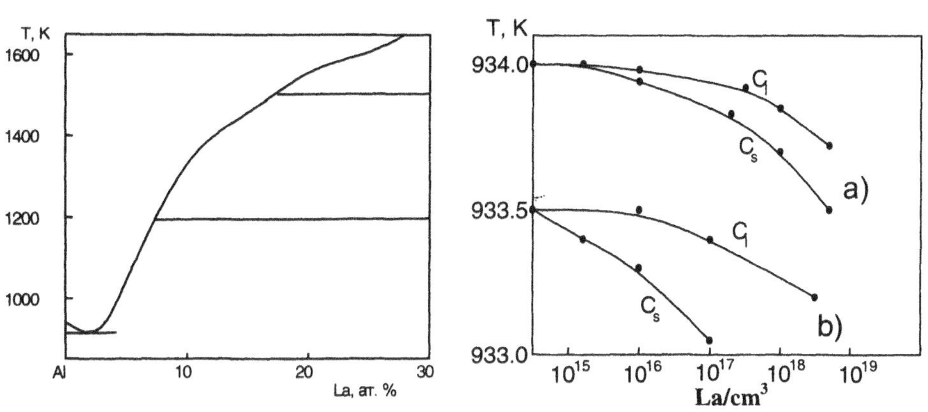

Figure 2. Phase diagrams for Al-La at a) a=1g, b) a=8,500g.

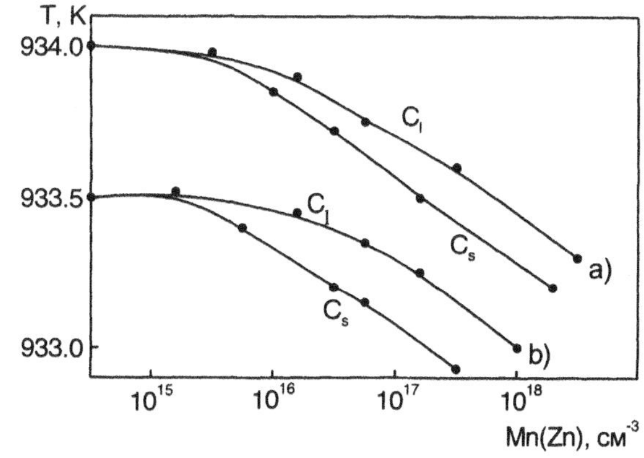

Figure 3. Phase diagram for Al-Mn and Al-Zn at a) a=1g, b) a=8,500g.

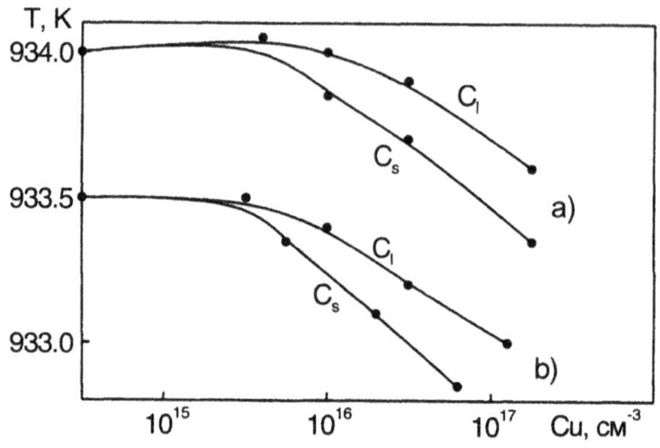

Figure 4. Phase diagram for Al-Cu at a) a=1g, b) a=8,500g.

PHASE DIAGRAMS OF LEAD

In the same way, we estimated the pressure-dependence of solubility of impurities in lead due to centrifugation at 8,500g. Figures 5 to 7 show the results the systems Pb-Sb, Pb-Bi and Pb-Ba. Thus we estimate that at 8,500g the melting temperatures of both Al and Pb increase by about 0.5K as a result of increasing the pressure in the melt. The distance between the solidus and liquidus curves also increase, causing the equilibrium segregation coefficient to increase. .

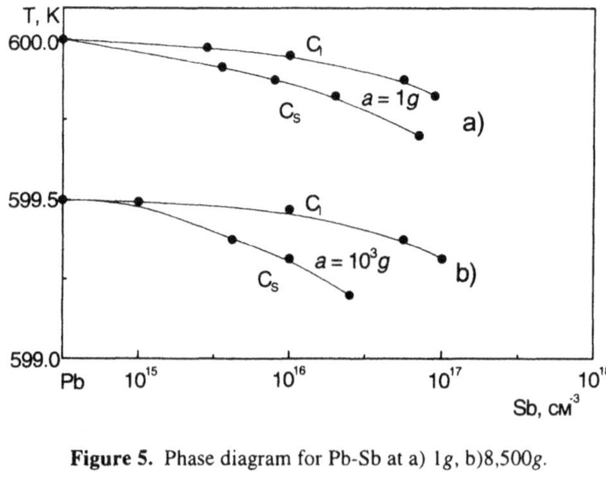

Figure 5. Phase diagram for Pb-Sb at a) 1g, b)8,500g.

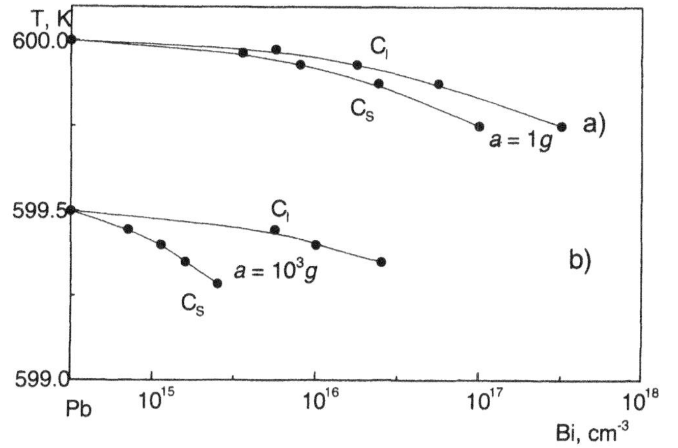

Figure 6. Phase diagram for Pb-Bi at a) $1g$, b)$8{,}500g$.

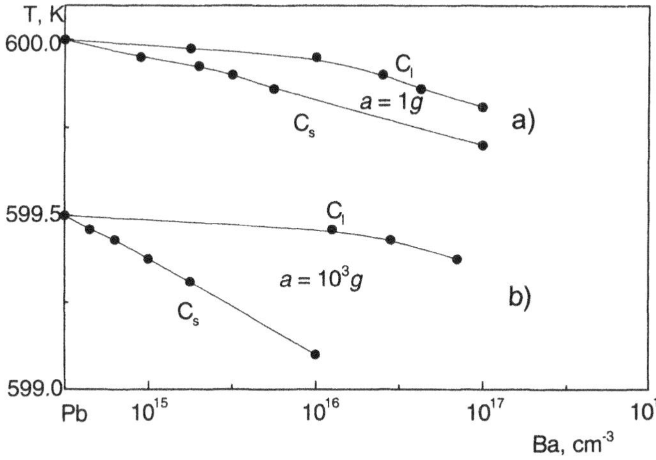

Figure 7. Phase diagram for Pb-Ba at a) $1g$, b)$8{,}500g$.

REFERENCES

1. Sh. Mavlonov, Equipment for division and cleaning of substances, *A. N. USSR* 21: 203 (1984).
2. Sh. Mavlonov, The phenomenon of crystallization in the centrifugal force fields and dynamo effect, *J. Cryst. Growth* 119: 167-175 (1992).
3. D. Shabanov, Sh. Mavlonov, B. Makhmudov, Crystallization phenomenon at the conditions of high gravity, in: *Abstracts for Symposium on Experimental Gravity*, Samarkand, Uzbekistan (1999).
4. M. Flemings, *Solidification Processing*, McGraw Hill, New York (1974).
5. Sh. Mavlonov, B. Makhmudov, T. Alimov, Some microdiagrams of the system Al-impurity, *Vistnik St. Petersburg University*, 3: 116-118 (1999).

DEFECT FORMATION IN TELLURIUM AT DIFFERENT GRAVITATIONAL LEVELS

L.L. Regel,[1] R.V. Parfeniev,[2] I.I. Farbshtein,[2] I.L. Shulpina,[2] S.V. Yakimov,[2] V.P. Shalimov,[3] and A.M.Turchaninov[4]

[1]Clarkson University, Potsdam, NY 13699-5814, USA.
[2]A.F. Ioffe Physico-Technical Institute, RAS, St. Petersburg, Russia.
[3]Baikov Institute of Metallurgy and Materials Science, RAS, 117334 Moscow
[4]RDPC "Kompozit," Korolev, Moscow region, Russia

ABSTRACT

We compared the influence of acceleration (μg, $1g$ and $5g$) during directional solidification of tellurium, both with and without seeding, on the formation of neutral and electrically-active defects. Microgravity conditions were achieved during experiments in the Crystallizator furnace on board the MIR space station. High gravity experiments were performed in the Meudon gradient-freeze furnace on centrifuge CF-18.

The concentration distributions of intrinsic defects such as electrically-active and neutral, were obtained from the density and mobility of free charge carriers measured along ingots in the temperature range 1.6 - 300K. The calculation of different contributions to hole scattering at low temperatures was made using an isotropic medium model. The results were compared with the properties of single crystals grown on earth. The ratio of the concentrations of neutral and electrically active defect was determined along the ingots and depends on the acceleration and other solidification conditions. The maximum concentration of neutral defects, $N_D \approx 10^{18}$ cm^{-3}, was determined for samples grown in μg and at $5g$ without a seed. The minimum value, $N_D \approx 10^{16}$cm^{-3}, was found for a polycrystalline sample grown at $1g$ with a seed. A Czochralski single crystal grown at $1g$ had a value an order-of-magnitude lower, 10^{15} cm^{-3}. The electrically-active defect density in the crystal with a seed was lower ($\leq 10^{15}$ cm^{-3}) than the above and decreased exponentially with increasing distance from the starting point of growth.

INTRODUCTION

This work was part of a program of investigation of the influence of acceleration on the solidification and resulting electrical properties of an anisotropic semiconductor, pure tellurium, and its compounds with selenium and silicon.[1] Among well-known elementary semiconductors (Si, Ge, Te, Se), Te and Se have extremely anisotropic structural and physical properties associated with a complex lattice structure, including spiral chains (Fig.1) packed into a hexagon, and with an anisotropic energy spectrum of free charge carriers.

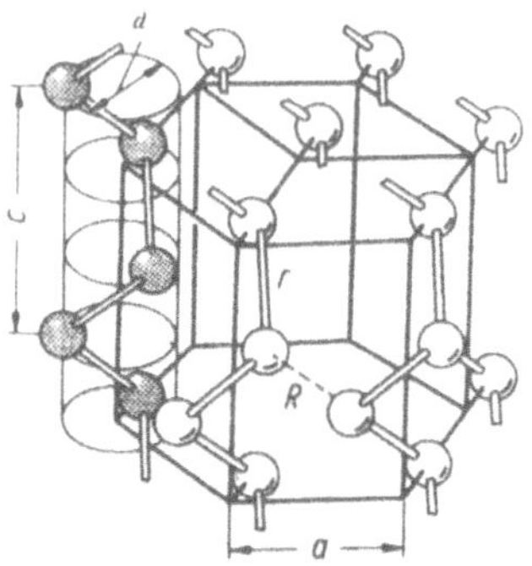

Figure 1. Crystallographic structure of tellurium. a, c, R, r and d are lattice parameters..

The hole concentration and hole mobility at low temperatures in pure tellurium are extremely sensitive to structural and point defects, mechanical stresses and dislocations. Different defects appearing in a Te crystal during solidification in ordinary conditions may cause p-type conductivity at low temperatures. That is why a program of solidification experiments from microgravity to high gravity was developed for pure Te. A gradient-freeze technique for growth of Te, Te-Se alloy, and $Te_{80}Si_{20}$ glass ingots at 1*g* and high gravity in the Meudon furnace had been created several years ago.[1] Later, three Te samples with different crystal structures were grown by the same technique in the Crystallizator furnace aboard the MIR space station in microgravity conditions.[2] Their structural and electrophysical properties were compared.[3] Here, we show that measurement of the galvanomagnetic properties of pure Te in the wide temperature range 1.6 - 300 K allowed us to determine the profiles not only of electrically active defects and impurities, but also of neutral defects by analyzing the temperature dependence of the hole mobility.

EXPERIMENT METHODS

Two series of experiments on tellurium melt solidification were fulfilled: with a seed to facilitate production of a monocrystal (S series) and without a seed (W series). The seeds were cleaved from a single crystal of pure Te along the threefold C-axis, which is the preferable direction of Te growth. In all experimental runs the seed was placed in quartz ampoule with a polycrystalline ingot of Te having a hole concentration $p_{77K} \sim 10^{14}$ cm^{-3}. This ampoule, in turn, was placed in a quartz container and sealed in a vacuum.

Directional solidification of tellurium in the Crystallizator and Meudon furnaces was accomplished using the gradient freeze technique, which prevented mechanical vibrations that would arise from translating the sample (which is also possible in Crystallizator). Each seeded S series ampoule was placed in a temperature gradient zone (40°C/cm) so that the solid-liquid interface position after heating of the furnace was estimated to have been within the seed, 2-3 mm from the union with the polycrystalline ingot. Each W series ampoule was placed so that complete melting of the ingot occurred after heating of the furnace. An experiment consisted of three stages: 1) heating of the furnace to melt the ingot; 2) homogenization of the melt, 3) slow programmed cooling to cause directional solidification.

In the Meudon furnace, experiments were performed on solidification of the samples: S-1*g*, S-5*g*, and W-5*g*. An experiment began with heating the hot zone up to 500°C and the cold zone up to 350°C to establish a gradient of 40°C/cm. This was followed by a 2-hour soak period at constant temperature for homogenization . The growth was carried out by cooling the hot zone at 75°C/h, to provide an estimate freezing rate of 1.8 cm/h. For the centrifuge experiments, the centrifuge was switched on during melting and solidification, and off again after cooling. The resultant acceleration vector was always parallel to the axis of the ampoules.

Both seeded and un-seeded experiments were also carried out using the Crystallizator furnace on board the MIR space station (S-μ*g* and W-μ*g*) and on earth without centrifugation (W-1*g*). An experiment began with heating the hot zone up to 600°C and the cold zone up to 350°C to establish a gradient of 40°C/cm. This was followed by 2 hours of homogenization at constant temperature. Growth was carried out by cooling the hot zone and the cold zone at 100°C/h, to provide a freezing rate estimated to be 4 cm/h.

The ingots grown in the microgravity conditions could be easily removed from the ampoules, in contrast to those grown at high gravity and on the earth.

The removed ingots were etched in a polishing solution of CrO_3 - HCl - H_2O to remove surface imperfections, and then selectively in 30% HNO_3 to reveal the morphology and microstructure visually. In samples of the S series, the grains were oriented mainly in the direction of the axis of growth, except for sample S-5*g*, in which there were some blocks with different orientations of the C-axis. The X-ray analysis of W-series samples showed that sample W-μ*g* had a microblock structure with random orientations of the C-axis in the crystal grains. The grains were 0.6-1 mm in the initial part (I) and 5-50 μm in the final part (II) of an ingot. All samples after solidification had nearly the same hole concentration ($p_{77K}=10^{14}$-10^{15}).

EXPERIMENTAL RESULTS

The upper part of Fig. 2 shows photographs of the ingots after removal from the ampoules. The electrical conductivity and Hall effect were measured over a wide temperature range, from 1.6K to 300K. The lower part of Fig. 2 shows conductivity profiles in these polycrystalline ingots measured at 77 K. The Hall coefficient was measured in a number of cross-sections along the ingots in a magnetic field of 0.5T. All samples were p-type at 77K. Figure 3 shows the variation of hole concentration determined from the Hall coefficient and Hall mobility $R\sigma$ at 77 K. Curves 3a and b are for Te grown with seeds at different gravity levels while 3c and d are for those without seeding. Figure 4 shows the hole concentration versus distance down the ingots. In our opinion, the observed high hole concentration in the part where solidification began and its subsequent decay arose from a high initial concentration of electrically-active structural defects whose number exponentially decreased with distance. As a result, the hole concentration decreased in the initial part of the ingot and then increased towards the end when the concentration of extrinsic defects (impurities) prevailed due to segregation at the freezing interface.[4]

The above results serve to answer the question about the possibility of producing a uniform crystal by decreasing gravity. The hole mobility $R\sigma$ was constant in the main part of the 1*g* Te sample (Fig. 3b). In the samples grown at lower and higher accelerations, the Hall mobility $R\sigma$ appeared to be smaller. This indicates the presence of the structural defects which contributed to the hole scattering at low temperatures.

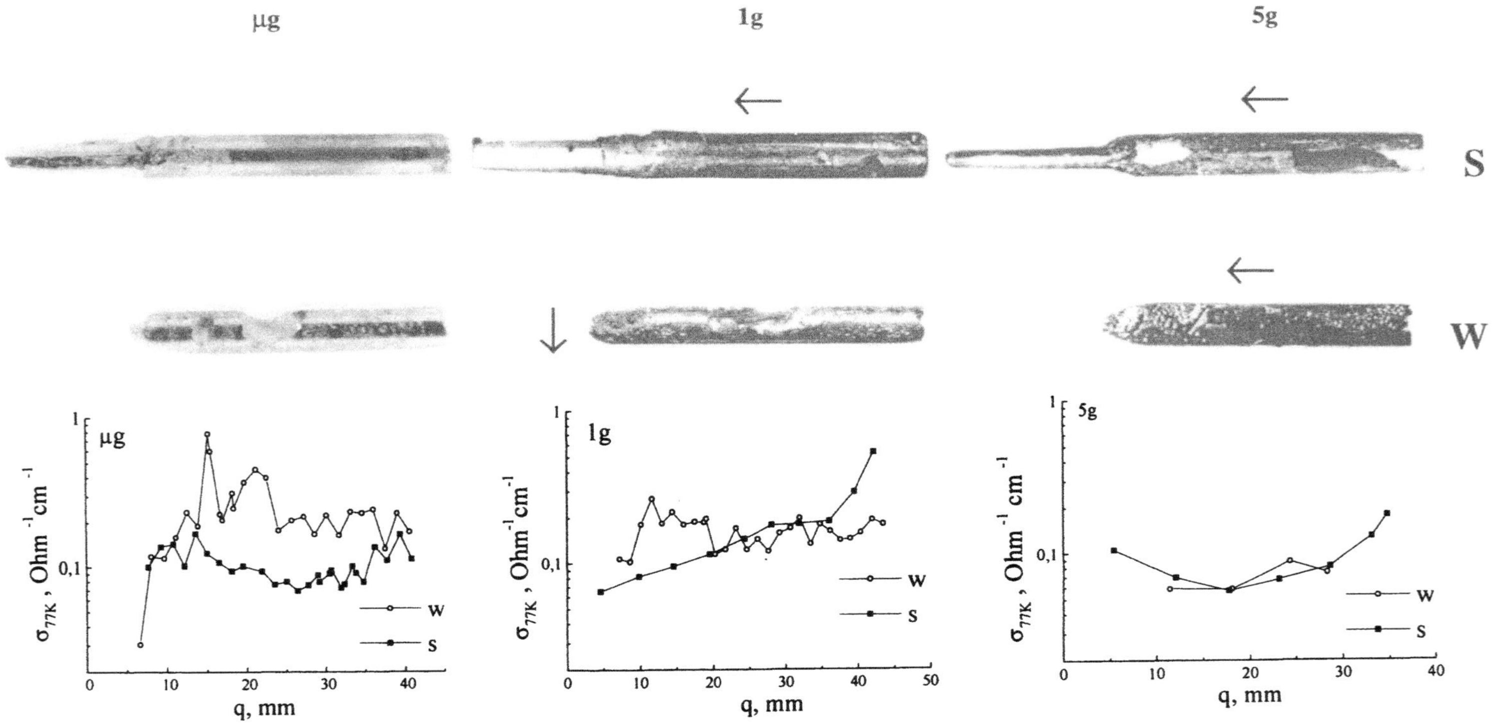

Figure 2. Samples grown in different gravitational conditions with a seed (S) and without a seed (W).
Upper: photos. Arrows indicate the gravity direction during the experiment.
Lower: variation of conductivity at 77 K along the ingots.

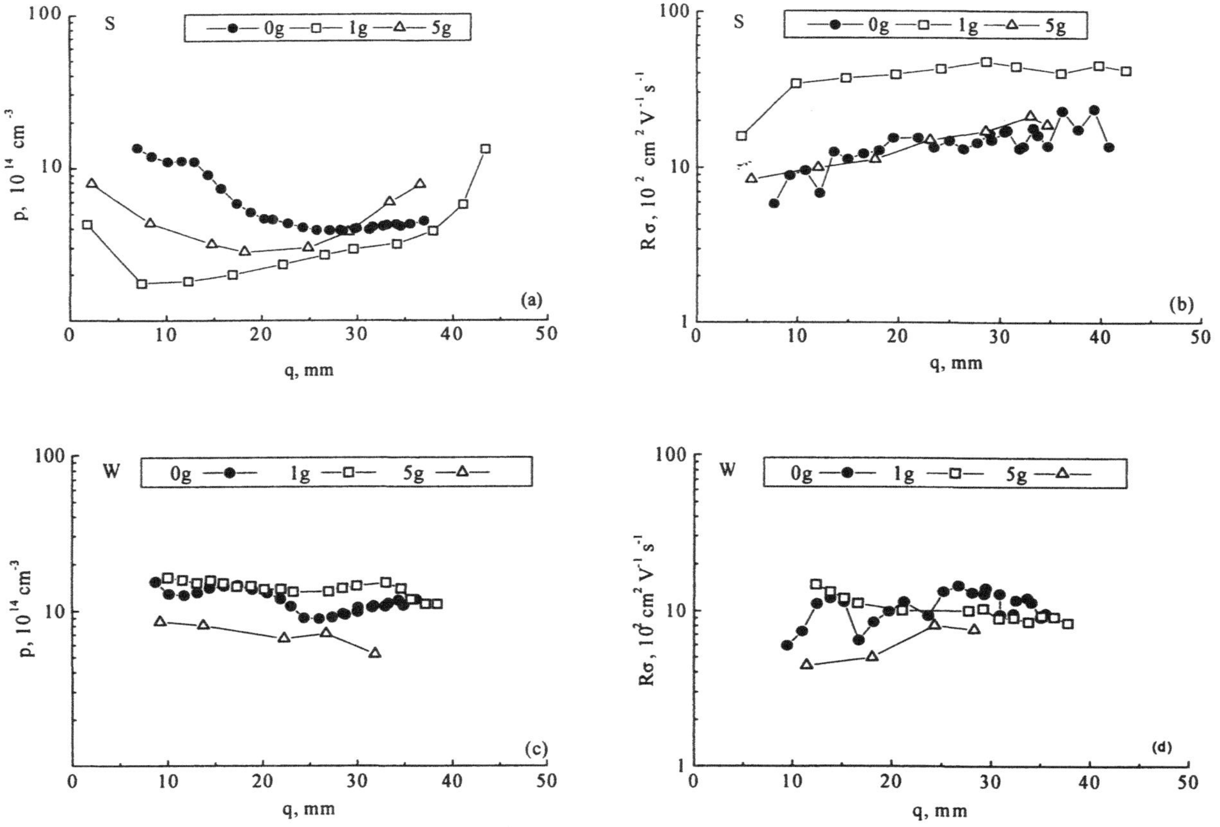

Figure 3. The hole density profiles p (a, c) and hole mobility Rσ (b, d) at 77.4 K along the Te crystals grown with seeds (a, b) and without seeds (c, d) in space (μg), on earth (1g), and in the centrifuge (5g). Here p=A/Rec, where A is Hall factor, and q is the distance from the starting point of crystallization.

Nevertheless, the mobility rose along the crystal length in the μ*g* and 5*g* samples. For measuring the temperature dependence up to helium temperatures, 8 potential contacts were placed on the samples. The results are divided into 2 regions of the samples: Part I close to the starting point of crystallization and Part II farther away. The temperature dependence of Rσ became stronger in the final part of II.[3] That corresponds to decreasing the concentration of structural imperfections in the freezing direction. So, the hole scattering by intrinsic defects became weaker in comparison with the scattering by acoustical phonons at 77K. A similar dependence of crystal perfection on the gravity level had been observed in experiments on the directional solidification of PbTe<Ag> with centrifugation.[5]

Figure 4 shows that the hole concentration profile p(q) at 77K depended on the relation between the concentrations of impurities (C_{imp}) and lattice defects (C_{def}) along the ingot. When $C_{def} > C_{imp}$, the effective impurity segregation coefficient k_{eff} cannot be determined from the free-hole density distribution along the sample. In this case, the minimum of p(q/l) shifts along the ingot axis away from the seed - ingot boundary, as was observed in the polycrystalline μ*g* and 1*g* ingots.

All of these results support an inference about the improvement of the crystal structure towards the end of Te crystals grown by the directional solidification under different gravity levels.

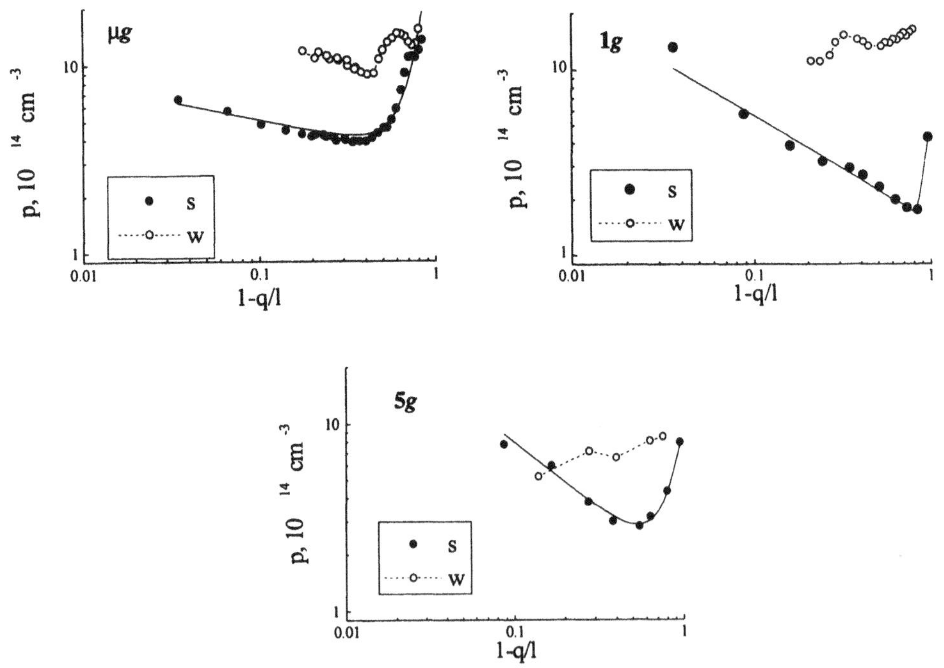

Figure 4. The dependence of hole density on the fraction solidified q/l in Te ingots grown with seeds (S) and without seeds (W) under different gravity levels. Points are experimental data and the solid curves were calculated by Eq. (1) using the parameters in Table 1.

ANALYSIS AND DISCUSSION

Figure 4 shows that Te ingots grown by directional solidification under different gravity levels had different hole density profiles measured at low temperatures. We now interpret those results.

The dopant concentration C in a solid at distance q from the initial growth interface with partial liquid mixing can be given by:

$$C = k_{eff} C_0 (1 - q / l)^{k_{eff} - 1} \tag{1}$$

where C_0 is the initial dopant concentration in the melt, l is the total length of the ingot, and q/l is the fractional distance down the ingot. The effective distribution coefficient k_{eff} is defined as the ratio of the impurity concentration in the solid to the average value in the melt, and has been assumed to be constant. The stagnant-film model to account for partial mixing yields the dependence of k_{eff} on freezing rate f:

$$k_{eff} = \frac{k_0}{k_0 + (1 - k_0) e^{-f\delta/D}} \tag{2}$$

where k_0 is the equilibrium distribution coefficient, D is the diffusion coefficient in the melt, and δ is the fictitious film thickness, which decreases with increasing convection in the melt and generally is a function of f and D.

The experimental data were fit to:

$$p(q) = k_{eff} C_0 (1 - q / l)^{k_{eff} - 1} + C_{e.a}^{0} \exp(-q / q_0) \tag{3}$$

in which two effects were taken into account: impurity segregation via Eq. (1) and the exponential decay of the intrinsic structural defect concentration in the initial transient with characteristic length q_0. Table 1 shows the parameters obtained by fitting the experimental data to Eq. (3). Figure 4 compares the experimental hole density distributions from Fig. 3 with calculated curves. Note, that the direct influence of gravity level on the distribution of electrically active intrinsic defects and impurity dopant for these directionally solidified Te ingots grown with a seed (S) and without a seed (W). Note that k_{eff} decreased with increasing acceleration.

Table 1. The values of the parameters in Eq. (3) for investigated ingots.

	C_0 (10^{14}cm^{-3})		k_{eff}		$C^0_{e.a.}$ (10^{15}cm^{-3})		q_0/l	
a	S	W	S	W	S	W	S	W
μg	4	11	0.8	0.55	6	8	0.09	0.09
$1g$	3.55		0.42		0.6		0.05	
$5g$	5.48		0.27		0.85		0.17	

For pure Te we can separate the contributions of the impurity and intrinsic defects to the free-hole density distribution. Figure 5 shows the temperature dependence of the galvanomagnetic phenomena in different samples, including the single crystal (SC) partly grown in μg. Figure 6 shows the data for inverse hole mobility. The temperature dependence of the mobility is attributed to three factors:

i) Scattering on phonons predominated at temperatures higher than 30K($1/u_L \sim T^{-3/2}$).
ii) Scattering on charge centers prevailed at low temperatures ($1/u_i \sim T^{-3/2}/N_i$).

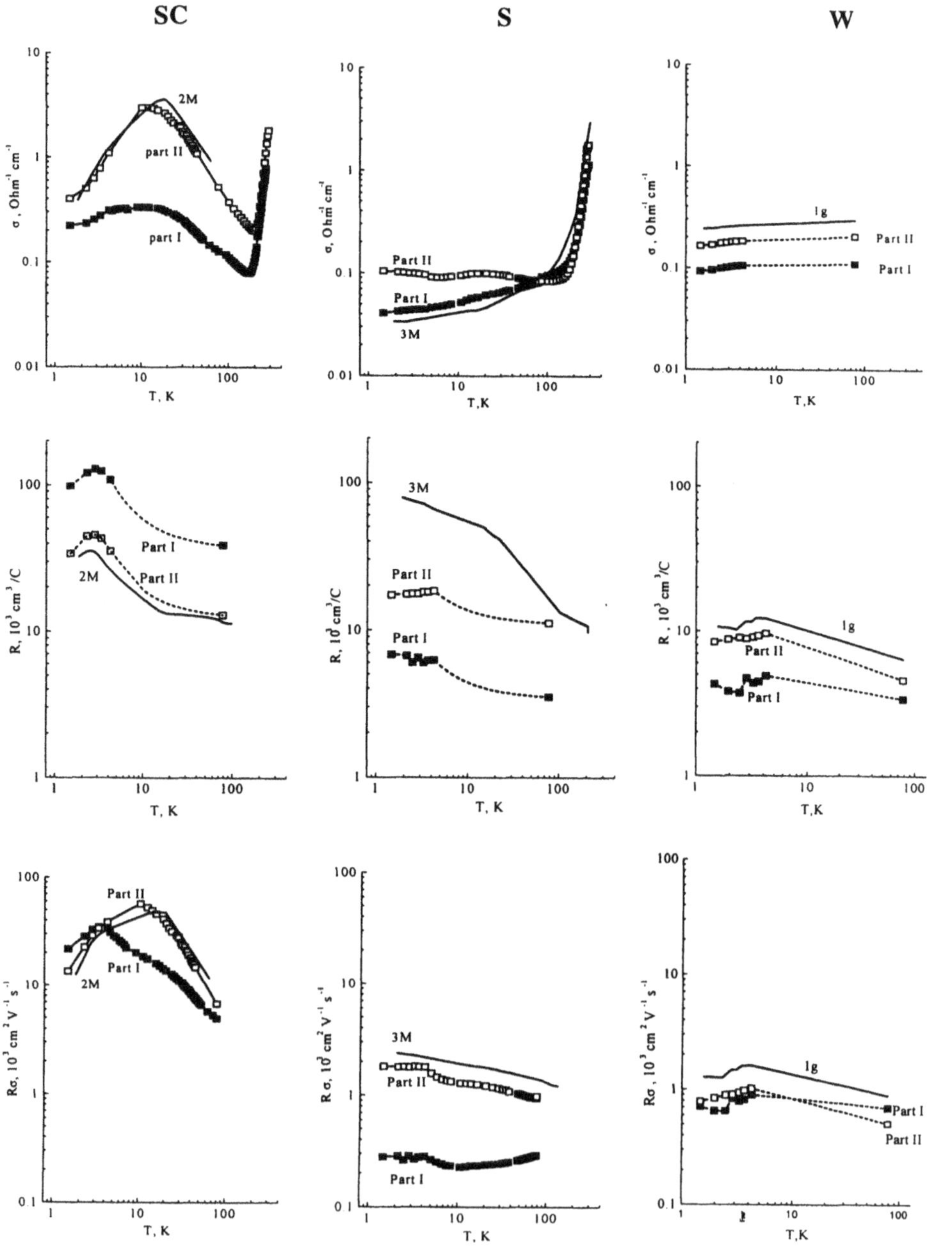

Figure 5. Temperature dependence of conductivity σ, Hall coefficient *R*, and mobility *R* σ of samples grown with different accelerations under otherwise identical conditions.
SC – single crystal, S – grown with a seed, W – grown without a seed.
Part I – close to the beginning of solidification, Part II – far from the beginning of solidification.

Scattering on defects which limits the maximum value of the hole mobility ($1/u_D \sim N_D$, supposedly independent of temperature). The resulting inverse mobility can be represented by:[8]

$$\frac{1}{u_p} = \frac{1}{u_L} + \frac{1}{u_i} + \frac{1}{u_D} = AT^{\frac{3}{2}} + BN_iT^{-\frac{3}{2}} + CN_D \qquad (4)$$

where $N_i(T) \cong p(T)$. The values of A and B were determined by fitting the data for the SC sample, as shown in Table 2. Figure 6 shows the inverse mobility data for a few ingots together with curves calculated with Eq. (4) using the parameters indicated in Table 2.

Table 2. The values of parameters in Eq. 4 for various samples.

Sample	**A** $\frac{V \cdot s}{cm^2 \cdot K^{3/2}}$	**B** $V \cdot s \cdot cm \cdot K^{3/2}$	$\mathbf{C_{ND}}$ $\left(\frac{cm^2}{V \cdot s}\right)^{-1}$	$\mathbf{N_D}$ cm^{-3}	**p** cm^{-3}
SC (II)	$3 \cdot 10^{-7}$	$4.9 \cdot 10^{-19}$	$9.3 \cdot 10^{-6}$	$7.7 \cdot 10^{15}$	$4.1 \cdot 10^{14}$
S (II)	$3 \cdot 10^{-7}$	$4.9 \cdot 10^{-19}$	$7.6 \cdot 10^{-4}$	$6.3 \cdot 10^{17}$	$5.6 \cdot 10^{14}$
W (II)	$3 \cdot 10^{-7}$	$4.9 \cdot 10^{-19}$	$1.0 \cdot 10^{-3}$	$8.3 \cdot 10^{17}$	$1.8 \cdot 10^{15}$

From Table 2 it can be seen that the differences in mobility and their temperature dependence can be explained by differences in C_{ND} and $N_i \equiv p$. We use the formula for a temperature-independent charge-carrier relaxation-time τ for scattering by hydrogen-atom type defects:[9]

$$\frac{1}{\tau} = \frac{20 \cdot \varepsilon_0 \cdot \hbar^3}{m^{*2} \cdot e^2} \cdot N_D \qquad (5)$$

In Table 2 the N_D values were calculated for averaged values of ε_0 and m* of the anisotropic components of the dielectric constant: for $\varepsilon_{11}(0) = 30$, $\varepsilon_{33}(0) = 43$; $\varepsilon_0 = \sqrt[3]{\varepsilon_{11}(0)^2 \varepsilon_{33}} = 33.82$; and effective mass of holes $m^*/m_0 = \sqrt[3]{m_\perp^2 \cdot m_\parallel}$; for $m_\perp/m_0 = 0.126$ and $m_\parallel/m_0 = 0.24$; $m^*/m_0 = 0.156$.

Figure 7 shows the neutral defect distribution estimated using Eq. (5). In the W-ingots the mobility was rather low and correlated with the value of N_D, which was rather high. The μg W sample had grown with some supercooling before nucleation and growth occurred. X-ray investigations revealed that on the whole this ingot was polycrystalline, with the different grain sizes in part I (near the starting point of growth) and part II (far from the starting point). The large grains (0.6 – 1 mm) in part I were highly stressed. Part II had grains from 5 to 50 μm, which led to a high value of the neutral-defect density N_D (Fig. 7 and Table 2) calculated from the hole mobility.

Measurements at liquid-helium temperatures showed that the mobility strongly depended on temperature only in the pure SC from 1.6 to 77.4 K. The mobility depended weakly on temperature in S-μg and S-5g samples, which had block structures, and W-μg, which was polycrystalline. The hole mobility was mainly defined by scattering by defects. The calculations gave $N_D = 6.3 \cdot 10^{17}$ cm^{-3} for region II of S-μg and $N_D = 8.3 \cdot 10^{17}$ for region II in W-μg.

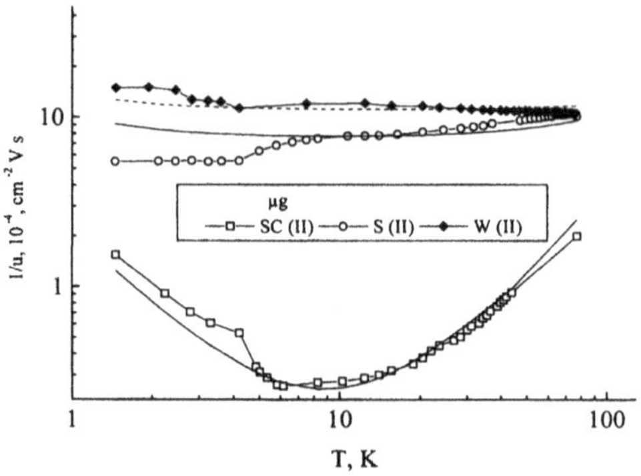

Figure 6. Temperature dependence of inverse hole mobility for samples solidified in space, far from the beginning of solidification (part II). —□— SC - single crystal feed partially melted and solidified, —O— S - solidified with a seed, —◆— W - solidified without a seed.

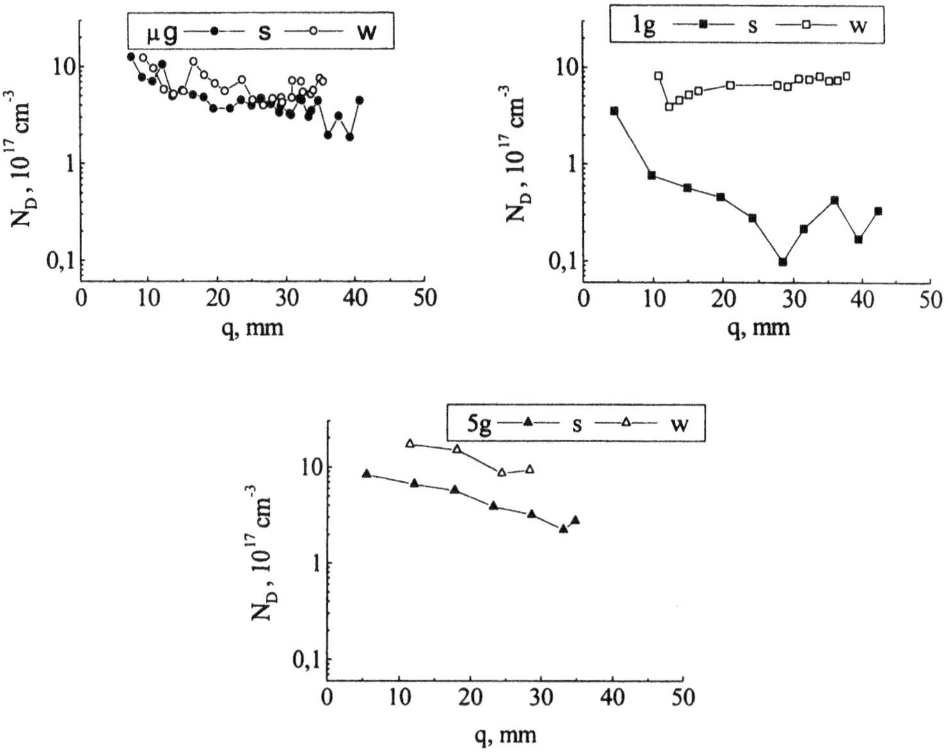

Figure 7. Neutral defect profiles in samples grown under different gravity levels without a seed (w) and with a seed (s).

In W-μg the N_D concentration varied along the ingot. Spontaneous detached solidification[10] of W-*μg* seemed to take place with supercooling of the melt. Marangoni convection may have taken place at the meniscus adjacent to the freezing interface, thereby modifying impurity segregation and crystalline orientation. Fewer dislocations and grain boundaries are also typical when detachment occurs.[10] Figure 8 demonstrates that the grains were perpendicular to the surface in the detached region of SC-*μg*. So, it is not surprising that this region had a two-times higher resistivity. (The hole resistivity in Te along the C-axis at 77 K is half that in the transverse direction.[11]) Figure 9 shows that in W-μg the N_D profile correlated with the resistivity profile at 300 K, where carrier scattering on phonons prevailed over scattering on point defects. In W-μg, fluctuations of the grain orientation may have arisen from contact of the growth interface with the ampoule wall. In this region the C-axis of a microcrystalline grain is normally directed to the internal ampoule walls (as in the case of tellurium whiskers growth on earth). So, the resistivity fluctuation along such an ingot is attributed to the increase in

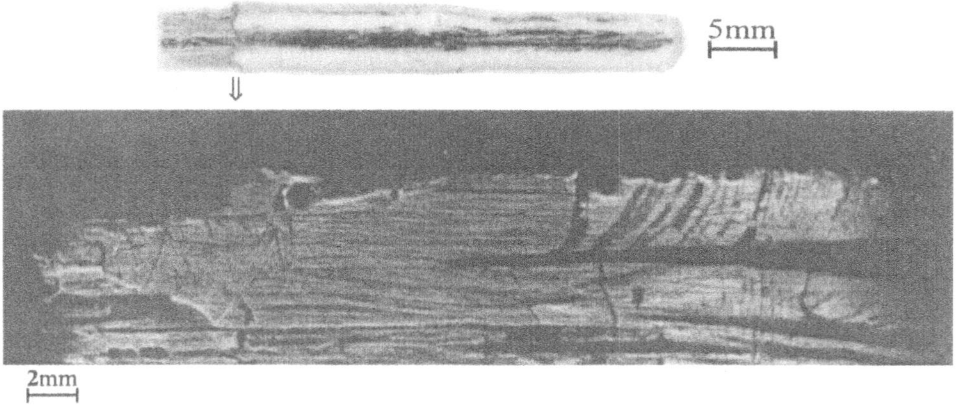

Figure 8. Te grown in space from single crystal (SC) feed by partially melting and refreezing.
Top: Photograph. Note the decreased diameter in the latter detached portion.
Bottom: X-ray topographic image of the cleavage surface $(10\bar{1}0)$using CuKα radiation and the $(20\bar{2}0)$ reflection with the C-axis directed horizontally. The arrow indicates where refreezing had begun.

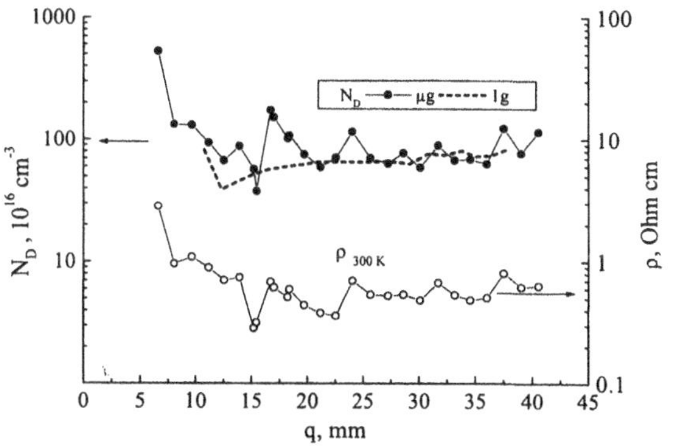

Figure 9. Variation of the electrical resistivity (right scale) at 300 K and the defect concentration (left scale) calculated from the temperature-independent part of the hole mobility along an ingot grown without a seed in space.

the contact region, the predominant role of inter-grain boundaries perpendicular to the current in resistivity, and of the anisotropy of the hole conductivity in Te. (The resistivity of tellurium for current perpendicular to the C-axis ($\rho_{\perp C}$) is approximately twice that for current parallel to the C-axis ($\rho_{\parallel C}$).) Such features were not observed in ingots grown on earth where the growing interface always touched the ampoule wall. Thus microgravity opens the possibility to prepare microcrystalline structures composed of anisotropic grains.

CONCLUSIONS

We infer from directional solidification experiments on pure Te at microgravity on the MIR space station, on earth at 1*g*, and at 5*g* on centrifuge CF-18 that:

i) The sensitivity of the electrical properties of Te at low temperatures to the distribution of both neutral and electrically active defects allowed us to determine the impurity and defect profiles along the ingot length at the level of 10^{-5} at%, which could not be achieved by other methods.

ii) The hole density profile in the initial part of all Te ingots did not correspond to the normal-freeze equation and was associated with an exponential decrease of the electrically-active intrinsic-defect concentration with increasing distance from the starting point of growth.

iii) A more than 1 order-of-magnitude smaller low-temperature hole mobility was observed in the μg sample compared to that for the single crystal partially melted and re-solidified in space under detached conditions, indicating the predominant role of neutral defects in scattering of holes. A high defect concentration, $N_D \approx 10^{18}$ cm^{-3}, was found in all ingots solidified without a seed.

iv) In the case of complete melting and re-solidification of Te in microgravity, the initial undercooling of the melt followed by spontaneous nucleation resulted in a microcrystalline structure with variable neutral-defect concentration along the ingot.

v) For the microgravity ingot grown detached, the variation of the intensity of scattering on neutral defects calculated for 77 K correlated with the non-monotonic resistivity profile measured at 300 K when carrier scattering on phonons predominated. That behavior was probably due to the influence of the growth interface contacting the ampoule walls on the crystallite orientation in the ingot. If in the contact region the C-axis of crystallites are oriented perpendicular to the ampoule wall, then resistivity will increase in this point due to the scattering of the charge carriers by the crystallite boundaries and crystal resistivity anisotropy ($\rho_{\perp c} > \rho_{\parallel c}$).

REFERENCES

1. L.L. Regel, I.V. Vidensky, A.V. Mikhailov, A.M. Turchaninov, R.V. Parfeniev, I.I., Farbstein, N.K. Shul'ga, and B.T. Melekh, Growth of semiconductor crystals of tellurium, tellurium-selenium alloy, tellurium-silicon and directed crystallization of the eutectic alloy under zero-gravity ("ALCUTEST-I, II" Program), in: *Proceedings of the International Astronautical Congress, Innsbruck, Austria*, Pergamon, New York (1986) IAF-86-283.
2. R.V. Parfeniev, I.I. Farbstein, S.V. Yakimov, V.P. Shalimov, A.M. Turchaninov, Solidification of anisotropic semiconducting compounds under microgravity conditions and their properties, *Joint Xth European and Vth Russian Symposium on Physical Sciences in Microgravity*, St. Petersburg, Russia (1997) Abstract 98.
3. R.V. Parfeniev, I.I. Farbstein, I.L. Shulpina, S.V. Yakimov, V.P. Shalimov, A.M. Turchaninov, A.I. Ivanov, and S.F. Savin, Tellurium recrystallization under microgravity conditions and the resulting properties of samples, *Physics of the Solid State*, 42: 244-252 (2000); Translated from *Fizica Tverdogo Tela* 42: 238-245 (2000).
4. D.H. Matthiesen, Growth of electronic materials in microgravity, in: *Proc. AIAA/IKI Microgravity Sci. Sympos., Moscow, USSR*, AIAA, Washington DC (1991) 110—118.
5. H. Rodot, L.L. Regel, and A.M. Turchaninov, Crystal growth of IV-VI semiconductors in a centrifuge, *J. Crystal Growth* 104: 280 (1990).
6. J.A. Burton, R.C. Prim and W.P. Slichter, The distribution of solute in crystal grown from the melt, *J. Chem. Phys.* 21: 1987-1991 (1953).
7. A.R. Regel and V.M. Glasov, *Physical Properties of Electronic Melts*, Nauka, Moscow (1980).
8. I.I. Farbstein, A.M. Pogarski, and S.S. Shalyt, Galvanomagnetic properties of tellurium and its valence band structure close to energy minimum, *Fiz. Tverd. Tela (Leningrad)* 7: 2383 (1965).
9. C. Erginsoy, Neutral impurity scattering in semiconductors, *Phys. Rev.* 79:1013 (1950).
10. L.L. Regel and W.R. Wilcox, Detached solidification in microgravity: a review, *Microgravity Sci. Technol.* 14: 152-166 (1999).
11. P. Grosse, *Die Festkörpereigenschaften von Tellur*, Springer Tracts in Modern Physics 48 (1969).

INFLUENCE OF CENTRIFUGATION ON FURNACE TEMPERATURE DISTRIBUTION AND FREEZING RATE OF InSb USING THE GRADIENT FREEZE TECHNIQUE

Illa L. Moskowitz, Liya L. Regel, and William R. Wilcox

International Center for Gravity Materials Science and Applications
Clarkson University, Potsdam, NY 13699-5814, USA

INTRODUCTION

For the gradient freeze technique, a tubular furnace is typically used that has an axial temperature gradient. Directional solidification of a molten material contained within this furnace is caused by slowly decreasing the temperature. Normally, the furnace is vertical with the temperature increasing with height. Since this thermal configuration produces little buoyancy-driven convection in a pure material, it is generally referred to as "thermally stable."

The freezing rate, temperature gradient and shape of the freezing interface are critical parameters in determining the properties of a solidified material. In the gradient freeze technique the freezing rate V_c is related to the axial temperature gradient G in the material and the rate of cooling dT/dt by:

$$V_c = (dT/dt)/G \qquad (1)$$

where it has been assumed that the freezing interface remains at a fixed temperature. In estimating the freezing rate, it is often assumed that the temperature gradient in the material is the same as it is in the furnace. However, one would expect the gradient to be significantly less in the material than in the furnace for the typical situation of a high thermal-conductivity semiconductor sealed in a low thermal-conductivity ampoule, which in turn is often surrounded by a metal cartridge. Arnold,[1] for example, calculated a temperature gradient in PbTe of about 1/3 that along the steel cartridge in the Meudon furnace used to study the gradient freeze technique during centrifugation. According to Eq. (1), this reduced gradient would give a freezing rate 3-times higher than estimated using the furnace gradient.

Few measurements have been made comparing the furnace temperature gradient with that in a molten material contained in a gradient freeze furnace. Only one such measurement appears to have been made previously during centrifugation. Chevy *et al.*[2] performed measurements in liquid tin. They surrounded the furnace with a water-cooled

Processing by Centrifugation, Edited by Regel and Wilcox
Kluwer Academic/Plenum Publishers, New York, 2001

jacket in order to avoid uncontrolled cooling of the furnace by the wind generated by centrifugation. In the thermally stable configuration, centrifugation to 10 g had negligible effect on the temperature gradient in the liquid tin. Prior to the present study there was no information on the influence of centrifugation on freezing rate.

EXPERIMENTAL

Tellurium-doped indium antimonide was used for all experiments, which are described in detail elsewhere.[3] The gradient-freeze furnace constructed by Derebail[4] was used on the HIRB materials processing centrifuge[5] at Clarkson University. The furnace was 28 cm long and 3 cm in internal diameter. Thermcraft fiber blanket insulation was placed around the ceramic core and the at the ends. The spacing between coils of the Kanthal A-1 18 gauge heater wire increased with increasing height, yielding a nearly constant gradient of about 20°C/cm in the position of the fused-silica ("quartz") growth ampoule containing the InSb. This furnace was mounted to the centrifuge arm by a hinge so that the resultant acceleration vector was always aligned approximately with the axis of the furnace.

K-type thermocouples in 0.5mm Inconel sheaths inside quartz capillary tubes were used to measure the temperature at various locations. A special ampoule permitting a thermocouple to be slid up and down in a capillary tube was used to obtain the temperature profile inside an InSb charge and along the center of the furnace beyond the charge. For profiling of the furnace, thermocouple capillary tubes were attached to the outside of the quartz furnace liner. The thermocouples were inserted to different lengths from the end of the furnace or around the periphery at the same length. A thermocouple for control of power was inserted from the side at 14 cm from the furnace bottom. Slip rings on the centrifuge were used to provide power to the furnace windings and to transmit the thermocouple signals to a recorder.

For centrifuge experiments, the power was ramped up until the control thermocouple reached 650°C and then held constant. Centrifuge rotation was started 2 hours after this temperature stabilized. Two experiments were conducted using this setup. In the first experiment, centrifuge rotation was gradually brought up to an acceleration of 3g over a period of about 30 minutes. The rotation rate was kept constant for 3 h and then gradually decreased to zero over a 30 minute period. In the second experiment, the acceleration was increased in steps of 0.5g up to 6g, and then decreased in steps until rotation was stopped. After each step, the centrifuge was allowed to rotate at constant speed for approximately 15 minutes.

Current interface demarcation (CID)[6-13] was used to reveal the freezing interface position at intervals throughout some solidification experiments. In this technique, electric current pulses are passed through the length of the material, causing sharp perturbations to the freezing rate and, thereby, to the impurity doping concentration. Proper etching produces striations that can be seen with optical microscopy. The freezing rate is calculated from the time between current pulses and the distance between striations. Three CID experiments were performed without centrifugation and one with centrifugation at 2 g.

RESULTS

Once the desired temperature was reached, the control thermocouple showed negligible variation in temperature during all experiments. No circumferential variation in temperature was detected at any acceleration. Figure 1 shows that the axial temperature gradient within the InSb was about the same as along the ampoule wall without centrifugation for both 9 mm and 12 mm ID ampoules.

Figure 2 shows the results of the first experiment described above for temperature measurements during centrifugation. Except for the control thermocouple, all thermocouples registered a rapid increase of several °C when centrifuge rotation was begun. Within a few minutes each thermocouple reading decreased to a new steady value. The final temperature gradient was less than before rotation. Figure 3 shows the results of the second experiment. Initiation of rotation also caused a sudden increase in the non-control thermocouple readings, and then decayed to give a decreased steady axial

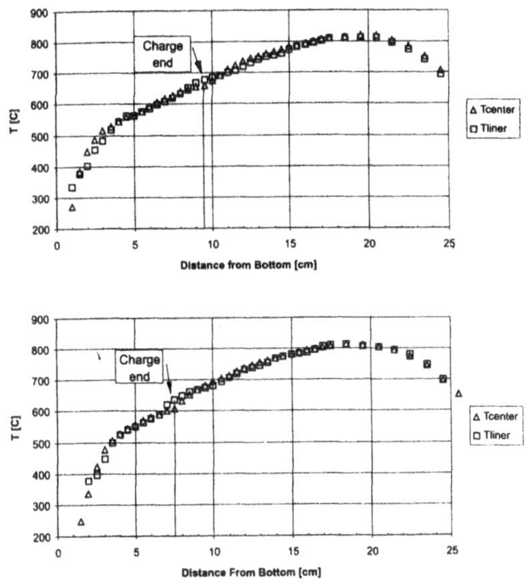

Figure 1. Temperature profiles measured along the quartz furnace liner and along the center of the furnace with the control thermocouple at 510°C. Top: With 9 mm diameter InSb charge in ampoule. Bottom: With 12 mm diameter InSb in ampoule.

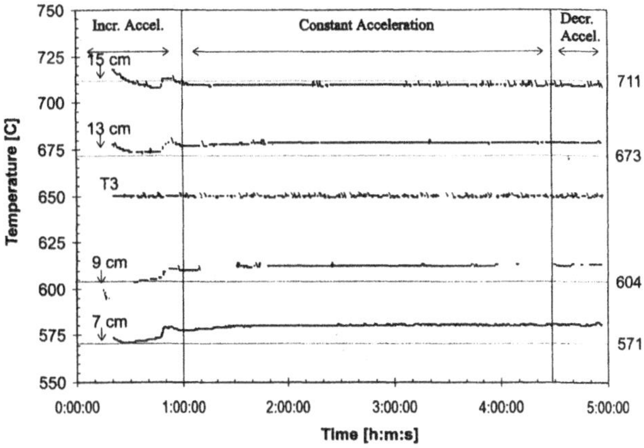

Figure 2. Temperatures along the quartz furnace liner in an empty furnace versus time with centrifugation at 3g (43 rpm). The distances from the bottom of the furnace are shown on the left. The horizontal straight lines represent the steady temperatures prior to centrifugation. The values of these temperatures are also shown on the right in °C. T3 is the control thermocouple.

temperature gradient. Smaller changes occurred in subsequent stepwise increases in g. Decreases in rotation rate caused no change in thermocouple readings in either experiment. This hysteresis may be attributed to a shift in furnace insulation due to centrifugation.

The Current Interface Demarcation (CID) experiments revealed that the freezing interface was faceted at some grains and changed direction abruptly at grain boundaries. Although irregular, on average it was relatively planar across the ingots. The measured growth rates are given in Figures 4 and 5, with the results summarized in Table 1.

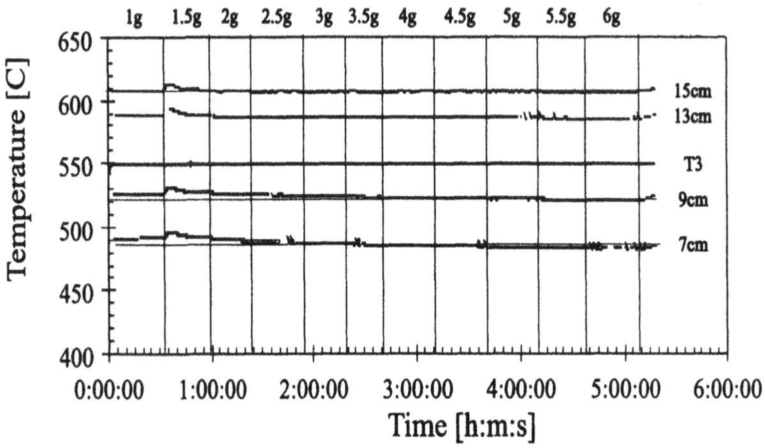

Figure 3. Temperatures along the quartz furnace liner in an empty furnace as the acceleration was increased in steps of 0.5g. The temperatures prior to centrifuge rotation are denoted by the horizontal lines. T3 is the control thermocouple.

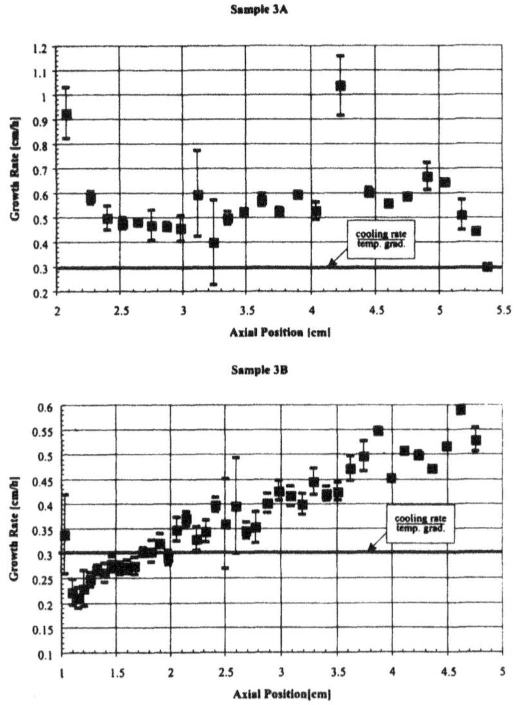

Figure 4. Growth rate versus distance down ingots from CID striations in experiments 3A and 3B.

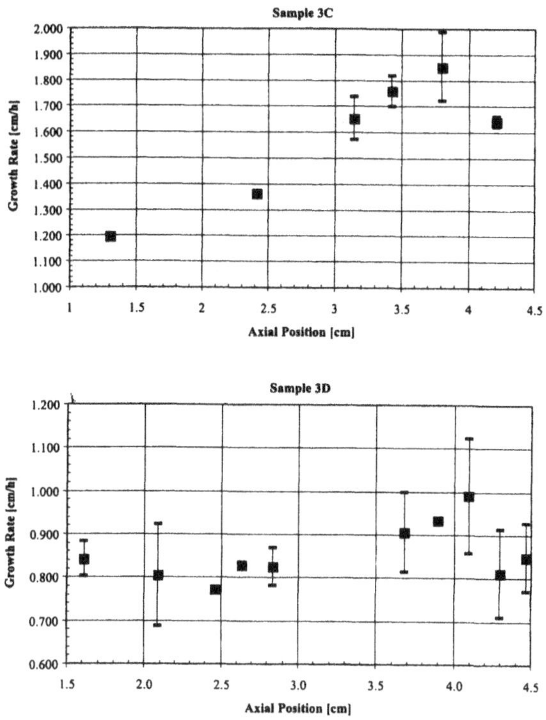

Figure 5. Growth rate versus distance down ingots from CID striations for experiments 3C and 3D.

Table 1. Summary of results of current interface demarcation (CID) experiments.

Expt.	Ampoule IDxOD	Cooling rate	Centrifuge rotation	Average CID growth rate	Cooling rate / T gradient
3A	9mm×11mm no quartz liner	6.25°C/h	none	0.6 cm/h	0.3 cm/h
3B	9mm×11mm	6.25°C/h	none	0.4 cm/h	0.3 cm/h
3C	12mm×16mm	6.25°C/h	none	1.6 cm/h	0.25 cm/h
3D	12mm×16mm	10.0°C/h	20 rpm (~2 g)	0.9 cm/h	0.4 cm/h

The growth rate in two of the 1 g samples varied significantly with position, while that grown with centrifugation had the most constant freezing rate. The measured growth rates were higher than those estimated from the ratio of the furnace cooling rate to the axial temperature gradient measured inside an ampoule containing InSb. The ratio of measured to estimated freezing rates ranged from 1.3 to 6.4. This difference tended to be larger for larger diameter ampoules, but was not reproducible. These results indicate that the freezing interface was not exactly at the melting point isotherm and that the temperature gradient inside the material varied in an unpredictable way during an experiment and from one experiment to another.

Acknowledgement

This research was supported by the National Science Foundation via grant number DMR-9414304.

REFERENCES

1. W.A. Arnold, *Numerical Modeling of Directional Solidification in a Centrifuge,* PhD Thesis, Clarkson University, Potsdam, New York (1993).
2. W. Chevy, P. Williams, M. Rodot, and G. Labrosse, Removal of convective instabilities in liquid metals by centrifugation, in: *Materials Processing at High Gravity,* L.L. Regel and W.R. Wilcox, eds., Plenum Press, New York (1994), pp 43-52.
3. I. Moskowitz, *The Influence of Centrifugation on the Temperature Distribution in a Gradient Freeze Furnace and Growth of Te-doped InSb with Current Interface Demarcation,* M.S. Thesis, Clarkson University (1998).
4. R. Derebail, *Directional Solidification of Indium Antimonide under High Gravity in Large Centrifuges,* Ph.D. Thesis, Clarkson University (1994).
5. R. Derebail, W.A. Arnold, G.J. Rosen, W.R. Wilcox, and L.L. Regel, HIRB - The centrifuge facility at Clarkson, in: *Materials Processing at High Gravity,* L.L. Regel and W.R. Wilcox, eds., Plenum Press, New York (1994) pp 203-212.
6. A.F. Witt and H.C. Gatos, Application of the Peltier effect for the determination of crystal growth rates, *J. Electrochem. Soc.* 115: 112 (1968).
7. M.J. Wargo, *The Effect of Direct Current on Crystal Growth from the Melt: InSb,* PhD Thesis, Massachusetts Institute of Technology, Cambridge (1982).
8. L.N. Brush, S.R. Coriell, and G.B. McFadden, Directional solidification of a planar interface in the presence of a time-dependent electric current, *J. Crystal Growth* 102: 725-742 (1990).
8. J. Zhou, *Accelerated Crucible Rotation Technique and Current Interface Demarcation during Directional Solidification of Te-doped InSb,* MS thesis, Clarkson University (1992).
9. J. Zhou, M. Larrousse, W.R. Wilcox, and L.L. Regel, Directional solidification with ACRT, *J. Crystal Growth* 128: 173 (1993).
10. Q. Li, H.N. Thi, and B. Billia, Quantitative analysis of the influence of Peltier interface demarcation on directional solidification, *J. Crystal Growth* 167: 277 (1996).
11. S. Corre, T. Duffar, M. Bernard, and M. Espezel, Numerical simulation and validation of the Peltier pulse marking of solid/liquid interfaces, *J. Crystal Growth* 180: 604-614 (1997).
12. L.L. Zheng, D.J. Larson, Jr., *J. Crystal Growth* 180: 293-304 (1997) .
13. L.L. Zheng, H. Zhang, D.J. Larson, Jr., and V. Prasad, *J. Heat Transfer* 120: 430-440 (1998) .

EFFECTS OF ROTATION ON TRANSPORT PROCESSES DURING CRYSTAL GROWTH BY SOLIDIFICATION

C. W. Lan

Department of Chemical Engineering, National Taiwan University
Taipei 10617, Taiwan, ROC

INTRODUCTION

The control of heat flow, dopant segregation, and the shape of the growth interface is an important task in bulk crystal growth.[1,2] A flat or slightly convex growth front is highly desired to minimize parasitic nucleation. Dopant uniformity, both radial and axial, is also a major concern. Therefore, better understanding of the melt flow and heat and mass transfer during crystal growth is important. In solidification the intrinsic coupling of transport processes to phenomena at the growth interface strongly influences crystal quality. For example, unstable flow can cause growth striations, sometimes even leading to periodic or chaotic back melting.

It is desirable to find an effective strategy for better control of solidification. Typically, the temperature profile in the furnace is tuned by furnace design to try to achieve the desired conditions, but this process is laborious, rather inflexible, and not always very successful. Thus, the use of external forces to assist in controlling crystal growth is becoming common.[3-19] Examples include magnetic fields,[4,5] accelerated crucible rotation (ACRT),[6] vibration,[7-12] centrifugation,[13-16] and reduced gravity.[17] Recently, a novel approach using steady ampoule rotation for flow and segregation control was proposed by Lan.[18] With a high enough rotation speed, diffusion-controlled segregation was achieved for the vertical Bridgman (VB) configuration. Yackel *et al.*[19] adopted this idea for the VB growth of ZnCdTe, and also showed that convection in the melt could be significantly suppressed by steady ampoule rotation. Radial segregation was increased due to the reduced mixing. Although for a moderate rotation speed, the flow suppression may not be adequate to reduce radial segregation, application in reduced gravity or in the cases having much weaker buoyancy-driven convection may be feasible.[20] In fact, our recent calculations show that centrifugal acceleration becomes important if the rotation speed is high enough.[21] The induced flow has a different flow direction and can, therefore, reverse the radial dopant distribution for a concave interface and lead to a minimum in segregation.

The effect of rotation on zone-melting (ZM) is also significant. Due to strong buoyancy flow, damping by steady ampoule rotation is not practical, although modification of the flow structure is possible. In a recent numerical study, Lan and Chian[22] showed that the local flow near the center of the growth interface could be easily modified by rotation.

Processing by Centrifugation, Edited by Regel and Wilcox
Kluwer Academic/Plenum Publishers, New York, 2001

As a result, a convex growth interface was obtained. (ACRT had a similar effect, but significant growth rate fluctuations, including back melting, were induced.) The Coriolis force significantly suppressed the onset of symmetry-breaking,[23] leading to three-dimensional (3D) flows. The nonlinear bifurcation structure could be changed dramatically by rotation as well; unsteady flows might be suppressed. In spite of the inherent bifurcation to 3D flows, ampoule rotation can reduce the effects of asymmetries arising from tilt of the ampoule or asymmetric heating.[22] More interestingly, rotation plays a critical role on the interfaces. During rotation, melting and growth can occur alternately, and the heat of fusion absorbed or released (Stefan effect) further reduces the asymmetry of the interface shape. This effect is particularly profound in the Pfann-Miller technique,[25-26] which uses slow rotation to reduce the zone asymmetry in horizontal ZM of organic compounds. Through 3D numerical simulation, Lan *et al.*[27] showed that the Stefan effect is the key mechanism for effective interface control. The angular convective heat transfer and the Coriolis force in such a configuration are not important. Side effects due to rotation on segregation and inclusion formation were also discussed.[27]

We see from the foregoing that the influence of ampoule rotation depends on the growth configuration and the convection level. Here, we give a brief review of our work on the influence of rotation on crystal growth, beginning with the relevant transport equations. Emphasis is the use of rotation for better control of vertical Bridgman (VB) and zone-melting (ZM) crystal growth, from a computational point of view. Extension to floating-zone crystal growth is discussed.

BACKGROUND

We consider the rotation of an ampoule containing melt and crystal about its own axis at a constant angular speed Ω. In dimensionless form the conservation laws for mass, momentum energy, and dopant can be expressed by the following equations in a rotating frame:

$$\nabla \cdot \boldsymbol{v} = 0, \tag{1}$$

$$\frac{\partial \boldsymbol{v}}{\partial \tau} + \boldsymbol{v} \cdot \nabla \boldsymbol{v} = -\nabla P + Pr \nabla^2 \boldsymbol{v} + \boldsymbol{F}, \tag{2}$$

$$\frac{\partial T}{\partial \tau} + \boldsymbol{v} \cdot \nabla T = \nabla^2 T \tag{3}$$

$$\frac{\partial C}{\partial \tau} + \boldsymbol{v} \cdot \nabla C = \frac{Pr}{Sc} \nabla^2 C, \tag{4}$$

The dimensionless body force $\boldsymbol{F}$ is:

$$\boldsymbol{F} = -Pr\, Ra_T (T - T_{ref}) [\boldsymbol{e}_g + rFr\boldsymbol{e}_r] + Ha^2 Pr\, \boldsymbol{v} \times \boldsymbol{v} \times \boldsymbol{B} - 2\, Ta^{1/2} Pr\, \boldsymbol{e}_z \times \boldsymbol{v}$$

and $\boldsymbol{v}$, P, T, and C are dimensionless velocity, pressure, temperature and dopant concentration, respectively; $\boldsymbol{e}_g$, $\boldsymbol{e}_r$, and $\boldsymbol{e}_z$ are the unit vectors in the gravitational, radial and axial direction, respectively; and r is radial distance, and B magnetic field strength. The dimensionless Rayleigh Ra_T, Hartmann Ha, Taylor Ta and Froude Fr numbers and their physical meaning are summarized by:

$$Ra_T = \frac{\beta_T g \Delta T L^3}{\nu_m \alpha_m} = \frac{\text{Buoyancy force}}{\text{Viscous force}} \qquad Ta = \Omega^2 L^4 / \nu_m^2 = \frac{\text{Coriolis force}}{\text{Viscous force}}$$

$$Ha = BL(\sigma / \mu_m)^{1/2} = \frac{\textit{Lorentz force}}{\textit{Viscous force}} \qquad Fr = \Omega^2 L / g = \frac{\textit{Centrifugal force}}{\textit{Gravity force}}$$

where L is the characteristic length, σ electrical conductivity, μ_m melt viscosity, and g gravitational acceleration. Additional dimensionless parameters are the Prandtl number $Pr = \nu_m/\alpha_m$ and the Schmidt number $Sc = \nu_m/D$, where ν_m is the kinematic viscosity and α_m and D are the thermal and dopant diffusivities, respectively. For molten oxides, Pr is about 1 and the thermal field can be affected easily by the flow. For molten semiconductors and metals, $Pr \ll 1$ and there is much less effect of flow on heat transfer and, thus, on the interface shape. For most dopants in melts, $Sc \gg 1$ and the dopant field, and thus segregation, are dominated by convection rather than by molecular diffusion.

In the above equations, the melt is assumed incompressible and the Boussinesq approximation is used. There is a controversy about the use of the fixed frame or the rotating frame in presenting the Boussinesq approximation. In the fixed-frame approach, the density variation is put in the gravity term only, and the effect of centrifugal force, which generates a pressure gradient in the radial direction, cannot be considered explicitly. As a result, at zero gravity, steady ampoule rotation should result in rigid-body rotation of the melt. In the rotating frame, the centrifugal force, $rFre_r$, resembles a gravitational acceleration, and the density variation due to a temperature gradient can be considered as shown above. Even at zero gravity, a thermal gradient in the axial direction, which is perpendicular to the centrifugal force, can induce flow. Normally, $g \gg r\Omega^2$ so that Fr is small and this effect can be neglected. On the other hand, when the centrifugal force is the dominant force, as in a high-g centrifuge or reduced gravity environment, the rigid-body rotation solution is not valid and Sweet-Eddington flow[28] may be induced. In obtaining the calculated results presented here, the centrifugal force was assumed to be small. It is reasonable to neglect this term with normal gravity ($Fr<10^{-3}$). For reduced gravity ($Fr\sim 1$), the centrifugal force term should not be neglected. Therefore, the recent prediction[20] of when centrifugal force can be neglected could be inaccurate in microgravity. Nevertheless, since the centrifugal force is axisymmetric,[20] the suppression of asymmetric flow due to residual acceleration may still be feasible. Further discussion of the centrifugal force can be found elsewhere.[21] Error due to the Boussinesq approximation could also be an issue, but this is beyond the scope of the present paper.

With a set of proper boundary conditions, the above equations can be solved numerically.[20-24] For crystal growth, the energy and solute balances at the interfaces are particularly important, and cannot be ignored. Taking the Bridgman configuration as an example, the balances (in dimensionless form with a fixed frame) at the growth front are:

$$Q_c - Q_m + \gamma St\,[\,(v_z - \frac{dh}{d\tau})\,\boldsymbol{n}\cdot\boldsymbol{e}_z + r\Omega\,\boldsymbol{n}\cdot\boldsymbol{e}_\phi\,] = 0 \qquad (5)$$

$$\boldsymbol{n}\cdot D\nabla C\,|_m + Sc\,(1-K)\,C[\,(v_z - \frac{dh}{d\tau})\,\boldsymbol{n}\cdot\boldsymbol{e}_z + r\Omega\,\boldsymbol{n}\cdot\boldsymbol{e}_\phi\,] = 0 \qquad (6)$$

Here, Q_c and Q_m are the heat fluxes at the crystal and melt sides, respectively, γ the density ratio of the crystal and the melt, St the Stefan number scaling the heat of fusion by the sensible heat of the melt, K the segregation coefficient of the dopant, $\boldsymbol{e}_\phi$ the unit vector in the azimuthal direction, and $\boldsymbol{n}$ is the unit normal vector, as illustrated in Fig. 1. If the interface shape is not axisymmetric ($\boldsymbol{n}\bullet\boldsymbol{e}_\phi \neq 0$), the freezing rate will oscillate and melt-back may even occur ($\boldsymbol{n}\bullet\boldsymbol{e}_\phi > 0$) during rotation. These boundary conditions are extremely important and cannot be ignored. As illustrated shortly, they play an important role in interface control and dopant segregation.

The above governing equations have been solved numerically by an efficient finite-volume method with multigrid acceleration.[29,30] Second-order accuracy was retained through deferred correction.

From the body force term, it can be seen that the effect of rotation depends on the relative contribution of the Coriolis force to the buoyancy force. If the buoyancy force is small, which is typical for the VB configuration, moderate rotation may significantly affect the melt flow. The resultant flow may not be enough to change much the interface morphology, but the dopant field can be affected dramatically. On the other hand, for the zone-melting configuration, due to radial heating the buoyancy force is usually dominant and the moderate rotation can only partly modify the flow structures. In the following sections, we will further discuss the effects of rotation on crystal growth through some examples. We focus on the VB and zone-melting configurations. Some applications in the floating-zone (FZ) configuration are also illustrated. This review is not extensive and includes mainly work done in the author's research group. Interested readers can find further discussion in the relevant references.

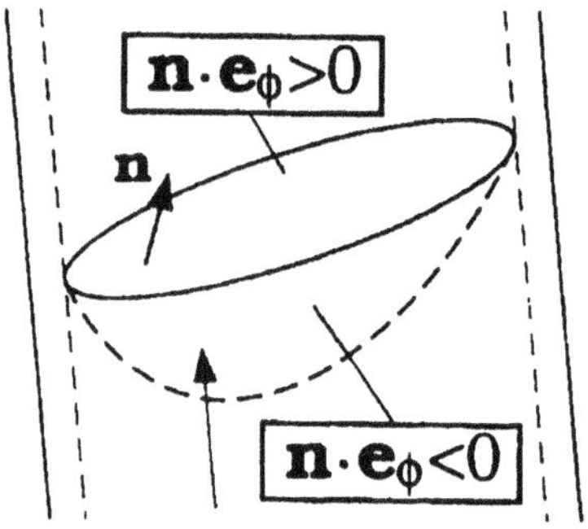

Figure 1. Schematic diagram of a growth interface showing its surface normal **n**.

SEGREGATION CONTROL IN VERTICAL BRIDGMAN CRYSTAL GROWTH

When the density decreases with height in the melt, buoyancy flow is weak and flow damping by ampoule rotation is usually effective. For comparison purposes, we take the growth of gallium-doped germanium in the Grenoble furnace as an example. This problem, without rotation, was investigated by Adornato and Brown.[31] Extensive benchmark numerical comparisons showed that both magnetic fields and steady ampoule rotation effectively suppress the melt flow, but their scaling laws are different.[32,33] Figure 2 shows that the flow intensity decreases linearly with rotation speed (or $Ta^{-1/2}$) and quadratically with magnetic field (or Ha^{-2}). The scaling laws apply to both concave and convex growth interfaces. Therefore, magnetic damping is much more effective than that by rotation. Nevertheless, in practice, rotation damping is much easier to implement and is not restricted to electrically-conducting melts. Figure 2 shows that for Ra_T = 2.489x 10^8 (at normal gravity), flow damping by 50 rpm rotation is slightly better than that by 0.5T, which is not a trivial magnetic field strength. To achieve diffusion-controlled doping segregation, the flow damping needs to be large enough to reduce the flow intensity to below the growth rate. A typical growth rate is marked by an arrow in Fig. 2. Therefore, at Ra_T = 2.489x 10^8 the magnetic field would have to reach 3T for damping, while more than 200 rpm of steady ampoule rotation would be necessary. Unfortunately, such a high rotation rate might cause problems, such as mechanical vibration. More importantly, the centrifugal force, which is neglected here, would become important and the scaling law would break down. For a concave interface, the flow damping can be even more effective because of the centrifugal force.[21]

Figure 3 shows typical flow structures and dopant fields for 0 and 100 rpm with an

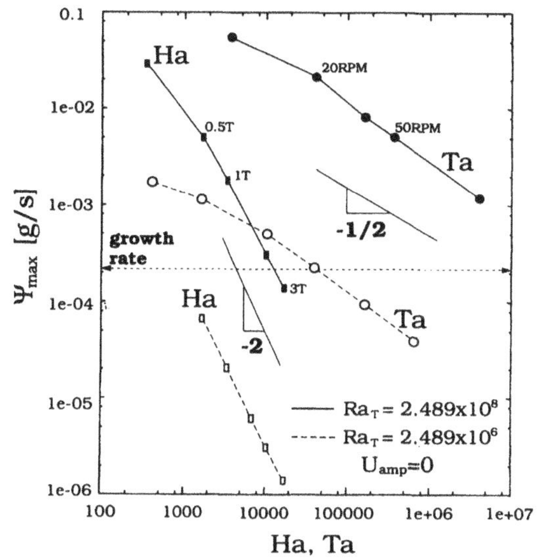

Figure 2. Effect of ampoule rotation and magnetic field on the maximum flow intensity in VB growth of Ga-doped Ge.[33]

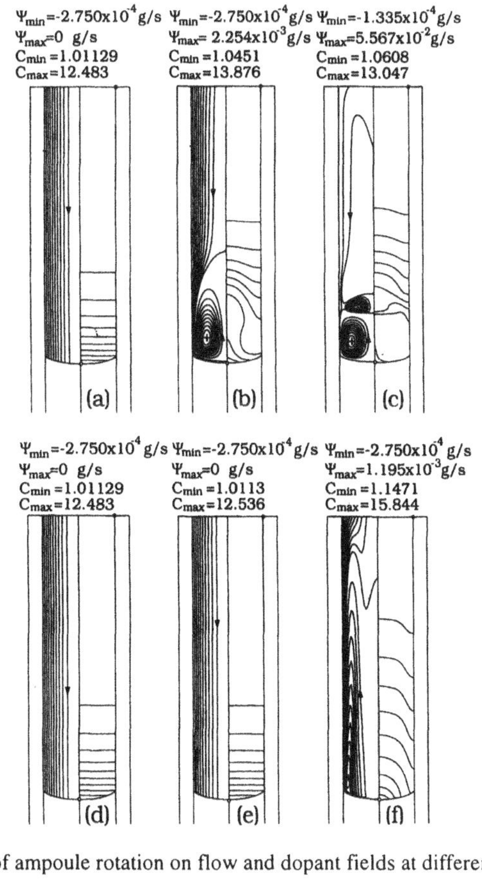

Figure 3. Influence of ampoule rotation on flow and dopant fields at different convection levels.[18]

ampoule lowering speed of $4x10^{-4}$cm/s. Figure 4 shows that the corresponding axial segregation is significantly influenced by ampoule rotation. Increasing the rotation rate pushes the concentration profile away from complete mixing behavior toward diffusion-controlled ($Ra_T = 0$). These results suggest that axial segregation could be further reduced by varying the rotation rate during solidification.

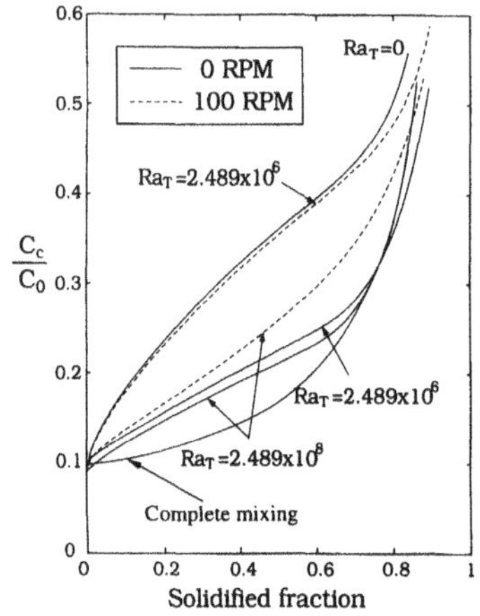

Figure 4. Influence of ampoule rotation on axial dopant segregation in VB growth.[18]

Another promising application for steady ampoule rotation is crystal growth in space. The residual acceleration (g-jitter), which may induce time-dependent segregation, can be significantly suppressed by ampoule rotation, even with a low speed. Figure 5 shows the effect of 5 rpm rotation on the flow and dopant fields in reduced gravity. Without rotation (Fig. 5a), the direction of the residual acceleration is important. Asymmetric 3D flow can be induced and lead to asymmetric doping in the grown crystal, as illustrated in the cross-sectional profiles on the bottom of Fig. 5. Figure 5b shows that 5 rpm rotation causes the flow and dopant fields to become nearly axisymmetric. Although the calculations in Fig. 5b did not consider centrifugal acceleration, the result should be similar to results using the full model. Because the centrifugal force is axisymmetric, once it dominates the melt convection, the effect of asymmetric residual acceleration is greatly reduced. Although ampoule rotation provides a solution to reduce the 3D buoyancy flow, it may not be desirable for an asymmetric growth interface. If the growth interface is not axisymmetric, which may be due to asymmetric heating, rotation would cause an oscillatory growth rate and induce severe rotational doping striations.[20]

HEAT FLOW CONTROL IN VERTICAL ZONE-MELTING

Because the density of the melt goes through a minimum with height, there is significant buoyancy-driven convection in vertical ZM. Figure 6a shows flow observed during ZM of $NaNO_3$ (diameter = 14.5mm). This flow, with the warmer melt flowing downward at the center, induces a gull-winged growth front, which is concave near the centerline and convex near the edge. Such an interface may render single crystal growth difficult in the core. Calculations predicted a similar interface shape for 1-in diameter GaAs.[22] Such a flow structure can be modified easily by steady ampoule rotation. Figure 6b shows the effect of rotation rate on interface convexity, zone length, flow structure and isotherms for GaAs.

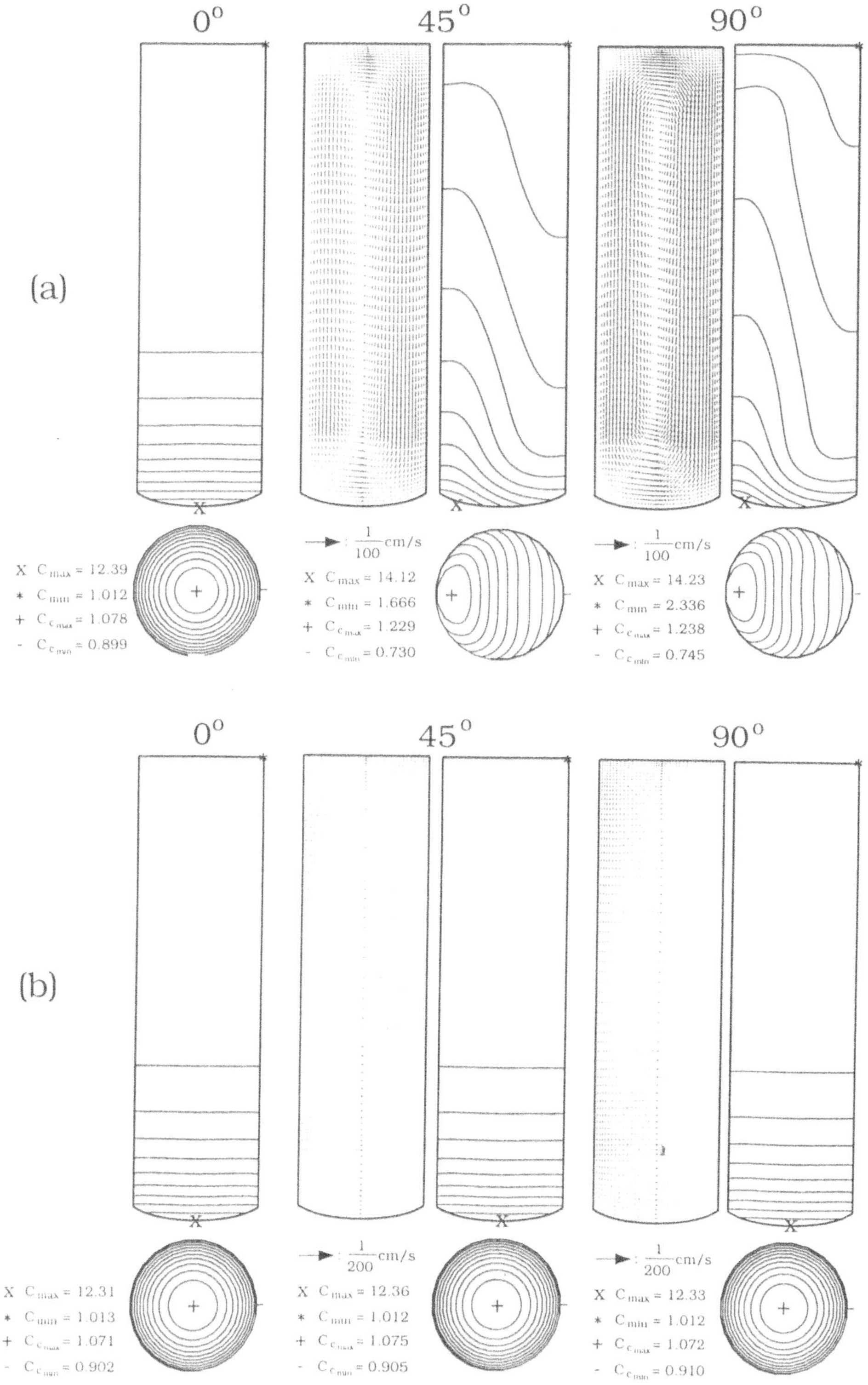

Figure 5. Influence of gravity orientation and rotation for $Ra_T=10^4$: (a) 0 rpm; (b) 5 rpm.[20]

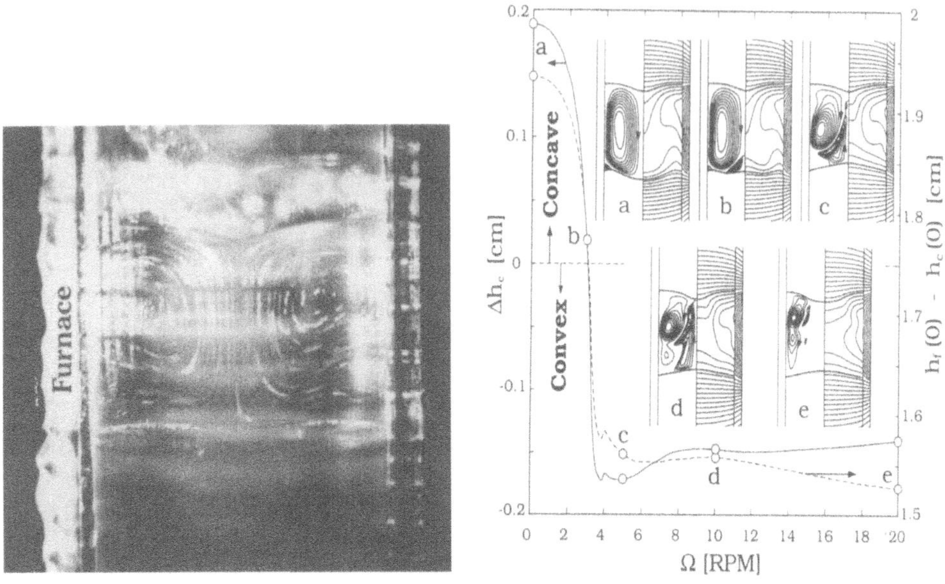

(a) (b)

Figure 6. Effects of flow on vertical zone melting: (a) flow visualization for $NaNO_3$; (b) influence of rotation on heat flow, interface convexity and zone length for GaAs.[22]

A dramatic change of interface shape, from concave to convex, occurred at about 3 rpm, where a secondary cell appeared at the zone center. There was little change in interface shape and zone length above 5 rpm rotation. At a higher rotation rate, about 20 rpm, the main flow intensity started to decrease. The damping effect was also effective near the melt core and the heat transfer there was changed from predominantly convective to conductive. A Taylor-Proudman column[34] near the melt core became obvious after 10 rpm. Such a change can also be seen from the distortion of the isotherms. Therefore, as far as interface control is concerned, only heat transfer near the growth front needed to be modified, and slow rotation seemed to be enough to achieve that.

In the previous calculations, axisymmetry was assumed. However, for the condition considered here, an unsteady 3D mode may prevail. A bifurcation diagram indicated that the melt flow at normal gravity should be oscillatory.[24] With 10 rpm rotation, an axisymmetric mode with a convex interface is possible.[23] The solution is the same as that obtained by the axisymmetric model. Near the center of the growth front, a convection cell is induced as well. A Taylor column near the centerline seems to prevent the flow from coming across and flow near the centerline is suppressed. As a result, axisymmetric flow becomes stable and the interface shape convex, both of which are favorable for the growth of single crystals.

Without ampoule rotation, the 3D unstable convective cells move back and forth, but no angular wave is found. If unsteady convection is not suppressed, a baroclinic instability, caused by coupling of the buoyancy and Coriolis forces, may appear.[35] Figure 7 illustrates a simple calculation, where the feed and growth fronts are fixed and flat. A thermal wave travels in the angular direction. One can get a clearer picture by visualizing the mid-plane isotherms. Note that the speed of the angular wave is slower than the rotation speed. If interface deformation is considered, one may expect similar dynamics at the growth front.

From the previous examples, it is seen that ampoule rotation may be used for control of the interface shape and the convection in vertical ZM. The 3D unstable flows may be suppressed as well. Even with tilting and asymmetric heating of the ampoule, rotation is also useful for reducing asymmetry.[23] The limiting case for asymmetry is the horizontal configuration. As illustrated below, the mechanism for interface control could be very different.

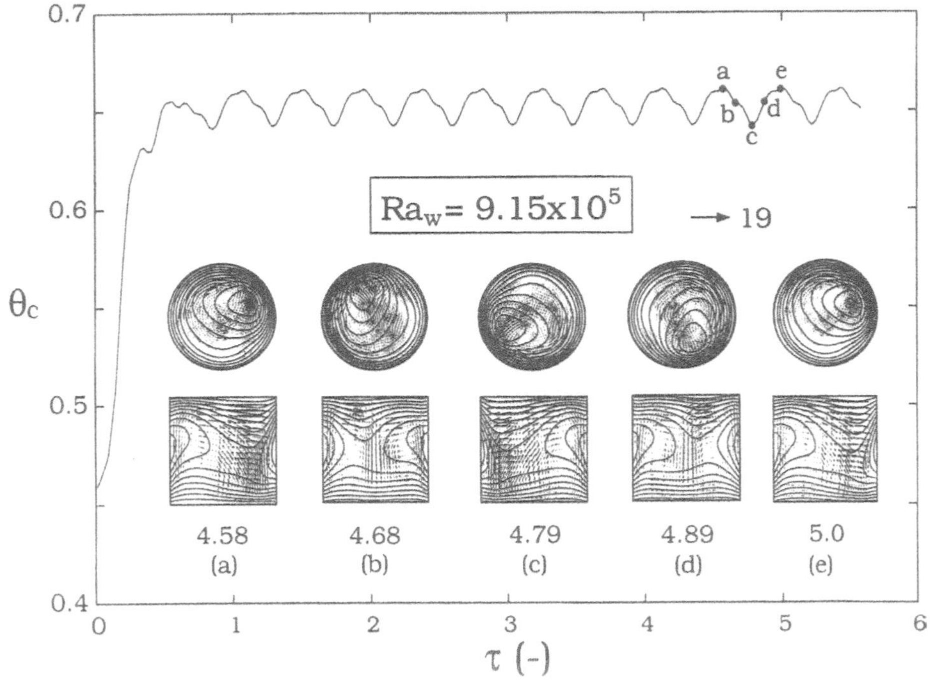

Figure 7. Calculated temperature response at the zone center for $Ra_w = 9.15\times10^5$ and Ta = 2056.49 for GaAs.[23]

INTERFACE CONTROL

As described above, ampoule rotation is also helpful for ameliorating the effects of asymmetric heating and poor ampoule alignment. The most interesting application of such an idea is to horizontal zone-melting (HZM) of low-Prandtl-number materials. Pfann *et al.*[25,26] first demonstrated the effectiveness of slow ampoule rotation on zone-shape control in HZM, the so-called Pfann-Miller or rotation-convection heating technique. With only 2.5 rpm rotation, a nearly axisymmetric zone as short as 20% of the ampoule diameter, was found possible for resorcinol. Nearly-flat interfaces could be produced as well. Since then, this technique has been used routinely in the HZM of organic and inorganic materials. The effectiveness of slow rotation on zone shape control has been attributed to convection smoothing out the interfaces.[25,26] Pfann *et al.*[26] claimed that to make the technique work, the flow pattern due to natural convection should not be significantly altered, in contrast to high speed rotation.[36]

In our recent flow visualization experiments using succinonitrile we found that although the zone shape was affected significantly by only 1 rpm rotation, the flow structure was completely altered, as shown in Fig. 8.[27] Without rotation, Fig. 8a shows that there was great asymmetry of zone shape and convection. Two flow cells were seen in this side view, with melt flowing up at the center and down near the interfaces. The upward flowing melt was heated at the top wall and flowed to the two sides to melt back the interfaces. As a result, the upper part of the zone became wider. Increasing heating power made the zone length longer, but the flow structure and the zone shape were not changed much. Therefore, the trapezoidal zone shape in Fig. 8a was dominated by natural convection. This observation is consistent with that reported by Pfann *et al.*[26] Figure 8b shows that 1 rpm ampoule rotation caused the zone to become much more symmetric and significantly changed the flow structure.

A fully 3D numerical simulation was carried out, as illustrated in Fig. 9.

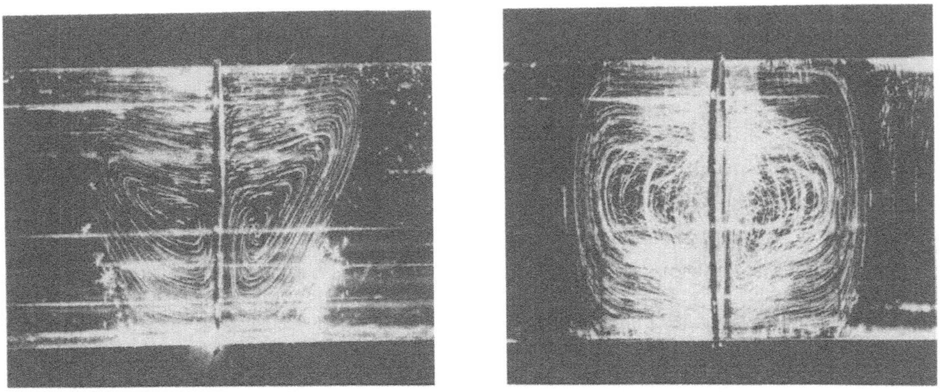

(a) (b)

Figure 8. Observed flow patterns and interface shapes for horizontal zone melting of succinonitrile containing particles. Photographs with 11s exposures. (a) 0 RPM; (b) 1 RPM[27].

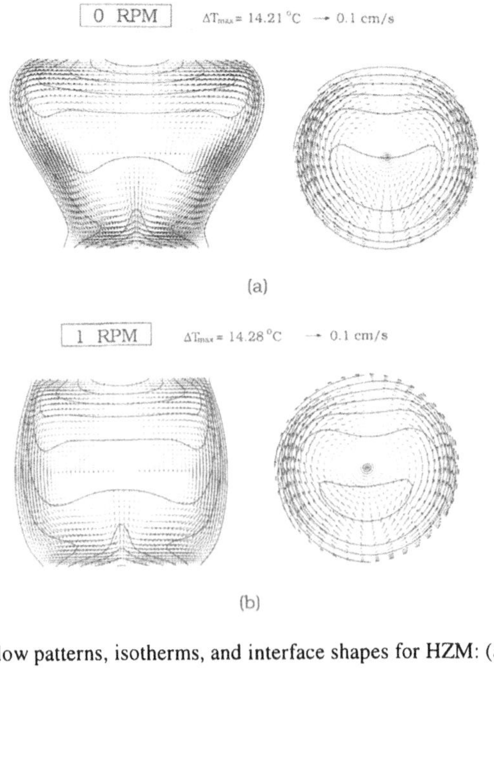

Figure 9. Calculated flow patterns, isotherms, and interface shapes for HZM: (a) 0 RPM; (b) 1 RPM[27].

The results are in good agreement with the experimental observations in Fig. 8. More importantly, without considering the deformable interfaces, the calculated results for 0 and 1 rpm rotation are near one another.[27] In other words, the contributions of the first two terms are small in Eq. (5), which represents the interface energy balance. The major contributions is from the last term, the so-called Stefan effect. Alternate regrowth ($\boldsymbol{n}\bullet \boldsymbol{e}_\phi<0$) and remelting ($\boldsymbol{n}\bullet \boldsymbol{e}_\phi>0$) occur. The more asymmetry is introduced, the more the Stefan effect is important.

The worst case was found in zone refining of an impure sample. Figure 10 shows that more inclusions appeared as the rotation rate was increased. This is attributed to an increased maximum growth rate with the same average growth rate when rotation causes alternate growth and melting.

The idea of reducing asymmetry by slow rotation can be applied to the floating zone (FZ) configuration as well. The most typical example is the growth of oxides. To illustrate that, we modeled FZ melting of a 4-mm $NaNO_3$ rod in a mono-ellipsoid mirror furnace (Fig. 11). The zone is very asymmetric without rotation, as shown in Fig. 12a. The side facing the lamp is much longer. The gray level indicates the temperature. The higher temperature gradient near the lamp also generates significant Marangoni convection, which makes the melt zone even longer and increases its asymmetry.

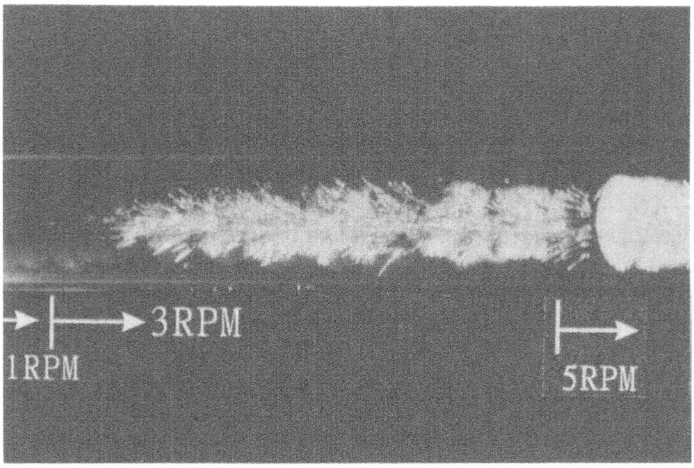

Figure 10. Inclusions in an impure material 9mm in diameter, after horizontal zone refining with different rotation speeds.[27]

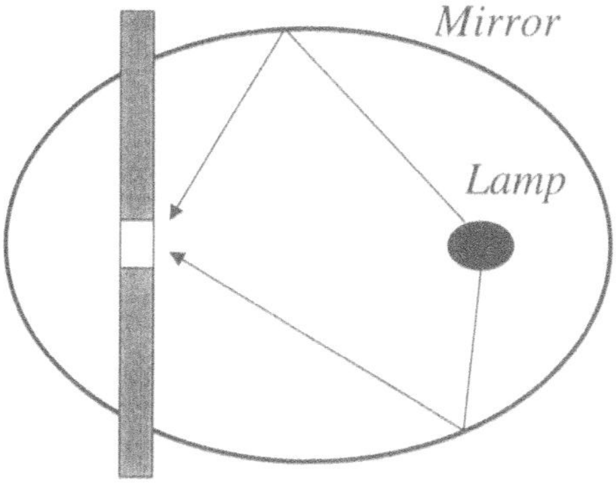

Figure 11. Schematic of mono-ellipsoid mirror furnace for floating-zone crystal growth

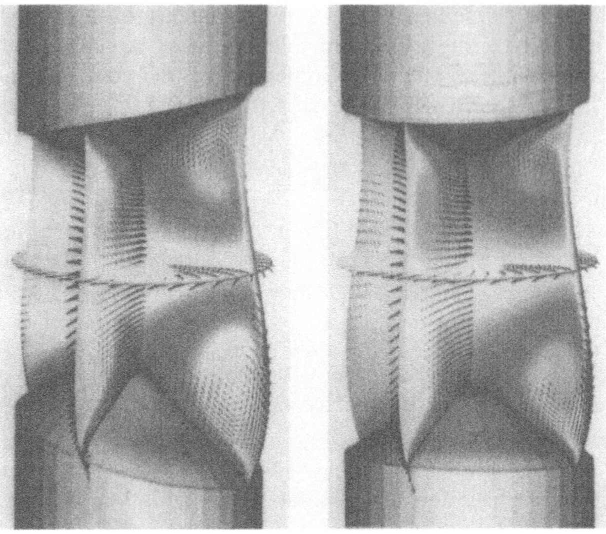

Figure 12. Calculated temperature field (gray), velocity vectors, and interface shapes for a floating-zone in 4-mm diameter $NaNO_3$: (a) 0 RPM; (b) 10 RPM.

The Marangoni flow can be seen from the velocity vectors in Fig. 12a, with significant flow from the hot to cold regions. Figure 12b shows that the zone becomes very uniform with 10 rpm co-rotation of the sample and crystal rods. Again, similar to HZM, the flow structure is not affected much by the rotation. The nearly axisymmetric interfaces are due to the rotational Stefan effect.

To further control the interface shapes, one may also use counter-rotation to generate forced convection that brings the hot melt directly from the surface to the zone center. As illustrated in Fig. 13 for 6mm diameter $NaNO_3$, flat interfaces are obtained by imposing 200 rpm counter-rotation.[38] From this flow visualization, we see that Marangoni flow is almost suppressed and forced convection is dominant. Increasing the rotation speed to 400 rpm caused the interfaces to become concave. More numerical illustrations can be found elsewhere.[39]

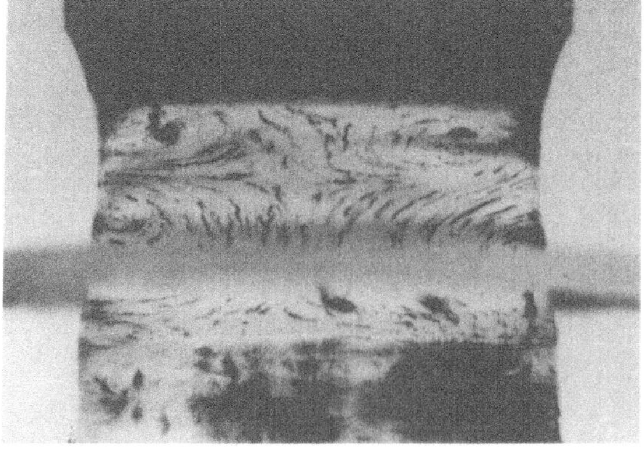

Figure 13. Flow visualization in a molten zone in a vertical 6-mm diameter $NaNO_3$ rod with 200rpm counter-rotation.[38]

CONCLUSIONS

Through flow visualization and numerical simulation, we see that there are many ways to use steady rotation to better control directional solidification. For situations with weak buoyancy-driven convection, the Coriolis effect can suppress the flow before centrifugal force becomes important. The flow suppression by rotation is similar to that by an axial magnetic field. As a result, we can control dopant mixing and the resulting axial segregation.

For situations with strong buoyancy-driven convection, such as in vertical zone-melting, interface control also can be achieved by modifying the flow. Rotation can also affect the interfaces through the so-called Stefan effect. However, the resulting oscillations in growth rate, perhaps including melt-back, can induce impurity striations, interface breakdown and inclusions formation. We showed that rotation can be used to form axisymmetric zones in oxides in a mono-ellipsoid mirror furnace, even though the Marangoni flow is not altered much by rotation. Counter-rotation can be used to further control the interface shape.

REFERENCES

1. R.A. Brown, Theory of transport processes in single crystal growth from the melt, *AIChE J.* 34: 881 (1989).
2. G. Müller and A. Ostrogorsky, Convection in melt growth, in: *Handbook of Crystal Growth 2b: Growth Mechanisms and Dynamics*, D.T.J. Hurle, ed., North-Holland, Amsterdam (1994).
3. H.P. Utech and M.C. Flemmings, Elimination of solute banding in indium antimonide crystals by growth in a magnetic Field, *J. Appl. Phys.* 37: 2021 (1966).
4. K.M. Kim, Suppression of thermal convection by transverse magnetic field, *J. Electrochem. Soc.* 132: 427 (1982).
5. D.H. Kim, P.M. Adornato, and R.A. Brown, Effect of vertical magnetic field on convection and segregation in vertical Bridgman crystal growth, *J. Crystal Growth* 89: 339 (1988).
6. H.J. Scheel, Accelerated crucible rotation: a novel stirring technique in high- temperature solution growth, *J. Crystal Growth* 13/14: 560 (1971).
7. A.V. Anilkumar, R.N. Grugel, R.N. Shen, and T.G. Wang, Control of thermocapillary convection in a liquid bridge by vibration, *J. Appl. Phys.* 73: 4165 (1993).
8. D.V. Lyubimov, T.P. Lyubimova, S. Meradji, and B. Roux, Vibrational control of crystal growth from liquid phase, *J. Crystal Growth* 180: 648 (1997).
9. W.S. Liu, M.F. Wolf, D. Elwell, and R.S. Feigelson, Low frequency vibrational stirring: a new method for radial mixing solutions and melts during growth, *J. Crystal Growth* 82: 589 (1987).
10. W. Yuan, M. Banan, L.L. Regel, and W.R. Wilcox, The effect of vertical vibration of the ampoule on the direction solidification of InSb-GaSb alloy, *J. Crystal Growth* 151: 235(1995).
11. V. Uspenskii and J.J. Favier, High frequency vibration and natural convection in Bridgman-scheme crystal growth, *Int. J. Heat Mass Transfer* 37: 691 (1994).
12. C.W. Lan, Effects of axial vibration on vertical zone-melting processing, *Int. J. Heat Mass Transfer* 43: 1987 (2000).
13. W.A. Arnold, W.R. Wilcox, F. Carlson, A. Chait, L.L. Regel, Transport modes during crystal growth in a centrifuge, *J. Crystal Growth* 119: 24 (1992).
14. G. Müller, G. Neumann, and W. Weber, The growth of homogeneous semiconductor crystals in a centrifuge by the stabilizing influence of the Coriolis force, *J. Crystal growth* 119: 8 (1992).
15. W.R. Wilcox and L.L. Regel, Influence of centrifugation on transport phenomena, *46th International Astronautical Congress*, Oslo, Norway (1995).
16. W.R. Wilcox, L.L. Regel, and W.A. Arnold, Convection and segregation during vertical Bridgman growth with centrifugation, *J. Crystal Growth* 187: 543 (1998).
17. A. F. Witt, H.C. Gatos, M. Lichtensteiger, M.C. Lavine, and C.J. Herman, Crystal growth and steady state segregation under zero gravity, *J. Electrochemical Soc.* 122: 276 (1975).
18. C.W. Lan, Effects of ampoule rotation on flows and dopant segregation in vertical Bridgman crystal growth, *J. Crystal Growth* 197: 983 (1999).
19. A. Yeckel, F.P. Doty, and J.J. Derby, Effect of steady ampoule rotation on segregation in high-pressure vertical Bridgman growth of cadmium zinc telluride, *J. Crystal Growth* 203: 87 (1999).
20. C.W. Lan, M.C. Liang, J.H. Chian, Influence of steady ampoule rotation on three-dimensional flows

and segregation in Bridgman crystal growth, *J. Crystal Growth* 212: 340 (2000).
21. C.W. Lan, Effects of centrifugal acceleration on the flows and segregation in vertical Bridgman crystal growth, *J. Crystal Growth*, submitted.
22. C.W. Lan and J.H. Chian, Effects of ampoule rotation on vertical zone-melting crystal growth: steady rotation versus accelerated crucible rotation technique, *J. Crystal Growth* 203: 286 (1999).
23. C.W. Lan, M.C. Liang, J.H. Chian, Suppressing three-dimensional unsteady flows in vertical zone-melting crystal growth, *J. Crystal Growth* 213: 395 (2000).
24. C.W. Lan, M.C. Liang, J.H. Chian, Three-dimensional simulation of vertical zone-melting crystal growth: symmetry breaking to multiple states, *J. Crystal Growth* 208: 327 (2000).
25. W.G. Pfann, *Zone Melting* (2nd End), John Wiley and Sons, New York (1966), p 97.
26. W.G. Pfann, C.E. Miller, and J.D. Hunt, New zone refining techniques for chemical compounds, *Rev. Sci. Instr.* 37: 649 (1966).
27. C.W. Lan, J.H. Chian, and T.Y. Wang, Interface control mechanisms in horizontal zone-melting with slow rotation, *J. Crystal Growth*, in press.
28. G. Buzyna and G. Veronis, Spin-up of a stratified fluid: theory and experiment, *J. Fluid Mech.* 50: 579(1971).
29. M.C. Liang and C.W. Lan, A finite-volume/Newton method for a two-phase heat flow problem using primitive variables and collocated grids, *J. Comp. Phys.* 127: 330 (1996).
30. C.W. Lan and M.C. Liang, Multigrid methods for incompressible heat flow problems with an unknown interface, *J. Comp. Phys.*, 152: 55 (1999).
31. P.M. Adornato and R.A. Brown, Convection and segregation in directional solidification of dilute and non-dilute binary alloy: effects of ampoule and furnace design, *J. Crystal Growth* 80: 155 (1987).
32. C.W. Lan and F.C. Chen, A finite volume method for solute segregation in directional solidification and comparison with a finite element method, *Comput. Meth. Appl. Mech. Eng.* 131: 191 (1996).
33. C.W. Lan, Comparison of flow and segregation control by ampoule rotation and magnetic fields for vertical Bridgman crystal growth, *J. Chin. Inst. Chem. Engrs,* to appear.
34. S. Chandrasekhar, *Hydrodynamic and Hydromagnetic Stability*, Dover, New York (1961).
35. H.P. Greenspan, *The Theory of Rotating Fluids*, Cambridge Univ. Press., Cambridge, UK (1969).
36. W. R. Wilcox, R. Friedenberg, and N. Back, Zone melting of organic compounds, *Chem. Rev.* 64: 187 (1964).
37. C.W. Lan, Three-dimensional simulation of heat flow and interfaces in floating-zone crystal growth, *J. Crystal Growth*, in preparation.
38. C.W. Lan, *Heat Transfer, Fluid Flow, and Interface Shapes in Floating-Zone Crystal Growth*, PhD Dissertation, University of Wisconsin at Madison (1991).
39. C.W. Lan and S. Kou, Effects of rotation on heat transfer, fluid flow, and interfaces in normal gravity floating-zone crystal growth, *J. Crystal Growth* 114: 517 (1991).

NUMERICAL MODELING AND FLOW VISUALIZATION IN THE GRADIENT FREEZE CONFIGURATION DURING CENTRIFUGATION

Peter V. Skudarnov,[1] Liya L. Regel, William R. Wilcox, and Goodarz Ahmadi

Clarkson University, Potsdam, NY 13699-5814, USA

ABSTRACT

A visualization system was used for observation of convection in water in the gradient freeze configuration during centrifugation with a 1.6-m radius arm. Without centrifugation, the usual axisymmetric flow pattern with a toroidal vortex near the bottom of the test cell was observed. With centrifugation at low rotation rates, the flow was primarily rotational about the cell's vertical axis. Centrifugation also modified the bottom toroidal vortex. As the rotation rate was increased, this vortex became asymmetric and was pushed toward the bottom. A flow pattern transition occurred at about 25 rpm with a flat bottom and at ~30 rpm with concave bottoms. There was a slight minimum in the horizontal velocity 3 mm above concave bottoms at about 35 rpm, and at ~20 rpm with a flat bottom.

Three-dimensional steady-state numerical modeling of the experiments was performed that included the centrifugal and Coriolis forces. Up to 20 rpm, the calculated flow patterns viewed in the tangential plane and a 3-mm horizontal plane agreed with those observed in the experiments. The numerical modeling predicted a minimum in mass flux in the vertical direction at about 10 rpm. The numerical modeling appeared to be invalid at rotation rates above 20 rpm, probably because the mesh was too coarse due to computer limitations.

The experimental velocities were insensitive to the curvature of the bottom, while the calculated velocities increased significantly with increasing concavity. This indicates that the bottom plastic surface was not isothermal as assumed in the theoretical model.

INTRODUCTION

This research was motivated by the unexpected gradient freeze results obtained by Regel and Rodot,[1-4] who reported uniform axial dopant distributions in crystals grown at a particular rotation rate in two different centrifuges. Although several simple models have been proposed to explain these results, none of them agree completely with all

[1] Current address: Florida International University, Miami, FL 33174.

Processing by Centrifugation, Edited by Regel and Wilcox
Kluwer Academic/Plenum Publishers, New York, 2001

experimental results and numerical simulations.[5]

In our previous papers,[6,7] we reported on visualization of convection in a simulated gradient freeze cell using HIRB, the Clarkson University materials processing and flow visualization centrifuge.[8] Neutrally-buoyant reflective particles were suspended in water and tracked in illuminated planar sheets. The velocities of these particles were determined using Particle Displacement Tracking Velocimetry in one vertical and two horizontal planes at rotation rates of 10, 15, 20, 30, and 40 rpm.[6,7,9] Quantitative flow patterns were obtained from the measured particle velocities using our computer code that implemented the interpolation procedure described by Ushijima.[10] The experimental and theoretical cells were 2.2-cm in diameter and 4.4-cm long, with a temperature difference between the top and the bottom of approximately 50°C. Two different plastic bottom end caps had been used in the test cell, one concave with a 1-mm depression and the other flat. Here, we present further experimental results obtained using the same techniques, as well as numerical simulations. The range of experimental centrifuge rotation rates was extended to 60 rpm. Two additional concave bottoms were used with 2 and 3-mm depressions, as well as additional experiments with a flat bottom.

Three-dimensional numerical modeling was performed using the computational fluid dynamics solver FLUENT, version 4.5.[11] Most of the modeling was done for an isothermal concave bottom with a 0.5-mm depression. The average flow velocity was calculated for the entire model cell, for its bottom part, and in one horizontal cross section. The mass flux in the axial direction was computed for three planes parallel to the bottom.

EXPERIMENTAL

The techniques used in this work were described in detail elsewhere.[6,7,9] As previously defined, the radial plane passes through the centrifuge arm and the axis of rotation. The tangential plane is perpendicular to the radial plane and passes through the ampoule's axis. Planes perpendicular to the radial and tangential planes are called horizontal, in the sense that they are perpendicular to the axis of the test cell.

Flow Patterns in a Horizontal Plane 3 mm Above the Bottom

Flow visualization was performed in a horizontal plane 3 mm above the outer edge of the bottom. Figures 1, 2 and 3 show the velocity field in this plane versus the rotation rate of the centrifuge for three different bottom curvatures. Without centrifugation the overall flow pattern was the customary axi-symmetric vortex, yielding the radially outward flow shown in Fig. 2a. For rotation rates up to 20 rpm, the flow pattern was the same for all bottom curvatures. This pattern consisted of a single circular vortex occupying the entire cell cross section and rotating in the direction opposite to that of the centrifuge (Figs. 1a & b, 2b & c, 3a & b). This rotation sense is referred to as clockwise, because the particles in this vortex moved clockwise on the video monitor (viewing from the bottom). The radial flow in this cross section was outward, i.e. toward the cell wall. At 30 rpm, with a concave bottom, two small secondary counterclockwise vortices appeared in the first and third quadrants of the cross section near the cell wall (Fig. 2d). With a flat bottom, a similar flow transition occurred at about 25 rpm (Fig. 3c). At 40 rpm, with a concave bottom the flow consisted of four unequal vortices, one in each quadrant of the cross section. In the first and third quadrant the vortices rotated counterclockwise, while in the second and fourth quadrants the vortices rotated clockwise (Fig. 2e). For concave bottoms, as the rotation rate was increased beyond 40 rpm, the vortices in the second and third quadrants became dominant and occupied more and more space in the cross section. The velocity in these vortices increased with increasing rotation rate, as indicated by the length of the arrows in Figs. 1d, 1e and 2f. For a flat bottom, at rotation rates beyond 40 rpm, four unequal vortices were observed similar to those at 40 rpm with a concave bottom (Figs. 3e & f).

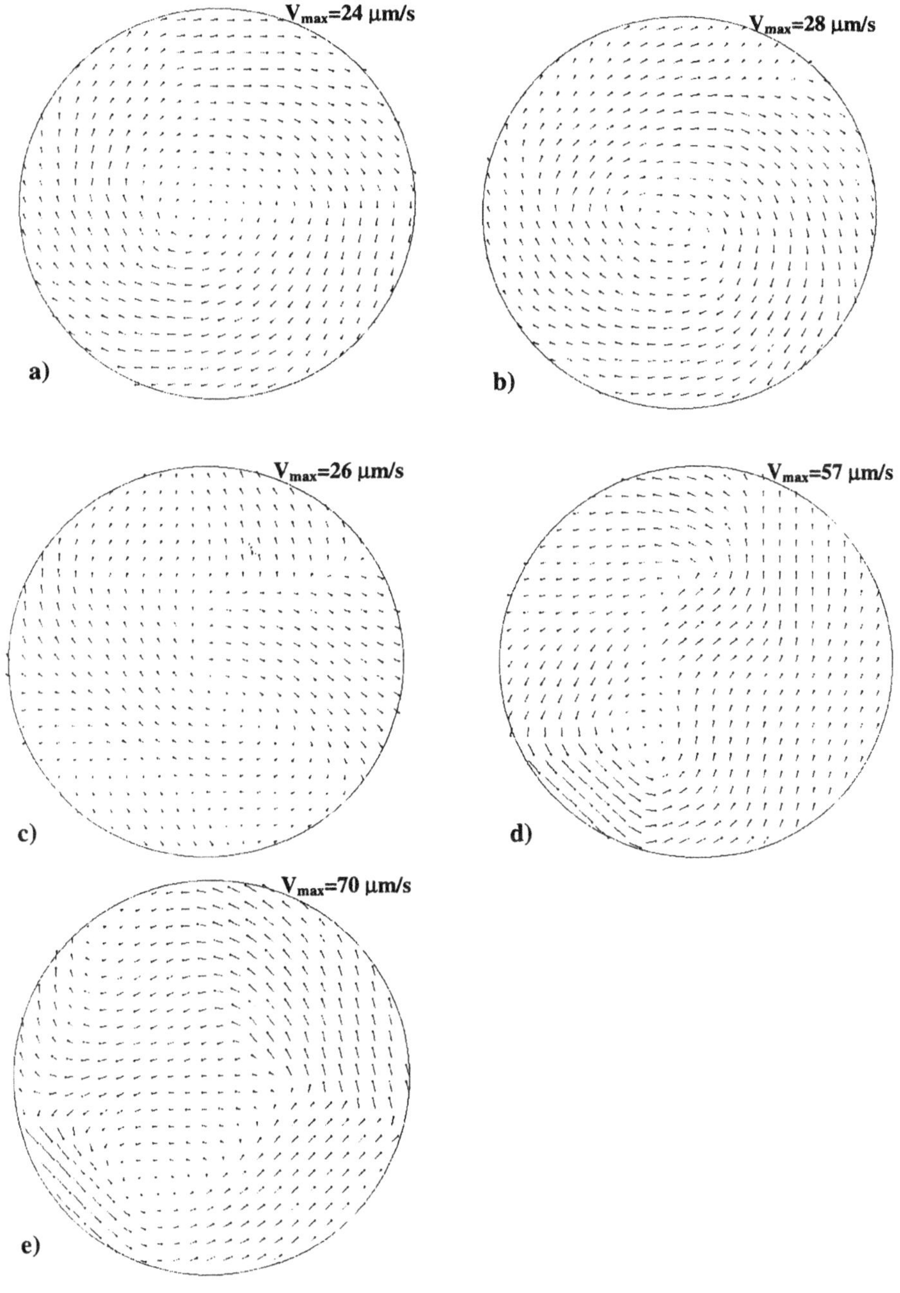

Figure 1. Flow patterns in a horizontal plane 3 mm above a concave bottom with a central depression of 2 mm. Velocities interpolated from measured particle velocities.
Rotation rates: a) 10 rpm; b) 20 rpm; c) 35 rpm; d) 45 rpm; e) 55 rpm.

Velocity Measurements

Particle velocities were measured in a plane 3 mm above the periphery of the bottom, with four bottom depressions: 0(flat), 1 mm, 2 mm, and 3 mm. Figure 4 shows the mean velocity and Fig. 5 the maximum velocity obtained from measured values.

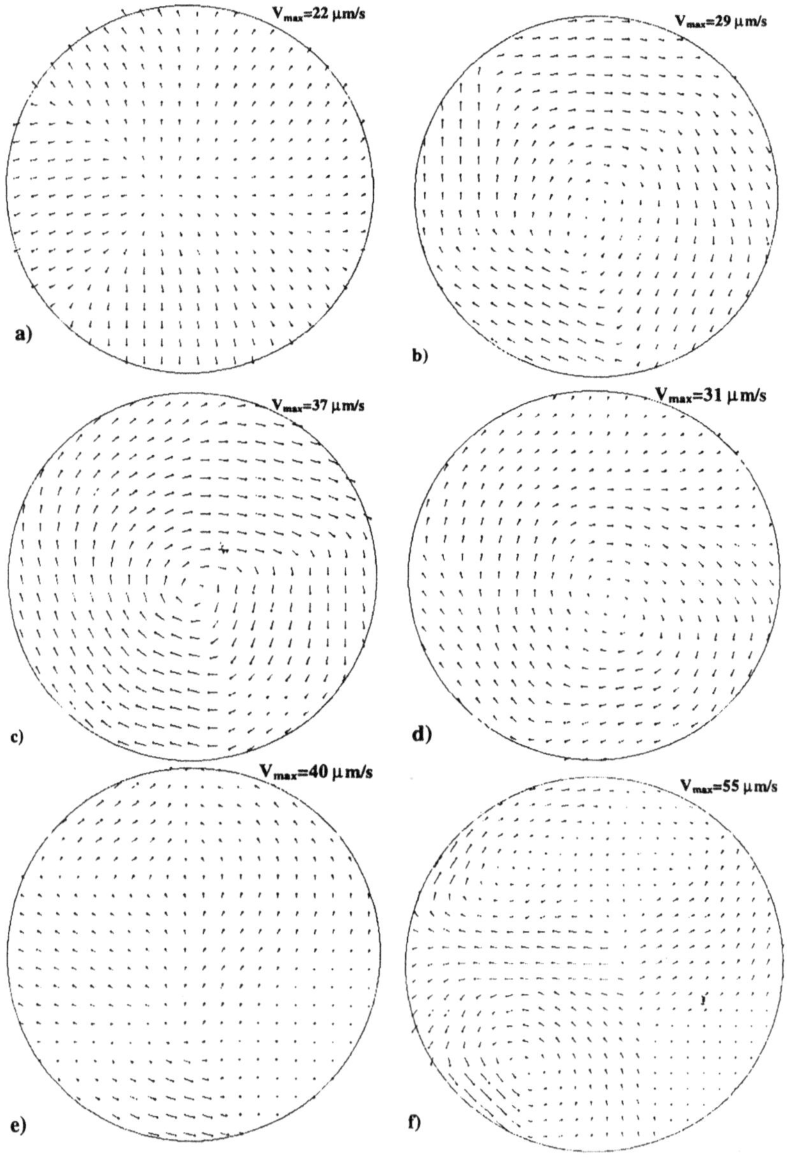

Figure 2. Flow patterns in a horizontal plane 3 mm above a concave bottom with a central depression of 3mm. Rotation rates: a) 0 rpm; b) 10 rpm; c) 20 rpm; d) 30 rpm; e) 40rpm; f) 55 rpm.

For curvature depressions of 2 and 3 mm, the mean velocity initially increased as the rotation rate increased, had a maximum at ~20 rpm, went through a minimum at ~35 rpm, and then increased again with increasing rotation rate up to the maximum studied rotation rate of 60 rpm.

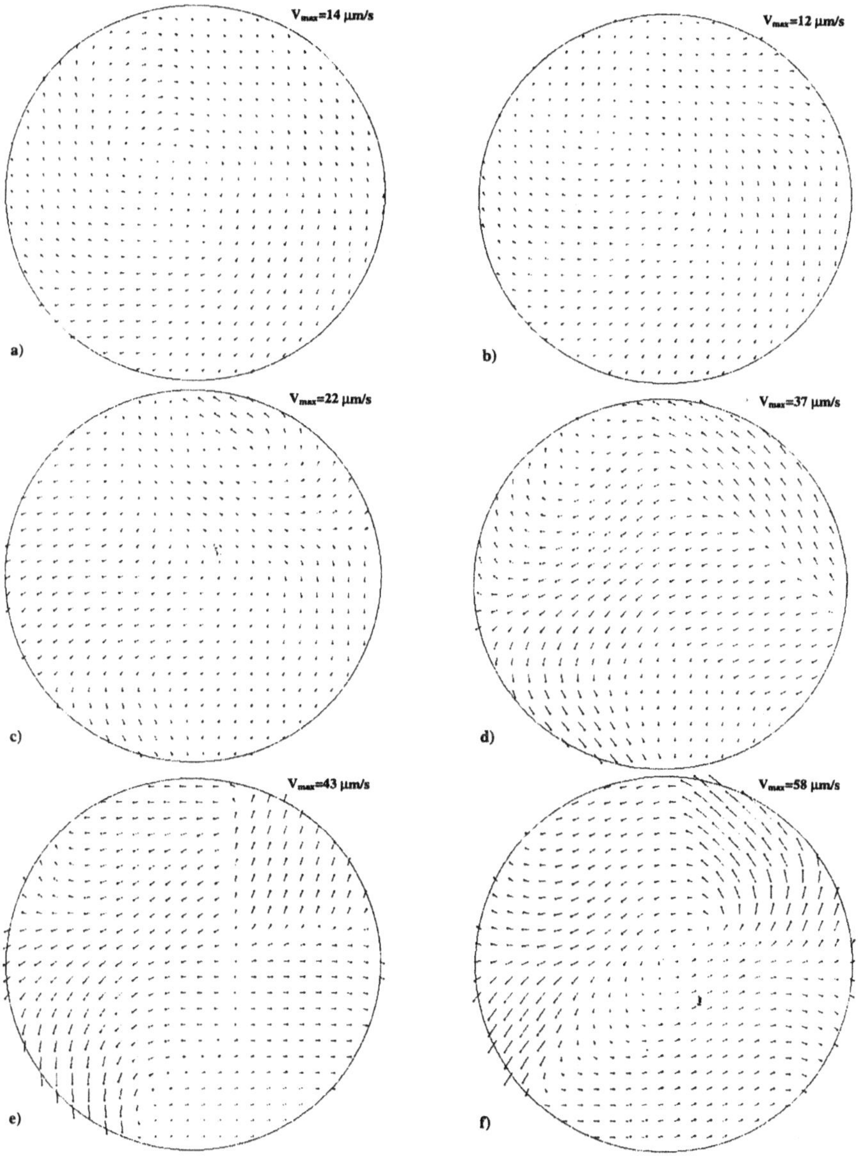

Figure 3. Flow patterns in the horizontal plane positioned 3 mm above a flat bottom end cap. Rotation rates: a) 10 rpm; b) 20 rpm; c) 25 rpm; d) 35 rpm; e) 45 rpm; f) 55 rpm.

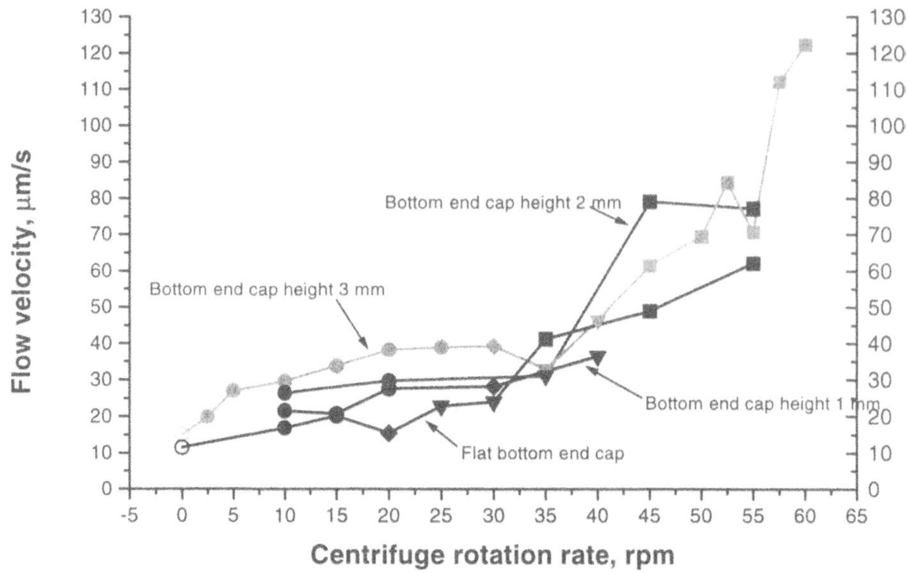

Figure 4. Mean velocity (µm/s) obtained from measured values in a horizontal cross section 3 mm above the outer edge of the bottom, for concave bottoms with 1, 2, 3 mm depression and for a flat bottom. Symbols indicate the flow pattern observed in the cross section:
❍ radial outward (without centrifugation) (see Fig. 2a);
● single clockwise vortex (see Figs. 2b and c);
◆ onset of flow transition. In addition to the main clockwise vortex, two secondary vortices are in the first and third quadrants of the cross section (see Fig. 2d);
▼ four vortices (see Fig. 3c);
■ two vortices (see Figs. 1d, e and f).

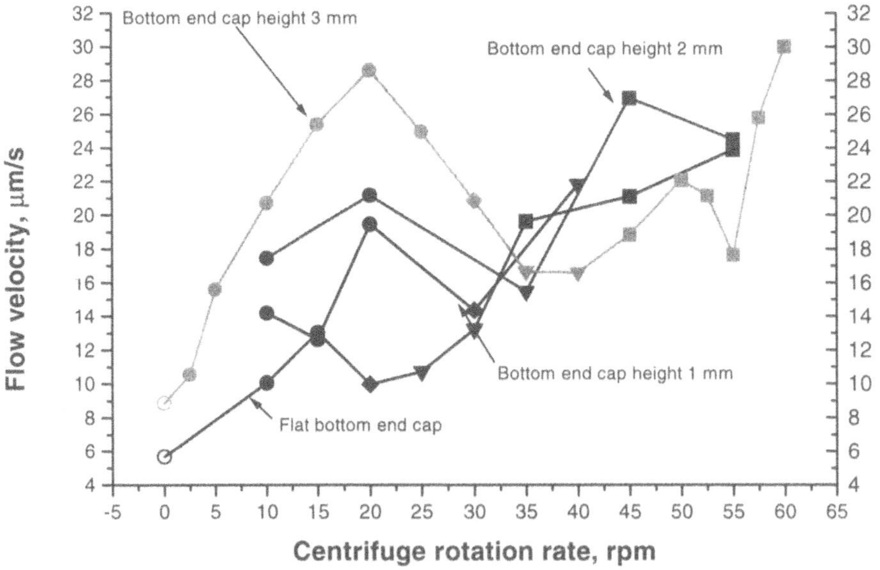

Figure 5. Maximum velocity (µm/s) obtained from measured values in a horizontal cross section 3 mm above the bottom, for concave bottoms of 1, 2, 3 mm depression and a flat bottom cap. Symbols show the type of flow, as in Fig 4.

For a flat bottom, similar behavior was observed with a maximum at ~15 rpm and a minimum at ~20 rpm. The maximum velocity increased slowly with increasing rotation rate up to 35 rpm (30 rpm for a flat bottom), beyond which it increased much faster.

NUMERICAL MODELING

The flow visualization experiments were modeled using FLUENT,[11] which solves simultaneously the conservation equations for mass, momentum, energy, and chemical species using a control volume based finite difference method. The governing equations are discretized on a structured body-fitting grid to enable computations in regions with complex geometries.

All of our modeling was three-dimensional for steady-state conditions, and included both the centrifugal force and the Coriolis force. The geometrical dimensions and the boundary conditions are summarized in Table 1.

Table 1. The geometrical dimensions and the boundary conditions used in the numerical simulations.

Parameter	Condition imposed
Centrifuge arm length	1.65 m
Test cell radius	11 mm
Test cell length	44 mm
Depression of top and bottom caps	0.5 mm
Temperature of the top	330 K (57°C)
Temperature of the bottom	280 K (7°C)

This model incorporated the dimensions of our test cell and centrifuge. Constant-temperature boundary conditions were applied at the end caps and a linear axial temperature profile on the cylindrical wall. The usual no-slip zero-velocity condition was applied at all walls. The net acceleration vector varied throughout the cell and was aligned exactly with the cell's vertical axis only at the center of the bottom. For each centrifuge rotation rate, the angle of tilt of the test cell's longitudinal axis with respect to the centrifuge's axis of rotation was calculated using MathCad 5.0 and then incorporated in the FLUENT numerical model. The temperature-dependent properties of water were used, with the density and viscosity approximated by fourth order polynomials and the thermal conductivity by a third order polynomial. The specific heat was taken to be constant.

Computations were performed on an IBM RS/6000 560 computer. A non-uniform grid of 37× 37×83 nodes was used, with a smaller spacing near the ends and a larger spacing in the middle of the cell. This grid was the finest with the available computer memory. At low rotation rates, a 19×19×83 grid yielded the same flow patterns, while deviations occurred at rotation rates of 30 rpm and above. Typical values of the initial normalized residuals were ~0.5 for velocity components, ~0.3 for pressure, and ~0.07 for enthalpy. Computations were stopped when the normalized residuals for the dimensionless velocity components remained ~0.01 for at least 200 iterations, the normalized residual for pressure ~0.001, and the normalized residual for enthalpy ~10^{-6}. At rotation rates of 30 rpm and above, the velocity vectors never ceased to fluctuate with each iteration.

Calculated Flow Patterns

Figure 6 shows the tangential plane of computed flow patterns for a concave bottom with a 0.5mm depression, both without rotation and with 10 rpm centrifuge rotation. Only the bottom portion of the cell is shown, because our main interest is in the flow near the bottom, i.e. the "freezing interface." The inner two vertical lines are an artifact of the simulation output and should be ignored. The flow pattern in the radial plane was essentially the same up to 10 rpm. Figure 7 shows that the flow patterns in the tangential and the radial planes differed only slightly at 20 rpm.

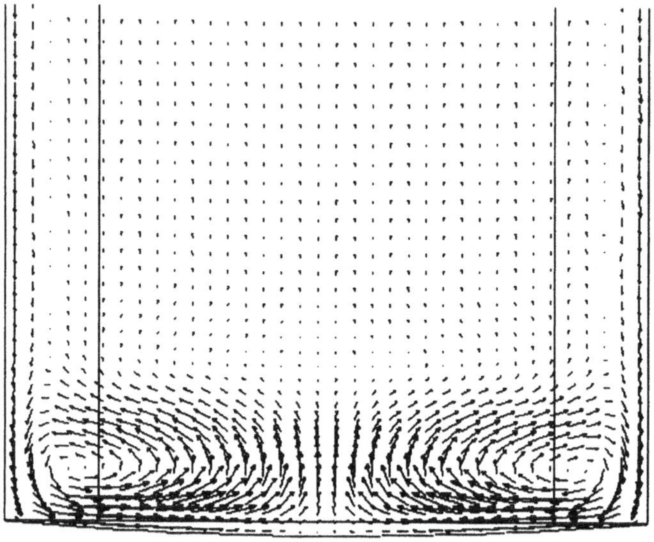

a)

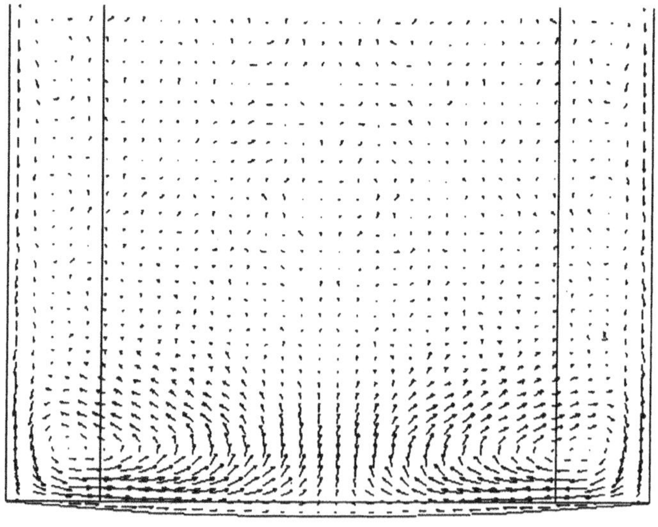

b)

Figure 6. Computed velocity field in the tangential plane. a: without rotation (1g). b: for 10 rpm

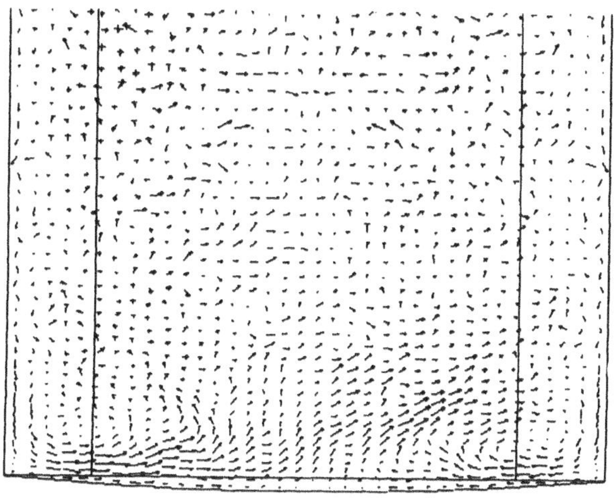

a

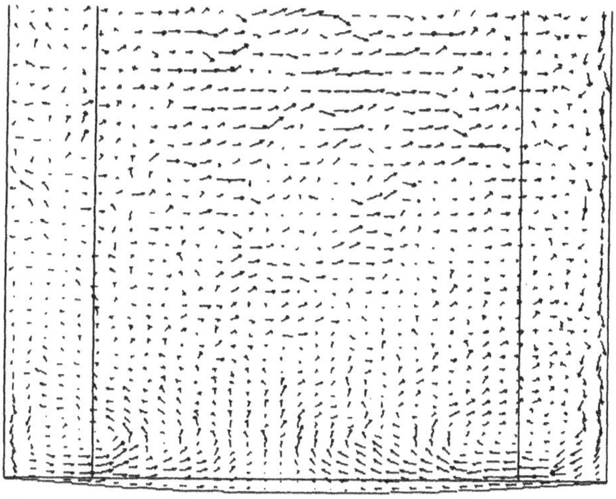

b

Figure 7. Computed velocity field at 20 rpm. a: viewed in tangential plane. b: radial plane.

The toroidal vortex near the bottom remained practically unchanged up to 20 rpm. At rotation rates beyond 20 rpm, the vortex in the tangential plane deformed: its height decreased, its left portion became confined to the left bottom corner of the cross section, and its right portion elongated in the radial direction. In the radial plane the right part of the bottom vortex became elongated in the radial direction, and its left part was pushed to the bottom left corner of the cross-section.

Figure 8 shows flow patterns in the horizontal plane 3 mm above the outer edge of the bottom. Without rotation, the flow in this plane was radially outward. At rotation rates up to 20 rpm, the flow consisted of a single clockwise rotating vortex. At a rotation rate of 30 rpm a secondary vortex emerged in the upper portion of the cross-section. At 40 and 50 rpm, two almost equal counter rotating vortices were present. The rotation sense of the vortices computed at 50 rpm was opposite to those in the experiments. The flow in horizontal planes above 3 mm was primarily rotational about the cell's vertical axis, and did not change appreciably with elevation.

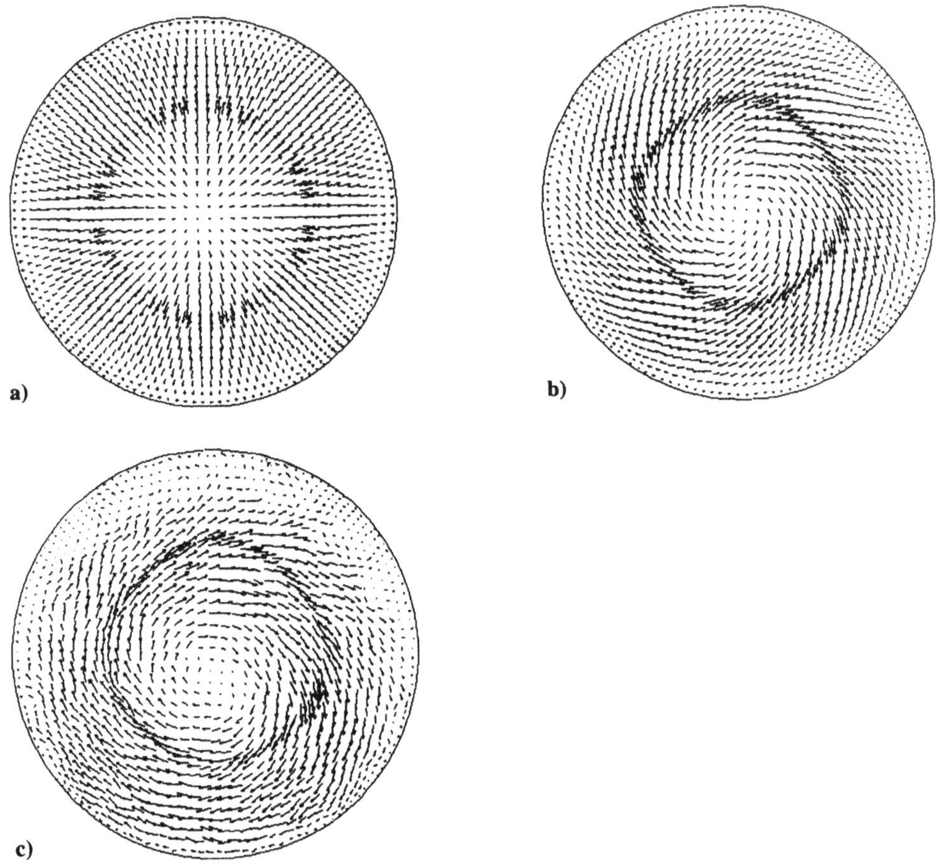

Figure 8. Computed velocity field in horizontal plane 3 mm above the bottom. a: without rotation. b: 10 rpm. c: 20 rpm.

Several numerical simulations were run using different boundary conditions to see how sensitive the flow is to these conditions. Tilting the cell by 1° resulted in only a slight distortion of the bottom toroidal vortex. Applying a slightly concave or a slightly convex temperature profile along the cell's cylindrical wall did not noticeably influence the flow pattern. In a plane 3 mm above the outer edge of the bottom, the average velocity for a slightly concave temperature profile was ~15% lower than for a linear profile. The average flow velocity in the same plane for a slightly convex temperature profile was ~15% higher than for a linear profile. Changing the bottom's depression from 0.5 mm to 3 mm did not affect the flow pattern, but did increase the average velocity by about 4X, both for the entire cell and for its bottom portion.

Computed Velocities and Mass Flow

The velocities in a plane 3 mm above the outer edge of the bottom were examined. Note that FLUENT computes the velocity in the center of each computational cell. To obtain average and maximum flow velocities in a plane we calculated the average and the maximum of absolute values of horizontal components of the velocity vectors in all computational cells that intersected this plane. Average and maximum velocities for this 3 mm plane are plotted in Fig. 9 versus the centrifuge rotation rate. Both velocities initially increased slowly with increasing rotation rate. Beyond 20 rpm, both increased faster.

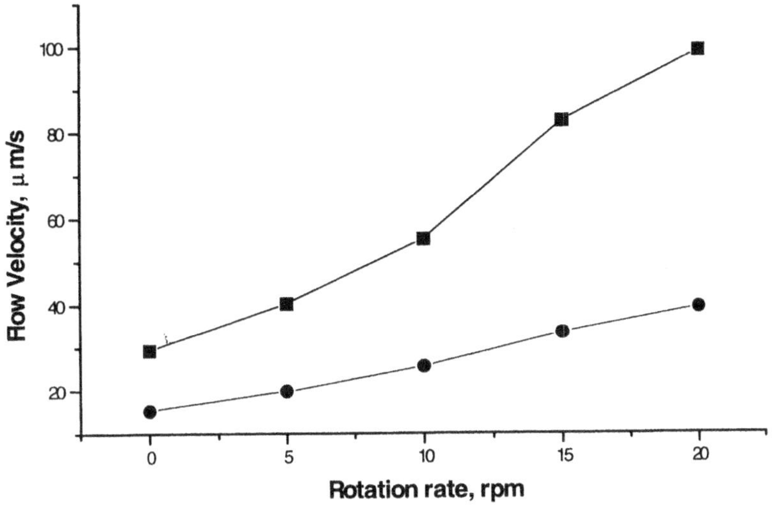

Figure 9. Average and maximum of the horizontal component of the computed velocities in a horizontal plane 3 mm above the outer edge of the bottom.
● Average velocity; ■ Maximum velocity.

Impurity segregation during solidification is influenced most by the vertical component of the flow near the freezing interface. Therefore, in order to understand the effect of centrifugation on segregation we computed the mass flow in the axial direction through three surfaces. The mass flow through a computational cell in a given direction is defined as the product of the flow velocity component in that direction, the face area of the cell perpendicular to that direction, and the density. For any closed volume problem like ours, at steady state the total integrated mass flow through any given plane must be zero, and so the mass flows into and out of horizontal planes must be equal. In our results the difference between mass flows into and out of a plane was about four orders of magnitude less than the mass flow in one direction. Figure 10 shows the mass flux (mass flow divided by the cross-sectional area) in one direction through three horizontal surfaces located 0.5 mm, 0.75 mm, and 1 mm from the periphery of the bottom. For all three, the mass flux had a gentle minimum at about 10 rpm. The smallest mass flux was through the surface closest to the bottom.

DISCUSSION AND CONCLUSIONS

Figure 11 compares the maximum velocity from the experiments with a 3mm bottom depression to the modeling results for 0.5 and 2mm depressions. The fairly good agreement between the 0.5mm depression simulations with 3mm experiments is most

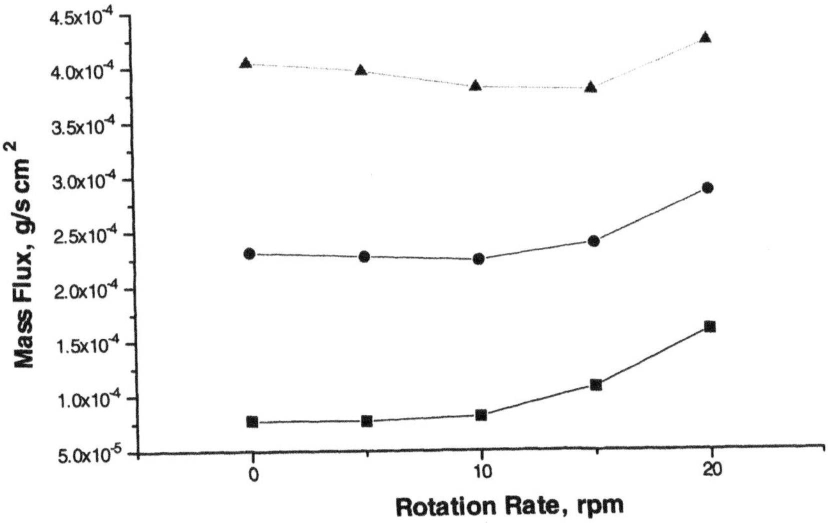

Figure 10. The mass flux in the vertical direction through the surfaces of computational nodes. ■ 0.5 mm above the bottom. ● 0.75 mm above the bottom. ▲ 1 mm above the bottom.

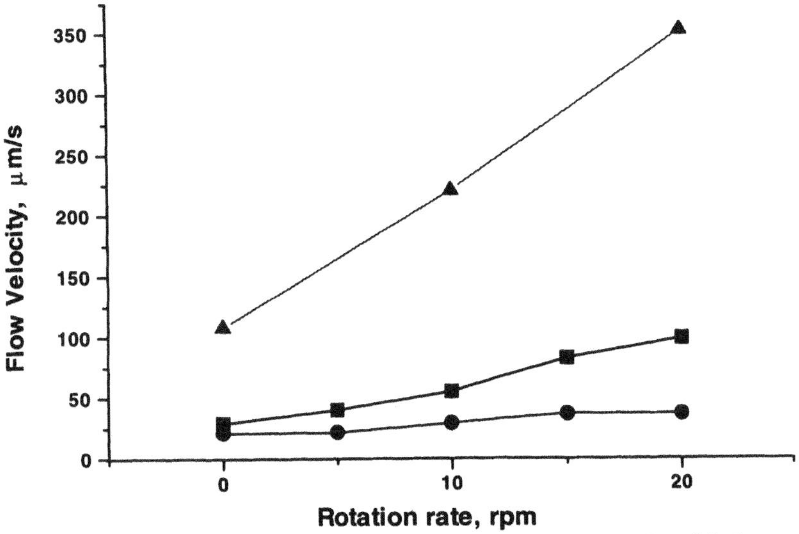

Figure 11. Maximum horizontal velocity component 3 mm above the outer edge of the bottom.
■ Calculated with a bottom depression of 0.5 mm.
▲ Calculated with a bottom depression of 2 mm.
● Experimental with a bottom depression of 3 mm.

likely because isothermal top and bottom caps was used in the modeling, while in the experiments this condition was probably not satisfied because of the low thermal conductivity of the plastic end caps. Thus, it is believed that the radial temperature gradient that drives the convection was about the same for experiments performed with a 3-mm depression as in the simulation at 0.5-mm depression. Furthermore, the experiments showed much less variation in velocity with change in depression than the numerical simulation, again reflecting the insensitivity of the radial temperature gradient in the experiments to the curvature of the end caps. Beyond 20 rpm, the experiments showed only a small variation in velocity while the numerical simulations predicted a steep increase of both average and maximum velocity. This, combined with the opposite rotation of the vortices noted earlier and the erratic velocity vectors, lead us to conclude that the numerical results were only valid up to 20 rpm, probably because the mesh size was too large. The single circular vortex in the horizontal plane at low rotation rates is attributed to the action of the Coriolis force on the basic radially outward flow observed in this plane without centrifugation. Experimentally, similar flow patterns and velocities were observed at 10, 15 and 20 rpm for all bottom curvatures. A flow mode transition was observed experimentally at ~30 rpm for concave bottoms, and at ~25 rpm for a flat bottom. As the rotation rate was increased, the bottom toroidal vortex started to rotate about the cell's vertical axis, was compressed down on the bottom surface of the cell, and became asymmetric. In the tangential plane its left side became smaller and was pushed toward the bottom left corner of the cross-section. The right side of the vortex became elongated in the radial direction, and at high rotation rates occupied practically the entire bottom region of the tangential cross-section. This asymmetry can be explained by the action of the Coriolis force as follows. Without centrifugation, liquid moves upwards in the center of the bottom toroidal vortex. At high rotation rates, the cell's axis is almost perpendicular to the centrifuge's axis of rotation. Thus, at high rotation rates, liquid in the center of the bottom vortex would move in the horizontal direction toward the centrifuge's axis, but the Coriolis force deflects it. The direction of the Coriolis force acting on liquid moving toward the centrifuge's axis is from left to right looking from above on a centrifuge rotating counterclockwise. Most of the velocity vectors are directed from left to right near the bottom.

The flow pattern in the vertical tangential plane was qualitatively similar to that calculated by Friedrich *et al.*[12,13] for a Ge melt. Unfortunately, they didn't give flow patterns in horizontal planes, making it impossible to compare with our results. The pattern found numerically by Arnold *et al.*[14,15] for PbTe grown at 5.2 *g* in an 18-m arm centrifuge agrees with our results for low rotation rates. They found that the usual axisymmetric flow pattern was only slightly modified by the Coriolis force. In our numerical simulations and experiments with low rotation rates, we also found that centrifugation only causes rotation about the cell's axis, leaving the vortex practically unchanged. Our flow velocities cannot be compared to those computed by Friedrich and Arnold because of the different material properties of semiconductor melts and water.

Changing the bottom surface curvature had little effect on the flow pattern in both numerical simulations and experiments. Similarly, tilting the cell 1° with respect to vertical did not appreciably change the computed flow pattern. This implies that a small misalignment of a gradient freeze ampoule from the net acceleration vector, as well as a small deviation of the temperature profile along the ampoule's cylindrical wall, should not result in dramatic changes of the flow.

Velocity measurements in the horizontal plane 3 mm above the bottom indicate a minimum at approximately 35 rpm for concave bottoms and 20 rpm for a flat bottom. This minimum may be related to the onset of the flow mode transition. The average velocity at this minimum was about 16 μm/s, or 58 mm/hour. Typical freezing rates used in the solidification experiments were on the order of 15 mm/h.[1-4] Thus, the flow velocity at the minimum was still about four times a typical freezing rate, which is not favorable for obtaining uniform "diffusion-controlled" doping. However, this velocity was measured 3 mm above the bottom. The velocity near the interface, and especially the vertical component of velocity, would have been much smaller.

Our experimental and numerical results suggest two possible explanations for the minimum in segregation reported in the literature for gradient freeze experiments. The first explanation is based on the numerically predicted minimum in the vertical mass flow at about 10 rpm. If the velocity at the minimum were much less than the freezing rate, this would lead to uniform axial doping, as observed experimentally. The Coriolis-induced rotation of this flow should yield very axisymmetric doping albeit with a radial variation. Unfortunately, experimental measurements of cross-sectional variations in doping have not been reported thus far.

The second possible explanation of minimum-segregation results is based on our experimentally observed minimum in flow velocity at moderate rotation rates (20 to 30 rpm), apparently related to the change in flow pattern. This minimum should yield an asymmetric cross-sectional dopant distribution in the resulting crystal, because the bottom toroidal vortex would be asymmetric.

Note that thus far, none of the numerical simulations reported in the literature have included solidification, which generates flow through the freezing interface.

Acknowledgements

This research was supported by the National Science Foundation under grant DMR-9414304. We are grateful to the Fluent Corporation for making their computer code available to us.

REFERENCES

1. L.L. Regel, G.V. Sarafanov, and A.M. Turtchaninov, State of the art and development prospects for research on crystal growth and metal solidification under conditions of elevated gravity, *Inst. Kosm. Issled. Akad. Nauk SSSR*, Preprint No. 907 (1984). Cited in L.L. Regel, *Materials Processing in Space*, volume 1, Consultants Bureau, New York (1990).
2. L.L. Regel, G.V. Sarafanov, I.V. Videnskii, A.M. Turtchaninov, and H. Rodot, Making lead telluride crystals doped with silver under conditions of elevated gravity, *Inst. Kosm. Issled. Akad. Nauk SSSR*, Preprint No. 908 (1984). Cited in L.L. Regel, *Materials Processing in Space*, volume 1, Consultants Bureau, New York (1990).
3. H. Rodot, L.L. Regel, G.V. Sarafanov, M. Hamidi, I.V. Videnskii, and A.M. Turtchaninov, Cristaux de tellurure de plomb elabores en centrifugeuse, *J. Crystal Growth* 79: 77 (1986).
4. H. Rodot, L.L. Regel, and A.M. Turtchaninov, Crystal growth of IV-VI semiconductors in a centrifuge, *J. Crystal Growth* 104: 280 (1990).
5. W.R. Wilcox, L.L. Regel, and W.A. Arnold, Convection and segregation during vertical Bridgman growth with centrifugation, *J. Crystal Growth* 187: 543-558 (1998).
6. P.V. Skudarnov, L.L. Regel, and W.R. Wilcox, Flow visualization study of convection in a centrifuge, in: *Centrifugal Materials Processing*, L.L. Regel and W.R. Wilcox, eds., Plenum Press, New York (1997) pp 75-87.
7. P.V. Skudarnov, L.L. Regel, and W.R. Wilcox, Visualization of convection in a simulated gradient freeze cell during centrifugation, *J. Crystal Growth* 193: 219-229 (1998).
8. R. Derebail, W.A. Arnold, G.J. Rosen, W.R. Wilcox, and L.L. Regel, HIRB - The centrifuge facility at Clarkson, in: *Materials Processing at High Gravity*, L.L. Regel and W.R. Wilcox, eds., Plenum Press, New York (1994) pp 203-212.
9. P.V. Skudarnov, *Flow Visualization and Numerical Modeling for the Gradient Freeze Configuration During Centrifugation*, Ph.D. Thesis, Clarkson University (1999).
10. S. Ushijima, H. Takeda, N. Tanaka, Image processing system for velocity measurements in natural convection flows, *Nuclear Engin. Design* 132: 265 (1991).
11. FLUENT (v4.51), *Fluid Flow Modeling*, Fluent Inc., Centerra Park, 10 Cavendish Court, Lebanon, NH 03766, USA.
12. J. Friedrich, J. Baumgartl, H.-J. Leister, and G. Müller, Experimental and theoretical analysis of convection and segregation in vertical Bridgman growth under high gravity on a centrifuge, *J. Crystal Growth* 167 (1996) 45.
13. J. Friedrich and G. Müller, Segregation in crystal growth under high gravity on a centrifuge: a comparison between experimental and theoretical results, in: *Centrifugal Materials Processing*, L.L. Regel and W.R. Wilcox, eds., Plenum Press, New York (1997) pp. 29-42.
14. W.A. Arnold, *Numerical Modeling of Directional Solidification in a Centrifuge*, Ph.D. Dissertation, Clarkson University, Potsdam, NY (1993).
15. W.A. Arnold, W.R. Wilcox, F. Carlson, A. Chait, and L.L. Regel, Transport modes during crystal growth in a centrifuge, *J. Crystal Growth* 119: 24 (1992).

PATTERN FORMATION IN NATURAL CONVECTION IN A ROTATING CONTAINER

Minerva Vargas, E. Ramos, F. Z. Sierra and A. A. Avramenko

Centro de Investigación en Energía-UNAM
Ap.P. 34, 62580, Temixco, Mor., México

ABSTRACT

We present an experimental and numerical study of the flow patterns that occur in natural convection inside a cylindrical cavity of aspect ratio (radius to height) of 0.25. The working fluid was water with a Prandtl number of 6.0. Observations made for a cavity fixed in the laboratory frame of reference at a Rayleigh number of 2.3×10^6 indicate that the convective pattern was steady and organized in a single asymmetric cell. Observations were also made with the cavity placed in a centrifuge. Experimental visualization with a rotating cavity was made with a Rayleigh number of 2.5×10^6 and a Taylor number of 1.7×10^7. With rotation, the flow became time-dependent with two differentiated modes. In the first mode, the flow was composed of regions that seemed to move in an almost independent way during tens of seconds. Then a general motion that lasted a few seconds set in. This sequence of events repeated with varying periods.

INTRODUCTION

It is now well established that melt motion in crucibles has a large influence on the properties of crystals grown using Czochralski-type techniques.[e.g.,1,2] Two specific points worth noting are:

a) The interaction of the convective flow with the material in the process of solidification at the liquid-solid boundary strongly influences the properties of the resulting solid.
b) Oscillations in the convective motion yield compositional striations in the crystal. Sound theoretical and experimental evidence indicates that under appropriate conditions the Coriolis force can stabilize the flow. This possibility has triggered an intense research effort to understand the interplay of the various physical mechanisms that determine the qualitative features of the natural convective flows in rotating systems.

A good frame of reference of the progress in this topic is provided by the proceedings of the International Workshops on Materials Processing at High Gravity, of which the present volume is the latest. In spite of the substantial progress achieved, several fundamental questions remain unanswered due to the inherent complicated nature of the phenomenon.

Processing by Centrifugation, Edited by Regel and Wilcox
Kluwer Academic/Plenum Publishers, New York, 2001

Natural convection in slender cavities at Ra = 10^6 or larger is a transition or fully turbulent flow that is beyond global prediction with analytical models and, thus, we must use experimental observations, approximate numerical solutions, or local analytical solutions to the transport equations. In contrast, due to geometrical constraints, it is expected that the flow patterns are limited and can be systematically classified in slender cavities.

Here, we present the results of numerical and experimental investigation of flow patterns for natural convection in a cavity with aspect ratio of 0.25. Experimental observations of the convective flow were made when the convective cell was fixed in the laboratory frame of reference and in a rotating system. The Rayleigh number was approximately 10^6 and the Taylor number was 1.7×10^7 . Information on the flow features was obtained by visualization, with two thermocouples placed inside the cell. Critical points or lines in the velocity field distribution were identified. In the appendix, we give the natural scaling indicated by Lie Group analysis of the equations governing non-rotating natural convection in a cylindrical cavity. This information will be useful in the description of the qualitative properties of these flows.

EXPERIMENTAL FACILITY

The flow patterns that occur inside a rotating cylindrical cavity were studied experimentally using a centrifuge with a 1.5 m radius arm that was designed to rotate at a maximum angular speed of 9 rad/s (86 rpm). A basket that carried the convective cell was located at the end this arm. A hinge allowed the basket to swing and orient itself in such a way that the vector sum of the centrifugal and gravity forces was always along the direction of the axis of the cylindrical cavity. The centrifuge carried two lead/acid, 60 W-h batteries and a DC/AC converter that provided electric power to a desktop computer located near the rotation axis of the centrifuge. This computer stored information collected during the experimental runs. The files containing experimental information were transferred via a network board to a laboratory server. The on-board computer could also control an experiment. A sketch of the centrifuge is given in Fig. 1, with a full description of the experimental apparatus given elsewhere.[3]

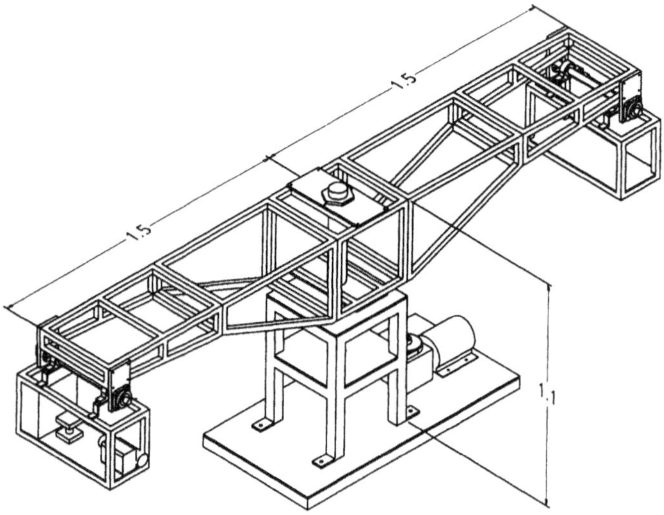

Figure 1. Sketch of the centrifuge; quotes are in meters.

Figure 2 shows the experimental equipment housed in the swinging basket. The convective cell was a cylindrical container of 1.7 cm internal diameter and 3.0 cm tall. Natural convection was generated by maintaining the bottom wall of the cavity at a higher temperature than that of the top wall, using two 7 W thermoelectric panels. Power for the heating (or cooling) systems was provided by the on-board batteries. The maximum possible temperature difference was 8°C. We used water as a working fluid at an average of room temperature, where the Prandtl number is about 6.0. The range of Rayleigh numbers that could be explored was 5.0 x 10^5 to 6.0 x 10^6, for which the flow was steady. Visualization of flow patterns was achieved by shining a sheet of laser light on the cavity to illuminate particle tracers made of liquid crystals, with a video camera normal to the light sheet. The video camera output was sent via an UHF antenna to a VCR located in the laboratory, allowing real-time visualization of experiments. The images obtained were processed to extract the features that qualitatively describe the major patterns. Using the video images, it was possible to reconstruct the velocity field within the light sheet, using PIV techniques. Two thermocouples were placed near the lateral wall, diametrically opposed and at a distance of 6.5 mm from the top wall. Their output was recorded by a PCLab data acquisition board and stored in the hard disk of the on-board computer.

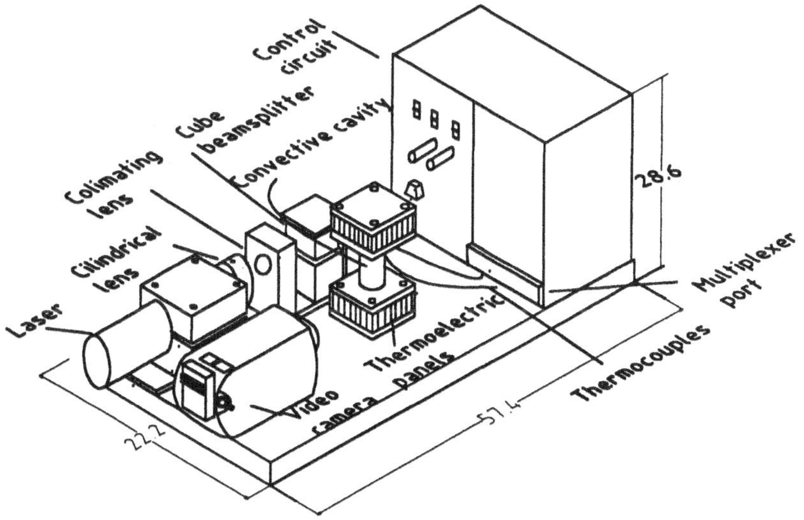

Figure 2. Sketch of the experimental swing basket containing the convective cell, and the monitoring and visualization equipment. Quotes are in cm

THEORETICAL ANALYSES

Formation of the flow pattern in the cylindrical cell was investigated by solving numerically the conservation equations with appropriate boundary conditions using the FLUENT software. In a non-rotating system with $10^4 < Ra < 10^6$ the flow displayed a single, non-axisymmetric cell. Roughly, the fluid ascended on one side of the cell, descended on the other side, and turned and changed its vertical direction near the top and the bottom regions. The specific orientation of the convective cell could not be determined *a priori*. The detailed structure of the cell depended on the Rayleigh number. For Ra ~10^4 the planes of symmetry of the convective motion were vertical. As the Rayleigh number was increased, a more complicated structure with two secondary cells appeared, as found elsewhere.[4,5] In preparation for future work, we present in the appendix the similarity variables found with a Lie group analysis of non-rotating natural convection.

EXPERIMENTAL RESULTS

The experimental observations were obtained using the following procedure. First, the desired temperatures of the top and bottom walls of the container were established and monitored with thermocouples located inside the cavity. Figure 3 shows a temperature record for Ra = 2.3 x 10^6 for an entire experimental run. The traces indicate that the flow displayed an initial transient state with a monotonic increase in velocity that lasted approximately 30 min. Then the system reached steady state. After 1850s rotation of the centrifuge was begun. Figure 4 shows that the pattern without rotation was a single, asymmetric convective cell including most of the volume. The maximum velocity was on the order of $2x10^{-4}$ ms^{-1}. Figures 5a, b and c show the steady flow calculated numerically for a Rayleigh number of 2.5 x 10^6, which is slightly higher than for the experimental conditions. The theoretical flow pattern and velocities display a qualitative correspondence with the experimental observations. Figure 5b shows the velocity field in the plane perpendicular to that in Fig. 5a. By integrating the information in Fig. 5, we see that the three-dimensional steady pattern was composed of an ascending region that conformed with the cylindrical geometry of the container. This ascending flow surrounded the descending flow in the upper region and was surrounded by the descending flow in the lower part, as discussed in detail elsewhere.[4]

Once the temperature record indicated that the transient was over, the centrifuge was set in rotation. The centrifuge angular speed was increased with a constant acceleration of 0.15 s^{-2} to reach a final rotation rate of 2.3 s^{-1} (22 rpm). Due to the effect of centrifugal acceleration and since the temperature between the top and bottom walls of the cell

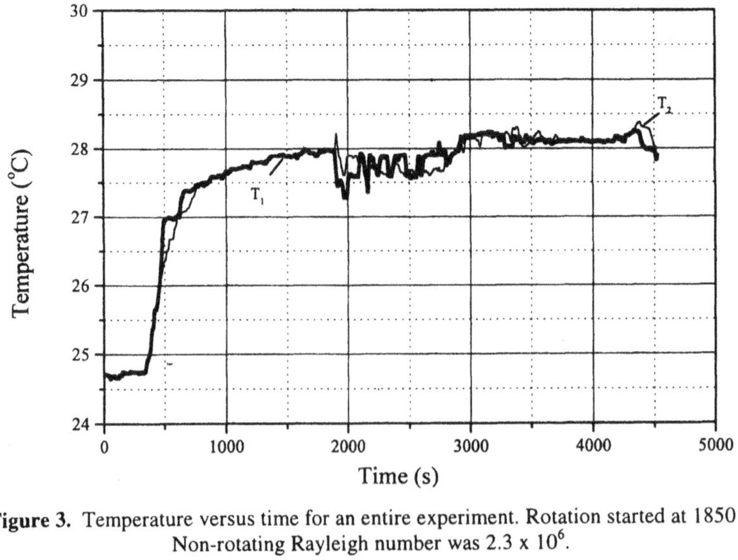

Figure 3. Temperature versus time for an entire experiment. Rotation started at 1850 s. Non-rotating Rayleigh number was 2.3 x 10^6.

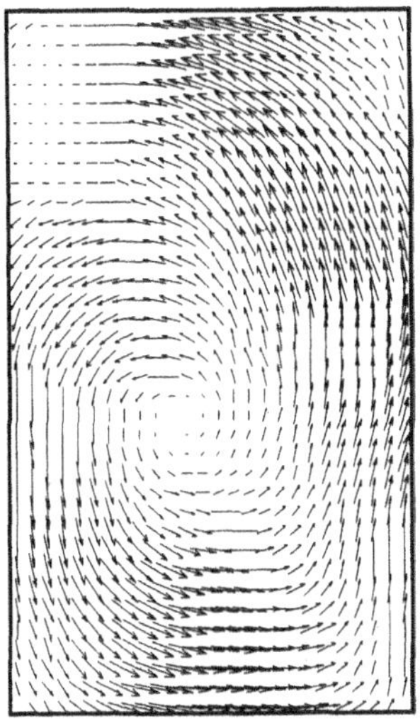

Figure 4. Flow pattern for non-rotating flow with Ra = 2.3×10^6.

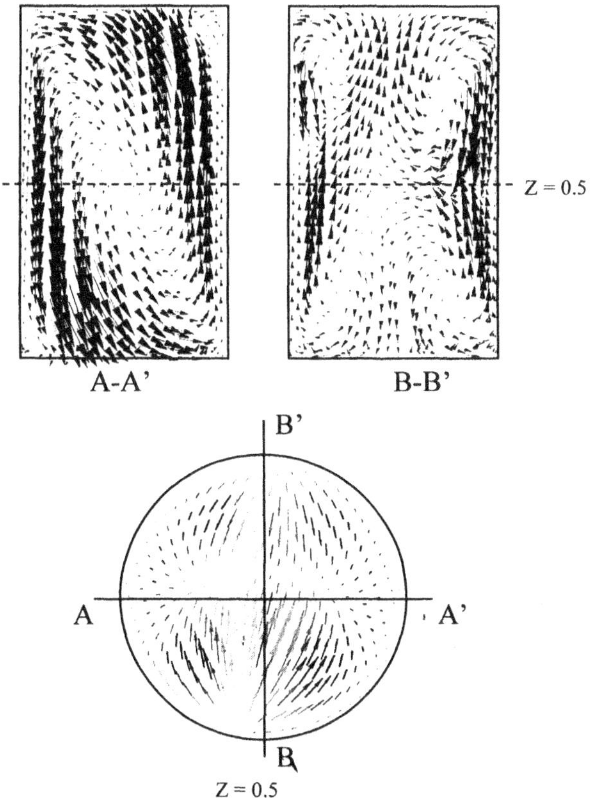

Figure 5. Computed velocity field for Ra = 2.5×10^6.
a) Vertical mid-plane with orientation defined by the line AA' in Fig. 5c.
b) Vertical mid-plane with orientation defined by the line BB' in Fig. 5c.
c) Horizontal mid-plane.

remained constant, the Rayleigh number for the rotating cell was $Ra_{rot} = 2.5 \times 10^6$. The Taylor number was 1.7×10^7 for the temperature record displayed in Fig. 3 for times greater than 1850s. Convection in the rotating frame was observed for 15 min.

Notable changes in the convection pattern were observed. First, the system never reached a steady state, but displayed an irregular, time-dependent behavior, akin to the transition to turbulence but with some clearly identifiable convective patterns. The change in qualitative behavior is featured in the temperature record as shown in Fig. 6, which is an amplification of Fig. 3. Oscillations were observed with periods ranging from 30s to 60s. The oscillations were not always present; sudden changes occurred between intervals of relatively steady behavior. Visual inspection of the video indicated that the flow displayed two modes. In the first mode, most of the fluid moved at a very small velocity ($< 10^{-5}$ ms^{-1}). Then a region of intense motion developed near the lower horizontal wall and probably also near the upper wall. This feature preceded the second mode, which consisted of a general downward motion in most of the observed plane. As commented above, the flow was unsteady and it is likely that other modes would develop. Figure 7 shows a sample of the velocity field found in the time-dependent flow. This picture is the time average of 80 individual frames captured at intervals of 1/30 s. This flow displayed two high-speed regions; one moving towards the right in the upper half near the vertical center, and the other towards the left at the lower right hand quarter. An interesting structure was a critical line extending from the upper left hand corner to the right wall, at the vertical center of the cell. This flow was three dimensional and velocities with a large component in the perpendicular direction of the plane analyzed are expected in the lower left quarter. Since the flow was time dependent, this picture is expected to change as the flow evolves.

DISCUSSION AND CONCLUSIONS

Natural convective flow in both rotating and non-rotating systems was observed experimentally. At Ra ~2×10^6, rotation caused the steady flow to become unsteady. This might be attributed to the experimental conditions for the non-rotating case being near to the flow transition, so that the increase of Rayleigh number due to centrifugal force triggered a time dependence. Also, since the observation time was relatively short (15 min), we cannot preclude the possibility that we observed a transient condition, although there was no indication that the oscillations were decaying.

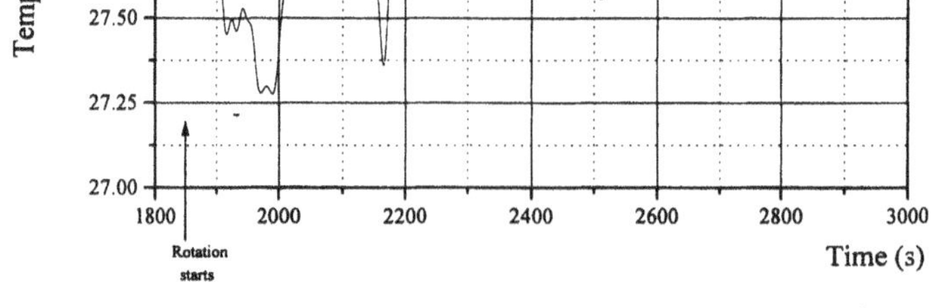

Figure 6. Temperature as a function of time in a rotating system with Ra = 2.5×10^6 and Ta = 1.7×10^7.

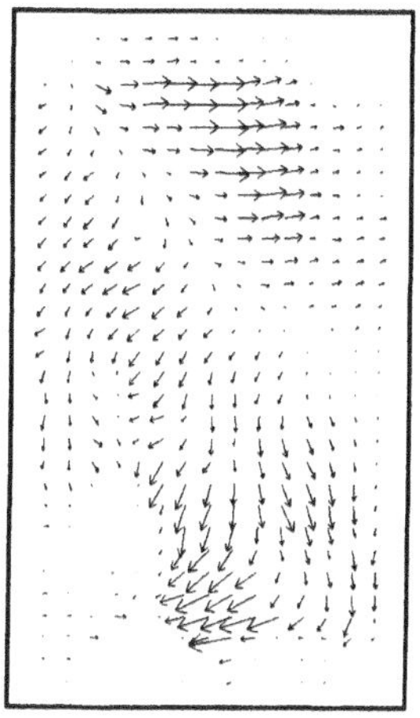

80 frames average

Figure 7. Flow Pattern for a rotating system with Ra = 2.5×10^6 and Ta = 1.7×10^7.

Acknowledgements

This investigation was partially supported by CONACyT (Mexico) G 0044E project. Financial support from CONACyT for A.A.A. is also gratefully acknowledged. Participation of Raúl Espejel, Gabriel Ascanio, Guillermo Hernández, Giovanni Esquivel, Juan M. García, Héctor Perales, Abraham Medina, José A. Córdoba, Leopoldo Ruiz, Alberto Caballero and Luis Velázquez in the various stages of the development of the centrifuge and image processing is appreciated.

REFERENCES

1. S.M. Pimputkar and S. Ostrach, Convective effects in crystals grown from melt, *J. Crystal Growth* 55: 614-646 (1981).
2. G. Müller, Convection and inhomogeneities in crystal growth from the melt, in: *Crystal Growth, Properties and Applications*, Vol. *12*, Springer (1988).
3. M. Vargas, E. Ramos, G. Ascanio, R. Espejel, G Esquivel and G. Hernández, A centrifuge for studies of fluid dynamics phenomena in rotating motion (submitted for publication).
4. A.A. Avramenko, F.Z. Sierra, M. Vargas and E. Ramos, Natural convection in a cylindrical cavity (submitted for publication).
5. G. Müller, G. Neumann and W. Weber, Natural convection in a vertical Bridgman configuration, *J. Crystal Growth* 70: 78-93 (1984).

NOMENCLATURE

Dimensional Quantities

a Net acceleration; vector sum of gravitational and centrifugal (m/s^2)
g Gravitational acceleration (m/s^2)
h Height of convective cell (m)
r Radius of the centrifuge (m)
T_H Temperature at the bottom of the convective cell (°C)
T_C Temperature at the top of the convective cell (°C)

Greek

α Thermal diffusivity (m^2/s)
β Thermal expansion coefficient (1/°C)
ΔT $T_H - T_C$ (°C)
ν Kinematic viscosity (m^2/s)
ω Angular velocity (rad/s)

Nondimensional Quantities

Ra Rayleigh number, $g\,\beta\,\Delta T\,h^3/\alpha\nu$
Ra_{rot} Effective Rayleigh number, $a\,\beta\,\Delta T\,h^3/\alpha\nu$
Pr ν/α
Ta $4\omega^2 h^4/\nu^2$

APPENDIX

Similarity variables for non-rotating natural convection in a cylindrical cavity can be found with Lie group analysis of the symmetries of the governing equations. The details of the derivation will appear in a forthcoming publication.

Considering scaling and rotation symmetries; the infinitesimal generator for natural convection in a cylindrical container is:

$$q = C_2\partial_\varphi + C_1 r\partial_r + (C_1 z + C_3)\partial_u - C_1 u\partial_u - C_1 v\partial_v - C_1 w\partial_w - 3C_1 T\partial_T - 2C_1 p\partial_p$$

where C_i are arbitrary constants. The self-similar independent variables are defined by:

$$\eta = \frac{r}{z+a}, \qquad \xi = r\exp(-b\varphi), \qquad \zeta = (z+a)\exp(-b\varphi)$$

where $a = C_3/C_1$, $b = C_1/C_2$. The self-similar velocities are:

$$(u,v,w) = \exp(-b\varphi)U(any:\eta,\xi,\zeta), \qquad (u,v,w) = \frac{U(any:\eta,\xi,\zeta)}{(z+a)},$$

$$(u,v,w) = \frac{U(any:\eta,\xi,\zeta)}{r}$$

Self-similar temperatures are:

$$T = \begin{bmatrix}\cos^3(n\varphi)\\ sin^3(n\varphi)\end{bmatrix}\vartheta(any:\eta,\xi,\zeta), \qquad T = \frac{\vartheta(any:\eta,\xi,\zeta)}{(z+a)^3}, \qquad T = \frac{\vartheta(any:\eta,\xi,\zeta)}{r^3}$$

Self similar pressures are:

$$p = \begin{bmatrix}\cos^2(n\varphi)\\ sin^2(n\varphi)\end{bmatrix}\Pi(any:\eta,\xi,\zeta), \qquad p = \frac{\Pi(any:\eta,\xi,\zeta)}{(z+a)^2}, \qquad p = \frac{\Pi(any:\eta,\xi,\zeta)}{r^2}$$

NON-AXISYMMETRIC CHIMNEY CONVECTION IN A MUSHY LAYER UNDER A HIGH GRAVITY ENVIRONMENT

D.N. Riahi

Department of Theoretical and Applied Mechanics
University of Illinois, Urbana, IL 61801, USA

ABSTRACT

Nonlinear buoyancy-driven chimney convection in arbitrary Prandtl number melts within mushy layers during alloy solidification is considered under a high-gravity environment, where the rotation axis is inclined at an angle γ to the high-gravity vector. The chimneys within the mushy layer produce undesirable freckles in the final form of the solidified material. These are imperfections that reduce the quality of the material. Methods of control aiming at eliminating or reducing significantly the convective flows in the chimneys also serve to eliminate or reduce significantly the existence and presence of the chimneys in the mushy layer, resulting in higher quality castings. Asymptotic and scaling examinations of the governing system for a non-axisymmetric and weakly unsteady mode of chimney convection resulted in new analytical expressions for the leading order magnitudes of the basic flow quantities as functions of the basic flow parameters. These results indicate that chimney convection can be reduced significantly in certain ranges of the parameter values under certain high gravity conditions and for prescribed and independent values of the radius of the chimneys.

INTRODUCTION

In the last two decades or so there have been a number of significant works on the problem of convection effects during alloy solidification and in the presence of a mushy zone of mixed solid and liquid phases.[1-10] For examples, Sample and Hellawell[1,2] did an experiment with aqueous NH_4Cl in a cylindrical mold with a chilled bottom surface where solidification was induced. They employed a rotation and tilting technique to change the orientation of gravity relative to the bottom surface of the cylinder. They found that for slow and steady rotation of the mold about the vertical axis, chimney formation and development was about the same as without rotation. However, for slow and steady rotation of a tilted mold, the number of chimneys was reduced. Fast rotation is not considered beneficial to crystal growers under normal gravity conditions as, for example, the experimental results of Kou et al.[11] indicated that segregates formed along a ring between the axis and the outer edge of the ingot with large rotational speed.

Regarding the governing system for a mushy zone, Hill et al.[3] developed a set of equations for a mushy zone and solved a reduced set of them approximately. Later, a more complete solution was given by Fowler.[5] Huppert and Worster[4] formulated a simple solidification model based on a global conservation relationship. Worster[6] provided a model for a region of dendritic or cellular growth. He followed Roberts and Loper's[12] treatment of the mushy layer as a porous medium and adopted Darcy's law in the momentum equation for the flow in the mushy zone. He included a general constitutive relation for the permeability of the medium as a function of the local liquid fraction. He then performed asymptotic and scaling analyses of the governing equations for nonlinear natural convection in a mushy layer in the limit of infinite Prandtl number Pr and for asymptotically very large solutal Rayleigh number R. He proposed a model in which there is upward steady convection in localized chimneys within the mushy layer. These chimneys were characterized as having negligible solid fraction and in a vertical direction. The model described above is consistent with experimental observation in an aqueous system[13] where Pr is not small. However, such a model may not be preferred for low Pr melt cases, such as metallic alloys, which are of commercial value. Riahi[14] investigated the structure of a convecting mushy layer and developed, in particular, a model relevant for low Pr.

Riahi and Sayre[15] studied strongly nonlinear compositional convection in a mushy layer of a solidification system for $Pr = \infty$ and under an external constraint of rotation where the rotation axis was inclined to the direction of the solidification velocity. They investigated only axisymmetric convection and did not include the Coriolis force terms in the governing equations. They applied asymptotic and scaling analyses to convective flow within the mushy layer and in cylindrical chimneys. They found that, for some particular moderate range of the rotation rate, where R was of the same order as the gradient acceleration parameter A, vertical convection in the chimneys was reduced significantly, while for $A >> R$ it was increased substantially. Riahi[16] extended the model due to Riahi and Sayre[15] to cases where the effects of finite Pr were taken into account in systems under rotational constraints. Again, Coriolis force effects as well as non-axisymmetric flow behavior were not taken into account. Riahi[17] studied nonlinear natural convection in chimneys for $Pr \rightarrow \infty$ and under an external constraint of inclined rotation. He applied asymptotic and scaling analyses to a weakly non-axisymmetric mode of convection with both centrifugal and Coriolis forces taken into account. Certain ranges in parameter space were determined where chimney convection would be significantly decreased. Riahi[17] assumed that the radius of any chimney is an independent parameter with a prescribed value. Riahi[18] extended the work to finite Pr and assumed that the radius of each chimney is a function of the other parameters. Riahi[19] carried out analysis for a more fully non-axisymmetric chimney convection and for finite Pr.

The present article considers previous theoretical studies[15–19], extends the earlier work[19] for weakly unsteady cases, where the radius a of any chimney is an independent parameter with prescribed value ($a << 1$), and obtains interesting results about chimney convection under certain ranges of parameter space.

FORMULATION

A layer of thickness h adjacent to a solidifying surface of a binary alloy melt is considered, where h is the mush-liquid interface elevation above $z = 0$ and the z-axis is perpendicular to the average location of the solidification front. A melt of constant composition C_0 and constant temperature T_∞ is solidified at a constant rate V_0, with the eutectic temperature T_e at the position $z = 0$ in a frame moving with the solidification velocity in the z-direction.

We assume that the solidifying system is rotating at some constant angular velocity Ω about a fixed axis at a constant angle γ with respect to the z-axis. The rotation axis is assumed to be anti-parallel to earth's gravity vector. The governing equations for the solidifying system are based on those for momentum, continuity, temperature and solute concentration, under the Boussinesq approximation, and in a moving coordinate system whose origin is centered on the solid-mush interface. The coordinate system rotates with the solidifying system at angular velocity Ω and translates with the solidifying front at speed V_0. The governing system is non-dimensionalized in the same way described by Riahi.[17–18] In the momentum equations the centrifugal force term is split into an average term, which is superimposed on the gravity term, and the deviation of the centrifugal force from this average term that is called the gradient acceleration term. The non-dimensional parameter R representing the modified gravity term can be significantly larger than the corresponding one due to earth's gravity at a significant rotation rate. Following Worster,[7] we treat the mushy layer as a porous medium where Darcy's law holds.

The governing equations are considered in cylindrical coordinates whose axial direction is along the z-axis. Nonlinear non-axisymmetric convective flow is considered both in a cylindrical chimney, whose axis coincides with the z-axis, and in the mushy layer in the asymptotic limits of strong compositional buoyancy force, negligible thermal buoyancy and sufficiently large Lewis number K/D, where K and D are thermal and solute diffusivity, respectively. The justifications for these limiting conditions are already provided in Worster[7] and will not be repeated here. The non-dimensional forms of the governing equations for momentum, continuity, temperature and solute concentration for the flow of melt in the liquid zone above the mushy zone as well as inside the chimney are:

$$\frac{1}{Pr}\left(\frac{\partial}{\partial t} - \frac{\partial}{\partial z} + \boldsymbol{u}\cdot\nabla\right)\boldsymbol{u} = -HR(\nabla P + S\hat{z}) + \nabla^2\boldsymbol{u} + HT\boldsymbol{u}\times\hat{\Omega} + HASG, \tag{1a}$$

$$\nabla\cdot\boldsymbol{u} = 0, \tag{1b}$$

$$\left(\frac{\partial}{\partial t} - \frac{\partial}{\partial z} + \boldsymbol{u}\cdot\nabla\right)\theta = \nabla^2\theta, \tag{1c}$$

$$\left(\frac{\partial}{\partial t} - \frac{\partial}{\partial z} + \boldsymbol{u}\cdot\nabla\right)S = \frac{1}{Le}\nabla^2 S. \tag{1d}$$

Here $\boldsymbol{u} = u\hat{r} + v\hat{\xi} + w\hat{z}$ is the velocity vector, u is the radial component of the velocity, $\hat{r}$ is a unit vector in the radial r-direction, v is the azimuthal component of the velocity, $\hat{\xi}$ is a unit vector in the aximuthal ξ-direction, w is the axial component of the velocity, $\hat{z}$ is a unit vector in the axial z-direction, P is the pressure, S is the solute concentration, θ is the temperature, $Pr = \nu/K$ is the Prandtl number, ν is the kinematic viscosity, $R = \beta\Delta c N_g K^2/(V_0^3\nu H)$ is the solutal Rayleigh number $N_g = (g^2 + \Omega^2 R_0^2)^{1/2}$, g is acceleration due to gravity, R_0 is the perpendicular distance from the center of gravity of the solidification basket to the rotation axis, $H = K^2/(V_0^2\Pi_0)$ is a non-dimensional parameter, whose value is generally large, [7] Π_0 is a constant reference value of the permeability $\Pi(\phi)$ of the porous medium, ϕ is the solid fraction of the mushy zone, $T = 2\Omega K^2/(V_0^2 \nu H)$ is the Coriolis parameter, $A = \beta \Delta C \Omega^2 K^2/(V_0^4 \nu H)$ is the gradient acceleration parameter, $\Delta C = C_0 - C_e, C_e$ is the eutectic concentration, β is the expansion coefficient for solute, $\hat{\Omega}$ is a unit vector along the rotation axis defined by:

$$\hat{\Omega} = \cos\gamma \hat{z} + \sin\gamma(\cos\xi \hat{r} - \sin\xi \hat{\xi}), \tag{1e}$$

t is time, and $\boldsymbol{G}$ is a position vector defined by:

$$\begin{aligned} \boldsymbol{G} = (r\cos^2\xi\cos^2\gamma + r\sin^2\xi - z\sin\gamma\cos\gamma\cos\xi)\hat{r} + \\ (z\sin\gamma\cos\gamma\sin\xi + r\sin\xi\cos\xi\sin^2\gamma)\hat{\xi} \\ +(z\sin^2\gamma - r\sin\gamma\cos\gamma\cos\xi)\hat{z}. \end{aligned} \tag{1f}$$

The non-dimensional form of the governing equations for the momentum, continuity, temperature and solute concentration in the mushy zone outside the chimneys are:

$$(1-\phi)\boldsymbol{u}/\Pi = -R(\nabla P + S\hat{z}) + T(1-\phi)\boldsymbol{u} \times \hat{\Omega} + AS\boldsymbol{G}, \tag{2a}$$

$$\nabla \cdot [(1-\phi)\boldsymbol{u}] = 0, \tag{2b}$$

$$[\frac{\partial}{\partial t} - \frac{\partial}{\partial z} + (1-\phi)\boldsymbol{u} \cdot \nabla]\theta = \nabla^2\theta + S_t\left(\frac{\partial}{\partial t} - \frac{\partial}{\partial z}\right)\phi, \tag{2c}$$

$$\begin{aligned} &\left(\frac{\partial}{\partial z} - \frac{\partial}{\partial t}\right)[(1-\phi)(C_r - S)] \\ &+(1-\phi)\boldsymbol{u} \cdot \nabla S = \frac{1}{Le}\nabla \cdot [(1-\phi)\nabla S]. \end{aligned} \tag{2d}$$

Here, $S_t = L/(\tilde{c}\Delta T)$ is the Stefan number, $\Delta T = T_L - T_e, T_L$ is the local liquidus temperature at $C = C_0$, C is dimensional solute concentration, $\tilde{c}$ is the specific heat per unit volume, L is the latent heat of solidification per unit volume, $C_r = (C_s - C_0)/\Delta C$ is a concentration ratio, and C_s is the composition of the solid phase forming the dendrites. As in Worster,[7] it is assumed that the liquidus relationship $\theta = S$ holds throughout the mushy zone.

The boundary conditions for (1)-(2) were originally given in Worster[7] and will not be repeated here since the analysis of (1)-(2) to be described in this article does not require explicit references to these boundary conditions. However, it turns out that the approximate solutions based on the asymptotic and scaling analyses to be described in this article are consistent with the actual boundary conditions for the dependent variables.

HIGH Pr CASE

The chimney under consideration is assumned to have an axis coincide with the z-axis of the coordinate system and its radius $a(\xi, z, t)$ is assumed to be small $(a << 1)$ with some prescribed order of magnitude. It is assumed that the orders of magnitude of r, ξ and z are, respectively, $a, 1$ and 1. Under the assumption that the magnitude of $(1-\phi)\boldsymbol{u}$ is of order one in the mushy zone, it is found from Eq. (2a) that to the leading term the pressure field in the mushy zone is unaffected by the flow velocity and $\theta = S$ is independent of r and ξ in this zone. Equations (2c) & (2d) then imply that $w = w_0(z,t)$, $\phi = \phi_0(z,t)$ and $\theta = \theta_0(z,t)$ at most, where w_0, ϕ_0 and θ_0 are the leading order terms for w, ϕ and θ, respectively. It is also assumed that $C_r >> \theta$ in the mushy layer.[7]

The analysis presented in this section is for the case where the nonlinear and time derivative terms in Eq. (1a) can be, at most, as significant as the viscous term in Eq. (1a). This condition leads to the following conditions:[14]

$$Pr \geq O(RHa^4), \quad \left(\frac{1}{w}\frac{\partial w}{\partial t}\right) \leq O(\sqrt{RHPr}), \tag{3a,b}$$

where w is the vertical velocity in the chimney.

Examining the non-axial terms in Eqs. (1b) or (2b), it is found[19] that a non-axisymmetric convection can be based on the following two conditions:

$$v \leq O(u), \quad \frac{\partial F}{\partial \xi} \leq O(F), \tag{4a,b}$$

where F can be any dependent variable.

Denoting an azimuthal average of any quantity F by

$$\langle F \rangle = \frac{1}{2\pi}\int_0^{2\pi} F d\xi, \tag{5}$$

assuming that any dependent variable and its derivatives are repeated in 2π interval in ξ, and taking the aximuthal average of Eqs. (1a) and (2a), it is observed[19] that the averaged form of these equations behave qualitatively similar to those studied in Riahi[17] under more restricted conditions.

We consider the chimney convection under Eq. (1) and assume[7] that $S \sim 1$ and $w >> 1$. Then Eq. (1) implies certain orders of magnitude[19] for $\langle w \rangle$, $\langle u \rangle$ and $\langle \psi \rangle$, where ψ is a stream function defined by the non-azimuthal components of $\boldsymbol{u}$. From Eq. (1c) or (2c) and a relevant condition[19] for HR, we found that the order of magnitude of the deviation of θ from θ_0 is smaller than one. Integrating the aximuthal average of Eq. (1c) in r from $r = 0$ to $r = a$ and following Worster,[7] we found the orders of magnitude of $\langle \theta - \theta_0 \rangle$, $\langle u \rangle$, $\langle w \rangle$ and $\langle \psi a \rangle$, where $2\pi\langle \psi a \rangle$ is the azimuthal average of the vertical flux in the chimney. These results were obtained for two distinct regimes, referred to as moderate axial flow regime (regime I) and large axial flow regime (regime II), and two corresponding parametries regimes in the parameter-space $(R, H, A_c, T, a, \gamma)$, where:

$$A_c \equiv 1 - (A/R)z\sin^2\gamma. \tag{6}$$

The results described above are simplified and are given below for certain relevant ranges of values for T and A_c. For $A_c = -0.001$ and $0.1 \leq T \leq 3$ or $T = 1$ and $0.001 \leq |A_c| \leq 0.1$, for regime I:

$$\begin{gathered}\langle u \rangle \sim HRA_c a^3, \quad \langle w \rangle \sim 1, \langle \psi a \rangle \sim RHA_c a^4, \\ a \sim a_0 + T\sin\gamma/(4A_c R ln a_0), a_0 \equiv 4[HR^2A_c^2 ln(HR^2A_c^2)]^{-1/4}.\end{gathered} \tag{7a-e}$$

For $A_c = 1.001(10^{-6})$ and $200 \leq T \leq 800$ or $T = 600$ and $1.0001(10^{-6}) \leq A_c \leq 1.001(10^{-6})$, for regime I:

$$\begin{gathered}\langle u \rangle \sim -RA_c/(T\sin\gamma), \langle w \rangle \sim 1, \langle \psi a \rangle \sim RA_c a/(T\sin\gamma), \\ a \sim \frac{(RA_c - 1)T\sin\gamma}{(RA_c)^2}\left\{1n\left[\frac{R^2A_c^2}{(RA_c - 1)T}\right]\right\}^{-1}\end{gathered} \tag{8a-d}$$

For $A_c = 1.0$ and $0.1 \leq T \leq 10$ or $T = 5.0$ and $0.001 \leq |A_c| \leq 1.0$, for regime II:

$$\begin{aligned} &\langle u \rangle \sim -HRa^3 A_c, \quad \langle w \rangle \sim (RHa^3)^2, \quad \langle \psi_a \rangle \sim RHA_c a^4, \\ &a \sim a_0 + T \sin\gamma/[4RA_c ln(\sqrt{H}/A_c)], \quad a_0 \equiv (A_c\sqrt{H})[ln(\sqrt{H}/A_c)]^{1/2}. \end{aligned} \qquad (9a\text{-}e)$$

For $A_c = 1.0$ and $5000 \leq T \leq 8000$ or $T = 5000$ and $1. \leq A_c \leq 5.$, for regime II:

$$\begin{aligned} &\langle u \rangle \sim -RA_c/(T\sin\gamma), \quad \langle w \rangle \sim (RHa^3)^2, \quad \langle \psi_a \rangle \sim RA_c a/(T\sin\gamma), \\ &a \sim [A_c/(RH^2)]^{1/16} + (RA_c^2)^{2/3}[1n(RH^2/A_c)]^{1/3}/(36T\sin\gamma H^{2/3}). \end{aligned} \qquad (10a\text{-}d)$$

The results given above are for small but prescribed values of the radius a of the chimney. It can be seen that certain ranges in the parameter space (R, H, T, A, γ) should be satisfied in order for the results in Eqs. (7)-(10) to be applicable. For moderate axial flow (regime I), chimney convection decreases with increasing A for $A_c > 0$ and increases with A for $A_c < 0$. Chimney convection is independent of T for small values of T and decreases with increasing T for large values of T. For large axial flow (regime II), similar results hold. Also Eqs. (7)-(10) indicate that chimney convection is independent of Pr but generally increases with R as is expected.

LOW Pr CASE

For liquid metals, Pr is small ($Pr << 1$). Here, we assume that the order of magnitude of the nonlinear inertial term in Eq. (1a) is larger than that of the viscous term, and the order of magnitude of the time derivative inertial term in Eq. (1a) is at most as large as the nonlinear inertial term. This leads to the following conditions:[14]

$$Pr << RHa^4, \left(\frac{1}{w}\frac{\partial w}{\partial t}\right) \leq O\left(\sqrt{RHPr}\right), \qquad (11a,b)$$

where w is the vertical velocity in the chimney.

We consider the chimney convection under Eq. (1). For low Pr, Eq. (1a) implies certain orders of magnitude[19] for $\langle w \rangle$, $\langle u \rangle$ and $\langle \psi \rangle$. These orders of magnitude satisfy the continuity equation (1b) and are based on a balance between the inertia, buoyancy and pressure gradient terms in Eq. (1a) to the leading terms in the asymptotic limit of sufficiently large R. From Eq. (1c) or (2c) and a relevant condition[19] for $HRPr$, we found that the order of magnitude of $(\theta - \theta_0)$ is smaller than one. Such order of magnitude is determined by applying the procedure described in the previous section for high a Pr. Similarly, the orders of magnitude of $\langle u \rangle$, $\langle w \rangle$ and $\langle \psi_a \rangle$ are found for both regimes I and II.

The results described above are simplified and are given below for certain relevant range of values for T and A_c. For $A_c = 0.001$ and $0.1 \leq T \leq 1.$ or $0.0001 \leq A_c \leq 0.001$ and $T = 0.8$, for regime I:

$$\left.\begin{aligned} \langle u \rangle \sim -a\sqrt{RHPrA_c}, \quad \langle w \rangle \sim 1, \quad \langle \psi_a \rangle \sim a^2\sqrt{RHPrA_c}, \\ a \sim a_0 - 2T\sin\gamma/[RA_c 1n(R^3A_c^3HPr)], \\ a_0 \equiv (RA_c)^{-3/4}(HPr)^{-1/4}[1n(R^3A_c^3HPr)]^{-1/2}. \end{aligned}\right\} \qquad (12a\text{-}e)$$

For $A_c = 1.0001(10^{-6})$ and $500 \leq T \leq 1000$ or $T = 800$ and $1.00001(10^{-6}) \leq A_c \leq 1.0001(10^{-6})$, for regime I:

$$\langle u\rangle \sim -RA_C/(T\sin\gamma), \quad \langle w\rangle \sim 1, \quad \langle\psi a\rangle \sim RA_C a/(T\sin\gamma),$$
$$a \sim T(RA_C - 1)/(R^2A_c^2)\ln\left[\frac{R^2A_c^2}{(RA_c-1)T}\right]. \tag{13a-d}$$

For $A_c = -0.02$ and $10 \le T \le 50$ or $T = 30$ and $0.02 \le |A_c| \le 0.2$, for regime II:

$$\langle u\rangle \sim -A_C a\sqrt{8RHPr}, \quad \langle w\rangle \sim HRa^2Pr, \quad \langle\psi a\rangle \sim A_C a^2\sqrt{8RHPr},$$
$$a \sim T\sin\gamma/(RA_c) + \exp[-\sqrt{HPr}/(2A_c^2\sqrt{2R})]. \tag{14a-d}$$

For $A_c = 0.01$ and $10^4 \le T \le 2(10^4)$ or $T = 20000$ and $0.005 \le |A_c| \le 0.05$, for regime II:

$$\langle u\rangle \sim -RA_c/(T\sin\gamma), \quad \langle w\rangle \sim HRa^2Pr,$$
$$\langle\psi a\rangle \sim aRA_c/(T\sin\gamma), \quad a \sim [A_c/(HPr)]^{1/2}. \tag{15a-d}$$

The results given above are for small but prescribed values of the radius a of the chimney. Again, as in the high Pr case, it is seen from Eqs. (12)-(15) that certain ranges in the parameter space (Pr, R, H, T, A, γ) should be satisfied in order for these results to be applicable. For moderate axial flow, chimney convection decreases with increasing A. Chimney convection is independent of T small values of T and decreases with increasing T for large values of T. Similar results also hold for large axial flow case. Also Eqs. (12)-(15) indicate that chimney convection increases with R, and it also increases with Pr in most of the ranges for T and A_c considered here.

CONCLUSIONS

(i) Chimney convection does not depend on Pr for sufficiently large Pr, while it depends strongly on Pr for sufficiently small Pr.
(ii) Chimney convection decreases with decreasing Pr for a low Pr melt.
(III) For high Pr, chimney convection decreases with increasing A for $Ah\sin^2\gamma < R$, where h is the height of the chimney. Chimney convection increases with A for $Ah\sin^2\gamma > R$.
(iv) For low Pr, chimney convection decreases with increasing A for $Ah\sin^2\gamma < R$.
(v) For high Pr, chimney convection is independent of the Coriolis force for small T, while it decreases with increasing T for large values of T.
(vi) For low Pr, the dependence of chimney convection on the Coriolis force is similar to those for high Pr.
(vii) Chimney convection increases with R.
(viii) Chimney convection usually increases with H.

REFERENCES

1. A.K. Sample and A. Hellawell, The effect of mold precession on channel and macro-segregation in ammonium chloride-water analog castings,*Met. Trans. B* 13: 495 (1982).
2. A.K. Sample and A. Hellawell, The mechanisms of formation and prevention of channel segregation during alloy solidification, *Met. Trans. A* 15: 2163 (1984).
3. R.N. Hill, D.E. Loper and P.H. Roberts, A thermodynamically consistent model of a mushy zone, *Quarterly J. Mech. Appl. Math.* 36: 505 (1983).

4. H.E. Huppert and M.G. Worster, Dynamic solidification of a binary melt, *Nature* 314: 703 (1985).
5. A.G. Fowler, The formation of freckles in binary alloy, *IMA J. Appl. Math.* 35: 159 (1985).
6. M.G. Worster, Solidification of an alloy from a cooled boundary, *J. Fluid Mech.* 167: 481 (1986).
7. M.G. Worster, Natural convection in a mushy layer, *J. Fluid Mech.* 224: 335 (1991).
8. M.G. Worster, Instabilities of the liquid and mushy regions during solidification of alloys, *J. Fluid Mech.* 237: 649 (1992).
9. H.E. Huppert, The fluid mechanics of solidification, *J. Fluid Mech.* 212: 209 (1990).
10. D.G. Neilson and F.P. Incropera, Effect of rotation on fluid motion and channel formation during unidirectional solidification of a binary alloy, *Int. J. Heat Mass Trans.* 36: 489 (1993).
11. S. Kou, D.R. Poirier and M.C. Flemings, Macro segregation in rotated remelted ingots, *Met. Trans. B* 9: 711 (1978).
12. P.H. Roberts and D.E. Loper, Towards a theory of the structure and evolution of a dendrite layer, in:, *Stellar and Planetary Magnetism*, A.M. Soward, ed., Gordon and Breach, New York (1983), p. 329.
13. S.M. Copley, A.F. Giamei, S.M. Johnson and M.F. Hornbecker, The origin of freckles in unidirectionally solidified casting, *Metall. Trans.* 1: 2193 (1970).
14. D.N. Riahi, On the structure of an unsteady convecting mushy layer, *Acta Mechanica* 127: 83 (1998).
15. D.N. Riahi and T.L. Sayre, Effects of rotation on the structure of a convecting mushy layer, *Acta Mechanica* 118: 109 (1996).
16. D.N. Riahi, Effect of high gravity on freckle formation and convection in a mushy layer in: *Centrifugal Materials Processing* L.L. Regel and W.R. Wilcox, eds., Plenum Publishing Corp, NY (1997), p. 169.
17. D.N. Riahi, Effects of centrifugal and Coriolis forces on chimney convection during alloy solidification, *J. Crystal Growth* 179: 287 (1997).
18. D.N. Riahi, High gravity convection in a mushy layer during alloy solidification, in: *Nonlinear Instabil ity, Chaos and Turbulence* L. Debnath and D.N. Riahi, eds., 1: 301 (1998), WIT Press, UK..
19. D.N. Riahi, Effects of rotation on a nonaxisymmetric chimney convection during alloy solidification, *J. Crystal Growth* 204: 382 (1999).

VISCOUS FINGERING IN A ROTATING POROUS MEDIUM

M.Z. Saghir,[1] H.H. Vaziri[2] and M.R. Islam[2]

[1] Ryerson Polytechnic University, Toronto, Ontario, Canada
[2] Dalhousie University, Halifax, Nova Scotia, Canada

ABSTRACT

Viscous fingering is an extremely important phenomenon in enhanced oil recovery, drilling underground liquid waste disposal, and porous bed reactors. It takes place when the viscous forces of a displacing fluid have greater momentum than those of the displaced fluid. While the onset and propagation of viscous fingers during liquid-liquid displacement is considered to have severe engineering consequences, little has been done to model the onset and propagation of a viscous finger. This is especially true for viscous fingering under combined thermal and chemical convection in a rotating porous medium. This constituted the subject of the present investigation.

The momentum-balance equation (Brinkman form) was solved along with the energy and mass-balance equations. The model consisted of a packed cylinder that rotated along the central axis with a given angular speed. The porous cylinder was initially filled with a viscous fluid. Water was injected through an inlet located at the top of the porous medium. Due to its lower viscosity, water formed miscible fingers that were affected by the rotation of the porous medium. The onset and transient growth of the viscous fingers were observed for various cases. In addition to studying viscous fingering under various hydrodynamic conditions, the role of the thermal gradient and double-diffusive convection were studied in detail. For each case, the evolution of the flow patterns was observed until steady state or quasi-steady state was reached. For some cases, however, steady state was never reached and flow instability remained throughout.

Even though no single finger dominated the flow, the tip of the displacement front consistently manifested a greater velocity than did the base of the front. The difference between the tip and the base of the displacement front was alleviated when hot water was injected into the system. The buoyancy force acted against the growth of fingers. Rotation of the bed about its axis distorted the viscous fingering formation and the flow was pushed toward the wall of the cylinder. Rotation reduced the adverse effects of viscous fingering for the cases studied.

INTRODUCTION

The study of heat and mass transfer in porous media has a large number of applications in environmental and geothermal engineering, secondary and tertiary oil recovery, fixed bed regeneration in chemical processing, hydrology and filtration.[1-8] Viscous fingering generally refers to the onset and evolution of instabilities that occur in the displacement of fluids in a porous bed. In most cases, this mechanism of instability is linked to the viscosity difference between the displaced and the displacing fluids. A fingering pattern may evolve when a less-viscous fluid (higher mobility) penetrates a more-viscous (lower mobility) fluid during a displacement process. This can occur for both miscible (no phase boundary or capillary force) and immiscible fluids. A similar instability can occur when a more-dense fluid displaces vertically downward a less-dense fluid in a porous medium, or even in a Hele-Shaw cell.[1] Fingering can also occur when a low viscosity Newtonian fluid, injected from a source, penetrates a Hele-Shaw cell (or a porous medium) filled with a miscible (or immiscible) non-Newtonian fluid. While there have been many studies reporting the effect of the above two factors, namely viscous forces and gravity forces, the effect of capillary forces with immiscible fluids is less understood.[9-12] Recent investigations show that heterogeneity plays an important role in defining the onset as well as the propagation of a viscous finger.

Numerical simulation of viscous fingering phenomena had been considered to be too difficult to be solved with computers. However, as computers became faster and more efficient in managing storage requirements, numerical simulation work started to be reported. For example, Zimmerman and Homsy[13,14] examined non-linear viscous fingering in miscible displacement in two-dimensions using a spectral method. Results indicated that the finger interaction mechanisms are independent of the concentration field and are governed fundamentally by the pressure field. In their three-dimensional simulation, they found that the advance of viscous fingering remained unchanged from two dimensions, suggesting that two-dimensional simulations are sufficient to capture the essential features of non-linear viscous fingering.[14] In their analysis, the buoyancy term was neglected, thereby eliminating the effect of gravity.

Christie *et al.*[15] simulated viscous fingering in both homogeneous and heterogeneous media for a single-phase flow using a three-dimensional finite-difference technique. Their main objective was to understand the mechanism that prevents viscous fingering from occurring or propagating. Even though they were credited with investigating three-dimensional viscous fingering, their findings indicated that there is little difference in macroscopic behavior in two and three-dimensional systems. They compared recovery curves for several two-dimensional and three-dimension simulations in their study by varying the water/gas ratios. In their analysis, strong gravity segregation was always present.

Tchelepi *et al.*[16] conducted experiments and numerical simulations for both stable and unstable displacements for millstone and limestone cores. Finite difference / particle tracking techniques were used in two-dimensional and three-dimensional models. The two-dimensional results showed reasonable agreement with their experimental observations of three-dimensional unstable flow in homogeneous and randomly heterogeneous media. They observed two regimes of flow for unstable displacements in homogeneous and randomly heterogeneous porous media. They reported the existence of an early-time regime of rapid-transition zone growth and a regime for which the concentration velocity reached nearly constant values. Heterogeneity of the limestone was the reason for the observed difference between the millstone and limestone results.[16] They also reported on behavior in microgravity, neglecting density effects.

The previous studies had several deficiencies. The first is that no one modeled the onset of viscous fingering, so that a systematic parametric study of both stable and unstable displacement processes had not been performed. The effect of gravity was neglected for most of the numerical studies. Consequently, this important aspect of the onset and propagation of viscous fingers remained unexplored. Finally, no one reported a relationship with thermally-induced fingering. It is conceivable that a temperature gradient would alter both the onset and the propagation of viscous fingers, but no study had been published coupling thermal and solutal effects. Some of these shortcomings were addressed here. Our objective was to examine numerically the onset of viscous fingering when one miscible fluid displaces another in a rotating porous medium. We were interested in two-dimensional axi-symmetric flows, and limited ourselves to the cases for which the orientation of the gravitational force is collinear with the direction of flow. The effect of temperature and rotation on the onset and propagation of viscous fingering was determined.

GOVERNING EQUATIONS AND NUMERICAL PROCEDURE

The Brinkman equation and the energy and the solutal conservation equations were solved numerically using a finite-element technique. Zongqin and Bejan[17] and Saghir and Islam[18,19] described the dimensional equations used in the model. In non-dimensional form, the following dimensionless variables were used:

$$U_r = \frac{u_1}{u_o}, U_z = \frac{u_2}{u_o}, U_\theta = \frac{u_3}{u_o}, R = \frac{r}{L}, Z = \frac{z}{L}, P = \frac{p}{\rho u_o^2}, u_o = \sqrt{g\beta_T \Delta T L},$$
$$\tau = \frac{t u_o}{L}, \vartheta = \frac{T - T_o}{\Delta T}, C = \frac{c - c_o}{\Delta c} \tag{1}$$

where u_o is a reference velocity, ΔT the initial temperature difference, Δc the initial concentration difference, T_o a reference temperature, c_o a reference concentration, β_T the thermal volume expansion, g the acceleration due to gravity and L a characteristic axial length. The governing equations follow. From a momentum balance (neglecting the Coriolis effect):

$$\frac{Re}{\phi}\frac{\partial u_i}{\partial \tau} + \frac{1}{Da} u_i = -P_{,i} + (u_{i,j} + u_{j,i})_{,j} + \frac{Gr}{Re}(\vartheta - NC) \tag{2}$$

The continuity equation is:

$$(u_i)_{,i} = 0 \tag{3}$$

Conservation of energy is expressed in non-dimensional form by (neglecting the heat of mixing):

$$G\frac{\partial \vartheta}{\partial \tau} + Re\,Pr\left[u_j \vartheta_{,j}\right] = F\left[\vartheta_{,j}\right]_{,j} \tag{4}$$

Conservation of a component in dimensionless form is:

$$Re\left[\phi\frac{\partial C}{\partial \tau}+u_j C_{,j}\right]=\frac{1}{Sc}(C_{,j})_{,j} \quad (5)$$

Here, $\mathrm{Re}=\frac{\rho u_o L}{\mu}$ is the Reynolds number, $\mathrm{Pr}=\frac{C_P \mu}{k}$ the Prandtl number, ρ the density of the fluid, μ its dynamic viscosity, C_p the specific heat, *k* the conductivity, $Da=\frac{\kappa}{L^2}$ the Darcy number, κ the permeability, $Ra_T=\frac{g\beta_T \Delta T L \kappa}{\alpha \nu}$ the thermal Rayleigh number, $Gr=\frac{g\beta_T \Delta T L^3}{\nu^2}$ the thermal Grashof number, α the thermal diffusivity for the porous medium $(\frac{k}{\rho C_p})$, $Sc=\frac{\nu}{\alpha_c}$ the Schmidt number, $Ra_c=\frac{g\beta_c \Delta c L \kappa}{\nu \alpha_c}$ the solutal Rayleigh number, and β_C the solutal volume expansion. The constant *G* is equal to $Re\,Pr\,\phi+\frac{(\rho C_p)_s}{\rho C_p}Re\,Pr(1-\phi)$, where ϕ is the porosity and the constant *F* is $\phi+\frac{k_s}{k}(1-\phi)$. The subscript *s* denotes the solid. Another dimensionless parameter, the relative buoyancy, is $N=\frac{\beta_c \Delta c}{\beta_T \Delta T}$. Here, *N* was set equal to 1. One of the most important parameters in double-diffusive phenomena is the Lewis number *Le*, which is the ratio of the thermal diffusivity to the solutal diffusivity, here taken equal to 100. Through their definitions, the thermal Rayleigh number Ra_T and the solutal Rayleigh number Ra_C are related by $Ra_C = Ra_T.\ N.\ Le$.

Figure 1 shows a schematic diagram of the situation analyzed, along with the boundary conditions. It consisted of a vertical cylinder with a diameter of 16 cm and a height of 10 cm. The cylinder contained an isotropic and homogeneous porous medium. Initially, the top 0.2 cm contained water and the remainder glycerin. (Note that water and glycerin are completely miscible.) A rotating shaft was located at the center of the cylinder. Pure water was continually injected from a 0.2 cm diameter inlet at the top center, and a glycerin/water solution was withdrawn from the bottom outlet, which was also 0.2 cm in diameter. At time $\tau = 0$, injection of water was initiated at a velocity of 0.776 cm/s. The water, in turn, pushed the glycerin. The initial concentration difference between the glycerin and the water was set equal to unity. Since the geometry was axi-symmetric, only half of the volume shown in Fig. 1 had to be considered. The radial velocity U_r was set equal to zero and the angular velocity $U_\theta = \omega r$ at all boundaries of the vessel. The vessel was adiabatic with no mass transfer to the environment (i.e., $\partial C/\partial n = 0$ on all walls). The axial velocity U_z was zero on the cylindrical wall. At the center axis of rotation, the rotational speed U_θ was set equal to zero (since r = 0).

Finite Element Formulation

The non-dimensional Eqs. (2)-(5) were solved using a finite-element technique. Quadrilateral elements were used with velocities, concentration and temperature at each node.[19] The glass beads constituting the packing were considered to have diameters $d = 5.25$ mm to produce a packing porosity of 0.39.

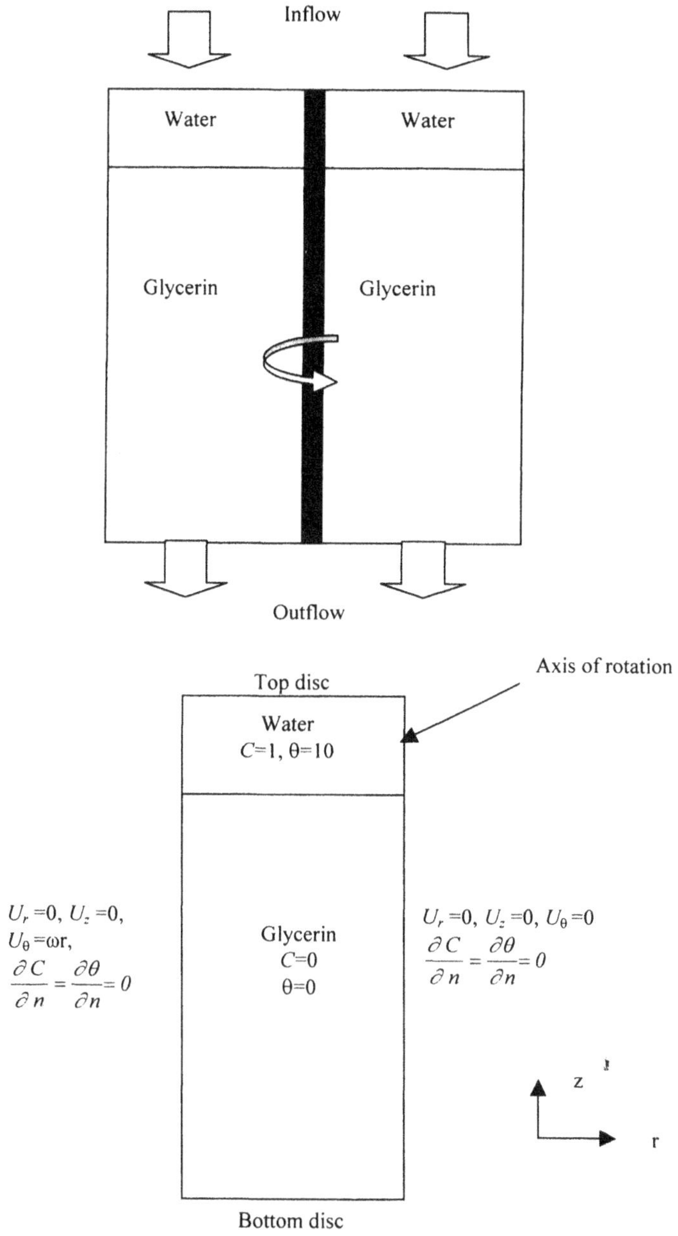

Figure 1. The system modeled. Top: physical arrangement. Bottom: boundary conditions.

Using the Kozeny-Carmen relation,[19] the permeability κ was:

$$\kappa=\frac{d^2}{174.3}\frac{\phi^3}{(1-\phi)^2} \tag{7}$$

This formulation resulted in $\kappa = 2.52 \times 10^{-4}$ cm^2. Since the characteristic length used in our model was L = 1 cm, the Darcy number $Da = 2.52 \times 10^{-4}$.

Table 1 gives the physical properties of the two liquids and the glass beads. The ratio of the thermal diffusivity in the horizontal and vertical directions was assumed to be equal to unity. A value of Da = 2.52×10^{-4} was used along with the other non-dimensional terms shown in Table 2.

Table 1. Physical properties of the fluids and the glass beads.

Material	Density (g/cm^3)	Viscosity (g/cm.s)	Conductivity (cal/cm.K)	Specific heat (cal/g.K)
Glycerin	1.0085	14.90	0.683×10^{-3}	0.5703
Water	1.000	1.003×10^{-2}	1.443×10^{-3}	0.9983
Glass beads	1.648	--	4.329×10^{-3}	0.1147

Table 2. Non-dimensional parameters used in the model.

Liquid	Ra_c	Ra_T	Le	Pr	Sc
Water	1.32×10^6	1.32×10^4	100	6.78	678
Glycerin	2.8	0.028	100	12.5	1250

Accuracy of Solutions and the Effect of Grid Distribution

A series of numerical runs were conducted for the base case in order to investigate the effect of the number of elements, as well as the distribution of grid blocks. At each time step, the four equations were solved simultaneously and the criteria for convergence were set on the four unknowns, namely the two velocities, the temperature and the concentration. Once all these variables, at two different time steps, reached a difference less than 10^{-4}, convergence was considered to have been achieved. Initially, all numerical solutions were obtained using a uniform grid in which 24 quadrilateral elements covered the horizontal direction and 32 quadrilateral elements covered the vertical direction. In order to improve the quality of depiction of physical phenomena, grid refinement was performed along the central vertical and top horizontal line. In total, 768 quadrilateral elements were used in the model. In another series of tests, numerical simulation was performed using 41 and 61 elements along the z-direction. Results were duplicated within 1% for the onset and within 5% for the propagation cases. Consequently, the chosen grid distribution was considered to be adequate for the modeling.

RESULTS AND DISCUSSION

Three different displacement cases were studied. In the first case, both the water and the glycerin were maintained at the same constant temperature. This is called the "isothermal-displacement case." In the second case, the water was injected with a temperature of 10°C above the glycerin temperature, thus giving a temperature difference of $\Delta\vartheta = 10$. This is designated as the thermal displacement case. Finally, the third case was a repetition of the thermal case but with rotation. This is called the "rotational-displacement case." The initial solutal and thermal Rayleigh numbers are those in Table 2.

Isothermal Displacement Case

Figure 2 shows the displacement front, which is marked with numerous fingers, at different dimensionless times. (In all such plots here, the horizontal scale is radial position, the vertical scale is vertical position, and the lines represent the boundary between glycerin and water at different times.) The overall shape of this front remained close to the elliptical shape that would be expected in the absence of viscous fingering and typical of Poiseuille flow in a porous media. This indicates that there was no dominant finger in the process. Experimentally, a dominant finger has always been observed, mainly because there is no perfectly isotropic, homogeneous porous media.[20,21] Since all naturally occurring porous media are heterogeneous, a single finger typically dominates the flow and leads to significant fluid bypass.

Note in Fig. 2 that the most dominant fingers arose near the center. Even though the flow rate for the overall system was kept constant, the velocity near the central region was the highest, leading to the formation of the greatest solutal and pressure gradients. These steeper gradients eventually led to liquid breakthrough near the central region (after a dimensionless time of 15 in this case). As soon as breakthrough occurred, the fingers near the central region become subdued. Also, the overall shape of the front became closer to a cone rather than an ellipsoid. As the displacement continued, the front became virtually free of viscous fingers.

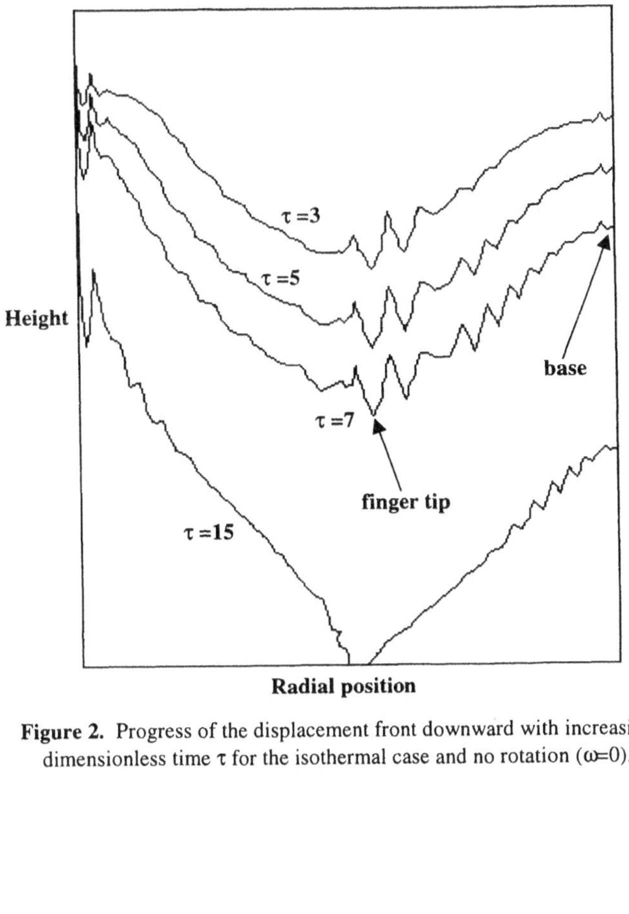

Figure 2. Progress of the displacement front downward with increasing dimensionless time τ for the isothermal case and no rotation (ω=0).

Figure 3 shows the dimensionless distance traveled by the base and the tip versus dimensionless time. Note that the base of the displacement front moved linearly at a relatively uniform pace, especially during the later stage. Note, also, that the viscous fingers practically disappeared after breakthrough at a dimensionless time of 15. Due to the dominance of viscous fingering in the early stages of the displacement, the propagation of the tip was much faster. The difference between the distance traveled of the base and the tip grew significantly with time, until breakthrough.

Thermal Displacement Case

In order to observe the effect of a temperature gradient on the onset and propagation of viscous fingers, a series of runs was performed. Figure 4 shows the displacement front for different dimensionless times. Despite the thermal changes, there appears to be little difference in the shape of the fingers from those shown in Fig. 2, while the displacement front moved approximately 10% slower. As a consequence, breakthrough took place later. The fingers that evolved were mostly locally contained. Consequently, their impact was not significant on macroscopic phenomena such as the temperature field.

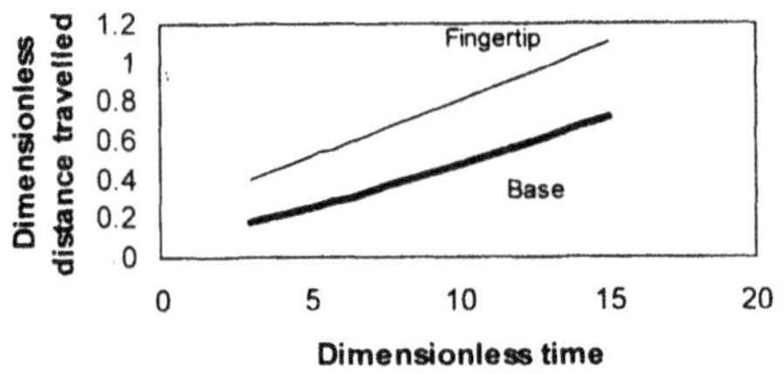

Figure 3. Distance traveled for the tip and the base of the fingers shown in Fig, 2.

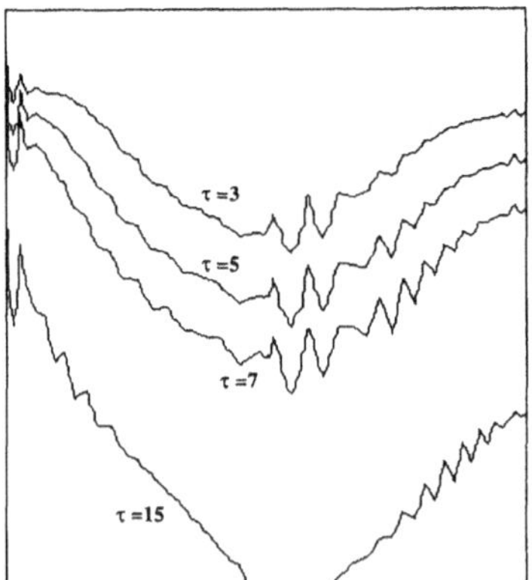

Figure 4. Displacement front at different times for the thermal displacement case without rotation.

Figure 5 shows the propagation of the tip and base of the displacement front for the thermal case. The displacement front of the thermal case propagates slower than for the isothermal case, for both the base and the tip. The difference between the two cases reached two maxima. The maximum difference occurred close to the breakthrough time. The difference between the two cases was not as intense for the propagation of the base and did not show a maximum. Instead, the difference between the two cases continued to grow. Because the thermal case adds water that is at a higher temperature (by 10°C), the density of the displacing phase is decreased. In this particular case, the thermal Rayleigh numbers Ra_T for the water and for the glycerin are 1.32×10^4 and 0.028, respectively. As a consequence, the buoyancy force acts against the growth of the viscous finger. This leads to a delayed movement of the tip of the viscous finger, for both the tip and the base of the displacement front.

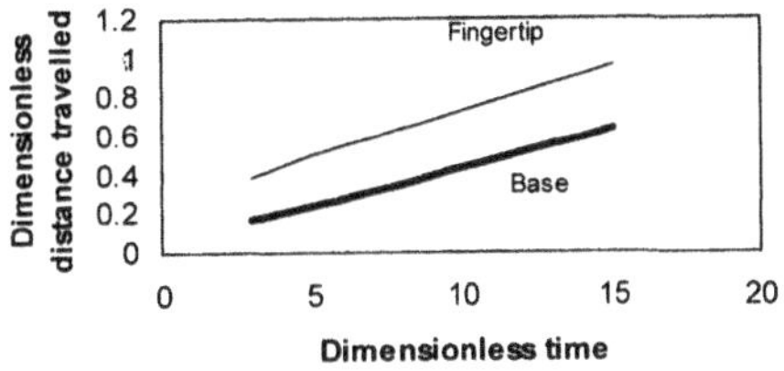

Figure 5. Distance traveled for the tip and the base of the fingers shown in Fig, 4.

Effect of Rotational Speed

In several numerical runs, the effect of rotation was examined for a speed from 2 rpm to 800 rpm. Figure 6 shows the development and propagation of viscous fingers at various times levels for a rotational speed of 2 rpm. It appears that the onset of viscous fingering was retarded by rotation while the shapes of the fingers were not significantly affected. Figure 7 shows the propagation of both the base and the fingertip. As time progressed, the baseline continued to show divergence from the fingertip line. The difference between the two remained approximately the same as for the previous thermal displacement case. Figure 8 shows the existence of viscous fingers at 10 rpm. The movement of the base of the displacement front continued to slow down as the rotational speed was increased. Figure 9 shows that the relative movement of the fingertip increased as the rotational speed was increased. Meanwhile, the viscous fingers continued to persist for all these cases. The first sign of diminishing fingering appeared at higher rotation rates, as shown in Fig. 10 for 80 rpm. Note that the base of the displacement front was significantly slowed. This aspect is discussed later. Figure 11 shows the propagation of the fingertip and the base. The base appeared to move very slowly while the finger tip moved faster than for the previous cases. However, the overall fingertip movement remained remarkably linear. In fact, after a dimensionless time of 7, both base and fingertip lines become parallel to each other. This behavior marks the beginning of diminishing viscous fingering. Figure 12 shows the displacement profile at 800 rpm, where viscous fingering did not exist. The anomalies on the left hand side are perturbations due to the high rotational speed and are not related to viscous forces. Breakthrough was delayed significantly, indicating that a lack of viscous fingers can increase recovery efficiency.

Figure 13 summarizes the effect of rotational speed on the propagation of the displacement front. The maximum difference between the fingertip and the base is plotted versus rotation speed. This graph shows the existence of two different regimes.

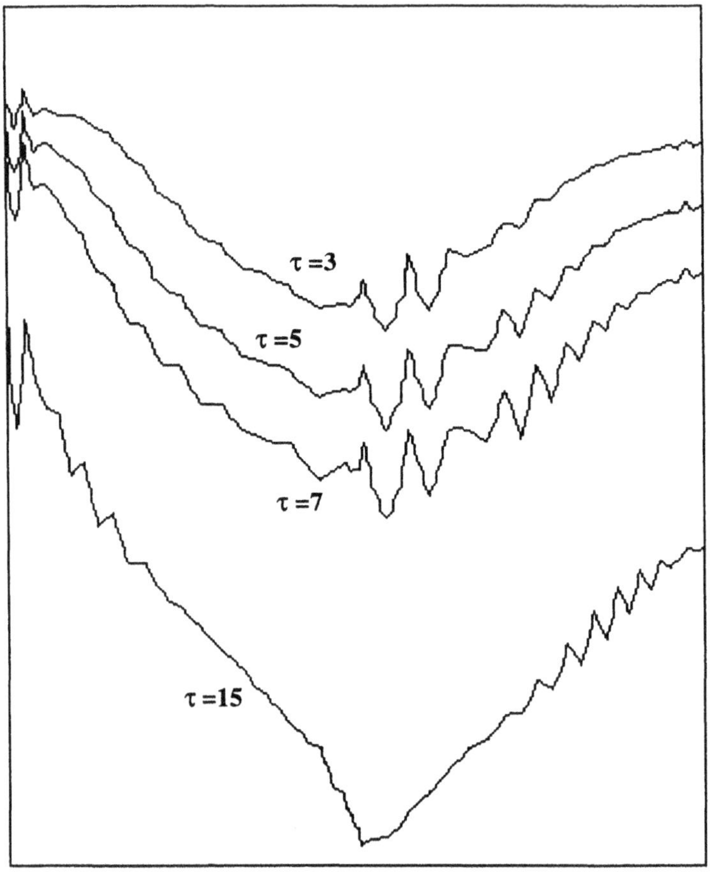

Figure 6. Evolution of the displacement front for the thermal displacement case with rotation at 2rpm.

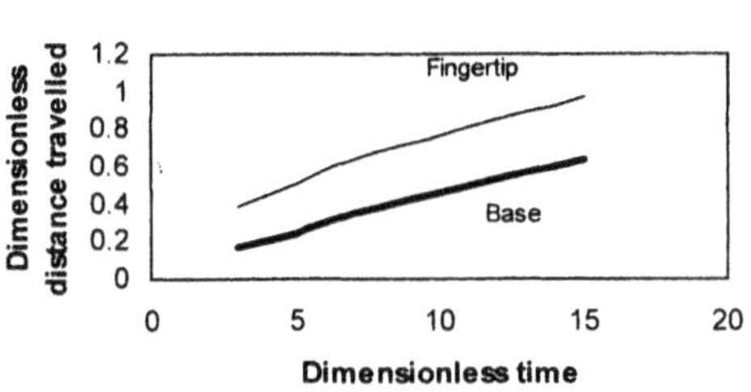

Figure 7. Distance traveled by the fingertip and base of fingers for the conditions of Fig. 6.

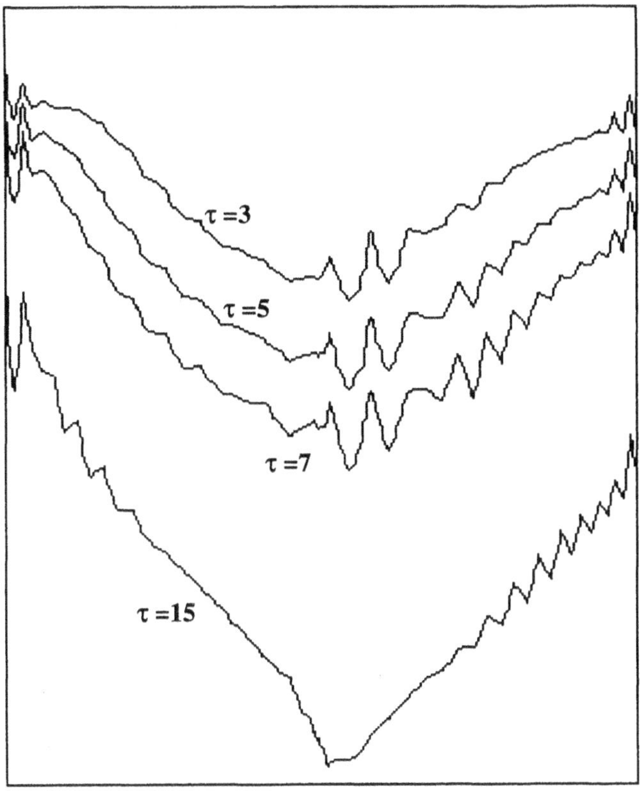

Figure 8. Viscous fingering for the thermal displacement case with rotation at 10rpm.

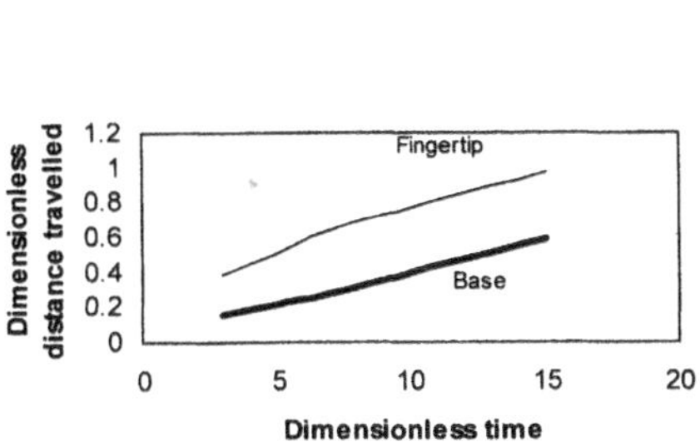

Figure 9. Distance traveled for the fingertip and the base of fingers under the conditions of Fig. 8.

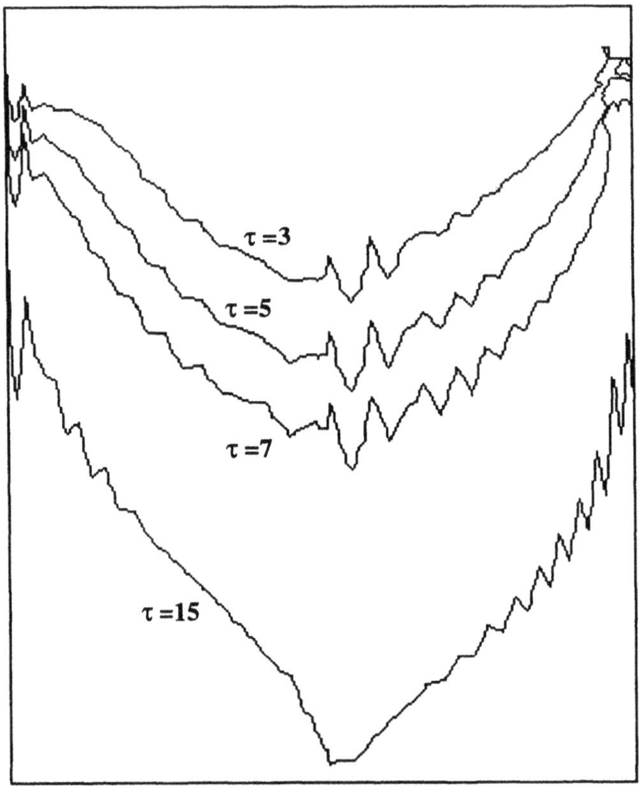

Figure 10. Viscous fingering at 80 rpm.

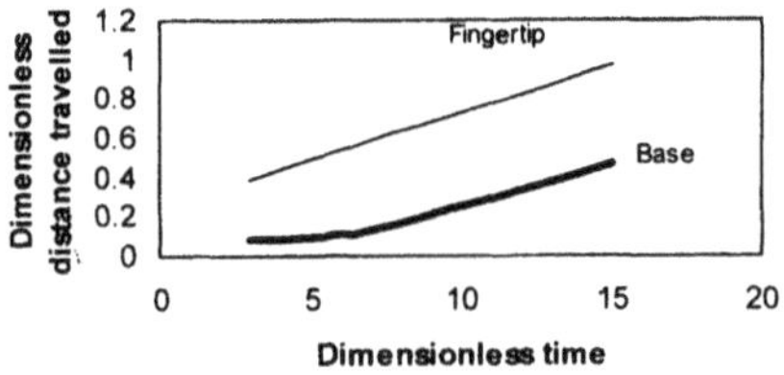

Figure 11. Distance traveled by the fingertip and base of viscous fingers at 80 rpm.

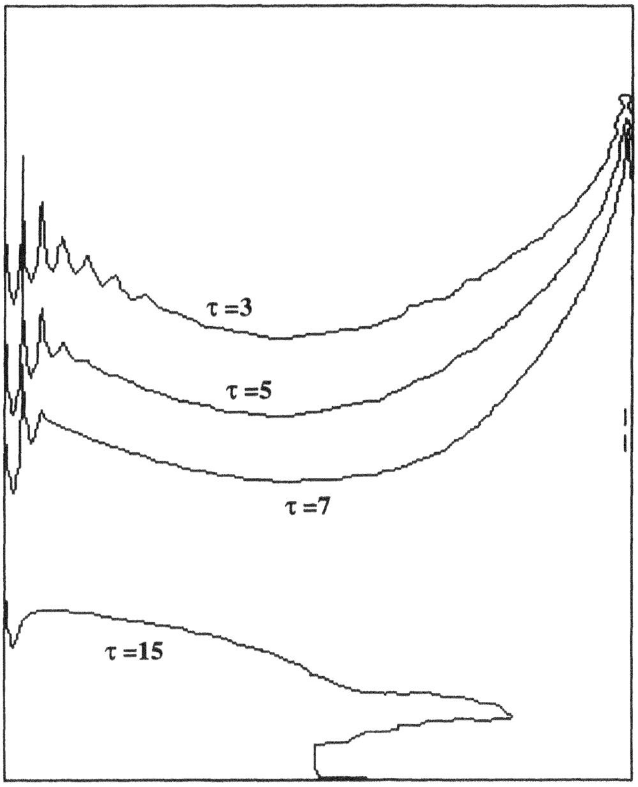

Figure 12. Evolution of the displacement front at 800 rpm.

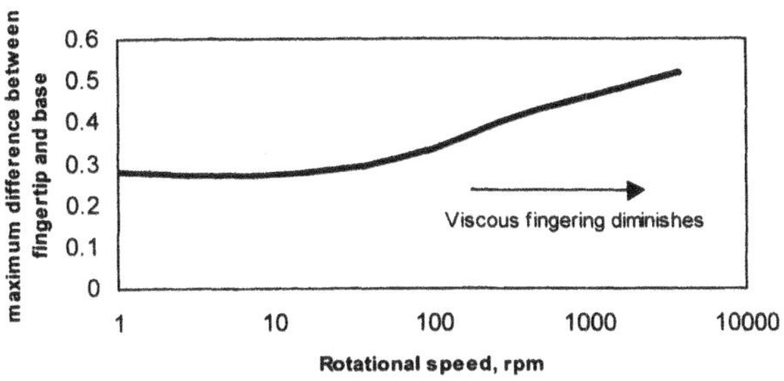

Figure 13. Influence of rotation on the evolution of viscous fingers.

The first shows little sensitivity to rotation speed and is dominated by the presence of viscous fingers. Interestingly, as the rotational speed was increased, the difference between the fingertip and the base of the front increased rapidly. As the rotation speed was increased, fewer viscous fingers formed and, as a consequence, the front became more conical. This behavior wouldn't be expected from conventional thinking in this field. However, it becomes clear when one considers that at higher rotation the centripetal acceleration prevents the formation and propagation of viscous fingers. In the absence of viscous fingers, the lateral dissipation of solute is minimized, leading to more flow in the axial direction. At the same time, a conical shape for the front means slower movement of the base of the front. This explains why the difference between the base and the front is significantly higher at higher rotational speed.

Acknowledgment

The authors would like to acknowledge the financial support of the Natural Science and Engineering Council (NSERC) of Canada.

REFERENCES

1. G.M. Homsy, *Ann. Rev. Fluid Mech.* 19: 271-311 (1987).
2. T.K. Perkins and O.C. Johnston, *Soc. Pet. Eng. J.*, 70-80 (1963).
3. E.J. Koval, *Soc. Pet. Eng. J.*, 307-311 (1963).
4. S. Vossoughi, J.E. Smith, D.W. Green and G.P. Willhite, *Soc. Pet. Eng. J.* 30-38 (1984).
5. M. Blunt and M. Christie, *Society of Petroleum Engineers Advanced Technology Series* 2, N.2, 383-390 (1994).
6. N. M. Chaudhari, *45th Annual Fall Meeting, Society of Petroleum Engineers*, paper SPE 2982 (1971).
7. J.R. Fanchi, G.D. Shank and R.L. Christiansen, Chaos: a source of miscible viscous fingering instabilities, *Petroleum Society of CIM/Society of Petroleum Engineers*, paper SPE 21587 (1990).
8. D.W. Peaceman and H.H. Rachford, Jr., Numerical calculation of multidimensional miscible displacement, *35th Annual Fall Meeting of the Society of Petroleum Engineers* (1960).
9. R.G. Bentsen, A new approach to instability theory in porous media, *Soc. Pet. Eng. J.* 25: 765-779 (1985).
10. R.G. Bentsen, Use of capillary pressure data to estimate the pseudointerfacial tension for analysis of stability of fluid flow in porous media, *AOSTRA J. Research* 2: 213-222 (1987).
11. M.R. Islam and R.G. Bentsen, A dynamic method for measuring relative permeability, *J. Can. Pet. Tech.* 25: 39-50 (1985).
12. M.R. Islam and R.G. Bentsen, Effect of different parameters on two-phase relative permeability, *AOSTRA J. Research* 3: 69-90 (1987).
13. W.B. Zimmerman and G.M. Homsy, Nonlinear viscous fingering in miscible displacement with anisotropic dispersion, *Phys. Fluids A* 3: 1859-1872 (1991).
14. W.B. Zimmerman and G.M. Homsy, Three-dimensional viscous fingering: a numerical study, *Phys. Fluid A* 4: 1901-1914 (1992).
15. M.A. Christie, A.H. Muggeridge and J.J. Barley, 3D simulation of viscous fingering and WAG schemes, *Technical Report SPE 21238, SPE 11th Symposium on Reservoir Simulation* (1991).
16. H.A. Tchelepi, F.M. Orr, N. Rakotomalala, D. Salin and R. Woumeni, Dispersion, permeability heterogeneity, and viscous fingering: acoustic experimental observations and particle-tracking simulations, *Phys. Fluid A* 5: 223-234 (1993).
17. Z. Zongqin and A. Bejan, The horizontal spreading of thermal and chemical deposits in a porous medium, *Int. J. Heat Mass Transfer* 30: 2289-2303 (1987).
18. Z. Saghir and M.R. Islam, Double diffusive convection in dual-permeability, dual-porosity porous media, *Int. J. Heat Mass Transfer* 42: 437-454 (1998).
19. A. Saghir, M. Hennenberg, and M.R. Islam, Double diffusive convection with Marangoni effects in a multi-cavity system, *Int. J. Heat Mass Transfer* 41: 2157-2174 (1998).
20. H.K. Sarma and B.B. Maini, An experimental evaluation of viscosity grading for controlling fingering in miscible displacement, *J. Can. Pet. Tech.* 32: 36-41 (1993)
21. Z. Saghir, O. Chaalal, O. and M.R. Islam, Experimental and numerical modelling of viscous fingering, *J. Pet. Sci. Eng.* (in press).

MAGNETOHYDRODYNAMIC EFFECTS IN HIGH GRAVITY CONVECTION DURING ALLOY SOLIDIFICATION

D.N. Riahi

Department of Theoretical and Applied Mechanics
University of Illinois, Urbana, IL 61801, USA

ABSTRACT

Nonlinear compositional convection in the melt and in cylindrical chimneys within a mushy layer during solidification of a binary alloy is investigated subjected to a strong magnetic field and under high gravity environment. Asymptotic and scaling analyses are applied to axisymmetric convection in the melt and in the chimneys. Three important nondimensional parameters are the solutal Rayleigh number R, the centrifugal acceleration parameter A, and the Chandrasekhar number (magnetic parameter) Q. Magnetohydrodynamic effects, due to the presence of a strong magnetic field, are stabilizing in the sense that convection and freckle formation can be reduced significantly. Such stabilizing effects can be enhanced by the presence of a moderate rate of rotation due to centrifugal force effects. In this stabilizing regime, chimney convection decreases with increasing either A or Q, even though it still increases with R.

INTRODUCTION

Rotational and/or magnetohydrodynamic effects on the convective flow in the melt adjacent to crystal-melt interface have been of interest to crystal growers for many years.[1] In industrial crystal growth processes it has been desirable to impose rotation and/or magnetic field upon the solidifing system in order to reduce the effects of instabilities or oscillatory convection that can reduce the quality of the solidified material. For example, Sample and Hellawell[2,3] did NH_4Cl-water experiments in a cylindrical mold with a chilled bottom surface where solidification was induced. They applied a rotation and tilting technique to change the orientation of gravity relative to the bottom surface of the cylinder. They found that for slow and steady rotation of the mold about the vertical axis, the channel formation and development was about the same as without rotation. Slow and steady rotation of a tilted mold substantially reduced the number of channels. Fast rotation is not considered useful to crystal growers, as, for example, the experimental results of Kou et al.[4] that indicated if

Processing by Centrifugation, Edited by Regel and Wilcox
Kluwer Academic/Plenum Publishers, New York, 2001

rotational speeds became too large, then segregates formed along a ring between the axis and the outer edge of the ingot.

The experimental results discussed above and the indication for possible usefulness of inclined rotational applied to solidifying systems ha led to further research and progress. A solidifying system under a high gravity environment provides a useful industrial example where inclined rotation can be beneficial.

The possible beneficial effects of rotation alone has already been explored by Riahi[5,6] for a solidification system. It was found that rotation can reduce the undesirable effects of convection, provided the axis of rotation is inclined with respect to the direction of the solidification. Furthermore, the value of the rotation rate should lie in certain domains. However, this latter condition may be difficult to meet in practice. Thus, it became desirable to explore the possible beneficial effects of a magnetic field in a solidification system as done recently by Riahi.[7] It was found,[7] in particular, that stabilizing effects of the field can be effective in reducing the strength of the chimney convection, provided the value of the magnetic field parameter Q (Chandrasekhar number) is sufficiently large. To provide flexibility in a flow controlling procedure and to achieve optimal conditions, it is quite desirable to investigate the combined effects of an externally imposed magnetic field and inclined rotation. The present paper shows such an investigation and reveals wider ranges in the parameter space (A, Q) where convection in the chimneys can be reduced significantly. One interesting result of the present study is that a strong magnetic field can widen the range of values of the rotation parameter where chimney convection can be weakened.

GOVERNING SYSTEM

The formulation for the solidification system in a high gravity environment used here follows most closely the approach of Riahi[8,9] in the absence of magnetohydrodynamic effects. We consider a layer of a binary alloy melt of some constant composition C_0 and temperature T_∞ being solidified at a constant rate V_0, with eutectic temperature T_e at position $z = 0$ held fixed in a frame moving with the solidification speed in the z-direction. The z-axis is assumed to be anti-parallel to the acceleration vector, as described in Riahi[9]. Within the layer of melt, there is a mushy layer adjacent to the solidifying surface and of thickness z_0. We assume that the solidifying system is placed in a centrifuge rotating at some constant angular velocity Ω about the centrifuge axis which makes an angle γ with respect to the z-axis. The centrifuge axis is assumed to be parallel to earth's gravity vector.

The magnetohydrodynamic aspect of the physical model is based on the Maxwell equations combined with the governing equations for the convective flow that are given in Chandra- sekhar[10]. The governing system of equations for the solidifying system, subjected to an external magnetic field $B\hat{z}$ in the z-direction of uniform strength B, is nondimensionalized using $V_0, K/V_0, K/V_0^2, \beta\Delta C\rho_0 gK/V_0, \Delta C, \Delta T$ and B as scales for velocity, length, time, pressure, solute, temperature and magnetic field, respectively. Here K is thermal diffusivity, $\Delta C = C_0 - C_e, \Delta T = T_L - T_e, C_e$ is eutectic concentration, T_L is local liquidus temperature at $C = C_0, \hat{z}$ is a unit vector in the positive z-direction, β is expansion coefficient for solute, ρ_0 is a reference (constant) density and g is acceleration due to earth's gravity.

Following Riahi[9], we will be concerned here mainly with convection in the melt and in a cylindrical chimney, whose axis is assumed to be parallel to the z-axi. We consider the governing equations in cylindrical coordinates with the axial direction being the z-axis. We assume axisymmetric flow and do not include the effect of Coriolis force.[11] The Coriolis effect would be included naturally in the fully non-axisymmetric flow case planned for investigation soon. As in the work treated in Riahi[9], the centrifugal acceleration term in the momentum equation is split into an average term, which is superimposed on the gravity term, and a so-called gradient acceleration term.[12] At a high rotation rate, the non-dimensional parameter representing the modified gravity term can be significantly larger than the corresponding one due to earth's gravity alone. We treat the mushy layer as a porous medium[13,14] in which Darcy's law is valid. For the analysis of convection in a cylindrical chimney, we assume that the chimney's axis coincides with the z-axis. The investigation is based on the limit of strong compositional buoyancy, negligible thermal buoyancy and large Lewis number K/D, where D is solute diffusivity.

The non-dimensional forms of the basic equations in the rotating frame for the magneto-hydrodynamic convective flow of the melt in the liquid zone above the mushy zone as well as inside the chimneys are given in their steady state form by

$$\left(-\frac{\partial}{\partial z}+\boldsymbol{u}\cdot\nabla\right)\boldsymbol{b}=\left(\frac{\partial}{\partial z}+\boldsymbol{b}\cdot\nabla\right)\boldsymbol{u}+\frac{1}{\tau}\nabla^2\boldsymbol{b}, \tag{1a}$$

$$\nabla\cdot\boldsymbol{b}=0, \tag{1b}$$

and by the steady forms of Eqs. (1a)-(1d) in Riahi[9], provided term $(HQ/\tau)(\frac{\partial}{\partial z}+\boldsymbol{b}\cdot\nabla)\boldsymbol{b}$ is added to the right-hand-side of Eq. (1a) in Riahi[9] and the Coriolis parameter T is set to zero. Here $\boldsymbol{b}$ is induced magnetic field, $\boldsymbol{b}=b_r\hat{r}+b_z\hat{z}$ under axisymmetry condition, $P=\tilde{P}/\rho_0+|(B\hat{z}+\boldsymbol{b})|^2\mu/(8\pi\rho_0)$ is modified pressure, μ is magnetic permeability, $\tilde{P}$ is pressure, $Q=\mu B^2K^2/(4\pi\rho_0\nu\eta V_0^2H)$ is Chandrasekhar number, $\tau=K/\eta$ is Roberts number and η is magnetic diffusivity. All the other variables and parameters are defined in Riahi[9].

The non-dimensional steady state form of the basic equations for the mushy zone outside the chimneys are given by

$$\left[-\frac{\partial}{\partial z}+(1-\phi)\boldsymbol{u}\cdot\nabla\right]\boldsymbol{b}=\left(\frac{\partial}{\partial z}+\boldsymbol{b}\cdot\nabla\right)[(1-\phi)\boldsymbol{u}]+\frac{1}{\tau}\nabla^2\boldsymbol{b}, \tag{2a}$$

$$\nabla\cdot\boldsymbol{b}=0, \tag{2b}$$

and by steady forms of Eqs. (2a)-(2d) in Riahi[9], provided term $(Q/\tau)\left(\frac{\partial}{\partial z}+\boldsymbol{b}\cdot\nabla\right)\boldsymbol{b}$ is added to the right-hand-side of Eq. (2a) in Riahi [9], and all the other parameters and variables here are defined in Riahi[9]. The boundary conditions for the solidification system are given in Worster[13] and in Riahi[8] and for the induced magnetic field can be of the form given in Riahi.[15] These boundary conditions are not repeated here since their explicit use is not needed in the present analysis though the resulting leading order solutions are consistent with such boundary conditions. Worster[14] indicates that $\theta=S$ in the mushy zone outside the chimneys, and such relation is also valid here.

In the next section we proceed with asymptotic and scaling analyses for the equations given in this section in the asymptotic limit of sufficiently large R, to determine the strongly nonlinear steady state, under axisymmetric condition, for the convective flow in the mushy zone and mainly in the chimneys for particular range in Pr where inertia terms in the momentum equation are negligible. It will be assumed that the Roberts number τ and the Chandraskhar number Q lie in the following ranges:

$$\tau\leq O(1), \tag{3a}$$

$$Q \geq O(R\tau). \tag{3b}$$

The range (3a) for τ is often satisfied in the laboratory experiments, while the range (3b) was determined recently[7] to be representative for the strong magnetic field case where stabilizing effects of the field becomes significant.

ANALYSIS AND RESULTS

We consider a chimney whose axis coincides with the z-axis, and with small radius $a(z)$ ($a << 1$). The maximum of r and z are a and 1, respectively. Assuming the magnitude of the flow velocity to be of order unity in the mushy layer, then Eq. (2a) in Riahi[9] implies that to the leading term the pressure field in the mushy layer is unaffected by the flow velocity and $\theta = S$ is independent of r. Equation (2a) here and Eqs. (2c)-(2d) in Riahi[14] for the assumed r-independent leading order variables $\theta_0(z), \phi_0(z), w_0(z)$ and $b_{z0}(z)$ then imply equations (4a,b) given in Riahi[6] plus the following equations:

$$-w_0' + (w_0 - 1)b_{z0}' = b_{z0} \cdot w_0' + \frac{1}{\tau} b_{z0}'', \tag{4a}$$

$$(1 - w_0)b_{z0}'' + w_0''(1 + b_{z0}) + b_{z0}'''/\tau = 0, \tag{4b}$$

when a prime denotes differentiation with respect to z. As in the earlier work,[14] $C_r >> \theta$ is assumed in the mushy zone.

Under the assumed axisymmetric condition, Eq. (1b) here and Eq. (1b) in Riahi[14] imply that stream functions $\psi(r, z)$ and $h(r, z)$ for the flow velocity and the induced field, respectively, can be introduced so that:

$$(u, w) = \left(-\frac{1}{r}\frac{\partial \psi}{\partial z}, \frac{1}{r}\frac{\partial \psi}{\partial r}\right), \tag{5a}$$

$$(b_r, b_z) = \left(-\frac{1}{r}\frac{\partial h}{\partial z}, \frac{1}{r}\frac{\partial h}{\partial r}\right). \tag{5b}$$

For the flow in the chimney, it is assumed that $s \sim 1$ and $w >> 1$. Then scalings of the form

$$w \sim HRa^2, \quad u \sim HRa^3, \quad \psi \sim HRa^4 \tag{6}$$

follow from Eq. (1a) in Raihi.[9] The assumption that the inertia terms in Eq. (1a) in Raihi[9] can be, at most, as large as the viscous terms in the same equation together with Eq. (6) imply that:

$$Pr \geq O(HRa^4) \tag{7}$$

This is the range under which the analysis of the present study is valid and is classified as the negligible inertia domain.[7]

By comparing of the magnetic field and buoyancy terms in the momentum equation under the assumptions that $b_z \leq O(1)$ and the nonlinear induced magnetic field term $\boldsymbol{b} \cdot \nabla \boldsymbol{b}$ is, at most, as large as the corresponding one due to external imposed field, we find (3b) and,

$$b_z \sim R\tau/Q, \quad b_r \sim aR\tau/Q, \quad h \sim a^2 R\tau/Q. \tag{8}$$

provided the hydrostatic condition is assumed in the mushy layer to the leading order term.[14] We designate $\theta_1(r, z)$ as the deviation of θ from θ_0. From Eq. (1) in Riahi[9] and the condition:

$$1/a^2 << HR << 1/a^4, \tag{9}$$

we find $\theta_1 << 1$. Using these results we simplify Eq. (1c) in Riahi [9], integrate it in r from $r = 0$ to $r = a$ and follow Worster[13] to get:

$$\theta_1 \sim \psi a \theta_0' 1nr, \tag{10a}$$

where

$$2\pi\psi a = \int_0^a 2\pi rwdr \tag{10b}$$

is the vertical flux in the chimney.

Using Eq. (3a) and designating $b_{z1}(r, z)$ to be the deviation of b_z from b_{z0}, then the simplified form of Eq. (1b) implies $b_{z1} << 1$. Integrating the simplified form of Eq. (1a) in r from $r = 0$ to $r = a$, we find:

$$b_{z1} \sim -r\psi_a' lnr. \tag{11}$$

Using Eq. (5), we find:

$$u \sim -\psi_a'/r \quad \text{as} \quad r \to a, \tag{12a}$$

$$b_r \sim -h_a'/r \quad \text{as} \quad r \to a, \tag{12b}$$

where

$$2\pi h_a = \int_0^a 2\pi r b_z dr \tag{12c}$$

is the vertical magnetic flux in the chimney.

Using the momentum equation for the mushy layer and Eq. (12), we find:

$$\Delta P \sim \psi_a' lna/R + Aa^2\theta_0/R + [Q/(\tau R)][(h_a')^2/(2a^2) - (1na)(1 + b_{z0})h_a''], \tag{13}$$

where Δp represents the radial pressure difference near the wall of the chimney. Using Eqs. (10) and (11), we find that the momentum equation in the mushy layer yields:

$$w \sim -R\psi a 1na[1 - Az\sin^2\gamma/R + \psi_a'' Q/(\psi a R)]. \tag{14}$$

This result holds near the wall of the chimney and implies the following condition in an average sense:

$$Q \leq O[\min(R, R/\lambda_1)], \tag{15a}$$

where

$$\lambda_1 \equiv |\langle \psi_a'' \rangle / \langle \psi a \rangle|, \tag{15b}$$

which is also a range of validity of the present analysis. Here, an angular bracket denotes a vertical average from $z = 0$ to the top of the chimney at $z = z_0$.

Following Worster[14] and Riahi,[16] we find that the scaled volume flux ψ_a in the chimney, due to the upward flow, is satisfied, to the leading terms, by the following equation:

$$\frac{\partial^2 \psi_a}{\partial z^2} + E^2 \psi_a = -16\lambda_2 R(1+\theta_0)(1 - Az\sin^2\gamma/R)/(Qlna), \tag{16a}$$

$$E \equiv 4/[a^2(-HQlna)^{1/2}], \tag{16b}$$

where λ_2 is a positive constant of order one. The condition $\psi_a' > O(\psi_a)$ then implies the following restriction on the parameter regime:

$$a^4 HQ|lna| < O(1), \tag{17}$$

which is assumed to be valid here. The solution to Eq. (16a), subjected to the initial conditions

$$\psi_a = \frac{\partial \psi_a}{\partial z} + 16\lambda_2 RB_0[1 - z_0 A\sin^2\gamma/(2R)]/(Qlna) = 0 \quad \text{at} \quad z = 0, \tag{18}$$

is found by standard methods[17] to be:

$$\psi_a = \left(\frac{-16R\lambda_2}{Qlna}\right)\{\ [1 - z_0 A\sin^2\gamma/(2R)]B_0 z + \left(\frac{1}{E}\right)\int_0^z \sin(Ez - E\eta) \\ [(1+\theta_0(\eta)).(1 - A\eta\sin^2\gamma/R) - E^2 B_0\left(1 - z_0\frac{A\sin^2\gamma}{2R}\right)\eta]d\eta\ \}, \tag{19}$$

where B_0 is a positive constant.

The leading term w_0 of the axial flow velocity in the mushy zone and outside the chimneys is given by: [13,16]

$$w_0 = -2\pi\psi_a N, \tag{20}$$

where N is the number density of all the chimneys in the mushy zone. This result is based on the principle of mass conservation that the downward flow through the mushy zone and outside the chimneys must be equal to the total upflow through all the chimneys per unit horizontal area.

The analysis in this section yields the following main results for a prescribed and small value of the chimney's radius a. Under the condition of Eq. (7), all the results were found to be independent of Pr to the leading terms. The present results are also independent of τ to the leading terms. Rotational effects are negligible for a non-inclined rotation where $\sin\gamma = 0$ or for an inclined rotation where $|\sin\gamma|$ is sufficiently small. For a weak rotation, $A << R$, rotational effects are negligible, and the volume flux in the chimney decreases with increasing Q. However, the volume flux increases with both R and H. For a moderate rotation,

$$O(1) = A\sin^2\gamma/R < 1, \tag{21}$$

the volume flux decreases with increasing A and Q. For a strong rotation,

$$A\sin^2\gamma/R > O(1), \tag{22}$$

the volume flux increases with A, but it still decreases with increasing Q. The variations of u given by Eq. (12a), and w given by Eq. (14), with respect to A, R, H and Q are similar, in general, to the way ψ_a varies with respect to these parameters.

CONCLUSIONS

(i) Chimney convection is independent of Pr for moderate or large values of Pr.
(ii) Chimney convection is independent of τ for small or moderate values of τ.
(iii) Rotational effects are negligible for the non-inclined rotation case where the axis of rotation is parallel to the direction of the solidification velocity.
(iv) For strong magnetic field, chimney convection decreases strongly with increasing Q regardless of the value of the rate of rotation.
(v) Rotational effects are negligible for weak rotation.
(vi) Chimney convection decreases with increasing A for moderate rotation.
(vii) Chimney convection increases with A for strong rotation.
(viii) Chimney convection generally increases with either R or H.

REFERENCES

1. P.J. Prescott and F.P. Incropera, Boinary solid-liquid phase change with fluid flow, in: *Advances in Transport Processes IX*, a.S. Mujumdar and R.A. Mashelkar, eds., elsevier, Science Publishers, Amsterdam (1993), p 57.
2. A.K. Sample and A. Hellawell, The effect of mold precession on channel and macrosegregation in ammonium chloride-water analog casting, *Met. Trans. B* 13:495 (1982).
3. A.K. Sample and A. Hellawell, The mechanisms of formation and prevention of channel segregation during alloy solidification, *Met. Trans. A* 15:2163 (1984).
4. S. Kou, D.R. Poirier and M.C. Flemings, Macrosegregation in rotated remelted ingots, *Met. Trans.B* 9:711 (1978).
5. D.N. Riahi, Effects of centrifugal and Coriolis forces on chimney convection during alloy solidification, *J. Crystal Growth* 179:287 (1997).
6. D.N. Riahi, Effects of rotation on a nonaxisymmetric chimney convection during alloy solidification, *J. Crystal Growth* 204:382 (1999).
7. D.N. Riahi, Effects of a vertical magnetic field on chimney convection in a mushy layer, *J. Crystal Growth*, in press (2000).
8. D.N. Riahi, Effect of high gravity on freckle formation and convection in a mushy layer in: *Centrifugal Materials Processing* L.L. Regel and W.R. Wilcox, eds., Plenum Publishing Corp. (1997), p 169
9. D.N. Riahi, Non-axisymmetric chimney convection in a mushy layer under a high gravity environment, in present volume.
10. S. Chandrasekhar, *Hydrodynamic and Hydromagnetic Stability*, Oxford University Press (1961).
11. D.N. Riahi and T.L. Sayre, Effects of rotation on the structure of a convecting mushy layer, *Acta Mechanica* 118:109 (1996).
12. W.A. Arnold, W.R. Wilcox, F. Carlson, A. Chait, and L.L. Regel, Transport modes during crystal growth in a centrifuge, *J. Crystal Growth* 119:24 (1992).
13. A. C. Fowler, The formation of freckles in binary alloy, *IMA J. Appl. Math.* 35:159 (1985).
14. M.G. Worster, Natural convection in a mushy layer, *J. Fluid Mech.* 224:335 (1991).
15. D.N. Riahi, Stability of the melt during the solidification of a binary alloy with a vertical magnetic field, *J. Math. Phys. Sci.* 26:429 (1992).
16. D.N. Riahi, On the structure of an unsteady convecting mushy layer, *Acta Mechanica* 127:83 (1998).
17. I. Stakgold, *Boundary Value Problem of Mathematical Physics*, Volume I, The Macmillan Co., London (1969).

BODY-FORCE-DRIVEN MULTIPLICITY AND STABILITY OF COMBINED FREE AND FORCED CONVECTION IN ROTATING CURVED DUCTS: CENTRIFUGAL FORCE

Liqiu Wang and Tianliang Yang

Department of Mechanical Engineering, The University of Hong Kong, Pokfulam Road, Hong Kong

INTRODUCTION

The present work focuses on the effect of centrifugal force on the bifurcation and stability of flow and heat transfer in curved ducts. The flow geometry is illustrated in Fig.1 with (r, y, θ) as the radial (normal), spanwise and streamwise directions, respectively. A viscous fluid is driven by a streamwise pressure gradient to flow through a square duct with a streamwise curvature and a uniform wall heat flux.

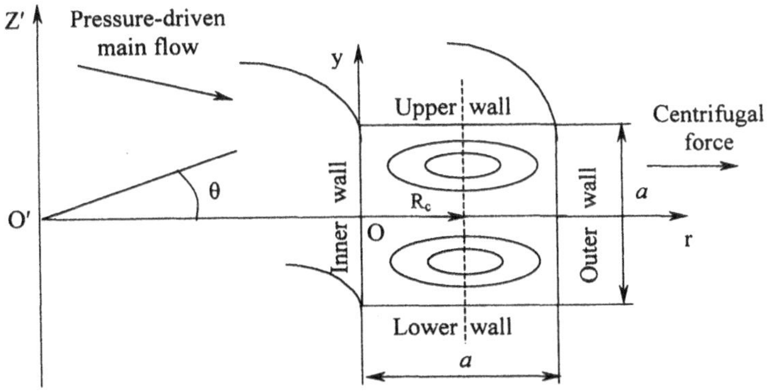

Figure 1. Physical problem and coordinate system

Centrifugal-force-driven flows are encountered in various practical processes. Flow and heat transfer in centrifugal materials processing, separation processes, heat exchangers and physiological systems represent some examples. Therefore, flow and heat transfer in the presence of centrifugal forces have drawn sustained attention. Dean[1] first formulated the problem in mathematical form under fully developed flow conditions, which is now

Processing by Centrifugation, Edited by Regel and Wilcox
Kluwer Academic/Plenum Publishers, New York, 2001

called the Dean problem. He demonstrated the existence of a pair of counter-rotating vortices as a secondary flow in a curved pipe. Subsequently, the Dean problem has been studied extensively. A full review can be found in Berger *et al.*,[2] Nandakumar and Masliya,[3] Ito and Nanbu,[4] and Berger.[5]

As the present work addresses the multi-solution structure and the stability of the flow and heat transfer inside a square curved channel, our brief literature review is mainly on the bifurcation and stability of the Dean problem. The Dean problem is characterized by a dimensionless parameter called the Dean number *De*, defined as a combination of the Reynolds number *Re* and the curvature ratio σ of the duct . It is used to measure the importance of the production of inertial and centrifugal forces relative to viscous forces. In a certain range of Dean numbers, the inertial, centrifugal and viscous forces are of comparable magnitude. Nonlinear effects can thus be very strong and a bifurcation appears. Cheng and Akiyama[6] first mentioned the multi-solution phenomenon in the Dean problem. They examined laminar-forced convection in a loosely coiled curved rectangular duct and found that there are dual solutions. Joseph *et al.*[7] predicted a four-vortex secondary flow structure in a curved square duct. Masliyah[8] found the same solution structure for a curved semicircular duct with flat outer wall. Nandakumar and Masliyah[9] studied dual solutions numerically. The two-vortex and four-vortex solutions were found to coexist irrespective of the tube's shape. The stability of the solution branches was, however, not fully addressed. Winters[10] presented a rather comprehensive study on the bifurcation structure of two-dimensional solutions to the laminar flow in curved rectangular ducts. The two-dimensional four-cell flow was found to be unstable. The symmetry-breaking bifurcation points and the asymmetric solutions were also computed. However, the calculation by Winters[10] was made only over a relatively small range of Dean numbers. Furthermore, the linear stability studied by Winters[10] was inconclusive. Daskopoulos and Lenhoff[11] conducted a flow bifurcation and stability study in the half region of the duct under the assumption of symmetry. The solution structure obtained was much more complicated than that by Winters.[10] In addition to the two-vortex and four-vortex flows, they found that symmetric six-vortex and eight-vortex flows co-exist in the range with a higher value of Dean number. The linear stability analysis made by Daskopoulos and Lenhoff[11] showed that the two-vortex, four-vortex and even six-vortex states could be stable to symmetric perturbations.

While a linear stability analysis is efficient in terms of the computation efforts required, it suffers two fundamental defects. First, it is not applicable to a finite disturbance. With a finite disturbance, a solution that is stable when based on linear stability analysis may not be always stable. Second, it provides no answer to questions related to the dynamic behavior of solutions, including how flows approach a stable solution after a disturbance, what happens to an unstable solution after a disturbance, whether all unstable solutions at a given set of parameters respond to disturbances in the same way, and whether the disturbances transform an unstable solution to a stable one at the same parameter value. A fully transient computation is necessary to examine the dynamic response of multiple solutions to finite random disturbances.

The bifurcation and stability of multiple solutions available in the literature are limited to Dean numbers below 250. However, solutions and their stability at high Dean numbers are more relevant for practical applications. Upon increasing the Dean number, a richer bifurcation structure and more intricate stability features are expected because of the stronger nonlinearity. New limit/bifurcation points and solution branches are expected. Because of the lack of a solution structure at high Dean numbers, there is a long-standing controversy over such solutions obtained by different methods that aimed at obtaining a solution without considering the multiplicity. To resolve this controversy, we need to find the bifurcation structure at high Dean numbers. Furthermore, it is also reasonable on physical grounds to expect phenomena related to a transition to turbulence at high Dean numbers, such as oscillations, period doubling, intermittency, and chaotic oscillation. These observations stimulated us to extend the previous works to higher Dean numbers and to examine the dynamic response of multiple solutions to finite random disturbances by direct transient computations.

The present work is a relatively comprehensive study of the bifurcation structure and stability of multiple solutions for laminar forced convection in a curved duct of square cross-section. The governing differential equations in primitive variables were solved for detailed bifurcation structure by a finite-volume/Euler-Newton continuation method with the help of a bifurcation test function, a branch switching technique, and parameterization of arc-length or local variable. Transient calculations were made to examine in detail the response of every solution family to finite random disturbances.

GOVERNING EQUATIONS AND NUMERICAL ALGORITHM

Governing Equations

We considered hydrodynamically and thermally fully developed laminar flow and heat transfer of a viscous fluid in a curved square duct with a uniform wall temperature at every streamwise position (Fig. 1). The properties of the fluid were taken to be constant. The flow is driven by a pressure gradient $\partial P/\partial\theta$ in the stream-wise direction. The velocity components are U, V and W in the directions of r, y and θ, respectively. The temperature of the fluid is T and T_w is the wall temperature. The streamwise temperature gradient is $\partial T/\partial\theta$. Under the assumption of both hydrodynamically and thermally fully developed flow, both $\partial P/\partial\theta$ and $\partial T/\partial\theta$ are constant. The continuity, momentum and energy equations governing the two-dimensional flow through a curved duct are non-dimensionalized by introducing the following dimensionless variables:[12-14]

$$u=\frac{U}{W_l}\qquad v=\frac{V}{W_l}\qquad w=\frac{W}{W_l}\qquad R=\frac{r}{a}\qquad Y=\frac{y}{a}$$

$$\Theta=\frac{R_c(T_w-T)}{Pr\,a\partial T/\partial\varphi}\qquad p=\frac{P}{\rho(\nu/a)^2}\tag{1}$$

where $W_1=-\dfrac{a^2}{R_c\rho\nu}\dfrac{\partial P}{\partial\theta}$ is a representative axial velocity; ρ, ν and Pr are the density, kinematic viscosity, and Prandtl number of the fluid; P is a pseudo-pressure that combines pressure, gravitational, and centrifugal forces;[12,15,16] a is the duct width/height; and R_c is the radius of curvature of the duct. This leads to the continuity equation:[12,17]

$$\frac{\partial}{\partial R}\{\{1+\sigma(R-0.5)\}u\}+\frac{\partial}{\partial Y}\{\{1+\sigma(R-0.5)\}v\}=0\tag{2}$$

The momentum equations are:

$$\frac{\partial u}{\partial \tau}+u\frac{\partial u}{\partial R}+\mathrm{v}\frac{\partial u}{\partial Y}-\frac{16\cdot Dk^2\cdot w^2}{\sigma[1+\sigma(R-0.5)]}=$$
$$-\frac{\partial p}{\partial R}+\{\frac{\partial^2 u}{\partial R^2}+\frac{\partial^2 u}{\partial Y^2}+\frac{\sigma}{1+\sigma(R-0.5)}\frac{\partial u}{\partial R}-\frac{\sigma^2 u}{[1+\sigma(R-0.5)]^2}\} \tag{3}$$

$$\frac{\partial \mathrm{v}}{\partial \tau}+u\frac{\partial \mathrm{v}}{\partial R}+\mathrm{v}\frac{\partial \mathrm{v}}{\partial Y}=-\frac{\partial p}{\partial Y}+[\frac{\partial^2 \mathrm{v}}{\partial R^2}+\frac{\partial^2 \mathrm{v}}{\partial Y^2}+\frac{\sigma}{1+\sigma(\mathrm{R}-0.5)}\frac{\partial \mathrm{v}}{\partial R}] \tag{4}$$

$$\frac{\partial w}{\partial \tau}+u\frac{\partial w}{\partial R}+\mathrm{v}\frac{\partial w}{\partial Y}+\frac{\sigma u w}{1+\sigma(R-0.5)}=\frac{1}{1+\sigma(R-0.5)}+$$
$$\{\frac{\partial^2 w}{\partial R^2}+\frac{\partial^2 w}{\partial Y^2}+\frac{\sigma}{1+\sigma(R-0.5)}\frac{\partial w}{\partial R}-\frac{\sigma^2 w}{[1+\sigma(R-0.5)]^2}\} \tag{5}$$

The energy equation is:

$$\frac{\partial \Theta}{\partial \tau}+u\frac{\partial \Theta}{\partial R}+\mathrm{v}\frac{\partial \Theta}{\partial Y}-\frac{4Dk\cdot w}{\sigma\, Pr[1+\sigma(R-0.5)]}=\frac{1}{Pr}\cdot[\frac{\partial^2 \Theta}{\partial R^2}+\frac{\partial^2 \Theta}{\partial Y^2}+\frac{\sigma}{1+\sigma(R-0.5)}\frac{\partial \Theta}{\partial R}] \tag{6}$$

Here, three dimensionless parameters were defined by:

$$\sigma=\frac{a}{R_c} \qquad Pr=\frac{\nu}{\alpha} \qquad D_k=\frac{\sigma a W_1}{4\nu} \tag{7}$$

It is customary to use the Dean number *De* with W_m as the characteristic velocity for flow in a curved channel. However, the employment of W_m results unavoidably in the appearance of an unknown dimensional parameter in the governing equations. Consequently, an iterative procedure should be applied, assuming some initial estimated values. This requires additional computation time. Therefore, we follow Wang and Cheng[12] in using a pseudo Dean number *Dk* with W_l as the characteristic velocity.

The boundary conditions (non-slip, impermeability and uniform peripheral temperature) may be written, in terms of dimensionless variables, as:

$$\begin{aligned} &u=\mathrm{v}=w=\Theta=0, \quad at \quad R=0,1 \quad for\ -0.5\le Y\le 0.5,\\ &u=\mathrm{v}=w=\Theta=0, \quad at \quad Y=-0.5,\,0.5 \quad for\quad 0\le R\le 1. \end{aligned} \tag{8}$$

In the present work, we fixed the curvature ratio of the duct σ at *0.02* (a typical value used in cooling systems of rotor drums and conductors of electrical generators) and *Pr* at *0.7* (the value for air) to study the effect of *Dk* by changing *Dk* from 0 to *800*.

The Numerical Algorithm for Bifurcation Structure

For the steady bifurcation structure, we removed the time-dependent terms in Eqs. (2)-(6). The governing Eqs. (2)-(6) were then discretized under the boundary conditions (8) by a finite volume method that was an adaptation of that in Patankar.[18] The main features of this method include a staggered mesh system, a power-law formulation for the combined effect of convection and diffusion terms, and a central-difference scheme for source terms.

After this discretization in the flow domain, we obtained a set of nonlinear algebraic equations, the so-called discretization equations, which approximate the governing equations. The individual solution branches were found by Euler/Newton continuation.[19,20] A test function proposed by Seydel[21,22] was used to detect bifurcation points during the continuation. The indirect method of branch switching technique developed by Seydel[21,22] was used to switch branches.

Grid-Dependence Check

Three pairs of grid sizes uniformly distributed in the flow domain were used to check grid dependence; *40×40*, *50×50* and *60×60*. The pair of numbers $L\times K$ represents the number of grid points used in the r- and z-directions, respectively. The bifurcation diagram showed no qualitative difference for all three grid sizes. The quantitative difference was also very small for most of the region of interest from *50×50* to *60×60*. The numerical data are shown in Table 1 for three cases with $Dk=100$, *550* and *800*, all at the primary solution branch S_1. For comparison, Table 1 lists three representative properties, namely, Dean number (De), maximum absolute value of secondary flow stream function ($|\psi|_{max}$) and maximum streamwise velocity (w_{max}), maximum temperature (Θ_{max}), as well as the *CPU* time for one continuation step. The computations were carried out on a *Digital* Personal Workstation *AU600*. The general trend of these results as the grid size was reduced indicates that the solutions for the *50×50* grid were accurate to within *1%*. Table 2 shows five representative properties used to verify the code further, together with the results of Wang and Cheng at $Dk=100$, $\sigma=0.02$ and $Pr=0.7$, where there is only one solution. The results are in good agreement, with less than 2% difference due to the different numerical methods used in the two studies. In order to have a balance between the cost of computer time and the accuracy of the solution, we carried out all computations with a *50×50* uniform mesh for square ducts.

Table 1. Variation of De, $|\psi|_{max}$, w_{max}, Θ_{max} and *CPU* time for different grids with $\sigma = 0.02$ and $Pr = 0.7$.

Dk	*Grid*	*De*	$\lvert\psi\rvert_{max}$	w_{max}	Θ_{max}	*CPU(s)*
100	40×40	76.5	5.488	0.04938	39.6	72.2
	50×50	76.5	5.528	0.04939	39.7	157.1
	60×60	76.5	5.546	0.04945	39.7	296.4
550	40×40	272.8	12.095	0.02867	78.7	72.4
	50×50	273.8	12.203	0.02884	79.4	158.2
	60×60	274.3	12.246	0.02891	79.9	310.7
800	40×40	362.8	14.622	0.02597	94.1	73.1
	50×50	364.6	14.787	0.02615	95.2	156.4
	60×60	365.6	14.857	0.02625	95.8	291.5

Table 2. Comparison of five representative results for $Dk=100$, $\sigma=0.02$, and $Pr=0.7$ obtained here with those by Wang and Cheng.[12]

Source	*Re*	*De*	$\lvert\psi\rvert_{max}$	w_{max}	Θ_{max}
Present work	541.0	76.5	5.528	0.0494	39.7
Wang & Cheng[12]	542.0	76.6	5.641	0.0496	39.9

Stability of the Solutions

The dynamic response of the steady solutions to finite random disturbances was determined. The governing equations with time-dependent terms were discretized using the finite volume method. The fully implicit method was used because of its superior numerical stability. Then the discretized equations were solved by the Euler/Newton continuation with τ as the continuation parameter. The initial condition at $\tau=0$, which also served as the starting point of the continuation scheme, was formed by the steady solution plus a finite random disturbance. The random disturbance was generated by the computer in the form of a vector of random values ranging from -1 to +1. The number of components of the vector equals the number of components of the steady solution vector of the discretized equations. The random disturbance was then formed by multiplying the random vector by the maximal percentage of disturbing value over the steady value (*4%*, *10%* and *15%* in the present study).

RESULTS AND DISCUSSION

Solution Structure and Flow Pattern

The solution structure is shown in Fig. 2 using the u-component of velocity at (0.9,0.64) as a scalar measure of the solution and Dk as the characteristic parameter. This spatial location was chosen because the flow profile changes most significantly in this region and both symmetric and asymmetric branches are made visible. Since the diagram is only a one-dimensional projection of the solution branches, some branches may appear to intersect with other branches at some points, but, in fact, they have no connection at all. In Fig. 2, the symmetric solution branches are named in ascending numerical order as S_n. The asymmetric solution branches are denoted in ascending numerical order as A_n. Six solution-branches were found. On each solution branch, there is at least one limit point. The limit points divide the branch into several sub-branches. Each sub-branch is labeled in ascending numerical order as $S_{n\text{-}m}$ or $A_{n\text{-}m}$, respectively.

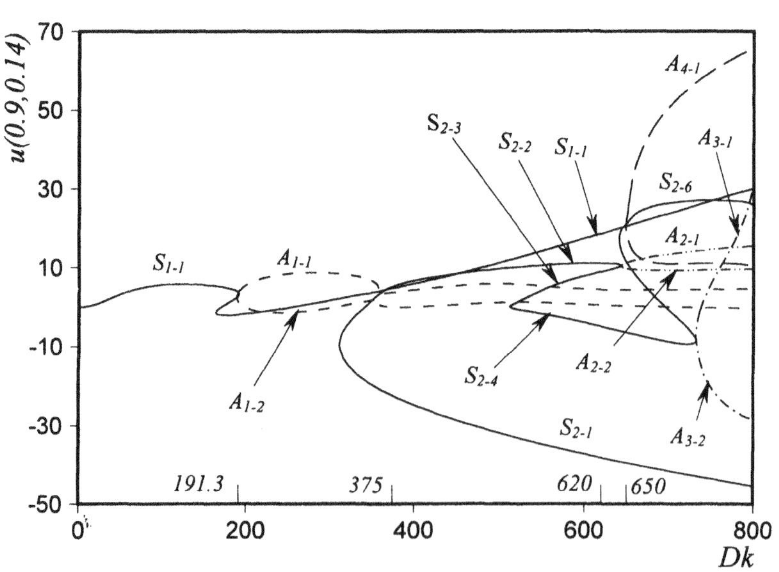

Figure 2. Solution branches for $\sigma = 0.02$ and $Pr = 0.7$.

Among the six solution branches, S_1, S_2 and A_1 were first reported by Winters.[10] Here, we confirmed and extended their findings up to $Dk=800$. While no new limit and bifurcation points were found along S_1 and A_1, four additional limit points[11] and three new symmetry-breaking bifurcation points were detected along S_2. In particular, three symmetry-breaking bifurcation points lead to three pairs of asymmetric solution branches, named A_2, A_3 and A_4, which have not been reported in the literature.

The secondary flows on various sub-branches are typified in Fig. 3. On $S_{1\text{-}m}$ ($m=1, 2, 3$), three solutions are the two-cell state (two Ekman vortices, Fig. 3(a)), the weak four-cell state with two Ekman vortices and two weak Dean vortices (Fig. 3(b)), and the four-cell state with two Ekman vortices and two Dean vortices (Fig. 3(c)), respectively. Figure 3(d) illustrates the secondary flow pattern at $Dk=300$ on the sub-branch $A_{1\text{-}1}$. Flows on $A_{1\text{-}1}$ are essentially an asymmetric 2-cell state. The vortex in the lower part of the duct stretches to the upper part near the outer wall. The solutions on $A_{1\text{-}2}$ can be formed by mirror images of the corresponding solutions on $A_{1\text{-}1}$ at the same Dk.

Figures 3(e) to 3(j) show some typical solutions for 6 sub-branches of S_2. The flows on $S_{2\text{-}1}$ are also four-cell states (Fig. 3(e)). However, this 4-cell structure differs from the one on $S_{1\text{-}3}$ (Fig.3(c)) mainly by the shape and size of the Dean vortices. Two Dean vortices stretch along the span direction rather than the radial direction. The spanwise distance between centers of two Dean vortices is noticeably larger than that of the 4-cell flow on $S_{1\text{-}3}$. The flows on $S_{2\text{-}2}$ (Fig. 3(f)) are a 2-cell state, which is qualitatively similar to that on $S_{1\text{-}1}$ but with a stronger secondary flow. The flows on $S_{2\text{-}3}$ (Fig. 3(g)) are a weakly 4-cell state with a pair of very weak Dean vortices. On the sub-ranches $S_{2\text{-}4}$, $S_{2\text{-}5}$ and $S_{2\text{-}6}$, the flow structure becomes complex. On $S_{2\text{-}4}$, we have 6-cell states with two pairs of Dean vortices along the outer wall (Fig. 3 (h)). The second pair appears because of the splitting of the original pair. It would be interesting to study, in the future, the relation between this vortex splitting and that due to the Eckhaus instability.[22] Flows on $S_{2\text{-}5}$ and $S_{2\text{-}6}$ are a 8-cell state with three pairs of Dean vortices (Fig. 3(i),(j)). The third pair is formed from the outer wall. This differs from the mechanism responsible for the appearance of the second pair.

The solution branch S_2 also has three symmetry-breaking bifurcation points, originating three asymmetric solution branches A_2, A_3 and A_4, respectively (Fig. 2). Each of these branches has one limit point dividing the branch into upper sub-branches $A_{2\text{-}1}$, $A_{3\text{-}1}$ or $A_{4\text{-}1}$ and lower sub-branches $A_{2\text{-}2}$, $A_{3\text{-}2}$ or $A_{4\text{-}2}$. Solutions on the lower sub-branches are the mirror images of the corresponding solutions on the upper sub-branches at the same Dk. Figures 3(k) to (m) detail flow structures on these three branches. The flows on the $A_{2\text{-}1}$ are an asymmetric 2-cell state (Fig. 3(k)) while they are asymmetric 7-cell structures on both $A_{3\text{-}1}$ and $A_{4\text{-}1}$. Figure 3(l) typifies the secondary flow pattern on $A_{3\text{-}1}$.

Friction Factor

For engineering applications, the friction factor is very important. Following the usual definition, the expression for friction factor can be written based on the local streamwise velocity gradient at the wall as:

$$f\,Re = \frac{2}{w_m}\left(\frac{\partial w}{\partial n}\right)_{wall} \tag{7}$$

Figure 4 shows the variation *fRe* on various solution branches as *Dk*. In the figure, *fRe* is shown on the basis of its value for the straight channel ($fRe_0 = 14.23$) to facilitate the comparison. It is observed that different solution branches have different *fRe* even for the same value of *Dk*.

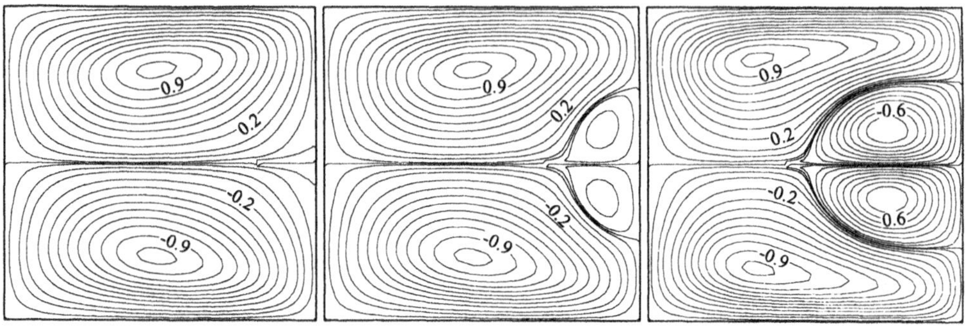

(a) $Dk=180$ on $S_{1\text{-}1}$; $|\psi|_{max}=7.440$ (b) $Dk=180$ on $S_{1\text{-}2}$; $|\psi|_{max}=7.059$ (c) $Dk=550$ on $S_{1\text{-}3}$; $|\psi|_{max}=12.20$

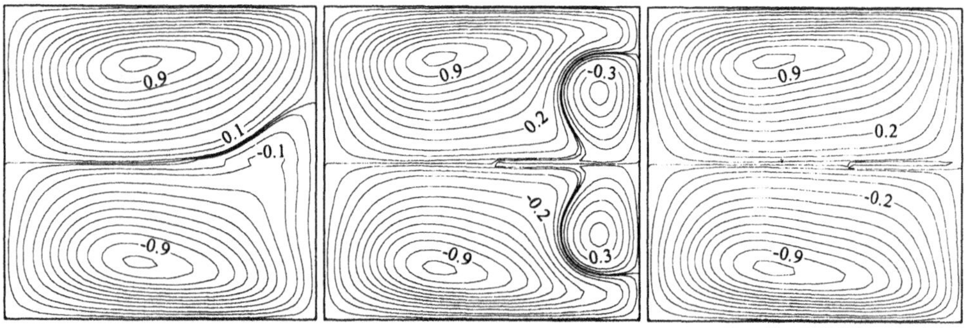

(d) $Dk=300$ on $A_{1\text{-}1}$; $|\psi|_{max}=9.601$ (e) $Dk=450$ on $S_{2\text{-}1}$; $|\psi|_{max}=10.35$ (f) $Dk=450$ on $S_{2\text{-}2}$; $|\psi|_{max}=11.70$

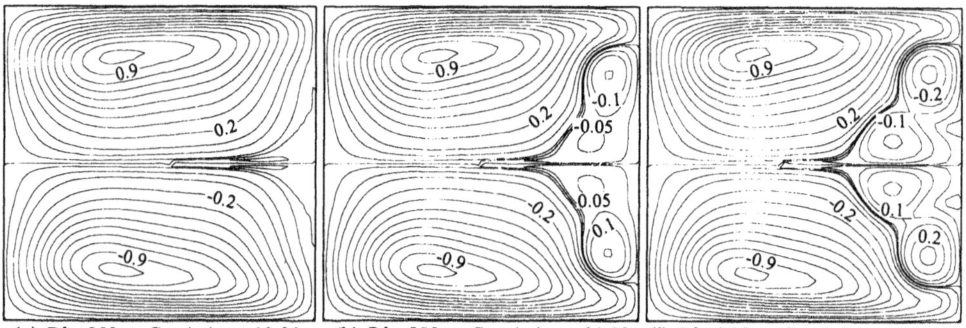

(g) $Dk=550$ on $S_{2\text{-}3}$; $|\psi|_{max}=12.30$ (h) $Dk=550$ on $S_{2\text{-}4}$; $|\psi|_{max}=11.53$ (i) $Dk=700$ on $S_{2\text{-}5}$; $|\psi|_{max}=12.68$

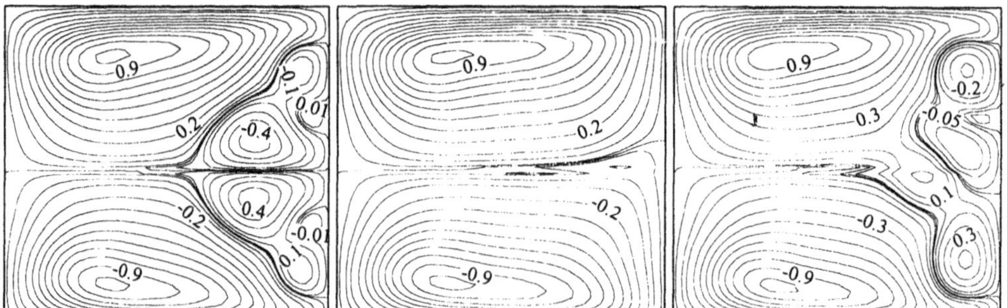

(j) $Dk=700$ on $S_{2\text{-}6}$; $|\psi|_{max}=13.00$ (k) $Dk=700$ on $S_{2\text{-}4}$; $|\psi|_{max}=14.02$ (l) $Dk=800$ on $S_{3\text{-}1}$; $|\psi|_{max}=13.48$

Figure 3. Typical secondary flow patterns on various solution sub-branches.

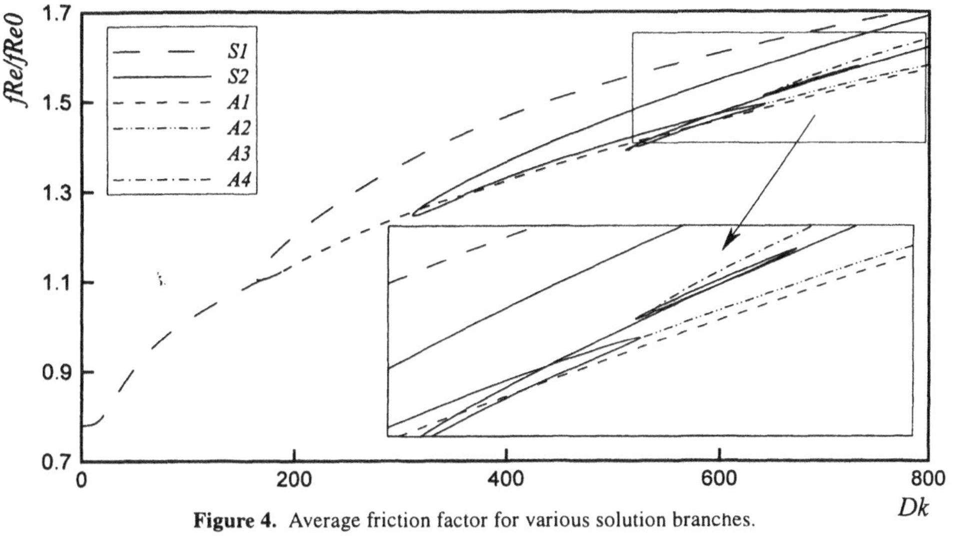

Figure 4. Average friction factor for various solution branches.

Stability of Solutions

Because there has been no study of dynamic responses of multiple solutions to finite random disturbances in the literature, a relatively comprehensive transient computation was made to examine the dynamic behavior. It was found that the final dynamic evolution after a short transient temporal period is independent of the initial disturbances for all solutions in the region $0 \leq Dk \leq 620$. At any fixed value of *Dk* in the range $0 \leq Dk \leq 620$, all steady solutions develop, after initial finite random disturbances, to the same final state, i.e., there is no co-existence of two or more stable states in this range within the scope of the present study. The stability of solutions on the sub-branch $S_{2\text{-}2}$ changes as *Dk* increases even without passing any bifurcation or limit point. In particular, the sub-branch is unstable in the range of $310.7 \leq Dk \leq 375$, stable in the range $375 < Dk \leq 620$, and unstable again if $Dk > 620$.

Five sub-ranges are identified with each having a distinct dynamic response to finite random disturbances. The first sub-range is for *Dk* from *0* to *191.3*. The stable solutions are the steady solutions on $S_{1\text{-}1}$. The final dynamic evolution at any fixed *Dk* is a 2-cell steady state on $S_{1\text{-}1}$ with the same *Dk*. Figure 5 shows the evolution process of the solution on $S_{1\text{-}2}$ for $Dk=180$. In this figure, the deviation of velocity components from their initial steady values is plotted against time τ at (*0.9, 0.14*), (*0.94, 0.1*) and (*0.96, 0.06*) for $Dk=180$. Both *u*- and *v*-velocity components are plotted for the first point (*0.9, 0.14*) while only *u*-velocity components are shown for the last two points. To facilitate the comparison, we use these four velocity components (either velocity itself or difference of the velocity from its initial steady value) in all figures that illustrate the dynamic response of the multiple solutions to finite random disturbances. The characteristic of the evolution is that four velocity components become constant and the perturbations vanish after a short period of time. Therefore, the steady solution is reached. The flow patterns and temperature profile of the transient solutions return to those on $S_{1\text{-}1}$.

The second sub-range is $191.3 \leq Dk \leq 375$ in which all steady solutions evolve to a unique temporal periodic solution at same value of *Dk*. The periodic feature of the dynamic evolution process is shown in Fig. 6 for $Dk=350$ on $S_{2\text{-}2}$. The period of the solution is about *0.14*.

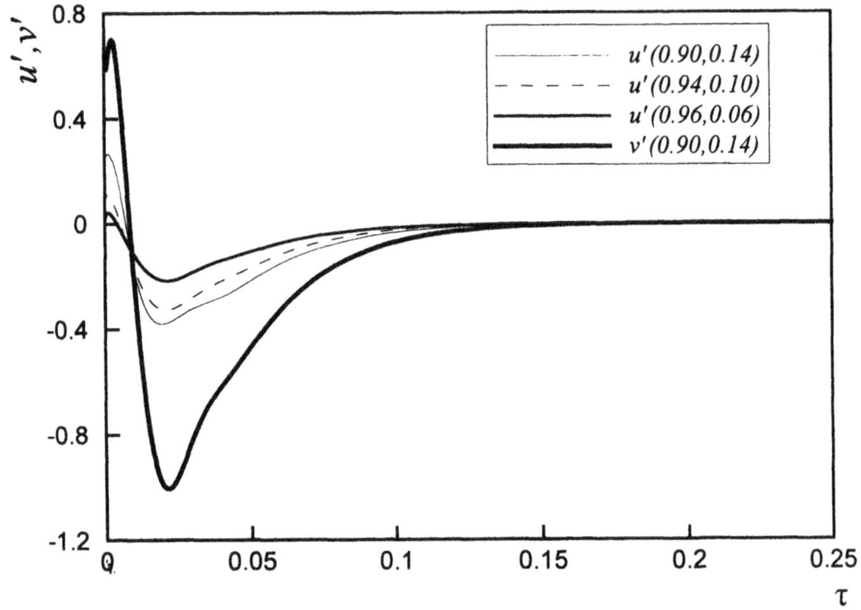

Figure 5. Dynamic response of solution *on* $S_{1\text{-}1}$ at Dk = *180* to finite random disturbances: evolution to original stable steady 2-cell state on $S_{1\text{-}1}$.

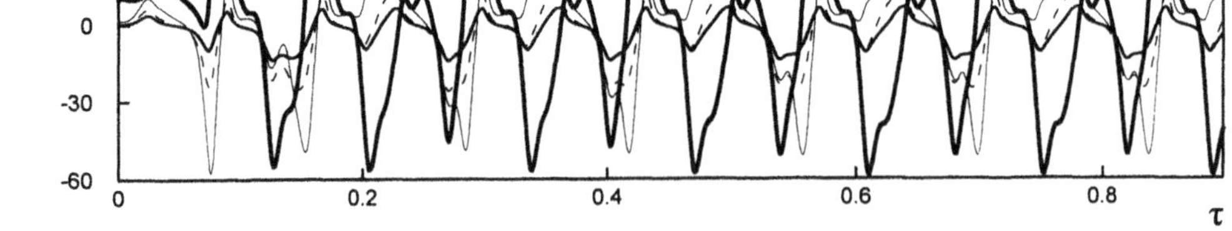

Figure 6. Dynamic response of solutions at Dk=*350* on $S_{2\text{-}2}$ to finite random disturbances: periodic oscillation (period=0.14)

The third sub-range is $375<Dk\leq620$, in which finite random disturbances lead all solutions to a 2-cell steady state. The stable solution is on $S_{2\text{-}2}$. Figure 7 shows the evolution of the transient solution on sub-branch $A_{1\text{-}1}$ at $Dk=380$. It is confirmed that the transient solution develops to steady solutions on $S_{2\text{-}2}$ at $Dk=380$. Figure 7 also shows the manner of evolution of unstable solutions to the stable steady solutions in this range. If there is a small perturbation, the unstable solution jumps to somewhere near the corresponding stable solution within a short period of time. Then it oscillates around the stable steady solution and finally reaches a stable solution.

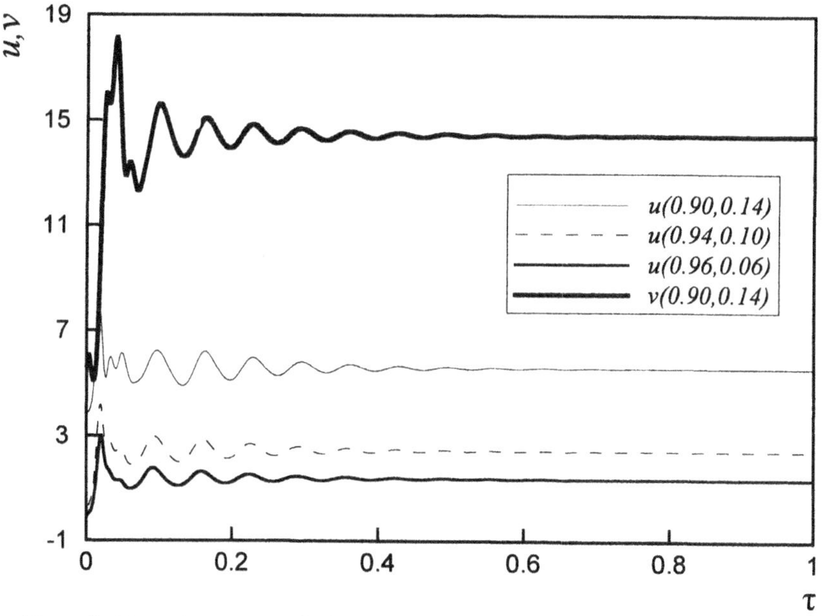

Figure 7. Dynamic response of solution at $Dk=380$ on $A_{1\text{-}1}$ to finite random disturbances: evolution to stable steady 2-cell state on $S_{2\text{-}2}$

The fourth sub-range is $620<Dk\leq650$ where the solutions respond to finite random disturbances in the form of a periodic oscillation with intermittency. Figure 8 shows a typical transient solution at $Dk=638$ on $S_{2\text{-}2}$. Most of the time, the transient solution oscillates periodically around the steady solution on $S_{2\text{-}2}$. The amplitude of the periodic oscillation increases with τ. When the amplitude reaches a certain value, the oscillation cannot be periodic anymore, and a burst starts. After a short period of time, the transient solution returns to periodic oscillation with relative smaller amplitude and another new periodic phase starts. As intermittency is one of basic ways for the onset of chaos,[21,23] the onset of chaotic flow and heat transfer in the curved duct is believed to be by intermittency. In the intermittent flow range, the number of bursts grows with increasing Dk, and the flow

becomes more chaotic. If *Dk* is high enough, the flow in the curved channel becomes completely chaotic.

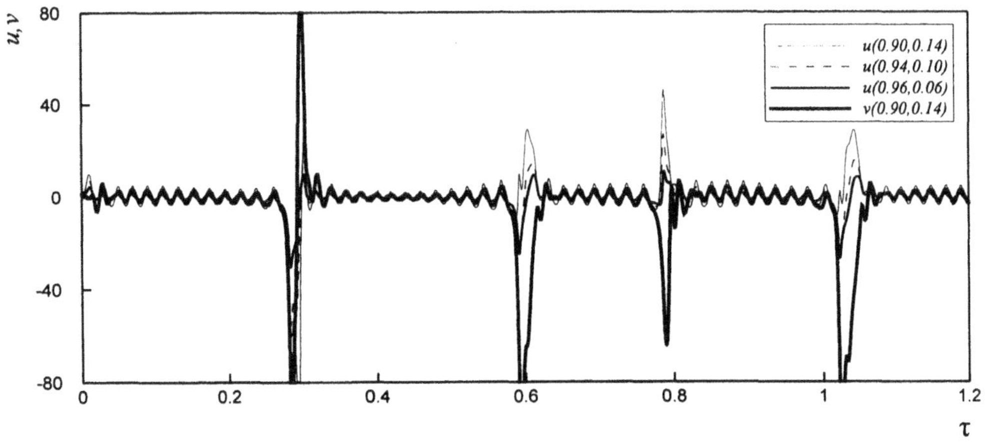

Figure 8. Dynamic response of solution at $Dk=638$ on $S_{2\text{-}2}$ to finite random disturbances: intermittency.

In the last sub-range, $Dk>650$, finite random disturbances convert the solutions to chaotically oscillating states. A typical chaotic transient solution is shown in Fig. 9. The steady solution at $D_k=700$ on $S_{2\text{-}6}$ plus a small perturbation is used as the initial condition to obtain the transient solution in Fig. 9. In this figure, intermittent bursts are still visible. Unlike the transient solutions in the intermittency region, however, the transient solution is not periodic anymore during the time period between two bursts. The flow oscillates irregularly.

CONCLUDING REMARKS

The governing equations were discretized by the finite volume method. The Euler-Newton continuation was used to solve the resulting system of nonlinear algebraic equations. An indirect method was employed to find the bifurcation points and to switch branches. The present work focussed on the bifurcation structure, the flow structure, and the stability of various solution branches for $\sigma=0.02$ and $Pr=0.7$.

Up to 6 solution branches were found within the *Dk* range of *0-800*. Three of these are new. At any fixed value of *Dk* in the range $0\leq Dk\leq 620$, all steady solutions develop, after the initial random disturbances, to the same final state. Finite random disturbances lead the steady solutions to a 2-cell stable state on $S_{1\text{-}1}$ in $0\leq Dk\leq 191.27$, periodic oscillation in $191.27<Dk\leq 375$, another 2-cell stable steady state on $S_{2\text{-}2}$ in $375<Dk\leq 620$, an intermittent oscillation in $620<Dk\leq 650$, and a chaotic oscillation in $650<Dk\leq 800$.

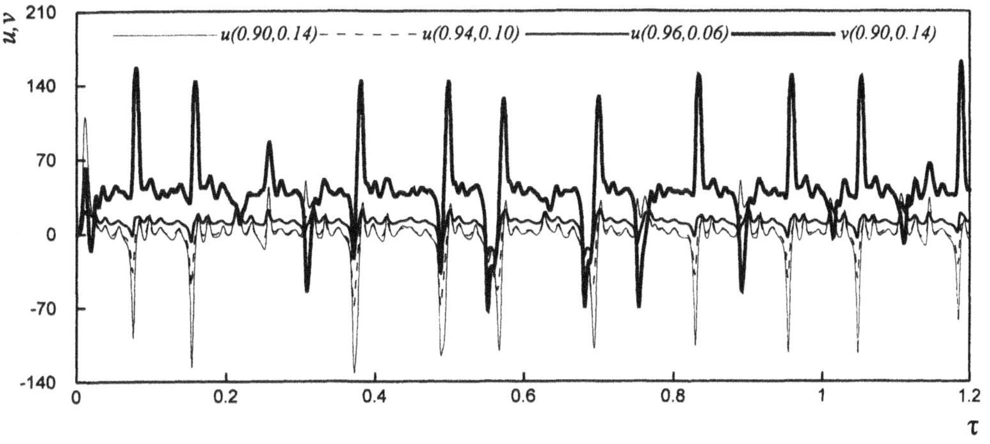

Figure 9. Dynamic response of the solution at $Dk=700$ on $S_{2\text{-}6}$ to finite random disturbances: chaotic oscillation.

Acknowledgement

This work was supported by a grant from the Research Grants Council of the Hong Kong Special Administrative Region, China (Project No. HKU7086/00E).

REFERENCES

1. W.R. Dean, The stream-line motion of a fluid in a curved pipe, *Phil. Mag.* 5:673 (1927).
2. S.A. Berger, L. Talbot, and L-S. Yao, Flow in curved pipes, *Ann. Rev. Fluid Mech.* 15:461 (1983).
3. K. Nandakumar and J.H. Masliyah, Swirling flow and heat transfer in coiled and twisted pipes, *Adv. Transport Proc.* 4:49 (1986).
4. H. Ito and K. Nanbu, Flow in rotating straight pipes of circular cross section, *J. Basic Engin.* 383 (1971).
5. S.A. Berger, Flow and heat transfer in curved pipes and tubes, *AIAA* 91-0030 1 (1991).
6. K.C. Cheng and M. Akiyama, Laminar forced convection heat transfer in curved rectangular channels, *Int. J. Heat Mass Transfer* 13:471 (1970).
7. B. Joseph, E.P. Smith and R.J. Adler, Numerical treatment of laminar flow in helically coiled tubes of square cross section, *AIChE J.* 21:965 (1975).
8. J.H. Masliyah, On laminar flow in curved semicircular ducts, *J. Fluid Mech.* 99:469 (1980).
9. K. Nandakumar and J.H. Masliyah, Bifurcation in steady laminar flow through curved tubes, *J. Fluid Mech.* 119:475 (1982).
10. K.H. Winters, A bifurcation study of laminar flow in a curved tube of rectangular cross-section, *J. Fluid Mech.* 180:343 (1987).
11. P. Daskopoulos and A.M. Lenhoff, Flow in curved ducts: bifurcation structure for stationary ducts, *J. Fluid Mech.* 203:125 (1989).
12. L. Wang, and K.C. Cheng, Flow transitions and combined free and forced convective heat transfer in rotating curved channels: The case of positive rotation, *Phys. Fluids* A6:1553 (1996).

13. R.K. Shah and A.L. Londong, Laminar Flow Forced Convection in Ducts, Academic, New York (1978).
14. S.W. Hong, S.M. Morcos and A.E. Bergles, Analytical and experimental results for combined forced and free convection in horizontal tubes, Proceedings of the Fifth International Heat Transfer Conference 3:154 (1974).
15. W.D. Morris, *Heat Transfer and Fluid Flow in Rotating Coolant Channels,* Research Studies Press, Wiley, Chichester (1981) p 27.
16. L. Wang, *Fluid flow and heat transfer in rotating curved channels,* Ph.D. Thesis, Department of Mechanical Engineering, University of Alberta (1995).
17. S.V. Patankar, *Numerical Heat Transfer and Fluid Flow,* Hemisphere Publishing Corporation, New York (1980) p 28.
18. H.B. Keller, Numerical solution of bifurcation and nonlinear eigenvalue problems, in: *Applications of Bifurcation Theory,* Academic Press, New York (1977) p 359.
19. W.C. Rheinboldt, Solution fields of nonlinear equations and continuation methods, *SIAM J. Numer. Anal.* 17:221 (1980).
20. R. Seydel, Branch switching in bifurcation problems for ordinary differential equations, *Numerical Mathematics* 41:91 (1983).
21. R. Seydel, *From Equilibrium to Chaos, Practical Bifurcation and Stability Analysis,* Elsevier Science Publishing Co., Inc., New York (1994) p 109.
22. Y. Guo and W.H. Finlay, Splitting, merging and wavelength selection of vortices in curved and/or rotating channel flow due to Eckhaus instability, *J. Fluid Mech.* 228: 661 (1991).
23. Y. Pomeau and P. Manneville, Intermittent transition to turbulence in dissipative dynamical systems, *Commun. Math. Phys.* 74: 189 (1980).

VISUALIZATION OF FLOWS IN CHANNELS WITH CURVATURE AND ROTATION

Liqiu Wang[1] and K.C. Cheng[2]

[1]The University of Hong Kong, Hong Kong
[2]University of Alberta, Edmonton, Canada T6G 2G8

ABSTRACT

Flows in channels with streamwise curvature and spanwise rotation were visualized in an end-view near the exit of test sections by injecting smoke. Two test sections were used; rectangular channels with aspect ratios of 1 and 10. The focus was on Dean and Coriolis vortices under the effects of secondary instabilities and flows in the region with a low negative rotation speed and a relatively high rotation speed. The results showed that: (1) low negative rotation leads to two unstable regions alternating with two stable regions (in the cross plane) in forms of multiple pairs of vortices; (2) the secondary instabilities cause Dean and Coriolis vortices to oscillate in various forms; (3) the flows at high rotation speeds are controlled by secondary instabilities rather than the primary instability. In particular, the secondary instabilities lead the flows to be unsteady and turbulent, somewhat like bursting flow in turbulent boundary layers.

INTRODUCTION

The physical model is given in the preceding theoretical paper.[1] Under the action of the pressure gradient along the channel axis, a viscous fluid is allowed to flow through a channel of rectangular cross section with width × height of ($a \times b$). The channel is uniformly curved around the axis $o'z'$ (streamwise curvature). At same time, the channel is rotating about that axis with a constant angular velocity Ω (spanwise rotation). The rotation can be *positive* or *negative* in terms of angular velocity vector. If positive, the rotation direction is in the direction of the relative velocity of the fluid inside the channel. When the rotation is negative, however, the rotation direction is in the direction opposite to the relative velocity of the fluid. The radial, spanwise and streamwise directions are r, y and θ, respectively.

The streamwise curvature and spanwise rotation of the channel introduce centrifugal and Coriolis forces in the momentum equations that govern the relative motion of the fluid with respect to the channel. Such body forces can stabilize the channel flow in some flow domains, and destabilize the flow in other domains. In particular, the stabilizing or destabilizing effects due to curvature and rotation may either enhance or counteract each other depending on the direction of the rotation. This can be intuitively understood through a simple analysis of force directions. In the plane of the cross section, the centrifugal force always acts outwards in the positive r-direction. However, the Coriolis force may act in either a positive or a negative r-direction depending on the rotation direction. If the rotation is positive, it will act along the positive r-direction. When the rotation is negative, however, it will act along the negative r-direction. Therefore, we may expect a change in flow pattern as the rotation direction and rate change.

For positive rotation with a low rotation rate ($Ro \sim o(\sigma)$ where Ro is the rotation number and $\sigma = a/R_c$ is the curvature ratio of the channel[1]), Miyazaki[2,3] numerically examined fully-developed laminar flow in circular and rectangular channels by a finite-difference method. The secondary flow consisted of a pair of counter-rotating vortices in a plane perpendicular to the axis of the channel. The interaction of the secondary flow with the pressure-driven main flow shifted the location of the maximum streamwise velocity away from the center of the channel and in the direction of the secondary velocity in the central region of the channel. The effects of rotation direction were not investigated by Miyazaki. Considering the channel to be a tube of circular cross-section with either positive or negative rotation, fully-developed laminar flow was investigated theoretically by Wang and Cheng.[4] The equations of motion were solved by a perturbation method for small curvature and low angular rotation rate of the tube. The inward Coriolis force due to negative rotation caused the direction of the secondary flow to reverse by overcoming the outward centrifugal forces in the plane of the cross section. The flow reversal occurred by passing through a multi-pair vortex flow region where, overall, the effect of the Coriolis force just neutralized that of the centrifugal force. Such flow reversal and multi-pair vortex flow were also found in rectangular channels by a finite-volume numerical analysis.[5] In the region with multiple pairs of vortices, the rotation number Ro (the inverse of the Rossby number) is of the order of the curvature ratio σ of the channel. Note that the mechanism for appearance of multiple pairs of vortices has been left unaddressed. Furthermore, it appears that no experimental work has been reported to show the flow in this region. This motivates the first part of the present study, i.e. to obtain qualitative information about the flows by flow visualization and to explore the mechanism of the appearance of the multi-pair vortex flow.

For moderate rotation in both positive and negative directions ($Ro \sim o(1)$), fully-developed laminar flow was examined by Daskopoulos and Lenhoff [6] for circular tubes and by Wang and Cheng[5] and Selmi *et al.*[7] for rectangular channels. The centrifugal and Coriolis forces caused centrifugal and Coriolis instabilities in the form of streamwise-oriented roll cells (Dean and Coriolis vortices). The interaction between the two instabilities resulted in a complex flow structure. In particular, the appearance of the Dean and Coriolis vortices led to a high inflectional profile of the streamwise velocity. Such a profile may be unstable with respect to streamwise wavy disturbances in accordance with Rayleigh's inflection point criterion.

For a channel with an infinite span, Finlay *et al.*[8] and Finlay[9] examined the secondary stabilities of Dean and Coriolis vortices with respect to streamwise wavy disturbances. They found that the instability leads to the development of undulating vortex flow and twisting vortex flow. In addition, the stability of Dean vortices and Coriolis vortices subjected to two-dimensional spanwise periodic disturbances (i.e., Eckhaus stability) was examined numerically by Guo and Finlay[10] for cross-section of infinite span . They found that the Eckhaus instability causes the vortex pairs to split apart or merge. This observation motivated the second part of the present study, i.e. to experimentally confirm these secondary instabilities and to expose other possible secondary instabilities.

The literature on high rotation rates ($|Ro| > 3$) appears to be rather limited. For positive rotation with a high rotation rate, Ludweig[11] and Hocking[12] employed a boundary-layer approximation to theoretically examine fully-developed laminar flow in a curved channel with either a square or a rectangular cross-section. They found that the secondary flow also consists of one pair of counter-rotating vortices in the cross plane. The streamwise velocity profile assumes a Taylor-Proudman configuration in the core region.[13] While Ludweig's[11] and Hocking's[12] papers appear to be the only publications on the high rotation case, the validity of the boundary-layer approximation has not been established when secondary flows are present. In fact, all of the numerical simulations failed to find stable flows at high rotation rates.[5-7,14] This motivated the third part of the present work, i.e. to experimentally visualize the flows in curved channels at a high rotation rate. Note that curved channels are usually operated at a high rotation rate in practical applications, so that the present results are believed to be not only of academic interest but also of practical importance.

For the present experiments, a smoke visualization apparatus was designed to reveal the flow in two curved channels rotating spanwise at low negative, moderate or high rotation rate. The work included three main parts. The first part was visualization of flow at a low negative rotation rate. The second part was visualization of flow under the effect of the secondary instabilities of Dean and Coriolis vortices. The third part was visualization of flow at a high rotation rate. In addition, a primary instability analysis was made. This led to the mechanism for appearance of multiple pairs of vortices and the conclusion that flows at a high rotation rate are controlled by a secondary instability rather than the primary instability.

EXPERIMENTAL APPARATUS AND TECHNIQUE

A schematic diagram of the experimental apparatus is shown in Fig. 1. It consisted of a test section, a rotating table with a rotating seal, an air supply system, and a smoke generator. The rotating table was driven by an electric motor with an adjustable-speed drive so that the range of rotating speeds was $n = 0 \sim 500$ *rpm*. The rotation speed was measured by using an optical slot switch running on a disc with 60 equally-spaced holes near its perimeter. The signal from the switch was fed to a Hewlett Packard HP 5314A Universal counter. With 60 holes in the disc, the frequency in *Hz* equaled the rotational speed in *rpm*.

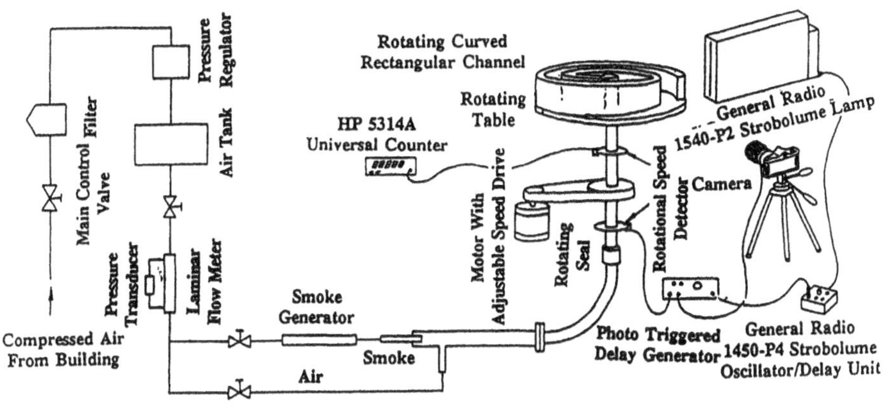

Figure 1. Schematic diagram of experiment apparatus.

A second disc with a single hole provided the signal for firing a General Radio 1540 Strobolume by way of a delay-generator/single-flash flip-flop to permit visual observation using a slit light source with one flash per rotation and also a single, properly timed flash for photographing the whole secondary flow field.

The building compressed air was used as the fluid. The air flow rate was measured by a Meriam flow element with a calibrated differential pressure transducer. Smoke was generated by burning Chinese incense sticks and was injected through a dispersing tube before the test section, as shown in Fig. 1. The very tiny smoke particles ($10^{-2} \sim 10^{-1}$ μm) were subjected to negligible gravitational and rotational buoyancy forces relative to the drag force, and so revealed the fluid velocity field. The smoke patterns were photographed ~0.8 *cm* from the exit of the test section to show the flow pattern. This provided an end-view of the secondary flow pattern for an observer looking upstream into the channel cross-section. A Nikon FM2 single lens reflex camera with Kodak T-Max black and white film P3200 was used.

The two test sections, denoted by *Ts-A* and *Ts-B*, are shown in Fig. 2. The top view and the exit cross-section of *Ts-A* are shown in Fig. 2(a). It consisted of an entrance spiral square channel with axial length 0.85 *m* and a curved square channel (270° bend with axial length of 1.2 *m*) with a constant radius of curvature R_c = 25.4 *cm*. The curved square channel had cross-section 5.08 cm × 5.08 *cm*, and was made from acrylic sheets. The air flowed through a rotating straight tube (with an inside diameter of 4 *cm*) along the axis of rotation, and then entered the spiral square channel before entering the test section. The flow near the exit of the channel is believed to have been fully developed because the ratio of length to width was 40.4. Test section *Ts-B* was also made from acrylic sheets and consisted of an entrance spiral rectangular channel with axial length 1.25 *m* and a curved rectangular channel (270° bend) with a constant radius of curvature R_c = 25.4 *cm*. The top view and the exit cross section for *Ts-B* are shown in Fig. 2(b). The air flowed through a rotating straight tube of 4-*cm* inside diameter along the axis of rotation and then entered the spiral rectangular channel before flowing into the curved channel. With an entrance spiral channel length of 1.25 *m* and a subsequent curved rectangular channel 1.2-*m* long with a constant radius of curvature, the flow near the exit of the channel is also believed to have been fully developed because the ratio of the length over the width of the channel was 96.5.

RESULTS AND DISCUSSION

End-view photographs will be shown for *Ts-A* and *Ts-B* with spanwise direction vertical and radial direction horizontal. In each photograph, the convex (inner) wall is on the left and the concave (outer) wall is on the right. For a given test section, the flow is characterized by two parameters, namely, Reynolds number *Re* and rotation number *Ro*, which are defined as $Re = \frac{W_m a}{\nu}$ and $Ro = \frac{\Omega a}{W_m}$, where W_m, ν, Ω are the mean streamwise velocity, the kinematic viscosity of air, and the angular rotation speed of the channel, respectively. The channel width *a* is used as the characteristic length in *Re* and *Ro*. For a given set of *Re* and *Ro*, several photographs, each taken at different instant, are usually presented to show the time variation of the flow qualitatively. No attempt was made to specify the exact time intervals.

Flows in Curved Channels with a Low Negative Rotation Rate

Examples of smoke patterns from the multi-pair vortex flow are illustrated in Fig. 3 for *Ts-A* and Fig. 4 for *Ts-B*.

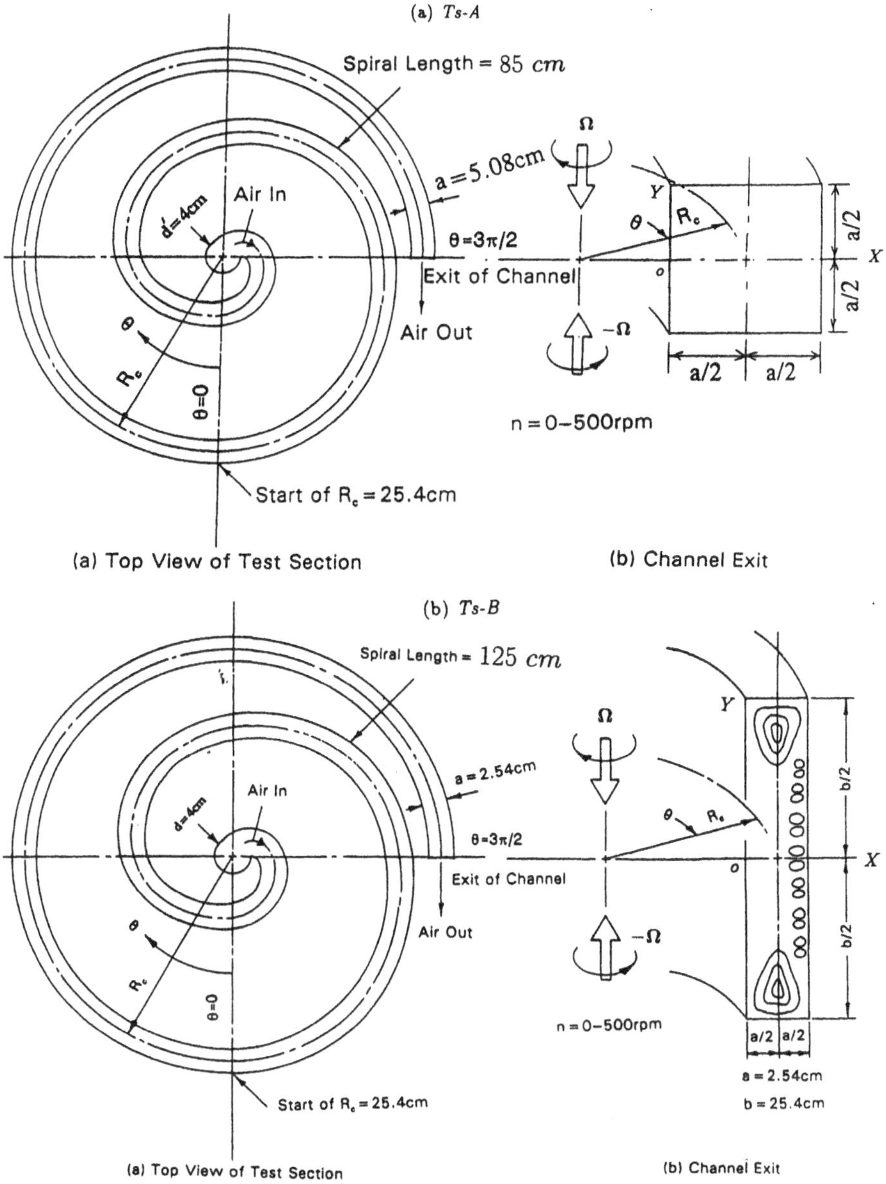

Figure 2. The two test sections.

For *Ts-A* these were obtained at Re=500, Ro=-0.330; Re=600, Ro =-0.274; Re=1000, Ro =-0.279 and Re=2000, Ro=-0.165. For *Ts-B* these were obtained at Re =110, Ro=-0.0792 and Re=452, Ro=-0.0914. The patterns are especially interesting because they show the existence of the multi-pair vortex flow.

In order to understand the mechanism, we use a displaced particle argument[15] to derive a stability criterion including both centrifugal and Coriolis force effects.

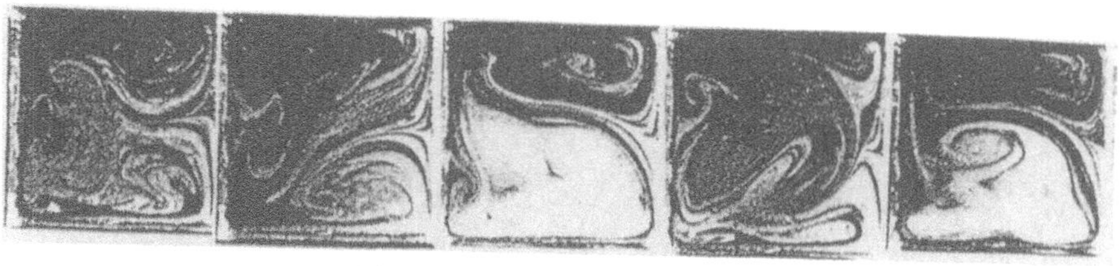

(a) $Re = 500, Ro = -0.330$

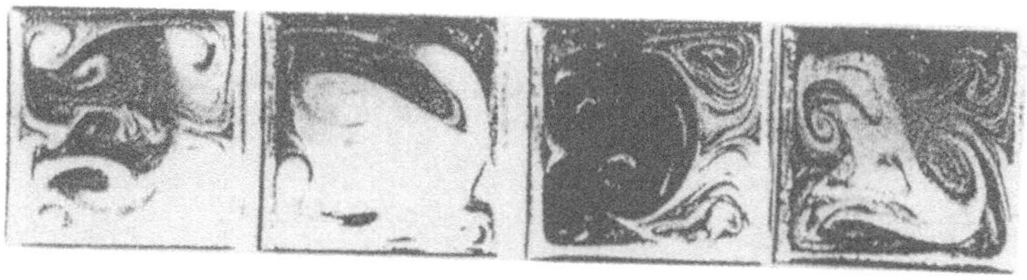

(b) $Re = 600, Ro = -0.274$

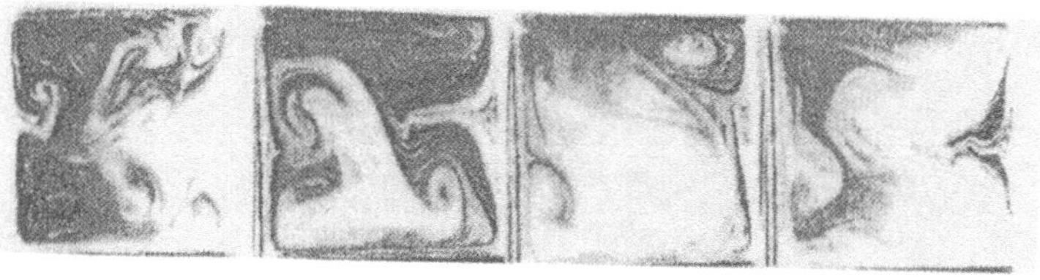

(c) $Re = 1000, Ro = -0.279$

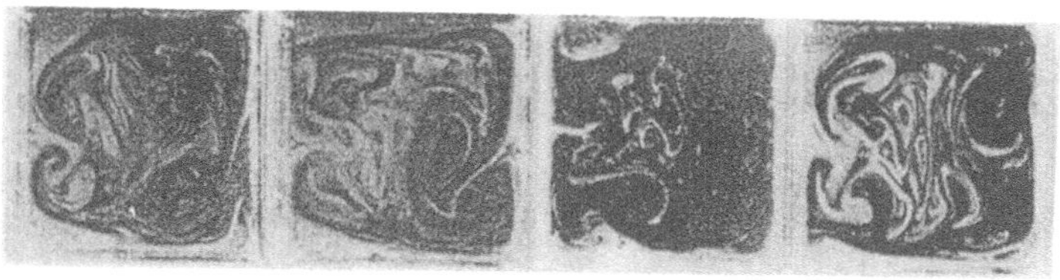

(d) $Re = 2000, Ro = -0.165$

Figure 3. Multi-pair vortex flows for *Ts-A*.

This can be regarded as a generalization of Rayleigh's theory to include both curvature and rotation effects. To simplify the analysis, consider an undisturbed flow in a rotating curved channel with an infinite span, as shown in Fig. 5. If a fluid particle moving with streamwise velocity w at a distance r from the center of curvature o is displaced by a radial disturbing force, the variation in moment of the momentum of the particle, taken around an axis perpendicular to w through o, can be written as:

$$\widetilde{W}(r+dr)-Wr=-2\Omega\int_0^{\Delta t} vrdt \tag{1}$$

by applying the principle of moment of momentum. Here, v is the displacement speed and Δt is the time required for the article to move from r to $r + dr$.

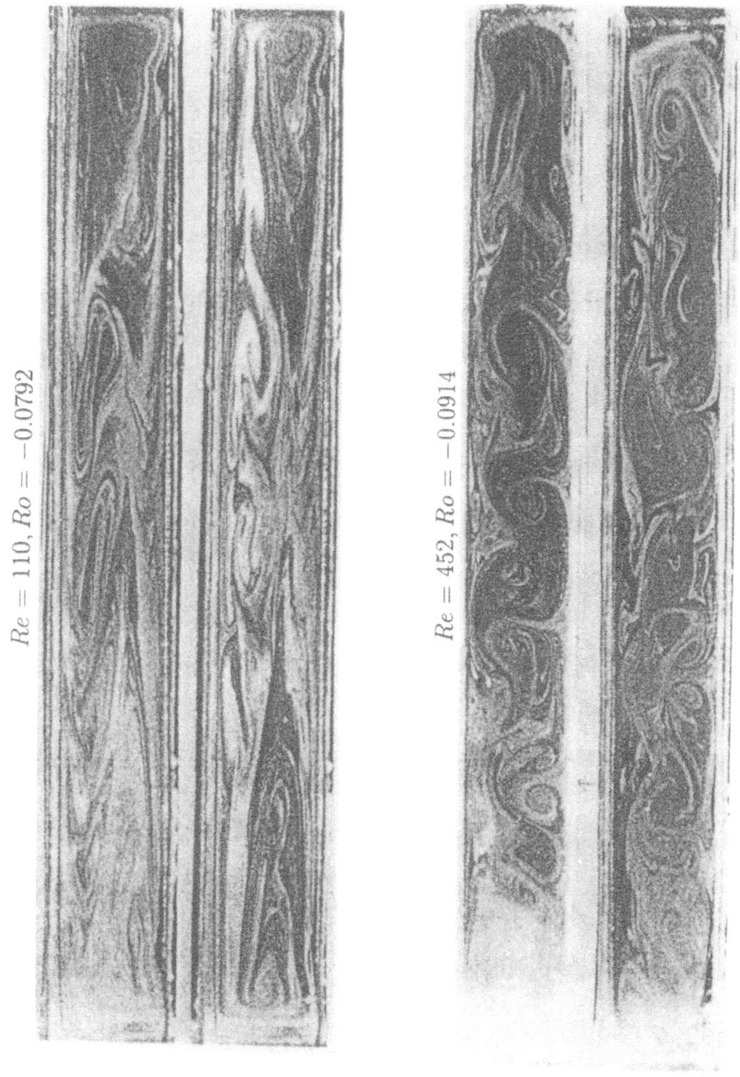

Figure 4. Multi-pair vortex flows for *Ts-B*.

This gives the streamwise velocity $\tilde{W}$ of the particle at the new position:

$$\tilde{W} = W - \frac{W\,dr}{r} - 2\Omega dr \tag{2}$$

to the first order in dr. With a streamwise velocity $\tilde{W}$, the particle at the new position is subjected to a centrifugal force F_1 and a Coriolis force F_2 along the radial direction:

$$F_1 = \frac{\rho \tilde{W}^2}{r + dr}, \qquad F_2 = 2\rho\Omega\tilde{W}. \tag{3}$$

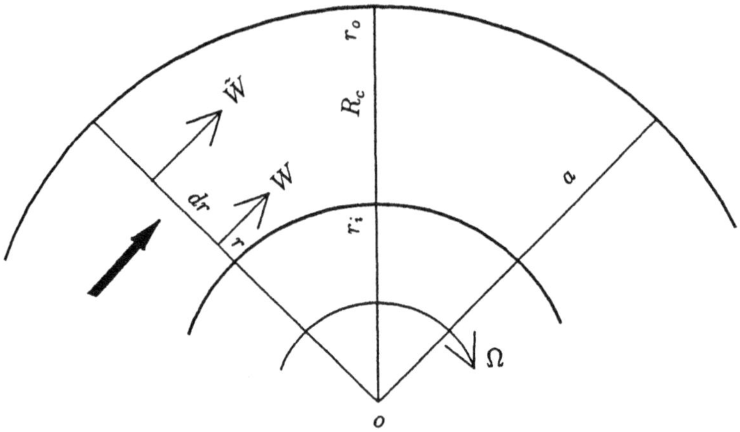

Figure 5. Sketch associated with the discussion of the displaced particle.

In the undisturbed flow, the centrifugal force and Coriolis force acting on a particle with a streamwise velocity $W + \frac{dW}{dr}dr$ at $r+dr$ is balanced by a pressure gradient in the radial direction, i.e.:

$$\left.\frac{\partial P}{\partial r}\right|_{r+dr} = \rho \frac{(W + \frac{dW}{dr}dr)^2}{r + dr} + 2\rho\Omega(W + \frac{dW}{dr}dr). \tag{4}$$

This pressure gradient also acts on the displaced particle.[15] Hence, the restoring force per unit volume acting on the displaced particle from Eqs.(3) and (4) is:

$$F = \rho \frac{(W + \frac{dW}{dr}dr)^2}{r + dr} + 2\rho\Omega(W + \frac{dW}{dr}dr) - \frac{\rho \tilde{W}^2}{r + dr} - 2\rho\Omega\tilde{W}\,. \tag{5}$$

Substituting Eq.(2) into Eq.(5) yields:

$$F = \rho\theta(r)dr + o(dr^2) \tag{6}$$

$$\text{where } \theta(r) = \frac{2W}{r}(\frac{dW}{dr} + \frac{W}{r}) + \frac{6\Omega W}{r} + 2\Omega(\frac{dW}{dr} + 2\Omega). \tag{7}$$

Nondimensionalizing $\theta(r)$, Ω, W and r by $(W_m/a)^2$, W_m/a, W_m and a, with W_m as the mean streamwise velocity across the channel ($W_m = \frac{1}{a}\int_{r_i}^{r_o} W(r)\,dr$), we have:

$$\Pi = \frac{\theta}{(W_m/a)^2} = 2[\frac{\sigma w}{1 + \sigma(x - 1/2)} + Ro][\frac{dw}{dx} + \frac{\sigma W}{1 + \sigma(x - 1/2)} + 2Ro], \tag{8}$$

$$\text{or } \Pi = \underbrace{\frac{2\sigma w}{1+\sigma(x-1/2)}\left(\frac{dw}{dx} + \frac{\sigma w}{1+\sigma(x-1/2)}\right)}_{1st\ term} + \underbrace{\frac{6\sigma w Ro}{1+\sigma(x-1/2)}}_{2nd\ term} + \underbrace{2Ro(\frac{dw}{dx} + 2Ro)}_{3rd\ term} \tag{9}$$

where $x = \dfrac{r - R_c}{a}, \quad w = \dfrac{W}{W_m}, \quad Ro = \dfrac{\Omega a}{W_m}, \quad \sigma = \dfrac{a}{R_c}.$ (10)

Note that Ro, the inverse of the Rossby number, represents the ratio of the Coriolis force to the inertia force. The parameter II may be regarded as a generalized Rayleigh stability criterion that includes the effects of both centrifugal and Coriolis forces. The intermediate term on the right side of Eq. (9) arises from the coupling between curvature and rotation, while the first term and last term are centrifugal instability-related and Coriolis instability-related effects, respectively. The flow is stable in the flow domain with II > 0, and unstable in the domain with II < 0.

For flow in a curved channel with rotation around the axis of curvature, the base flow is unaffected by rotation.[16] For a small gap ($\sigma << 1$), the solution for the base flow can be written as:[16]

$$w(x) = 6x(1 - x). \tag{11}$$

For a small gap ($\sigma << 1$) and low rotation rate ($Ro \sim o(\sigma)$), the generalized Rayleigh criterion II reduces to:

$$\text{II} = 2(\sigma w + Ro)\,\frac{dw}{dx}. \tag{12}$$

The condition II(x) = 0, which can be obtained either by $\sigma w + Ro = 0$ or by $dw/dx=0$, gives the boundary between the stable region and unstable region. Using $dw/dx=0$, we get:

$$x_{01} = \frac{1}{2} \quad \forall Ro. \tag{13}$$

Using $\sigma w + Ro = 0$, we obtain:

$$\begin{aligned} x_{02} &= \frac{1}{2}\left(1 + \sqrt{1 + 2Ro/(3\sigma)}\right) \\ x_{03} &= \frac{1}{2}\left(1 - \sqrt{1 + 2Ro/(3\sigma)}\right) \end{aligned} \tag{14}$$

for all values of Ro satisfying:

$$-1.5 \le \frac{Ro}{\sigma} \le 0. \tag{15}$$

The sign of II in different regions is listed in Table 1. Thus, the stable/unstable regions can be determined, as shown in Fig. 6. Although we show all the regions of Ro in Table 1 and in Fig. 6, the results are valid only for small $|Ro|$ with $Ro \sim o(\sigma)$.

Table 1. Sign of II for small gaps and low rotation rates.

Ro/σ	**Positive II**	**Negative II**
$Ro/\sigma \le -1.5$	$0 < x < ½$	$½ < x < 1$
$-1.5 \le Ro/\sigma \le 0$	$x_{03} < x < ½$ $x_{02} < x < 1$	$0 < x < x_{03}$ $½ < x < x_{02}$
$Ro/\sigma \ge 0$	$0 < x < ½$	$½ < x < 1$

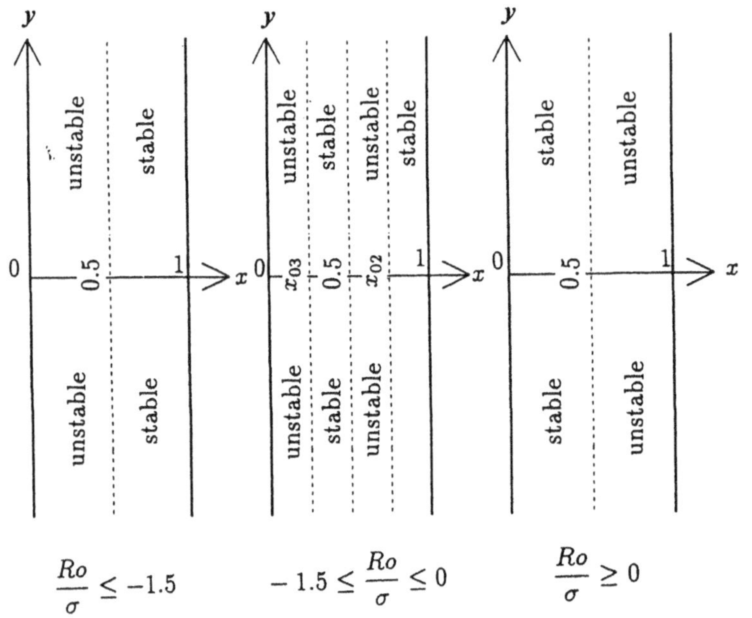

Figure 6. Stable and unstable regions for small gaps with a low rotation rate.

It is interesting that in the region $-1.5 < Ro/\sigma < 0$ there exist two potentially unstable regions separated by two stable regions. We might expect that competition between the destabilizing mechanisms in the two regions would lead to a complicated flow. The flow shown in Figs. 3 and 4 is in or around this range of parameters. Therefore, we conclude that the appearance of the multi-pair vortex flow comes from the existence of two potential unstable regions alternating with two stable regions in the cross-plane. The oscillating nature of the flow can be inferred from a comparison between photos taken at different times for each case in Figs. 3 and 4. Two mechanisms possibly contribute to this oscillating behavior, i.e. the secondary instability of the primary instability or invalidation of the principle of exchange of instabilities in this region. These flow visualization experiments were not appropriate to determine the exact cause. A detailed stability analysis is, thus, required in the future.

Dean and Coriolis Vortices

An important feature relating to stabilization and destabilization is the appearance of the Dean and Coriolis vortices. It is of special interest because they may cause a significant change in the flow structure and the wall heat transfer distribution. In addition, they have an important effect on the transition from laminar to turbulent flow and relaminarization from turbulent to laminar flow.

Dean and Coriolis vortices may be present as a result of centrifugal and Coriolis instabilities imposed on the flow by curvature and rotation. Experimentally, these vortices are identified by mushroom-shaped smoke patterns. Such patterns form because of spanwise variations of streamwise velocity due to the different inwash and outwash regions associated with pairs of counter-rotating vortices. The stems of individual mushroom shapes mark the region of flow away from the wall surface between the vortices that form one pair.

The different possible kinds of secondary instabilities may produce Dean and Coriolis vortices with different features in different flow regimes. The properties of the disturbances triggering the initial vortex formation (such as their steadiness, uniformity along the spanwise direction, etc.) may also lead to various Dean and Coriolis vortices. Consequently, we would expect different mushroom-shaped smoke patterns due to single or combined effects of the two causes (secondary instabilities and initial disturbances). In a fully-developed region, which the present work is mainly concerned with, the Dean and Coriolis vortices become fully developed and should be less sensitive to the initial disturbances because, once initiated, vortex behavior and development are strongly controlled by the centrifugal and Coriolis instabilities from the curvature and rotation of the channel. The different Dean and Coriolis vortices observed in our experiments are, therefore, believed to have been caused by their secondary instabilities.

The Dean and Coriolis vortices observed here can be divided into two groups: symmetric and asymmetric vortices with respect to the radial-streamwise plane. The symmetric Dean and Coriolis vortices usually appeared at a dynamic parameter close to the critical value for their onset. Such vortices were identified by the mushroom stems that formed the radial symmetry lines. The mushroom-shaped smoke patterns were mirror images on each side of the line. In other words, the two vortices in each pair were similar. The asymmetric Dean and Coriolis vortices were often observed at higher dynamic parameters. When viewed in cross (radial-spanwise) planes, such vortices were distorted generally by spanwise and/or radial unsteadiness in the flow such that no line of symmetry was observed.

Symmetric Coriolis Vortices. Figure 7 shows examples of smoke patterns from symmetric Coriolis vortex flow. They were obtained for *Ts-B* at *Re*=193.6, *Ro*=-1.171 and *Re*=452, *Ro*=-1.718. The patterns were especially interesting because they showed only small unsteadiness. No significant time variation on shape and size could be observed. The narrow mushroom stems indicate narrow outwash regions from the high pressure wall. Also the large mushroom petals imply large, spread-out sidewash and inwash regions. The fairly uniform size and shape at different spanwise positions indicate that the flow was fully developed, since otherwise a difference would result from the sensitivity of developing vortices to initial disturbances. A comparison of the smoke patterns in Fig. 7(a) with those in Fig. 7(b) shows that the shape of the symmetric Coriolis vortices changed with the operating parameter. This is consistent with numerical simulations[9] that indicated that the vortex shape changes with spanwise wavenumber, Reynolds number and rotation number.

Radially Oscillating Coriolis Vortex Flow. Figure 8 shows two examples of radially oscillating Coriolis vortices. They occurred at *Re*=600, *Ro*=-1.37 for *Ts-A* and at *Re*=193.6, *Ro*=-0.663 for *Ts-B*. At this stage of flow, very little spanwise oscillation was present and mushroom patterns seemed to move almost exclusively in the radial direction. When these oscillations occurred, they were mostly symmetric Coriolis vortices, and the radial unsteadiness decreased in magnitude as the overall average vortex height increased.

The vortex heights at different spanwise locations in Fig. 8(b) show relatively small variations in height and spacing. This indicates that the Coriolis vortices were fully developed along the streamwise direction and the oscillation was caused by the secondary instability of the Coriolis vortices.

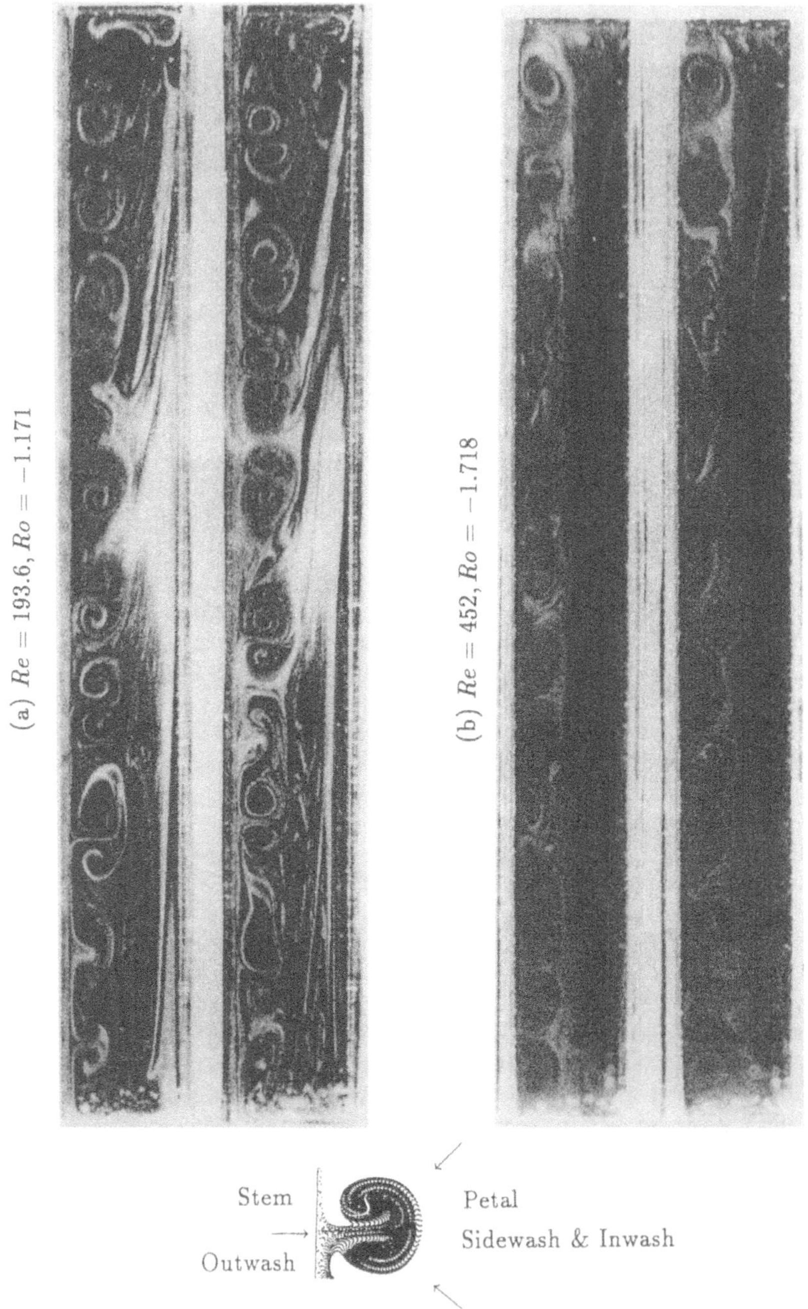

Figure 7. Symmetric Coriolis Vortices for *Ts-B*.

(a) $Re = 600, Ro = -1.37$ (Ts-A)

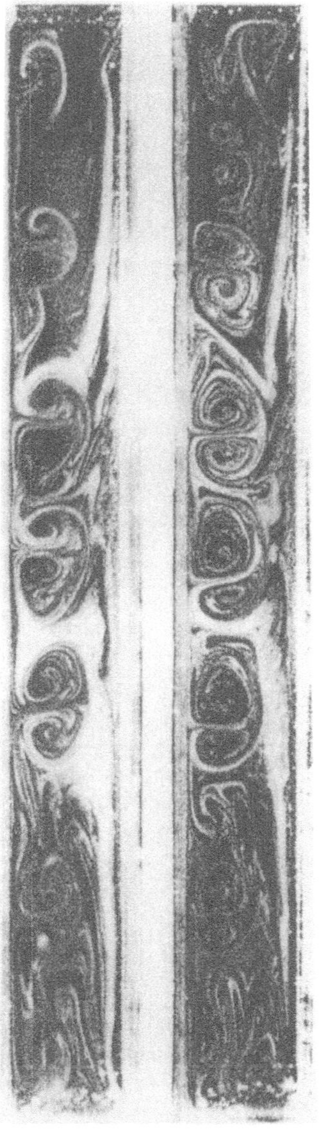

(b) $Re = 193.6, Ro = -0.663$ (Ts-B)

Figure 8. Radially oscillating Coriolis vortices.

Spanwise Oscillating Dean and Coriolis Vortex Flow. Figure 9 shows mushroom-shaped smoke patterns at *Re*=500, *Ro*=0 for *Ts-A* and at *Re*=581, *Ro*=-1.12 for *Ts-B*. The time variations in shape show that very little radial oscillation was present and the mushroom patterns seemed to move almost exclusively in the spanwise direction. When these oscillations occurred, they were mostly symmetric. The uniform height and spacing in spanwise direction provide evidence of fully developed vortices. As well, the oscillation is believed to be caused by the secondary instability.

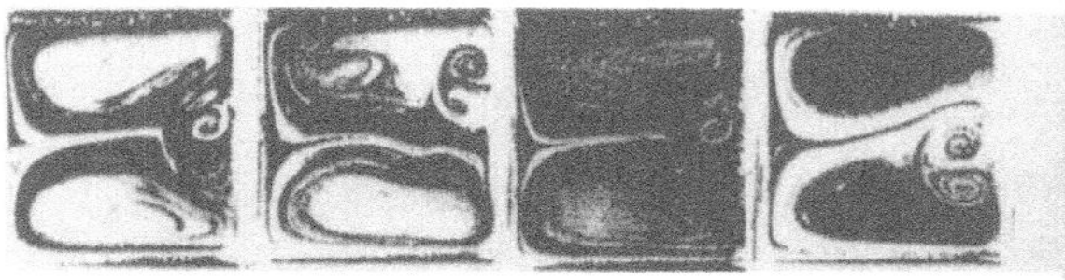

(a) Spanwise oscillating Dean vortices at $Re = 500, Ro = 0$ for $Ts\text{-}A$

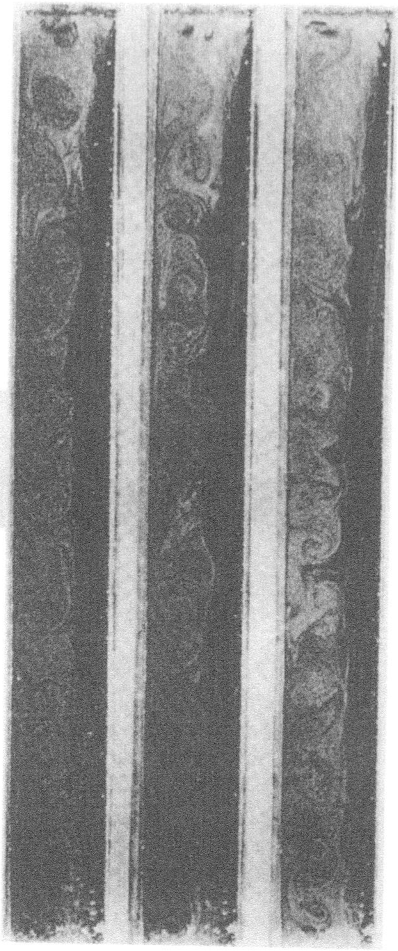

(b) Spanwise oscillating Coriolis vortices at $Re = 581, Ro = -1.12$ for $Ts\text{-}B$

Figure 9. Spanwise oscillating Dean and Coriolis vortices.

Simultaneous Spanwise and Radially Oscillating Dean Vortex Flow. The mushroom-shaped smoke patterns, oscillating simultaneously in the spanwise and radial directions could be observed without a twisting motion. One example is shown in Fig. 10 for *Ts-A* at $Re = 550$, $Ro = 0$. The vortices in this vortex flow are also nearly symmetric, and generally have larger-scale oscillation and greater unsteadiness than the smoke patterns identified with radial, span or twisting modes.

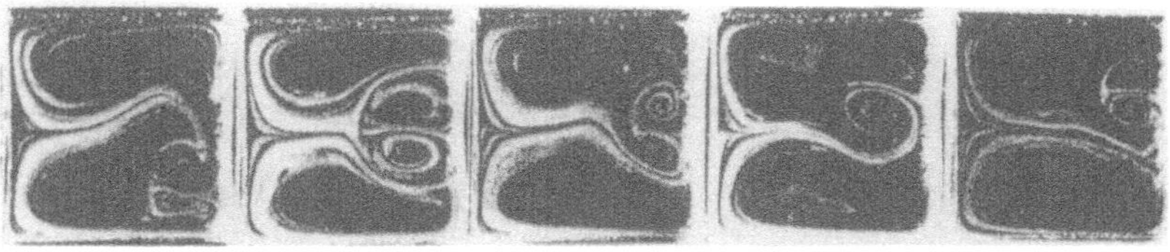

Figure 10. Simultaneous spanwise and radially oscillating Dean vortices at $Re = 550$, $Ro = 0$ for *Ts-A*.

Twisting Dean Vortex Flow. Curved and rotating channel flows have been simulated using three-dimensional incompressible time dependent Navier-Stokes equations.[8,9] Evidence was obtained that the Dean or Coriolis vortices develop two different types of traveling waves in the streamwise direction. The undulating Dean or Coriolis vortex flow contains long streamwise wavelengths and very small growth rates which make them difficult to observe experimentally. The twisting Dean or Coriolis vortex flow contains much shorter streamwise wavelength than the undulating Dean and Coriolis vortex flow. This makes them more readily observable.

Streamwise fully-developed traveling waves generate the same events at one time but at different streamwise locations, or at one streamwise location but at different times. With twisting, the vortex centers oscillate only a little in the spanwise direction. The motion occurs mostly in the radial direction as the vortices in one pair oscillate in shape and strength.

The smoke patterns for twisting Dean and Coriolis vortices were mushroom-shaped with a rocking type of motion. Examples of smoke patterns from twisting Dean vortices are presented in Fig. 11. They were obtained at Re=600, Ro=0 for *Ts-A* and at Re=581, Ro=0 for *Ts-B*. The twisting Dean vortices were asymmetric. The Reynolds number for the two cases with twisting Dean vortices ($Ro = 0$) was larger than $1.96Re_c$, where Re_c is the critical Reynolds number for the onset of Dean vortices. These results are consistent with numerical simulations.[8]

For *Ts-B*, the rocking motion was rarely observed alone. Rather it was usually observed in conjunction with at least one other mode of oscillation. In many cases, the rocking vortex pairs were adjacent to pairs with a different type of motion.

Small Vortex Pairs. Using a linear stability theory and a spectral method, Guo and Finlay[10] examined the Eckhaus stability of the Dean and Coriolis vortices. They found that the Eckhaus stability boundary is a small closed loop. Outside the boundary, the Eckhaus instability causes the vortex pair to split apart by the formation of a new vortex pair or merge together, and no vortex flow is stable to spanwise disturbances when $Re > 1.7Re_c$. When Re is not too high ($Re < 4Re_c$), the wavenumber of vortices is selected by the Eckhaus instability. Similar phenomena have also been reported by Finlay *et al.*,[8] Finlay,[9] and Bland and Finlay[17] in their numerical simulations of channel flows with curvature or rotation. Vortex splitting/merging and readjustment of the spanwise wavenumber were clearly evidenced here in the radial-span cross section by the presence of small vortex pairs at some instants of time. Examples are shown in Fig. 12 for *Ts-B* at Re=452, Ro=0; Re=452, Ro=-0.091; Re=323.3, Ro=-0.819; and Re=581, Ro=1.12.

Complex Oscillating Models. In some cases, usually with high dynamical parameters, the oscillation of the Dean and Coriolis vortices appeared to be so chaotic that the mushroom-shaped patterns were barely recognizable or could not be identified clearly.

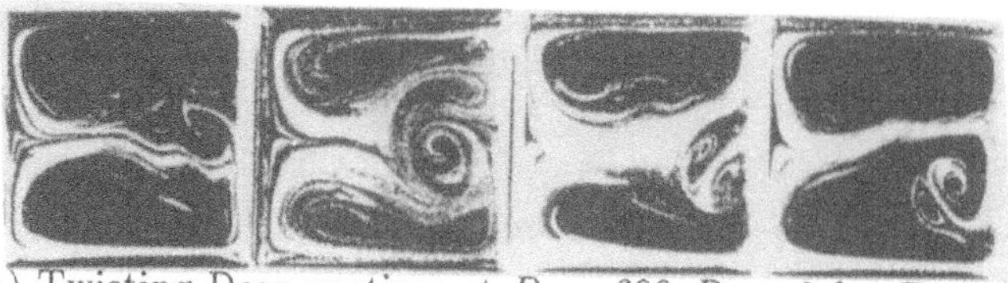

(a) Twisting Dean vortices at $Re = 600$, $Ro = 0$ for $Ts\text{-}A$

(b) Twisting Dean vortices at $Re = 581$, $Ro = 0$ for $Ts\text{-}B$

Figure 11. Twisting Dean vortices.

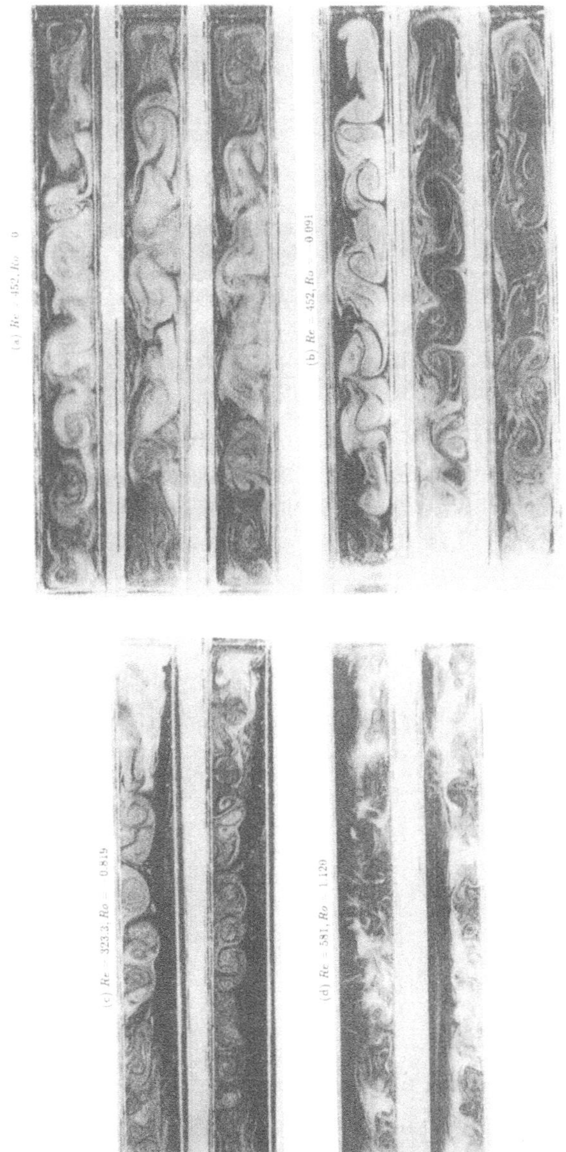

Figure 12. Small vortex pairs for *Ts-B*.

Photographs of such patterns are shown in Fig. 13 for *Ts-B* at Re=193.6, Ro=0.

Flows in Curved Channels with a High Rotation Rate

For a small gap ($\sigma << 1$) and high rotation rate ($Ro \sim o(1)$), the generalized Rayleigh criterion II can be approximated as:

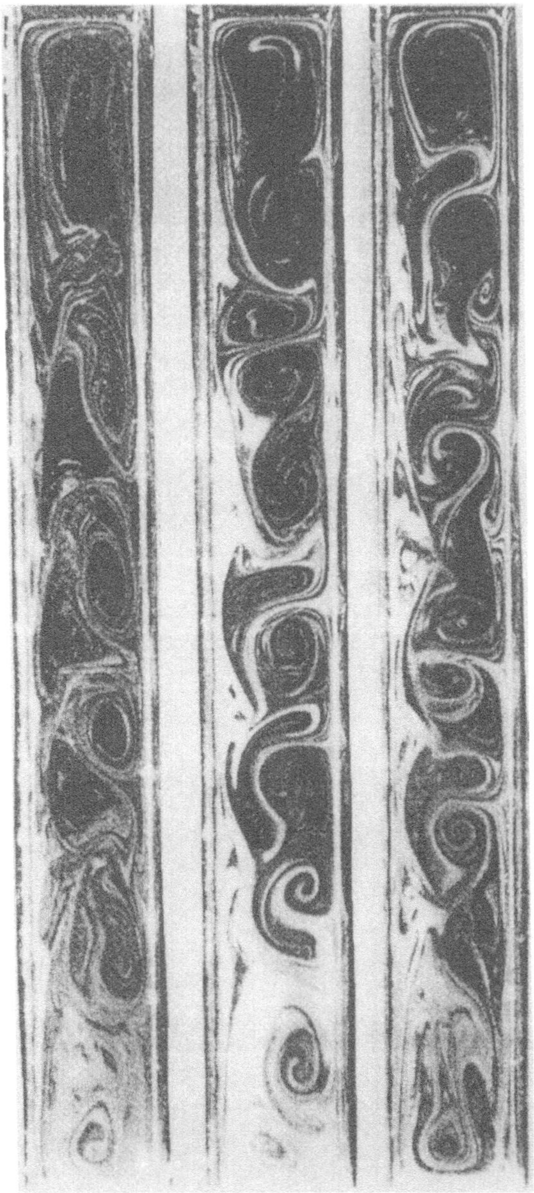

Figure 13. Complex oscillating Dean vortices at Re = 193.6, Ro = 0 for Ts-B.

$$\Pi(x) = 2Ro(\frac{dw}{dx} + 2Ro) \qquad (16)$$

which vanishes at a radial position x_0 satisfying:

$$\frac{dw}{dx} + 2Ro = 0. \tag{17}$$

This, with Eq. (11), yields:

$$x_0 = \frac{1}{2}(1 + \frac{Ro}{3}) \qquad \forall \ |Ro| \leq 3 \tag{18}$$

The sign of II in different regions is summarized in Table 2. Figure 14 shows the stable and unstable regions. Although Table 2 and Fig. 14 show the results for the whole range of *Ro*, they should only be used for high rotation rates. They show that the unstable regions decrease as $|Ro|$ increases, while high rotation with $|Ro| > 3$ always stabilizes the flow in the whole cross-section.

Table 2. Sign of II for small gaps and high rotation rates.

Ro	**Positive II**	**Negative II**
$Ro \leq -3$	$0 < x < 1$	
$-3 < Ro \leq 0$	$x_0 < x < 1$	$0 < x < x_0$
$0 \leq Ro < 3$	$0 < x < x_0$	$x_0 < x < 1$
$Ro \geq 3$	$0 \leq x \leq 1$	

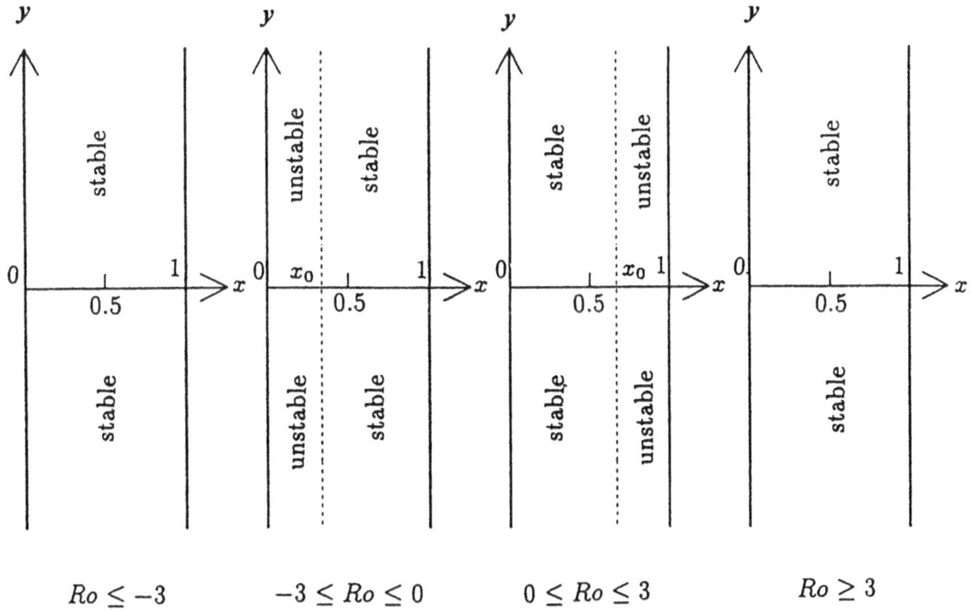

Figure 14. Stable and unstable regions for small gaps with a high rotation rate.

Examples of smoke patterns from flows in a region with $|Ro| > 3$ are illustrated in Fig. 15(a) for *Ts-A* and in Fig. 15(b) for *Ts-B*. For *Ts-A* these patterns were obtained at *Re*=200, *Ro*=21.42; *Re*=200, *Ro*=-10.05; *Re*=600, *Ro*=9.78; and *Re*=1000, *Ro*=-3.43.

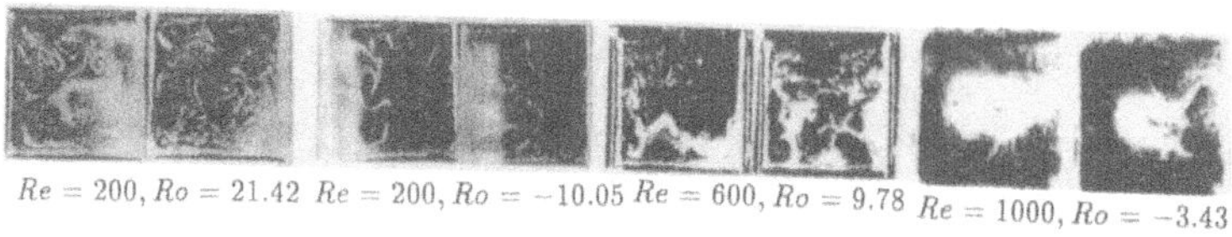

(a) Flows in a region with $| Ro |> 3 |$ for *Ts-A*

(b) Flows in a region with $| Ro |> 3 |$ for *Ts-B*

Figure 15. Flows in a region with $|Ro| > 3$.

For *Ts-B*: *Re*=110, *Ro*=-8.11; *Re*=110, *Ro*=10.60; *Re*=452, *Ro*=-3.59; and *Re*=452, *Ro*=3.45. The patterns are somewhat similar to bursting flow in turbulent boundary layers,[18] and are far from the one pair of counter-rotating vortices predicted by Ludwieg[11] and Hocking.[12] Therefore, the boundary layer approximation may not be valid for flow in curved channels with a high rotation rate. Also, the high unsteadiness of the flow implies that the flow is controlled by the secondary instability rather than the primary instability. In particular, a secondary instability can produce turbulent flows at low Reynolds numbers.

CONCLUDING REMARKS

An experimental apparatus was designed and constructed to visualize fully-developed flow in two kinds of channels with streamwise curvature and spanwise rotation, in terms of the end-view secondary flow near the exit of the test section. The work focused on Dean and Coriolis vortices under the effects of a secondary instability, and on flows with a low negative rotation speed and a relatively high rotation rate.

For the low negative rotation speed, there were two potentially unstable regions alternating with two stable layers in the cross-plane. The competition of the two destabilizing mechanisms caused the appearance of multi-pair vortex flows. The flow visualization showed that such flows were associated with oscillations. This suggests the need for further stability analysis.

Pairs of counter-rotating Dean and Coriolis vortices were observed as mushroom-shaped smoke patterns. The narrow mushroom stems indicated narrow outwash flow regions, while large mushroom petals implied large, spread-out sidewash and inwash regions. Fairly steady and symmetric Dean and Coriolis vortices were observed in some parameter ranges. Their size and shape changed with the operating parameters.

Secondary instabilities of the Dean and Coriolis vortices led to oscillation in various forms, even in streamwise fully-developed flow regions. The observed oscillatory modes included: (1) one with mostly radial motion; (2) one with mostly spanwise motion; (3) one with significant simultaneous radial and spanwise motion; (4) one with rocking motion. In addition, oscillating modes were also present with motion so complex that it was difficult to describe the flow visualization results. The experimental results need to be confirmed both theoretically and numerically in the future.

In the cross planes, twisting Dean and Coriolis vortices were evidenced by asymmetric mushroom-shaped smoke patterns with rocking motion. The region where the twisting Dean and Coriolis vortices were observed was in agreement with that predicted.[8,9] This kind of vortex flow differed from that with simultaneous radial and spanwise motion mainly in three aspects:

(1) The twisting vortices were asymmetric, but the simultaneous radial and spanwise oscillating vortices were symmetric.
(2) With twisting, the spanwise locations of vortex centers oscillated only a little. With simultaneous radial and spanwise oscillations, the spanwise locaιions of vortex centers could oscillate significantly.
(3) The simultaneous radial and spanwise oscillating Dean and Coriolis vortices usually had a larger-scale oscillation and greater unsteadiness than the twisting Dean and Coriolis vortices.

The vortex pairs sometimes seemed to appear and disappear, as indicated by small vortex pairs observed temporarily in the cross planes under some experimental conditions. This is believed to be caused by the Eckhaus instability discussed by Guo and Finlay.[10]

The flow at large $|Ro|$ appeared to be controlled by the secondary instability rather than the primary instability. In particular, the secondary instability caused the flow with large $|\mathrm{Ro}|$ to be unsteady and turbulent, somewhat like bursting flow in a turbulent boundary layer. This appears to produce low Reynolds number turbulent flow, and one may conclude that the boundary-layer approximation is not valid for this type of flow.

Acknowledgements

This work was supported by the Research Grants Council of Hong Kong, the CRCG at the University of Hong Kong, and the Natural Sciences and Engineering Research Council of Canada.

REFERENCES

1. L. Wang and T. Yang, Body-force-driven multiplicity and stability of combined free and forced convection in rotating curved ducts: centrifugal force, in: *present volume*.
2. H. Miyazaki, Combined free and forced convective heat transfer and fluid flow in a rotating curved circular tube, *Int. J. Heat Mass Transfer* 14:1295 (1971).
3. H. Miyazaki, Combined free and forced convective heat transfer and fluid flow in a rotating curved rectangular tube, *J. Heat Transfer* 95:64 (1973).
4. L. Wang and K. C. Cheng, Flow transitions and combined free and forced convective heat transfer in a rotating curved circular tube, *Int. J. Heat Mass Transfer* 39:3381 (1996).
5. L. Wang and K. C. Cheng, Flow transitions and combined free and forced convective heat transfer in rotating curved channels: the case of positive rotation, *Phys. Fluids*, 8:1553 (1996).
6. P. Daskopoulos and A. M. Lenhoff, Flow in curved ducts: part 2, rotating ducts, *J. Fluid Mech.*, 217:575 (1990).
7. M. Selmi, K. Nandakumar and W. H. Finlay, A bifurcation study of viscous flow through a rotating curved duct, *J. Fluid Mech.* 262:353 (1994).
8. W. H. Finlay, J. B. Keller and J. H. Ferziger, Instabilities and transition in curved channel flow, *J. Fluid Mech.* 194:417 (1988).
9. W. H. Finlay, Transition to oscillatory motion in rotating channel flow, *J. Fluid Mech.* 215:209 (1990).
10. Y. Guo and W. H. Finlay, Splitting, merging and wavelength selection of vortices in curved and/or rotating channel flow due to Eckhaus instability, *J. Fluid Mech.* 228:661 (1991).
11. H. Ludwieg, Die ausgebildete kanalstromung in einem rotierenden system, *Ing.-Arch. Bd.* 19:296 (1951).
12. L. M. Hocking, Boundary and shear layers in a curved rotating pipe, *J. Math. Phys. Sci.* 1:123 (1967).
13. H. P. Greenspan, *The Theory of Rotating Fluids*, Cambridge University Press, London (1968).
14. K. Nandakumar, H. Raszillier and F. Durst, Flow through rotating rectangular ducts, *Phys. Fluids A* 3:770 (1991).
15. D. J. Tritto and P. A. Davies, Instabilities in geophysical dynamics, in: *Hydrodynamic Instabilities and the Transition to Turbulence (2nd edn.)*, H. L. Swinney and J. P. Gollub ed., Springer-Verlay, New York, 229 (1985).
16. S. Chandrasekhar, *Hydrodynamic and Hydromagnetic Stability*, Oxford University Press, London (1961).
17. S. B. Bland and W. H. Finlay, Transitions towards turbulence in a curved channel, *Phys. Fluids A* 3:106 (1991).
18. S. J. Kline, W. C. Reynolds, F. A. Schraub and P. W. Runstadler, The structure of turbulent boundary layers, *J. Fluid Mech.* 30:741 (1967).

EFFECT OF HIGH GRAVITY ON THE SOLID-LIQUID INTERFACIAL FREE ENERGY

C. Patuelli and R. Tognato

Dipartimento di Fisica and Istituto Nazionale di Fisica della Materia, Alma Mater Studiorum, Università di Bologna, 40127 Bologna, Italy

INTRODUCTION

The aim of this paper is to evaluate the effect of high gravity on the solid-liquid interfacial free energy per unit area, σ_{SL}, of face-centered cubic metals. There is considerable ambiguity about the location of the solid-melt interface on an atomic scale. The atoms in the melt have a packing that is not too different from the corresponding solid and the densities of the two phases are comparable. Moving from solid to liquid through the interface, a region of gradual transition from solid to liquid is expected.[1] It is reasonable to assume that the interfacial free energy per unit area may be due to the melt region considering that for a diffuse interface σ_{SL} is zero.[2]

GENERAL CONSIDERATIONS

A flat liquid-solid interface present in volume $V_{gram\ atom}$ occupied by a gram atom is considered. Because crystal and liquid regions are three dimensional, the interface region is two dimensional since its height $h \ll (V_{gram\ atom})^{1/3}$. The interface region may be composed of a single layer of height b, which is the nearest neighbor distance. Its total energy is given by:

$$\Gamma = Gb^3Y[\ln(Y/3)]/4\pi \tag{1}$$

where G is the shear modulus and Y is an integer, such that Yb is the perimeter of a single layer expressed in terms of the nearest neighbor distance. The surface area exposed to the crystal phase is:

$$\langle A\rangle \cong (Y^2b^2\sqrt{3}/24) \tag{2}$$

and will contain $n = 7(Y/9.464)^2$ atoms [3]. The corresponding interfacial free energy per unit area is:

Processing by Centrifugation, Edited by Regel and Wilcox
Kluwer Academic/Plenum Publishers, New York, 2001

$$\Gamma/\langle A\rangle \cong Gb[\ln(Y/3)]/Y \tag{3}$$

Since $\ln(Y/3) << Y \cong (V_{gram\ atom})^{1/3}/b$, one may write:

$$\Gamma/\langle A\rangle \cong Gb^2/(V_{gram\ atom})^{1/3} \tag{4}$$

For face-centered cubic metals $Gb^2 \cong 10^{-9}$ N and $(V_{gram\ atom})^{1/3} \cong 10^{-2}$ m, so that $\Gamma/\langle A\rangle \cong 10^{-7}$ J/m^2. This value is not reasonable when compared with the experimentally measured $\sigma_{SL} \approx 10^{-1}$-$10^{-3}$ J/m^2.

Surface-energy values comparable with experimental data are obtained if, instead of a single layer, one takes into account a number n_s of superimposed single layers. Suppose that single layer energies may be added. Then the total energy of a poly-layer is given by $n_s\Gamma$, while the surface area $\langle A\rangle$ remains constant. However, the value of n_s required for agreement between theory and experiment ranges between 10^6-10^4. This is not a reasonable result considering that reported interface heights usually require $n_s \cong 10^2$-10^1.[4]

Domains in an Interface

It is possible to decrease the value of n_s if one supposes that the interface region is composed of domains having perimeters $Y_n b$ and surface areas exposed to the crystal phase of $\langle A_n\rangle \cong (Y_n^2\, b^2\sqrt{3}/24)$.[5] The number of such domains is given by the ratio between the surface area of a single layer and the surface area of a domain, $n_D = \langle A\rangle / \langle A_n\rangle$. The total energy of each domain is $\Gamma_n = Gb^3Y_n[\ln(Y_n/3)]/4\pi$ and each domain contains $n_n = 7(Y_n/9.464)^2$ atoms.

The presence of domains in a two-dimensional system has been demonstrated by Monte Carlo simulation.[6] The domain theory can be derived from the dislocation model for solid-liquid phase transition.[7] If one assumes that the domain energies are additive, it is possible to write:

$$\sigma_{SL} \cong {n_D n_s \Gamma_n}/{\langle A_n\rangle} = \frac{\langle A\rangle}{\langle A_n\rangle}\, {n_s\Gamma_n}/{\langle A_n\rangle} \tag{5}$$

After substitution:

$$\sigma_{SL} \cong Gb\left[{24\left({Y}/{Y_n} \right)^2 n_s \ln\left({Y_n}/{3} \right)}/{Y_n 4\pi\sqrt{3}} \right] \tag{6}$$

We have applied the previous considerations to a face-centered cubic metal, aluminum, using $G = 1.73\times10^{10}$ N/m^2, $T_m = 933$ K, $V_{gram\ atom} = 11\times10^{-6}$ m^3, and $b = 2.92\times10^{-10}$ m. In order to obtain the σ_{SL} value of 9.3×10^{-2}J/m^2 for $n_s = 10$,[8] Eq. (6) requires $Y_n \cong 3\times10^6$ and $Y \cong 8\times10^7$. It is preferable to express σ_{SL} as a function of the number of atoms. So using the expressions relating Y and Y_n to n and n_n:

$$\sigma_{SL} \cong Gb\left[{\left({n}/{n_n} \right) n_s \ln(4n_n)}/{13 n_n} \right] \tag{7}$$

For $n_s = 10$, Eq. (7) requires $n \cong 1x10^{15}$ and $n_n \cong 1x10^{12}$. The necessity of domains has thus been proven.

A More Refined Approach

The solid-liquid interfacial free energy may be written as:[9]

$$\sigma_{SL} = -TS_{melt} \tag{8}$$

where T is a generic temperature and S_{melt} is the spatial negative entropy per unit area, and is given by:

$$S_{melt} = -k_B \left(\frac{n_{interf}}{< A_{interf} >} \right) \int_0^1 W(Z) \ln W(Z) dZ \tag{9}$$

Here, k_B is Boltzmann's constant, n_{interf} is the number of atoms in the interface region, $<A_{interf}>$ the surface area of the interface region, $Z = z/c$ is a dimensionless coordinate normal to the crystal phase surface, z is the length of the interfacial region which extends from z= 0 to z= c, c is a cut-off length, $W(Z) = \rho(z)/\rho$ is a dimensionless density and ρ is the bulk liquid density. For a random distribution W(Z)= 1 and S_{melt} vanishes, whereas for a non-random distribution the integral of Eq. (9) is positive and the spatial entropy is less than zero, i.e. $\sigma_{SL} > 0$. The entropy deficit of the melt region may be associated with lower disorder than that of the bulk liquid. It is possible to express Eq. (9) in a more analytical form.[10-13]

We are interested in finding a difference between the melt region entropy and the liquid phase entropy. It is possible to write for a single atom of the melt region:

$$s'_{melt} \cong k_B \left(\frac{1}{b^2} \right) \left[\ln \left(\frac{1 - \exp\left(-{m_a ab}/{kT} \right)}{{m_a ab}/{kT}} \right) - \left(\frac{{m_a ab}/{kT}}{\exp\left({m_a ab}/{kT} \right) - 1} - 1 \right) \right] \tag{10}$$

where m_a is an atomic mass, a is acceleration due to gravity and b is the atom height, with the atom being treated as a hard sphere. Taking the atomic mass of aluminum, $m_a \cong 4.48x10^{-26}$ kg and a= g one finds $s'_{melt} = 0$, which is a reasonable result: gravity should not influence melt atomic entropy deficit. For a domain containing n_n atoms and height n_s:

$$s''_{melt} = k_B \left(\frac{1}{n_n b^2} \right) \left[\ln \left(\frac{1 - \exp\left(-{n_n m_a a n_s b}/{kT} \right)}{{n_n m_a a n_s b}/{kT}} \right) - \left(\frac{{n_n m_a a n_s b}/{kT}}{\exp\left({n_n m_a a n_s b}/{kT} \right) - 1} - 1 \right) \right] \tag{11}$$

Assuming that domain contributions may be summed up, it is possible to write:

$$S_{melt} \cong k_B \frac{n}{n_n^2 b^2} \left[\ln\left(\frac{1-\exp\left(-n_n m_a a n_s b / kT\right)}{n_n m_a a n_s b / kT} \right) - \left(\frac{n_n m_a a n_s b / kT}{\exp\left(n_n m_a a n_s b / kT\right) - 1} - 1 \right) \right] \qquad (12)$$

The number of domains in this case is given by n/n_s, where the value of n has been obtained by means $(V_{gram\text{-}atom})^{2/3}/\pi b^2$. After substitution, Eq. (8) becomes:

$$\sigma_{SL} = -T k_B \frac{n}{n_n^2 b^2} \left[\ln\left(\frac{1-\exp\left(-n_n m_a a n_s b / kT\right)}{n_n m_a g n_s b / kT} \right) - \left(\frac{n_n m_a a n_s b / kT}{\exp\left(n_n m_a a n_s b / kT\right) - 1} - 1 \right) \right] \qquad (13)$$

Note that $\sigma_{SL} > 0$ only if the term between the square brackets is a negative quantity. For $a = g = 9.805$ m/s^2 and T = 933 K, Eq. (13) gives $\sigma_{SL} \cong 9.3\text{x}10^{-2}$J/m^2 for $n_s = 9$ (here, $n_s b = 26.3\text{x}10^{-10}$ m), $n_n \cong 8\text{x}10^7$ and $n \cong 5\text{x}10^{15}$.

For body-centered cubic metals, the hypothesis has been advanced that a solid-liquid interface might be 5 to 10 atomic distances wide.[14] The n_s and n values are in agreement with those obtained by means of Eq. (6), while the n_n value is much lower. This result is more reasonable; the model has a high degree of refinement so it gives more reliable predictions. Again the necessity of domains has been proved.

There is also another important result: the use of Eq. (13) allows evaluation of the effects of gravity on the interfacial free energy. In order to maintain the value of $\sigma_{SL} \cong 9.3\text{x}10^{-2}$J/m^2 at $a = 10g = 98.05$m/s^2 it is necessary for $n_s b < b$, which is not a reasonable result: a solid-liquid interface must always have a finite width. As a consequence, we hypothesize that gravity may influence the interfacial energy. In particular, an increase in acceleration should increase the interfacial free energy and decrease the interface thickness. This is due to the fact that accelerations increase the order of the melt region with respect to the liquid phase (an external field decreases the entropy by breaking some elements of randomness of a system [13]), so there are increases of the melt deficit. This may be obtained by decreases of the interface height. Since we are treating differences of entropy between the solid and the melt it is reasonable to suppose that the present results do not depend on the orientation of the interface with respect to the acceleration vector. Predicted behavior is shown in Figs. 1 and 2. To obtain the results of Fig. 1, it was assumed that the interface thickness should never be less than b (at any acceleration an interface of finite thickness must be present). Consequently, the interfacial free energy changes from $9.3\text{x}10^{-2}$ J/m^2 at 9.805 m/s^2 and $n_s b = 26.3\text{x}10^{-10}$ m to $6.5\text{x}10^{-1}$ J/m^2 at 9805 m/s^2 and $n_s b = 2.92\text{x}10^{-10}$m. Variations of n_n over a reasonable range have a negligible influence on the final results.

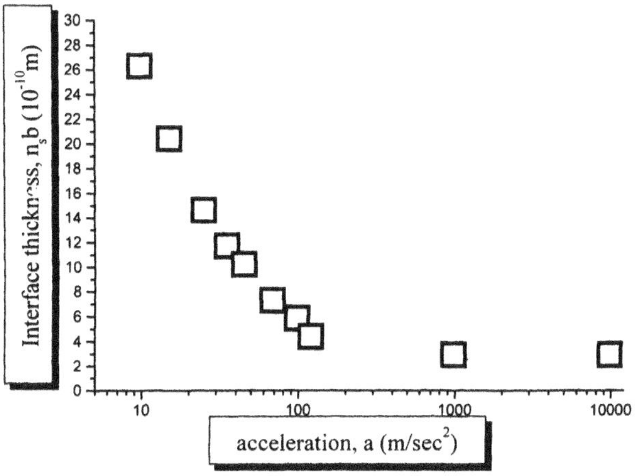

Figure 1. Interface thickness, $n_s b$, versus acceleration.

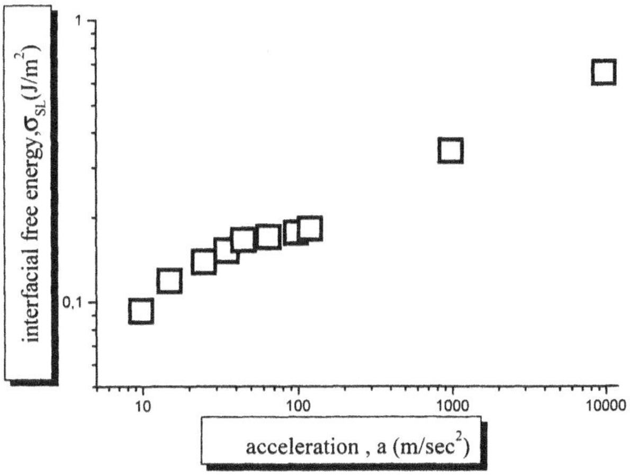

Figure 2. Interfacial free energy, σ_{SL}, versus acceleration.

These results may influence the kinetics of processes in which interfacial free energy plays an important role, such as nucleation and growth of a new phase. These two processes are also governed by mass transport.[15] If we hypothesize that an interface under high acceleration behaves as an interface under high pressure,[13] then high acceleration should increase the activation energy because of decreased activation volume.[16] These decreases of activation volume[17] are consistent with interface height decreases in the presence of high acceleration.

FINAL CONSIDERATIONS

A flat solid-liquid interface is composed of domains. The presence of domains allows us to obtain, at earth's gravity for a well-defined domain height, interfacial free energy values comparable with experimental data. The presence of domains permits us to conclude that increasing acceleration decreases the domain height and increases the interfacial free energy .Interfaces under high acceleration are thought behave similarly to those at high pressure. The results may be applied to centrifuges.

Acknowledgements

This research was carried out with grants from Agenzia Spaziale Italiana (ASI) grants.

REFERENCES

1. J.W. Cahn and R. Kikuchi, *Phys Rev. B* 31: 4300 (1983).
2. Q. Jang, H.X. Shi and M. Zhao, *Acta mater.*47: 2109 (1999).
3. R. Tognato, *Phys. Stat. Sol (a)* 96: 451 (1986).
4. D.P. Woodruff, *The Solid-Liquid Interface*, Cambridge University Press, Cambridge (1973).
5. R. Tognato, *Surface Science* 199: 282 (1988).
6. S.W. Koch, *Dynamic of First Order Phase in Equilibrium and Nonequilibrium Systems*, Springer Verlag, Berlin (1984).
7. T.V. Ramakrishnam, Liquid to solid transition, *Material Science Forum* 3: 57 (1985).
8. W. Kurz and D.J. Fisher, *Fundamental of Solidification*, Trans Tech Publications, UK (1986).
9. R.H. Ewing, *Phil. Mag.* 25: 779 (1972).
10. C. Patuelli and R. Tognato, *Trans. Mat. Res. Soc. Japan* 19A: 205 (1994).
11. C. Patuelli and R. Tognato, *Space Bound 1997*, Canadian Space Agency, St. Hubert, Canada (1997).
12. R. Tognato, *Material Science Forum* 203: 197 (1996).
13. R.E. Salvino, *Phys. Rev. A* 41: 4236 (1990).
14. D.W. Oxtoby and A.D.J. Haymet, *J. Chem. Phys.* 76: 6262 (1982).
15. L.L. Regel and W.R. Wilcox, eds., *Material Processing in High Gravity*, Plenum Press, New York (1994).
16. C.A. Swenson, Physics at High Pressure, *Solid State Physics* 11: 135, Academic Press, New York (1960).
17. J.P. Hirth and J. Lothe, *Theory of Dislocations*, McGraw Hill, New York (1968).

MECHANICAL BEHAVIOR OF ENERGETIC MATERIALS AT HIGH ACCELERATION

Y. Lanzerotti[1] and J. Sharma[2]

[1]U. S. Army TACOM ARDEC, Picatinny Arsenal, NJ 07806-5000, USA
[2]Naval Surface Warfare Center, Silver Spring, MD 20903, USA

INTRODUCTION

We have been studying the mechanical behavior of energetic materials during high acceleration by using an ultracentrifuge.[1-5] Energetic materials are of significant interest for scientific and practical reasons in the extraction (mining) industry, structure demolition, space propulsion, and ordnance. In these applications the materials can be subjected to high acceleration, fluctuating and/or sustained. The nature of the fracture process of such materials under high acceleration is of particular interest, especially in ordnance and propulsion applications. For example, explosives in projectiles are subjected to setback forces as high as 50,000 *g* during the gun launch. These high setback forces can cause fracture and premature ignition of explosives.

Fundamental understanding of the behavior of energetic materials subjected to high acceleration is a key to better practical ordnance designs that solve the problems of abnormal propellant burning and premature ignition of explosives during gun launch. The pressure gradient that is experienced by an explosive during acceleration in a gun and under *g*-loading in an ultracentrifuge is unique and produces different kinds of behavior and failure than under other material test conditions. The present work is particularly relevant to the future development of insensitive energetic materials to be used in devices with higher acceleration.

Previously,[1-5] we used an ultracentrifuge to study the fracture behavior at 25°C of TNT (trinitrotoluene), Composition B [59% cyclotrimethylenetrinitramine (RDX), 40% TNT, 1% wax], four types of Octol [70% cyclotetramethylenetetranitramine (HMX), 30% TNT; 75% HMX, 25% TNT; <75% HMX, 25% TNT, <1% HNS (hexanitrostilbene); and 83% HMX, 17% TNT], cast TNAZ (1,3,3-trinitroazetidine), pressed TNAZ, pressed plastic-bonded explosives [LX-14 (95% HMX, 5% Estane), Composition A3 Type II (91% RDX, 9% polyethylene), PAX-2A [85% HMX, 9% BDNPF (bis-dinitropropyl acetal formal), 6% CAB (cellulose acetate butyrate)], and PAX-3 (85% HMX, 9% BDNPF, 6% CAB/25% aluminum). In the present work, we studied the fracture behavior at high acceleration in an ultracentrifuge at -10°C of cast TNAZ, pressed TNAZ, and pressed plastic-bonded explosives [LX-14, Composition A3 Type II, PAX-2A and PAX-3].

Processing by Centrifugation, Edited by Regel and Wilcox
Kluwer Academic/Plenum Publishers, New York, 2001

TECHNIQUE

A Beckman ultracentrifuge model L8-80 with swinging-bucket rotor model SW 60 Ti was used to rotate the samples up to 60,000 rpm (about 500,000 *g*). The distance of the specimen from the axis of rotation ranged from 6 to 12 cm.

The samples were machined into the shape of the frustum of a cone. The large diameter was typically 11 mm and the small diameter 9 mm. The angle between the base and the side was 80°. Each sample was fitted into a 5-mm long, 11-mm o.d. aluminum cylinder. At one end the i.d. was 11 mm and at the other end 9 mm. The angle between the inner and outer sides of the sleeve was 10°. The 9-mm diameter top of the sample faced away from the axis of rotation.

For an experiment the acceleration was increased up to a maximum and then held there until the total elapsed time reached 5 min. The sample was then decelerated smoothly. In a series of experiments on a material, the initial maximum acceleration was less than that causing fracture. The maximum acceleration was then increased systematically in each successive 5-min run. The sample fractured when the shear or tensile strength of the material was exceeded, causing particles to break loose from the exposed surface and transfer to the closed-end of the tube. A hemispherical fracture surface resulted.

RESULTS

TNAZ

Polycrystalline melt-cast TNAZ fractured at grain boundaries at about 93 k*g* at -10°C. The cast had been made with 70% liquid TNAZ and 30% 40 μm maximum size TNAZ particles. This melt-cast TNAZ was 97.3% of its theoretical maximum density (TMD) of 1.84 g/cm^3. Pressed TNAZ fractured at about 55 k*g* at -10°C and its density was 98.4% of this TMD.

Pressed Plastic-Bonded Explosives

Table 1 compares the fracture behaviors of melt-cast TNAZ and of pressed TNAZ with those of LX-14, PAX-2A, PAX-3, and Composition A3 Type II. The fracture acceleration of LX-14 was greater than 140 k*g* at -10°C. This LX-14 had been pressed to 95.9% of its TMD of 1.849 g/cc. The fracture acceleration of PAX-2A is greater than 140 k*g* at -10°C. This PAX-2A was pressed at 99.4% of its TMD of 1.79 g/cc. PAX-3 has been found to fracture at about 123 k*g* at -10°C. This PAX-3 was pressed at 97.4% of its TMD of 1.955 g/cm^3. Composition A3 Type II fractured at about 55 k*g* at -10°C. It had been pressed to 98.1% of its TMD of 1.662 g/cm^3.

The fracture acceleration of cast TNAZ was greater than that of pressed TNAZ at -10°C and 25°C.[4] Since the fracture acceleration of cast explosives is inversely related to the grain size,[2] the high fracture acceleration of this cast TNAZ may therefore be due to its very small grain size (< 0.1 mm). The fracture acceleration of Composition A3 Type II was less than that of PAX-3 at -10°C and 25°C. LX-14 and PAX-2A did not fracture at -10°C in these experiments.

The fracture acceleration of the pressed plastic bonded explosives PAX-3 and Composition A3 Type II at -10°C were about twice that at 25°C. The fracture acceleration of pressed TNAZ and melt-cast TNAZ at -10°C was about 10% less than at 25°C.

Table 1. Fracture acceleration of pressed explosives in an ultracentrifuge at -10°C

Explosive	Fracture Acceleration	Density (%TMD)
LX-14	>140 k*g*	95.9
PAX-2A	>140 k*g*	99.4
PAX-3	123 k*g*	97.4
TNAZ (melt-cast)	93 k*g*	97.3
Comp A3 Type II	55 k*g*	98.1
TNAZ	55 k*g*	98.4

REFERENCES

1. Y.D. Lanzerotti and J. Sharma, Brittle behavior of energetic materials during high acceleration in an ultracentrifuge, *App. Phys. Lett.* 39: 455 (1981).
2. Y.D. Lanzerotti and J. Sharma, Mechanical behavior of energetic materials during high acceleration ultracentrifuge, in: *Grain Size and Mechanical Properties – Fundamentals and Applications,* M.A. Otooni, R.W. Armstrong, N.J. Grant and K. Ishizaki, eds., Materials Research Society, Pittsburgh 362: 131 (1995).
3. Y.D. Lanzerotti, L.V. Meisel, M.A. Johnson, A. Wolfe and D.J. Thomson, Fracture surface topography of energetic materials using atomic force microscopy, in: *Atomic Resolution Microscopy of Surfaces and Interfaces,* D.J. Smith, ed., Materials Research Society, Pittsburgh 466: 179 (1997).
4. Y.D. Lanzerotti and J. Sharma, Mechanical behavior of energetic materials during high acceleration in an ultracentrifuge, in: *Shock Compression of Condensed Matter – 1997,* S.C. Schmidt, D.P. Dandekar, J.W. Forbes, eds., American Institute of Physics (1998).
5. L.V.Meisel, R.D. Scanlon, M.A. Johnson, and Y.D. Lanzerotti, Self-affine analysis on curved reference surfaces: self-affine fractal characterization of a TNT surface, in: *Shock Compression of Condensed Matter – 1999,* M.D. Furnish, L.C. Chhabildas, and R.S. Hixon, eds., American Institute of Physics (2000).

INDEX

GPSR Compliance

The European Union's (EU) General Product Safety Regulation (GPSR) is a set of rules that requires consumer products to be safe and our obligations to ensure this.

If you have any concerns about our products, you can contact us on ProductSafety@springernature.com

In case Publisher is established outside the EU, the EU authorized representative is:

Springer Nature Customer Service Center GmbH
Europaplatz 3
69115 Heidelberg, Germany

Batch number: 10755805

Printed by Printforce, the Netherlands